Amateur Telescope Making 1

ALBERT G. INGALLS
Editor

Published By

Willmann-Bell, Inc.
Publishers and Booksellers Serving
Astronomers Worldwide Since 1973

Published by Willmann-Bell, Inc.
P.O. Box 35025, Richmond, Virginia 23235

Copyright ©1996 by Jeremy Graham Ingalls and Wendy Margaret Brown

All rights reserved. Except for brief passages quoted in a review, no part of this book may be reproduced by any mechanical, photographic, or electronic process, nor may it be stored in any information retrieval system, transmitted, or otherwise copied for public or private use, without the written permission of the publisher. Requests for permission or further information should be addressed to Permissions Department, Willmann–Bell, Inc. P.O. Box 35025, Richmond, VA 23235.

Printed in the United States of America

First Printing 1996
Second Printing 1998

Library of Congress Cataloging-in-Publication Data.
Amateur telescope making / edited by Albert G. Ingalls.
   p.   cm.
   Originally published under title: Amateur telescope making. advanced. New York : Munn, 1937, a sequel to Amateur telescope making.
      Includes indexes.
      ISBN 0-943396-48-4 (v. 1). – ISBN 0-943396-49-2 (v. 2). – ISBN 0-943396-50-6 (v. 3).
      1. Telescopes–Design and construction–Amateur's manuals. I. Ingalls, Albert G. (Albert Graham) II. Title: Amateur telescope making, advanced.
QB88.A62   1966
681'.212–cd20   96-32194
                                                            CIP

98 99 00 01 02 03 04 05 9 8 7 6 5 4 3 2

How It Came About

SHEER ACCIDENT. That is how this book and the amateur telescope making development which has resulted from its publication originally came to be. If you are bitten by the bug of the amateur telescope making hobby, you may pretty nearly blame the fortuitous fumbling of one man's thumb. Thus closely are things that are worthwhile in life linked up with the most trifling circumstances. Some years ago the editor of this volume, while in a public library, was half-consciously thumbing over a bound volume of *Popular Astronomy,* and by merest chance caught sight of the intriguing words "The Poor Man's Telescope." These words, it proved, formed the title of an arresting article (November 1921) by Russell W. Porter, which told how the author made the concave mirror for his own reflecting telescope. A second article (March 1923) related how a group of Vermont villagers under the same writer's instruction made their own telescopes and became amateur astronomers. The "poor man's" telescope, it was set forth, was not the more familiar refracting kind but the reflector. It was called the poor man's telescope because even a poor man, if he did not begrudge hard labor, might possess one by making it himself.

After reading these articles, an attempt was made at once to find detailed treatises on telescope making and, since the book resources of the whole vast New York Public Library were immediately at hand, it was fully anticipated that an armful of works on that art in the English language would be found readily available. Now, it is a rare thing in these days

Russell W. Porter (left of center holding mirror) and the group of Vermont villagers that made, under his instruction, their own telescopes and which was described in the November 1925 issue of Scientific American *(Photo circa 1923).*

of plentiful books concerning everything under the sun, when one cannot easily lay hands on at least a dozen works about even an obscure subject; generally, in fact, one's first task is to eliminate all but the best of the lot. Nevertheless, it turned out that in the whole English-speaking world there was only one book on telescope making for the amateur, and even that was not available in American book stores. This was *The Amateur's Telescope,* by the Rev. William F.A. Ellison, Director of Armagh Observatory in Northern Ireland and a veteran maker of telescope mirrors. A copy of that book was obtained from London and it proved to be a gold mine. With its aid work was started on a modest mirror of 6-inch diameter.

At this juncture Russell W. Porter, author of the articles on the poor man's telescope, was personally discovered and proved willing to lend ready ear to certain frantic appeals for practical advice, and in course of time the mirror was completed and installed in a most unpretentious mounting of wood.

Then a larger idea took shape. Why not, with the book by the Rev. Ellison and the immediate assistance of Russell W. Porter, and with the *Scientific American* as a ready medium of access to large numbers of scientifically-minded persons, attempt to popularize amateur telescope making as a widespread hobby? Would it make appeal? Would it? No one knew. To test the potential "reader interest," if any, in the subject, an article was published in that magazine (November 1925), describing a night spent with the group of Vermont amateurs which Mr. Porter had

Amateur Telescope Making 1   v

His curiosity aroused by the November 1921 issue of Popular Astronomy *describing the "poor man's" telescope, Albert G. Ingalls in June 1925 travelled to Springfield, Vermont and to Stellafane to see first hand what was going on and to possibly write about it in* Scientific American. *Russell W. Porter captured the event with this sketch.*

fostered, at their stargazing mountain-top clubhouse-observatory near the village of Springfield. In response, 368 of the readers of that article wrote to the editor of the *Scientific American* urging the publication of practical instructions for making telescopes such as the Vermont amateurs had made and used.

This looked like an auspicious beginning for so specialized a hobby, and Mr. Porter was accordingly invited to prepare two such articles. These two articles (January and February 1926), brief and inadequate as any mere article or two on such a subject must necessarily be, aroused so much interest that the publication of a book of instructions, more detailed in nature, was at once decided upon and a request for the right to reprint *The Amateur's Telescope* in America was cabled to the Rev. Ellison in Northern Ireland. This book, or most of it, and the two *Scientific American* articles by Porter were combined with other matter to make a modest volume of 102 pages, the first thin edition of the present work.

As time went on, the telescope-making hobby enlisted the interest and keen enthusiasm, sometimes almost fanatical, of more and more of the readers of the *Scientific American*. Descriptions of telescopes actually made

In the November 1925 issue of Scientific American *Albert G. Ingalls wrote about Stellafane and "How a Group of Enthusiasts Learned to Make Telescopes and Become Amateur Astronomers."*

were published in every issue of the magazine after 1926, and clubs of enthusiastic amateur telescope makers and astronomers were formed in many of the larger communities. Through correspondence and travel their members became mutually acquainted and, all over the nation and, indeed, all over the world, wherever the *Scientific American* circulated—in the mountains of Java, in South Africa, the Argentine, Australia and New Zealand, India, Japan, Canada, and elsewhere—amateurs interested in science and refined mechanics found themselves engaged in rubbing one piece of glass on another to make a telescope mirror and, as soon as this was completed, eagerly starting larger and larger ones. The first edition, some 3400 copies, of the little 102-page book was gone by 1928. A second edition, enlarged

Amateur Telescope Making 1 vii

During the late 1920s and well into the 1940s Albert G. Ingalls frequently returned to Stellafane. Here, after breakfast and before the table has been cleared Ingalls (seated) confers with David O. Woodbury, author of The Glass Giant of Palomar *a fascinating account of how the 200-inch Hale telescope came about. Immediately above Ingalls is a sundial painted on the wall of Stellafane's "Pink House."*

to 285 pages by the addition of new matter, was prepared that year, and the 5400 copies of that edition had vanished by 1932. The present edition contains the same matter, with trifling alterations and deletions, and with some 200 pages added.

Still the hobby goes marches on. Thousands of telescopes have been labored over by eager workers young and old, skilled and less skilled, men and women (several of these), "poor" men and rich men too. Telescope making is a scientific hobby and it appeals doubtless because it exacts intelligence; requires patience and sometimes dogged persistence in order to whip the knotty but fascinating problems which arise; demands hard work—is not dead easy; and compels the exercise of a fair amount of handiness—enough to exclude the born bungler but no more than is possessed by the average man who can "tinker" his car or the household plumbing, or dissect and wreck a watch. Some use of the brain is also called for, but one need not be an Einstein. The hobby also appeals because the worker derives something of a thrill while shaping the refined curve of the glass as he realizes that, with scarcely any special tools but chiefly with the aid of an elementary test which greatly magnifies minute irregularities on the curve, he is able to work to within almost a millionth of an inch of absolute perfection. Finally, it may legitimately make appeal because the end-product, the telescope, is not only a tangible evidence, visible to all, of the worker's possession of the several virtues cited above, but is a valuable scientific instrument which places him on the threshold of astronomy and astrophysics, perhaps the most romantic branch of modern science.

The reader doubtless will discover that this book is a mine of practical information but that the same information is not arranged in a single sequence—he must mine it out. This is because the various parts were written by many different authors and at different times. Like Topsy, the book "just grew" or, as is sometimes said of the British Empire, it is "a fortuitous, unsystematized agglomeration of ill-assorted entities acquired at different times by opportunism and otherwise." However, like that very practical commonwealth, it works—thousands who have used it can testify to that. To organize its contents thoroughly, so that the reader might march straight through a logical sequence without jumping about, would require that it be rewritten entirely and by a single writer. But then it would lose most of its claim to authoritativeness, simply because it would thereby lose most of its numerous contributing authorities; one cannot eat one's pie and have it too. So the diligent worker will be forced to make the best of this disability, reading the volume twice or more while he works, and using the index to correlate cognate phases of the work. . .

Albert G. Ingalls,
Associate Editor
Scientific American
New York, November, 1932

Or So Ingalls Thought...

The book you are now holding in your hands and its two companion volumes is a rearrangement of the original three volume work. Insofar as possible similar items are now grouped together; each as a Chapter within a descriptive Part. Chapters are usually divided into sections, subsections, etc. While the Table of Contents for a typical "ATM" of old was 2 pages it is now 3 to 5 times larger. The Index has similarly been expanded. The objective was not to *rewrite* but to *logically rearrange* the text into a more user-accessible format. Spelling, hyphenation and usage have been standardized, and errors-in-fact corrected, but the original text remains fundamentally unchanged. Most of the original text from the last printings of the original ATMs is included here but the day of vacuum tube power supplies for photomultiplier tubes is past—likewise pendulum clocks. What remains are nearly 1,800 pages devoted to praticle telescope making and is arranged as follows:

Amateur Telescope Making 1.

    Part A Newtonian Telescope Mirror Making

    Part B Optical Testing

    Part C Workshop Wisdom

    Part D Observatory Buildings

Amateur Telescope Making 2.

    Part A Refractor Telescopes

    Part B Telescope Mechanics

    Part C Telescope Adjustments

    Part D Binoculars

    Part E Schmidt Cameras

    Part F Optical Flats

Amateur Telescope Making 3.

    Part A Optical Production Methods and Machinery

    Part B Eyepieces and Small Lenses

    Part C Optical Coatings and Coating Equipment

    Part D The Eye and Atmosphere

    Part E Other Optical Instruments

    Part F Instrumentation for Solar Observations

In the early 1990s notice was given in *Sky & Telescope* magazine that a re-publication of ATM was planned and a request was made for corrections. Mel Bartels, Robert P. Brown, William Herbert, Virgil Johnson, Keith G. Mogridge, Philip Moniot, and Gordon Rayner responded to this request.

Seventy years after Albert Ingalls travelled to Springfield, Vermont up Breezy Hill to Stellafane the tradition continues. Here, on a Thursday evening in August 1995, are gathered many of the Springfield Telescope Makers who have spent most of the week in preparation for the 1995 Stellafane Convention.

I am thankful for their comments, many of which have been incorporated into this edition.

Just before publication and over a three month period John D. Koester, Kenneth J. Launie, and Harold R. Suiter spent most of their spare time checking the final manuscript. I cannot emphasize too much the value of their help. In large measure these books "ring true" because they found the discordant phrase, word or equation and brought to it our attention. One of these readers, Harold R. Suiter is the author of the highly acclaimed *Star Testing Astronomical Telescopes* which I also published. When he returned the first volume of this work I saw that he had made notes at the end of some chapters which if expanded would follow the tradition established by Ingalls who regularly made such comments throughout the text. His initial response was that these were more-or-less similar to the side comments made between us during the production of his book and that some readers might think it presumptuous that one might tamper with the work of Ingalls, Porter and the other contributors. I did not think this to be the case. It seemed to me that the perspective they provided would add to the work rather than detract from it and would be welcomed by the readers if his comments were made at the end of the chapters and their source identified. Messrs Koester, Launie and Suiter have made these better books in many unseen ways and I am fortunate that their dedication to this avocation has been shared with you and me. What ever errors remain are mine and I would appreciate learning about them so that subsequent printings might be corrected.

Perry Willmann Remaklus
President and Publisher
Willmann-Bell, Inc.
Richmond, August 1996

Table of Contents

A Newtonian Telescope Mirror Making — 1

A.1 Mirror Making for Reflecting Telescopes — 3
- A.1.1 Introduction to the Reflecting Telescope — 3
- A.1.2 Grinding the Mirror — 5
- A.1.3 Making the Pitch Lap — 6
- A.1.4 Polishing the Mirror — 7
- A.1.5 Testing the Mirror — 8
- A.1.6 Figuring the Mirror — 11
- A.1.7 Silvering the Mirror — 17
- A.1.8 Biographical Note by the Editor — 21
- A.1.9 Note Added in 1996 — 23

A.2 The Beginner's First Telescope — 25
- A.2.1 Editor's Note — 25
- A.2.2 Introduction — 26
- A.2.3 The Poor Man's Telescope — 28
- A.2.4 The Main Thing is the Mirror — 30
- A.2.5 And So— — 33
- A.2.6 Now We Start — 35
- A.2.7 Keep Hogging — 36
- A.2.8 Checking Progress — 38
- A.2.9 No More Long Strokes — 40
- A.2.10 Scratches — 42
- A.2.11 Polishing — 43
- A.2.12 All Set to Begin Polishing — 44
- A.2.13 Mounting the Poor Man's Telescope — 48
- A.2.14 Collimation — 53
- A.2.15 Albert G. Ingalls Editor's Note — 56
- A.2.16 Notes Added in 1996 — 62

A.3 The HCF Lap — 63
- A.3.1 Advantages of HCF — 63
- A.3.2 Making and Using an HCF Lap — 64
- A.3.3 Using HCF for Zonal Correction — 66

A.4 The Amateur's Telescope — 69
- A.4.1 Introductory — 69
- A.4.2 Literature — 71
- A.4.3 Tools and Materials — 72
- A.4.4 Rough Grinding — 74
- A.4.5 Testing; Foucault's Shadow Test — 79
- A.4.6 Polishing — 83
- A.4.7 Figuring — 87
  - A.4.7.1 Parabolizing by Long Stroke — 88
  - A.4.7.2 Parabolizing by Graduating Facets — 89
  - A.4.7.3 Parabolizing by the Small Polisher System — 90
  - A.4.7.4 Parabolizing by Overhang — 90
  - A.4.7.5 Working Uphill — 91
- A.4.8 Editor's Note — 94
- A.4.9 How to Recognize the Paraboloid, Zonal Testing — 95
- A.4.10 Silvering — 101
  - A.4.10.1 To Polish the Film — 101
  - A.4.10.2 A Few Hints on Silvering — 102
  - A.4.10.3 Care of the Film — 103
- A.4.11 Mounting the Mirror — 104
- A.4.12 A Last Word to Beginners on Insufficient Grinding — 107

A.5 An Amateur's View of Mirror Making — 111
- A.5.1 From One TN to Another — 111
  - A.5.1.1 Pitch — 111
  - A.5.1.2 Abrasive Action, Rolling and Stationary — 111
  - A.5.1.3 The Spit Test for Radius of Curvature — 112
  - A.5.1.4 Tool Effect — 113
  - A.5.1.5 Thermal Effect — 113
  - A.5.1.6 Evaporation Effect — 114
  - A.5.1.7 Friction Effect — 116
  - A.5.1.8 Tool Deformation — 117
  - A.5.1.9 Tool Plowing — 118
  - A.5.1.10 The Clock Stroke — 118
  - A.5.1.11 The Blending Overhand Stroke — 119
  - A.5.1.12 The Semistroke — 120
- A.5.2 Backwoods Technique — 120

	A.5.2.1	The Handle	120
	A.5.2.2	The Grip	120
	A.5.2.3	Rough Grinding	121
	A.5.2.4	Fine Grinding	123
	A.5.2.5	The Lap	124
	A.5.2.6	Polishing	127
	A.5.2.7	Correcting	128
	A.5.2.8	Figuring	134
A.5.3	The Second Mirror		135
A.5.4	In Retrospect		136

A.6 Subdiameter Tools · 139
A.6.1 Large Mirrors and Subdiameter Tools · 139
A.6.2 Use of Subdiameter Tools on a 12-inch Mirror · 140
A.6.3 Use of Subdiameter Tools on a 10-inch Mirror · 141
A.6.4 Construction of Subdiameter Tools · 144

A.7 The Prism or Diagonal · 147
A.7.1 Editor's Notes · 151
A.7.2 Sizing a Newtonian Diagonal · 151
    A.7.2.1 Addendum, 1948 · 154

A.8 Prism Diagonals-Axial Aberration Effects · 157
A.8.1 Axial Spherical Aberration · 158
A.8.2 Axial Chromatic Aberration · 159
A.8.3 Effect on Definition · 160
A.8.4 Extra-Axial Aberrations · 161
A.8.5 Prism Glasses · 162

A.9 How to Make a Diagonal for a Newtonian · 163
A.9.1 Making the Blank · 163
A.9.2 Making a "Surround" · 164
A.9.3 Grinding · 165
A.9.4 Polishing · 166
A.9.5 Testing for Flatness · 166
A.9.6 Mounting the Diagonal · 167

A.10 The Building of a 19-Inch Reflecting Telescope · 169
A.10.1 The Mirror · 170
A.10.2 Mounting · 178
A.10.3 The Clock · 184
A.10.4 Note Added in 1996 · 186

B Optical Testing · 187

B.1 Curves Found During Figuring · 189

B.2 Where Is The Crest Of The Doughnut? · 197
   B.2.1 A Study in Shadows 197

B.3 Accuracy in Parabolizing a Mirror · 201
   B.3.1 Editor's Note 206
   B.3.2 Note Added in 1996 207

B.4 The Ronchi Test for Mirrors · 209
   B.4.1 Note Added in 1996 214

B.5 Hindle's Method in the Knife-edge Test · 215

B.6 The Slit Test · 219

B.7 Shadow Appearance · 223
   B.7.1 Doughnut Mathematics 225
   B.7.2 Shadow Behavior 231
   B.7.3 The Error of Observation 235
   B.7.4 Accuracy of the Knife-edge Test 238
   B.7.5 Testing Equipment 240
   B.7.6 Testing Routine 243
   B.7.7 Note Added in 1996 248

B.8 The Caustic Test · 251
   B.8.1 Caustic Testing Procedure 262
   B.8.2 Interpretation 265
   B.8.3 Sample Calculations 266
   B.8.4 Test Rigs 268
      B.8.4.1 Test Rig for Second Method 269
      B.8.4.2 Test Rig for Third Method 269
   B.8.5 Accuracy 276
   B.8.6 Nonparaboloidal Surfaces 280
   B.8.7 Editor's Note 281
   B.8.8 Note Added in 1996 283

B.9 Quantitative Optical Test for Telescope Mirrors · 285
   B.9.1 Quantitative Failure of Test 285
   B.9.2 Apparatus and Procedure 286
   B.9.3 Geometry of Test 288

B.9.4	Appearances of Common Forms of Aberration	289
B.9.5	Editor's Note	289
B.9.6	Notes on the Ronchi Band Patterns	290
B.9.7	Note Added in 1996	292

B.10 Gregorian Secondary Test — 293
    B.10.1 Editor's Note, 1948 297

B.11 The Hartmann Test — 299

B.12 Notes on the Optical Testing of Aspheric Surfaces — 305
    B.12.1 The Ellipsoid 305
    B.12.2 Testing The Paraboloid On Near Objects 308
    B.12.3 Quantitative Test of Hyperboloidal Mirrors 310
    B.12.4 Note Added in 1996 314

B.13 Null Test for Paraboloids — 315
    B.13.1 Editor's Note 319

B.14 Testing Convex Spherical Surfaces — 321

B.15 A Bilateral Slit Mechanism — 325

B.16 Small Pinholes — 329
    B.16.1 Determining the Optimum Size 329
    B.16.2 Illuminating the Pinhole 331
    B.16.3 Making the Pinhole 332
    B.16.4 Differences in Usage Between Small and Large Pinholes . 332
    B.16.5 Note Added in 1996 334

B.17 Astigmatism — 335

B.18 Optical Bench Testing — 341
    B.18.1 Introduction 341
    B.18.2 Lens Characteristics 341
    B.18.3 A Nodal Slide Optical Bench 343
        B.18.3.1 Illuminator 344
        B.18.3.2 Telescope-microscope 345
        B.18.3.3 The Optical Bench Proper 347
        B.18.3.4 The Slides X and S 347
    B.18.4 Alignment 347
    B.18.5 Testing Methods 349
        B.18.5.1 Tests on the Axial Image 350

B.18.5.1.1 Equivalent Focal Length 350
B.18.5.1.2 Back Focal Length 351
B.18.5.1.3 Working Distance 351
B.18.5.1.4 Flange Focal Length 352
B.18.5.1.5 Axial Critical Aperture Ratio or
 Aperture Tolerance 352
B.18.5.1.6 Axial Chromatic Aberrations 353
B.18.5.2 Axial Spherical and Zonal Aberrations 353
B.18.5.3 Tests Using Extra-Axial Images 354
B.18.5.3.1 Curvature of Image Field 354
B.18.5.3.2 Lateral Chromatic Aberration 357
B.18.5.3.3 Distortion 357
B.18.5.4 Comatic, Lateral Spherical or Sinical Aberration . 358
B.18.5.5 Astigmatism 358
B.18.5.6 Testing Complete Telescopes 359
B.18.5.6.1 Convergent Tubes 362
B.18.5.6.2 General Caution 363
B.18.5.7 Approximate Refractometry 363
B.18.5.7.1 Radius Measurement in General 365
B.18.6 Summary 366
B.18.7 References 367

B.19 Interference of Light 369
B.19.1 Newton's Fringes 374
B.19.2 Haidinger's Fringes 378
B.19.3 Low-reflection Coatings 380
B.19.4 Diffraction 383
B.19.5 Image Spikes 383
B.19.6 Focal Diffraction 387
B.19.6.1 Edge Diffraction 387
B.19.7 Refractor vs. Reflector 392
B.19.8 References 392
B.19.9 Notes Added in 1996 393

C Workshop Wisdom 395

C.1 Advice From TN's 397
C.1.1 General 397
C.1.1.1 Choosing a Mirror Size 397
C.1.1.2 Walking Around the Barrel 397
C.1.1.3 The Correct Level for Grinding and Polishing . . . 398

Amateur Telescope Making 1

C.1.1.4	Flexure	398
C.1.1.5	Warping of Wood on which Tool is Mounted	398
C.1.1.6	Uniform Working Temperature	399
C.1.1.7	Keep a Log Book	399
C.1.1.8	Sagitta equals $r^2/2R$	400
C.1.1.9	When $r^2/2R$ is not exact enough	400
C.1.1.10	Scratches	402
C.1.1.11	Mirror Breakage	403
C.1.1.12	Workplace Cleanliness	403
C.1.1.13	Shipping a Mirror	404
C.1.1.14	Take Courage	404
C.1.1.15	Three in One	405
C.1.2	Mirror Substrates	405
C.1.2.1	Fused Quartz	405
C.1.2.2	Pyrex	406
C.1.2.3	How Plate Glass is Made	407
C.1.2.4	Optical Glass	407
C.1.2.5	Glass Mirror Substitutes	408
C.1.2.6	Rotating Mercury Mirror	412
C.1.2.7	Cellular Mirror	413
C.1.2.8	Cemented, Built-Up Disks	415
C.1.2.9	Suction Mirror	416
C.1.2.10	Magnesium Oxychloride Mirror	416
C.1.3	Cutting Circular Disks and Holes	417
C.1.4	Tools and Fixtures	418
C.1.4.1	Inverting Tool and Mirror	418
C.1.4.2	Inverting Device	419
C.1.4.3	Limiting Devices	419
C.1.4.4	Making Templates	420
C.1.4.5	Metal Template	420
C.1.4.6	Removable Handle for Mirror Disks	421
C.1.4.7	Grinding Stands	421
C.1.5	Grinding Abrasives	421
C.1.5.1	Designations for Carbo Grain Sizes	421
C.1.5.2	Crushed Steel and Pyrex	422
C.1.6	Grinding	422
C.1.6.1	Who Discovered the Method of Concaving a Glass Disk?	423
C.1.6.2	Nature of Grinding	424
C.1.6.3	Wets	424
C.1.6.4	One-third Strokes	424
C.1.6.5	Strokes in Grinding	425

C.1.6.6	Why Disks Grind Concave	426
C.1.6.7	Refusal of the Mirror to Become Concave	426
C.1.6.8	Amount of Glass Removed for Each Stage	427
C.1.6.9	Chamfering Disks	428
C.1.6.10	Getting Overall Contact while Grinding	428
C.1.6.11	Bad Central Contact in Final Grinding	429
C.1.6.12	Estimating Fineness of Grinding	430
C.1.6.13	Streak of Rouge Test for Contact in Fine Grinding	431
C.1.6.14	Water Drop Test for Contact in Fine Grinding	431
C.1.6.15	Sticking Mirror	431
C.1.6.16	Mirror Sticks to Tool	432
C.1.6.17	Get a Sphere before Beginning to Polish	433
C.1.6.18	When is Fine Grinding Finished?	434
C.1.6.19	Finishing With Emery	434
C.1.6.20	Searching for Tiniest Pits	435
C.1.6.21	Futility of Attacking Big Pits with Rouge	435
C.1.6.22	Test for Center of Curvature During Grinding	437
C.1.6.23	Keeping Track of the Radius of Curvature While Grinding	437
C.1.6.24	Very Exactly Finding the Radius of Curvature	438
C.1.6.25	Bubbles Between Mirror and Tool	438
C.1.7	Pitch Laps	438
C.1.7.1	Pitch	438
C.1.7.2	Making the Pitch Lap	439
C.1.7.3	Modifications of the Plain Pitch Lap.	444
C.1.7.4	Straining Pitch	445
C.1.7.5	Speckling of Lap	445
C.1.7.6	Pitch Flammability	446
C.1.7.7	Channels in Laps	446
C.1.8	Nonpitch Laps	446
C.1.8.1	The HCF Lap	446
C.1.8.2	The Paper Polishing Lap	449
C.1.8.3	Laps in Hot Places	450
C.1.8.4	Substitutes for Pitch	450
C.1.9	Polishing Agents	450
C.1.9.1	Rouge Size	450
C.1.9.2	How to Make Rouge	450
C.1.9.2.1	Editor's Note	452
C.1.10	Another Method for Making Rouge	453
C.1.11	Breaking-up Rouge	453
C.1.12	Polishing	453
C.1.12.1	Polishing, Theory	453

	C.1.12.2	Grabbing	459
	C.1.12.3	Sleeks	460
	C.1.12.4	Scratches From Rouge	460
	C.1.12.5	Stuck Disks	460
	C.1.12.6	Test for Complete Polish	461
	C.1.12.7	Detecting the Most Minute Pits and Scratches	461
	C.1.12.8	Cold Pressing	461
	C.1.12.9	Prolonged Cold Pressing	461
C.1.13	Figuring		462
	C.1.13.1	Correcting Turned-up Edge	462
	C.1.13.2	Correcting a Hole	462
	C.1.13.3	Correcting a Hyperbola	463
	C.1.13.4	Forestalling Turned-down Edge	463
	C.1.13.5	Turned-down Edge	463
	C.1.13.6	Turned Edge	463
	C.1.13.7	Test for Slight Turned Edge	464
	C.1.13.8	Diffraction Effects	465
C.1.14	Testers		465
	C.1.14.1	A Cool Pinhole	465
	C.1.14.2	Permissible Distance Differential of Pinhole and Knife-edge along the Axis	466
	C.1.14.3	Electric Lamp for Knife-edge Test	466
	C.1.14.4	Tiny Pinhole for Advanced Workers	467
	C.1.14.5	Slow Motion Devices for Testing	467
	C.1.14.6	Reversing Knife-edge	467
	C.1.14.7	Testing Tunnel	468
C.1.15	Testing		468
	C.1.15.1	First Announcement of Foucault Test	469
C.1.16	Conic Sections		471
C.1.17	Record Keeping		473
C.1.18	Correct Paraboloidal Shadow		475
	C.1.18.1	Measuring Zones	476
	C.1.18.2	Testing without Masks	477
	C.1.18.3	Interpreting Shadows	478
	C.1.18.4	Precision in Reading Knife-edge Shadows	479
	C.1.18.5	The Inside-and-outside Test	480
		C.1.18.5.1 Editor's Note	481
	C.1.18.6	Learning to Understand the Knife-Edge Test	482
	C.1.18.7	Avoiding Fatigue in the Knife-edge Test	483
	C.1.18.8	Making Focograms	484
	C.1.18.9	Diffraction Ring (Star) Test	486
C.1.19	Abnormalities		493

C.1.19.1	Warped Mirror	493
C.1.19.2	Astigmatized Mirror	494
C.1.19.3	Striae	494
C.1.19.4	Strains in Glass and Their Detection	494
C.1.19.5	Editor's Note	497

C.1.20 Notes on the Eyepiece 498
C.1.21 Telescope Mechanics 502
    C.1.21.1 Tubeless Telescope 502
    C.1.21.2 Hardening Brass 502
    C.1.21.3 Blackening Brass 502
    C.1.21.4 Paint for Inside of Tube 502
    C.1.21.5 An Adapter Tube 502
    C.1.21.6 Finders 503
    C.1.21.7 More about Finders 504
C.1.22 Telescope Designs 505
    C.1.22.1 Ritchey-Chrétien 505
    C.1.22.2 Cassegrainian Notes 507
    C.1.22.3 The Herschelian Telescope 514
C.1.23 Literature of Interest to the TN 515
    C.1.23.1 Observatories 515
    C.1.23.2 Herschel's Mirrors 516
    C.1.23.3 Properties of Pitch 517
    C.1.23.4 Wassell and Blacklock Letters 517
C.1.24 Graduating Setting Circles 517

D Observatory Buildings   519

D.1 Telescope Housings   521
D.1.1 The Warmed Observing Room 521
D.1.2 Turret Telescopes 523

D.2 The Amateur's Observatory   527
D.2.1 Observatory Size 530
D.2.2 Weatherproofing 531
D.2.3 Rollers 531
D.2.4 The Sliding Roof Observatory 532
D.2.5 Domes 534
    D.2.5.1 Shutters 543
    D.2.5.2 Revolving the Dome 544

D.3 Thermal Effects of Observatory Paints — **549**
- D.3.1 Introduction 549
- D.3.2 Test Procedures 549
- D.3.3 Interior painting 552

Part A

Newtonian Telescope Mirror Making

Chapter A.1

Mirror Making for Reflecting Telescopes[†]

A.1.1 Introduction to the Reflecting Telescope

In the reflecting telescope, *the mirror's the thing*. It is the heart of the instrument, and is usually completed before the other parts of the telescope are begun. The tube and mounting are then built to match its focal length, which cannot be precisely predetermined.

We are concerned in this chapter with the shaping of the telescope mirror. This consists solely in giving one side of it a concave, polished surface. This surface is to be so very nearly spherical that we shall first attempt to make it precisely so; and at the very last we shall alter it to the kind of surface known among the highbrows as a paraboloid of revolution.

Such an automobile headlight has the property of throwing out from a concentrated source of light placed at a focal point near it, a beam of parallel rays. (See Figure A.1.1.) We shall, however, use this reflector the other way around, that is, by receiving parallel rays of light from a distant object (star); and by reflecting them from a properly curved mirror we shall bring them to a point of focus (F, Figure A.1.1).

Our curve, however, is so small a portion of this widely sweeping parabola (the black area represents the mirror) that it is extremely shallow, and so it nearly coincides with the superimposed spherical curve. At first, therefore, we shall seek to hollow out a spherical curve, later deepening it very slightly into the paraboloid.

Since the angle of reflection of a beam of light is equal to the angle of incidence, the parallel, arriving rays will be reflected approximately to a

---

[†]By Russell W. Porter, M.S. Formerly Optical Associate, Jones and Lamson Machine Co., Springfield, Vermont. Associate in Optics, The California Institute of Technology.

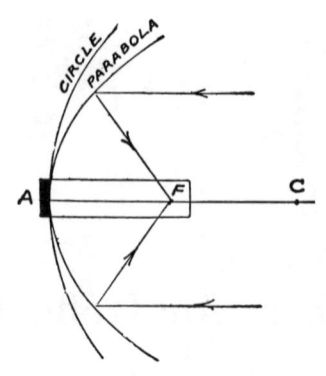

Fig. A.1.1: Theory of the Mirror. Many find it difficult to understand why the focal length is only one-half of the radius or distance to the center of curvature, while in the shadow test the light is focused at the center of curvature. In the first case, the rays are coming from a star, at almost infinite distance, and are therefore virtually parallel, while the rays that reach the mirror from the pinhole are divergent (radii). In this diagram, let us imagine we could grasp the two parallel rays indicated and actually pull their right-hand ends together until they touched the point C. As we drew them in, the angle at which they would now meet the mirror's surface would change, and since light is reflected away at the same angle at which it strikes a mirror, the reflected rays would shift at the same time from F to C, at double the distance of F.

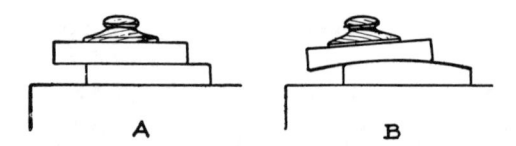

Fig. A.1.2: Why the Curves Develop. The upper disk tends to hollow out, because at extremities of the strokes the abrasive effect on both disks is increased. This is due to the overhang and to the consequently increased pressure on the central portion of the upper disk, as well as the marginal part of the lower.

focus whose length may be regarded as one-half of the radius of curvature, C–A, Figure A.1.1.

Enlarging the mirror of Figure A.1.1, A, we have in Figure A.1.3 the essentials of the Newtonian, reflecting telescope. Light from a distant object falls down the tube to the mirror, and normally would, by reflection, produce an image at the focus, F. The converging rays are, however, intercepted at D by a small diagonal mirror or prism that delivers them to a lens called an eyepiece at the side of the tube, where the image is examined.

A.1.2. Grinding the Mirror 5

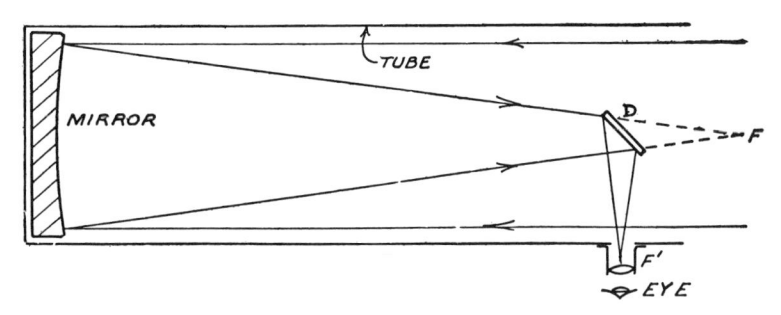

Fig. A.1.3: Why a Diagonal is Needed. *Without it, the rays would come to a focus at F. But then the observer's head would eclipse the light from the object. The diagonal mirror, or a prism, reflects them to F'.*

A.1.2 Grinding the Mirror

I will take as our standard a mirror 6 inches in diameter, having a 4-foot focal length. The beginner is not advised to essay a larger mirror for his first effort, since his difficulties will be found to multiply quite disproportionately as the diameter increases. If two flat glass disks (*A*, Figure A.1.2) are ground together, one over the other, with an abrasive between, lo and behold!—the upper one becomes concave, the lower one convex. This is because the pressure per unit area, and therefore the amount of abrasion, is increased on the central portions of the upper disk and outer portions of the lower one when the upper disk overhangs as in *B*.

A straight, back-and-forth stroke, in which a given point on the upper disk moves across one-third the diameter of the lower, has the property of holding the two surfaces spherical. This is due to the fact that spherical surfaces are the only ones which remain in continuous contact at every point when moved over each other in any direction. This fact is a veritable godsend to the amateur—and to the professional, too, for that matter—for he may go confidently forward through the different stages of grinding and polishing with the knowledge that his mirror will come out nearly as it will be when it is finally deepened into a paraboloid.

The depth of the curve increases with grinding, and it is gauged with a template of the proper radius. Since by our rule, the radius, *A–C*, Figure A.1.1, of the curve of the glass is twice its focal length *A–F*, a template is made from tin, with a radius of twice 48 inches, or 96 inches. Therefore a stick of wood (not a string, which would be elastic) should be tacked to the floor at one end so as to pivot, and a knife point held at the opposite end, or a sharpened nail driven through at the proper distance, should be used to scratch the desired curve to which the tin should be cut. For our 6-inch mirror the hollow will come to about 0.05 inch deep.[1]

The lower disk of glass is fastened to a pedestal or to a weighted barrel so that one can walk around it in grinding, or it may be held between one removable and two fixed buttons on the corner of a stout bench or table. Using melted pitch, a round handle is attached to the upper disk, which is first heated slowly in water to a slightly unpleasant warmth for the hand, taking care that no cold water drops fall on the warmed disk, for they might break it.

The grinding is done by placing wet Carborundum grains of successively finer sizes between the two disks, care being taken after each size is used to wash all parts of the work entirely free of the larger sized grains, which would otherwise scratch the disk. The strokes are straightforward and back, the center of one disk crossing that of the other. The glass also rotates bit by bit in the hands, in order to present a new direction for each stroke; and from time to time, in order to prevent the wearing of the glass unsymmetrically, the worker shifts positions around the pedestal; or, if working on a bench, he turns the lower disk, called the "tool" (we shall discard this tool at the end) to a new position.

Each grade of abrasive is used long enough to remove the coarser pits left by the preceding grade, and it will save much time and labor in the polishing if a small quantity of washed 6F ("sixty minute") emery is used after the No. 600 Carborundum.

All the preceding work is covered in great detail by Rev. W.F.A. Ellison in *The Amateur's Telescope*, also reproduced in the Section A.4, and therefore there would be no point in going again over the same ground here.

The bench and both disks are now thoroughly washed in order to remove all traces of grit, preparatory to polishing.

A.1.3 Making the Pitch Lap

Pitch is melted over a stove. It is tempered by adding (not over the fire) sufficient turpentine until a cooled sample placed between the teeth will just "give" slowly without crumbling, or will show a slight indentation of the thumbnail under moderate pressure. The pitch is poured (Figure A.1.4) over the tool, which has been warmed in water, and dried, and when it is partly cool, the glass is wetted (in warm water) and pressed down on the pitch until perfect contact is obtained between glass and pitch. V-shaped channels 1 inch apart are now cut across the pitch at right angles to each

---

[1] This method of using a template to gauge the depth of the curve has fallen from favor among modern TNs. A variation of the "spit" test in Section A.5.1.3 is much easier to do.

A.1.4. Polishing the Mirror

Fig. A.1.4: Preparing the Pitch Lap. *Melted pitch is being poured on the convex, upper face of the tool. Note the temporary collar of wet paper, which acts as a retaining wall for the pitch until it cools. Tool and mirror should previously have been placed in lukewarm water. If pitch is poured on a cold tool it will "set" so rapidly that there will be little time to make it conform to the curve of the mirror. But if the two disks are somewhat warm, there will be about ten minutes' time in which to make a lap that will preserve good contact. Thus the worker may "take it easy" and do it correctly. Keep cold drafts away from the job. Warm water striking cold glass is not likely to break it, but cold water striking warm glass may.*

other, to allow free access of the rouge and water to all parts of the glass. Do not center this system of channels or you may produce zones in the mirror. Decenter it $1/4$ inch in one direction, $1/2$ inch in the other (Figure A.1.6).

A.1.4 Polishing the Mirror

Rouge mixed with water is now substituted for the Carborundum and the polishing is carried on to completion, using the same strokes as in grinding. The time thus far consumed in grinding should be about five hours; polishing may require nine hours, divided into "spells." Through all these operations Ellison goes with care in Section A.4, anticipating the pitfalls into which the tyro inevitably falls. Were I to emphasize one caution over another, it would be the care required in preserving complete contact between the glass and the pitch-lap surfaces while polishing.

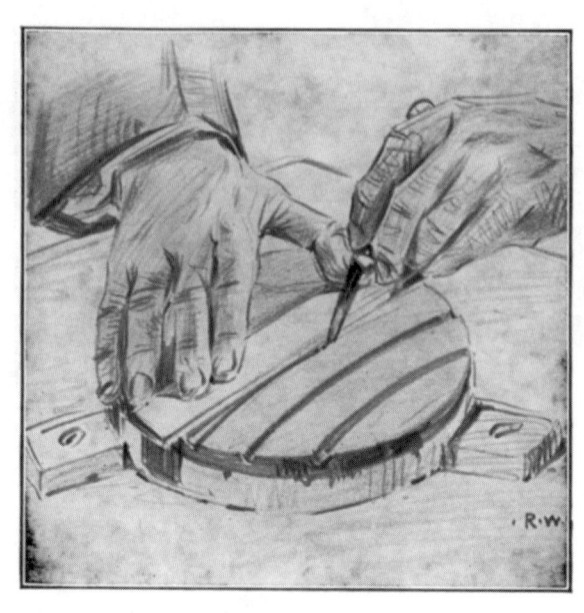

Fig. A.1.5: *Cutting Channels in the Pitch Lap.(Right) Use a flexible straightedge and a sharp knife. Keep everything wet to minimize sticking of the pitch. In spacing the channels, precision serves no particular purpose. Do not center them, in any case. After the lap is formed and the channels are cut, leave the mirror on the lap until the tool, pitch, and mirror have regained uniform room temperature. It should then be "cold pressed," or weighted, to ensure the establishment of an even contact, which may have been disturbed during the cooling process.*

A.1.5 Testing the Mirror

If one-third strokes have been maintained in grinding and polishing, the surface of the glass will be nearly spherical. How shall we find out? The method I shall now describe is one of the most delicate and beautiful tests to be found in the realm of physics. By it, imperfections of a millionth of an inch on the glass can be detected, and all the tools required are a kerosene lamp and a safety razor blade! This method of testing mirrors, called the Foucault knife-edge test, was unknown until about 1850; before that time mirror makers were groping in the dark. Even the great Herschel—father of the reflecting telescope—did not know when his mirrors were right, except by taking them out and trying them on a star.

If an artificial star made by a tiny pinhole (use a needle point) in a tin chimney on a kerosene lamp (an electric lamp will also be suitable) were placed at the center of the sphere of which the mirror's curve is a very small part, all of that portion of the light that emerges from the pinhole

A.1.5. Testing the Mirror

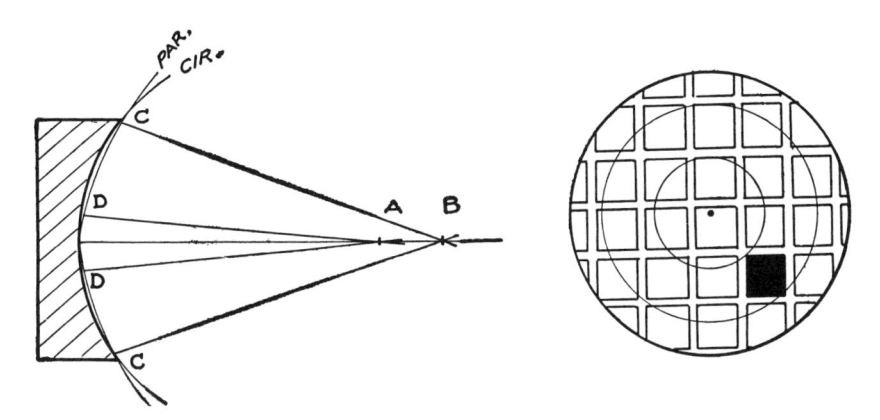

Fig. A.1.6 (Left): How Much Have We Parabolized? The radius of a parabola shortens as its vertex is approached. Therefore, the zone of the parabola near the edge C may be regarded (in practice) as part of a sphere with radius C–B. The central zone is regarded as part of a smaller sphere (shorter radius) with radius D–A. In the shadow test, we can actually measure the distance A–B with a scale, and from this we can work out the amount that we have deepened or parabolized the center of our spherical mirror.

(Right) A Typical Pitch Lap for a 6-inch Mirror. The black square represents a facet removed from the lap in an effort to treat a depressed zone. Thus, there would be less abrasion over the path traveled by this region as the mirror was rotated in polishing, and a zone (see rings on drawing) would tend to be raised above the general level of the glass.

and strikes the mirror is reflected back to the pinhole; for these light rays are all radii of the sphere, and by reflection they must return as radii back to their source, the pinhole.

In practice, the pinhole is pushed over a little to the right of the center of curvature so that the cone of reflected light may clear the chimney and enter the eye, as shown in Figures A.1.7, A.1.8, and A.1.9. The mirror is placed on its edge on some suitable support, at table height, in a fairly darkened room. The lamp and the knife-edge (mounted on a block of wood) are placed on a table as shown, and about 8 feet from the mirror, viz., at its center of curvature. The lamp remains stationary.

At first, considerable difficulty may be encountered in picking up with the eye the reflected cone of light. One way is to replace the tin chimney with a glass one, walk away from the lamp, keeping it in line with the mirror, when the image of the lamp will be seen in the mirror itself. Then bring the eye forward slowly, keeping the lamp image in view, and move the knife-edge to the right until it cuts off half of the image. The tin chimney is then put on and the image of the pinhole may be picked up somewhere

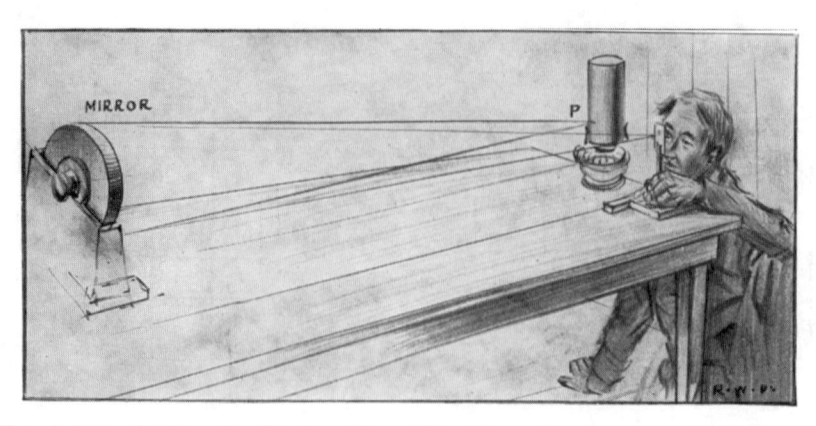

Fig. A.1.7: Making the Shadow Test. *The mirror does not necessarily have to rest on the same surface with the lamp and the knife-edge, but all three should rest on stable supports which will not vibrate after the hand is removed from the knife-edge.*

near the edge of the safety razor blade. As the eye approaches the position shown in the figures, this pinhole image begins to expand until a position is reached where the mirror is flooded with light over its entire surface—almost dazzling. See shadowgraph A, Figure A.1.10. An alternate method is to use a piece of ground glass, which can be prepared by rubbing it with Carborundum, to explore the neighborhood of the lamp, picking up the bright spot of light on it.

Now comes the remarkable knife-edge test. The razor blade is moved in from the left until it cuts into the reflected cone of rays. If at a, Figure A.1.9, that is, *inside* of the center of curvature, a shadow will come in on the mirror from the left, as might be expected (shadowgraph B). If, however, it cuts the rays at c, Figure A.1.9, that is, *outside* the center of the curvature, the shadow will advance over the mirror from the right, giving appearance the reverse of shadowgraph B (or as B appears with the page turned upside down.) But *at* the center of curvature, b, the mirror, if spherical, darkens simultaneously over its entire surface, becoming evenly gray like shadowgraph C, Figure A.1.11. As the knife-edge is moved farther, the shadow quickly vanishes. This is the simple test for a spherical surface, but it would be sheer luck if one's mirror appeared thus at the first test.

Viewed as just described, the surface of the curved mirror does not seem curved, but has the strange illusion of being flat. The observer *knows* it actually has a section like Y, under shadowgraph C, but it *appears* flat, like apparent section X, same place.

A.1.6. Figuring the Mirror   11

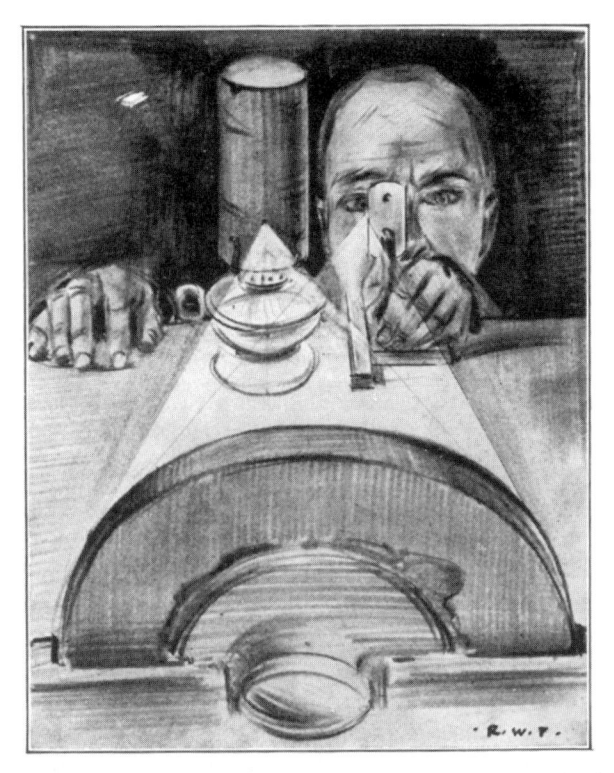

Fig. A.1.8: Making the Knife-edge Test. *The semicircle in the foreground is the back of the mirror, with its handle, set in a simple wooden framework, which can be made of a packing box cut down. Beyond is the lamp with metal chimney pierced by a needle hole; also, the knife-edge. The latter consists simply of a dulled safety-razor blade or any strip of metal set in a split stick of wood which is driven into a hole in a block of wood. This crude equipment serves as well as if it were elaborated with more complicated devices.*

A.1.6 Figuring the Mirror

The surface having been brought to a sufficiently fine polish and to a spherical curve, the remaining work on the mirror, known as the "figuring," consists in slightly deepening this spherical surface into a paraboloidal surface, and this is done by polishing away the center faster than the edge. The final goal is to make the mirror appear, when the razor blade is beginning to cut off the light, like the shadowgraph E, F, (Figure A.1.12) or some intermediate depth, depending on the focal length, which need not be exact.

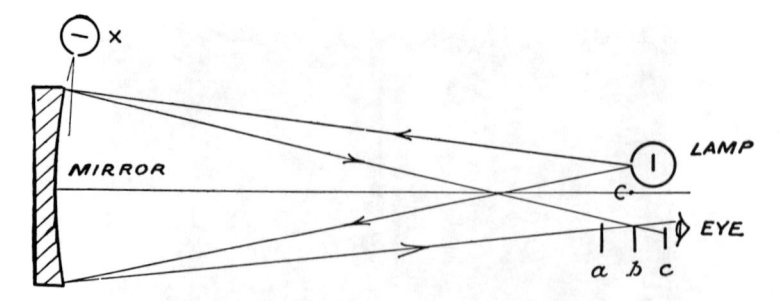

Fig. A.1.9: *Finding the Center of Curvature of a Spherical Mirror. This is comparatively easy if the mirror is a true sphere. This point, b, can be located quite precisely. If, however, the mirror is parabolized, we speak of the "mean center of curvature," that is, halfway between that of the outer zone, regarded locally as a section of a sphere, and the central zone, similarly regarded.*

A common imperfection will be a raised or depressed zone, appearing like G and H, Figure A.1.13, whose true (lower) and apparent (upper) sections are shown beneath them. In the case of the raised zone the shadow has all the reality of a flat surface on which is a raised portion in the shape of a ring, the left slopes of shadowgraph G, being in the shade, and the right slopes being in the light, *as though* the mirror were illuminated by a lamp placed on the opposite side of the glass from the knife-edge, as at X, in Figure A.1.9. Figure A.1.14 is an attempt to show how this imaginary lighting, at grazing incidence, *would* produce these shadows. Here the shadow of the man's fingers is superposed over the knife-edge shadow of a paraboloid. Conversely, a depressed zone (shadowgraph H) will have its lights and shades reversed.

Other characteristic shadowgraphs shown indicate curved surfaces well known to geometers under mouth-filling names. I would refrain from repeating them here for fear of throwing the novice into a panic of discouragement, but they must, nevertheless, be labeled for purposes of identification. Perhaps it will refresh the student's memory to note again the relations of these curves as shown in conic sections.

We have already considered the sphere whose section gives a circle (near top of cone, Figure A.1.15). Its neighbors below are the ellipse, the parabola (whose plane lies parallel to one edge of the cone), and the hyperbola. When rotated the ellipse produces a solid, the ellipsoid. This term, which is synonymous with prolate spheroid, when used in mirror making refers to the region of the vertex. The parabola, when detached from the cone and rotated, produces the paraboloid and *this is the surface of a perfect telescope mirror*. The hyperbola is steeper and on the other side of the parabola. The shadows of these three curves are all alike in shape and size, though they differ somewhat in depth, and therefore in or-

A.1.6. Figuring the Mirror

der to distinguish them it is necessary to take measurements by means of the knife-edge. The same curves are not even alike in depth of shadow for all mirrors; a mirror of short focus gives a stronger shadow (shadowgraph E) and a mirror of long focus a fainter shadow (shadowgraph F). The oblate spheroid (shadowgraph D) comes from the side, not the vertex, of the ellipse.

There is something uncanny about these shadows and shadowgraphs. As before mentioned, they should all be interpreted *as though* illuminated by light coming in from the right. But if one can force one's self to imagine these shadows as produced by light coming from the left, they will give an impression exactly the obverse. For example, in the case of the zone (shadowgraph G), one can change its appearance from a bas-relief to an intaglio, like shadowgraph H, by imagining it lighted from the left; and with a little experience one can make it perform in either manner at will. The rule is to consider the light coming from a direction opposite to the knife-edge. Ellison is almost unique among mirror workers in placing the light on the left and the knife-edge on the right.

Now all of these possible surfaces into which one's mirror may develop, are to be treated in the same way—the apparently raised portions are worn down to an apparently flat surface. There are several ways of accomplishing this result and all are described by Ellison (see Section A.4), at greater length than the present space could possibly permit. In general, a zone may be reduced by removing a part of the pitch lap, for it is evident that a square of pitch removed as shown in Figure A.1.6 (right) would tend to raise a zone on the mirror. The danger here is in producing unexpected zones, and the drawback of having to remake the lap (always a fussy job) if the altered pitch fails to correct the glass. Suffice it to say that, as explained in Section A.4.7, there are several strokes and positions of the glass overhanging the tool that will bring almost any surface to that of the desired sphere, ready for the slight deepening into a paraboloid, without changing the lap.

This is the hardest, but at the same time the most fascinating, part of mirror making. Any one of these surfaces is so close to the sphere that no mechanical means could detect a difference between them. And yet, under the knife-edge, each type stands out glaringly with its own characteristic shadow—never to be forgotten when once seen.

Let us now assume that the mirror has been brought spherical—that it appears flat, under test. The curve now to be sought belongs to type E, F (shadowgraphs). This is very close to the sphere—so close that but a few moments' polishing with a long stroke, or by letting the glass overhang the tool sidewise, will produce it. *Frequent testing is therefore essential during this crucial work of figuring the mirror.*

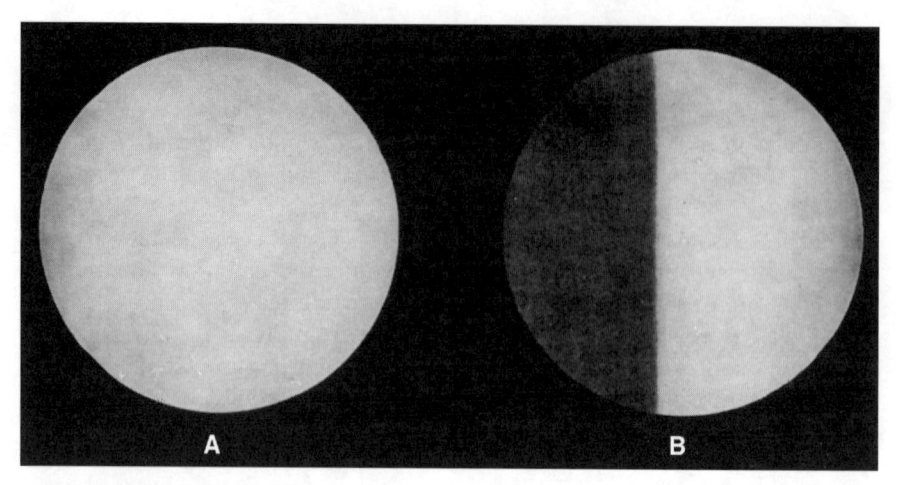

14   *Chapter A.1.   Mirror Making for Reflecting Telescopes*

Fig. A.1.10: *Left: Appearance of mirror before the knife-edge cuts the reflected light. Right: Knife-edge inside the center of curvature of the mirror.*

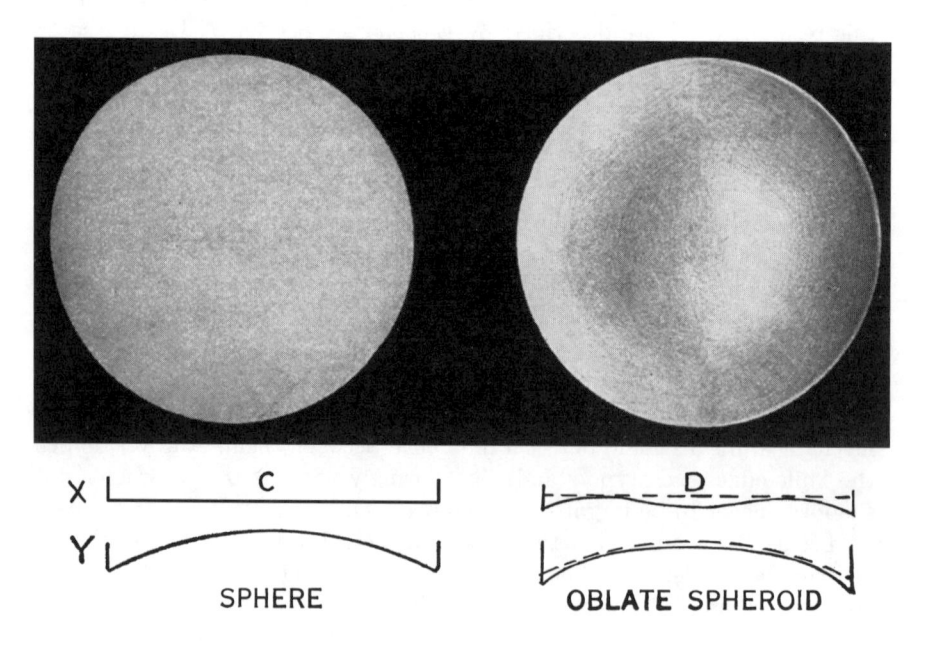

Fig. A.1.11: *Left: Sphere, Right: Oblate Spheroid.*

A.1.6. Figuring the Mirror

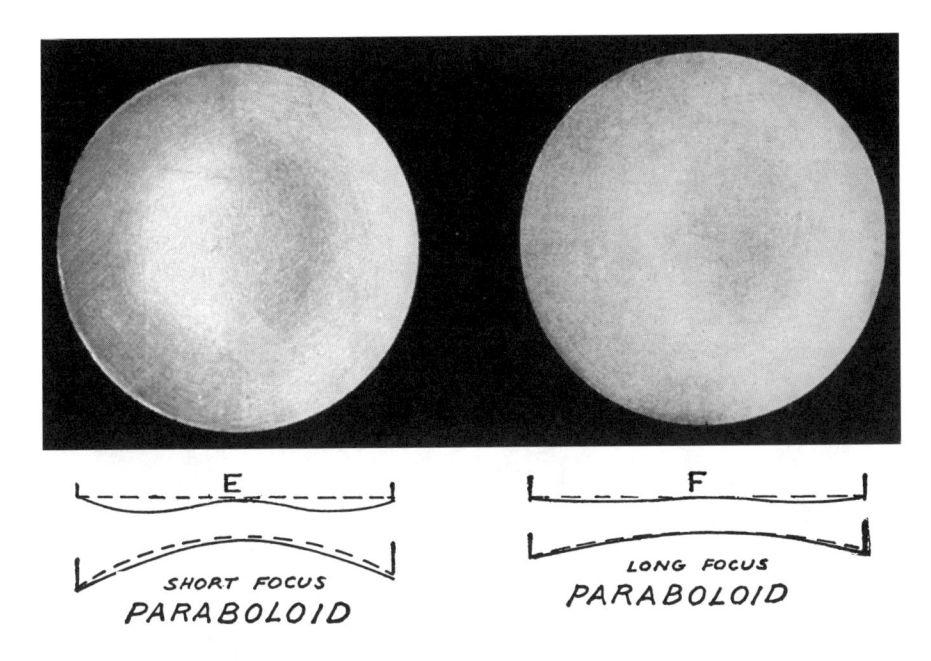

Fig. A.1.12: Left: Short Focus Paraboloid, Right: Long Focus Paraboloid.

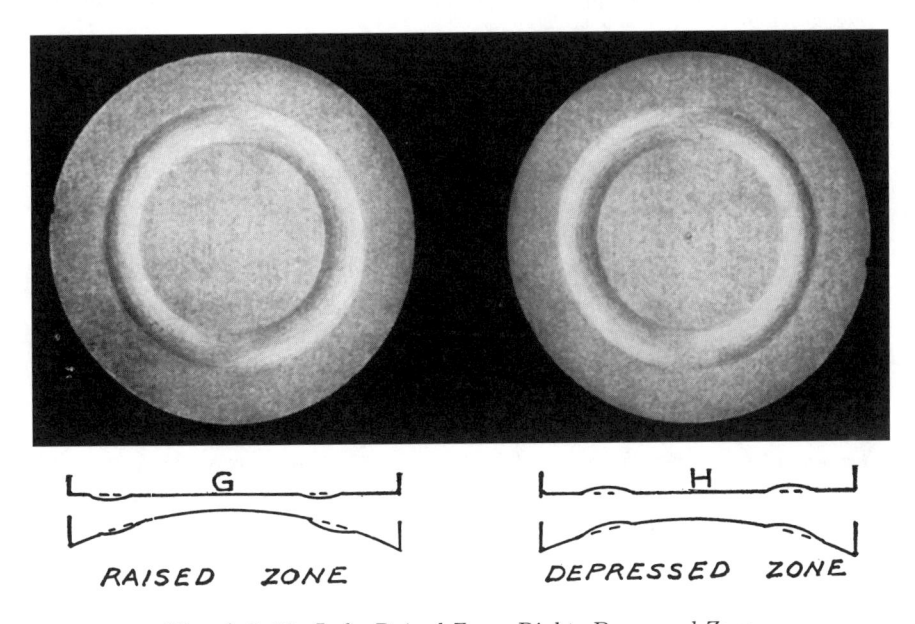

Fig. A.1.13: Left: Raised Zone, Right: Depressed Zone.

16   Chapter A.1. Mirror Making for Reflecting Telescopes

Fig. A.1.14: *An Effort to Explain the Illusion of the Knife-edge Shadows. The real source of light in the shadow test is the pinhole, which is in the front of the mirror. But the mirror appears as though it were being illuminated from one side, grazingly, as in this sketch.*

In Figure A.1.6, the two curves represent sections of a sphere and of a paraboloid. It is evident that the parabolic curve is flatter at the margins, C, C, of the glass than at the central portion, D–D. Therefore light reflected from the pinhole will bring the rays from C and C to a point at B, on the axis of the mirror, further away than the point where the deeper part of the curve, D–D will focus them.

The distance between A and B is given from the equation, AB equals the square of the radius of the mirror, divided by its radius of curvature. Substituting for our 6-inch mirror of 4-foot focal length, we have, AB equals $(3)^2$ divided by 96, or $9/96$, which is about $1/10$ inch.

We now diaphragm out all of the mirror except $1/2$ inch around the margin, and mark on the table the position of the knife-edge *when the light darkens equally over the exposed portions.* All of the mirror is then covered except a central portion 2 inches in diameter, and the knife-edge test is again applied similarly (see Fig. A.4.6). This time, if the surface is correctly parabolized, we shall have to move the knife edge toward the mirror $1/10$ inch, as above determined. In both of the above tests, what we are really doing is to select limited parts of the paraboloid and regard each

A.1.7. Silvering the Mirror

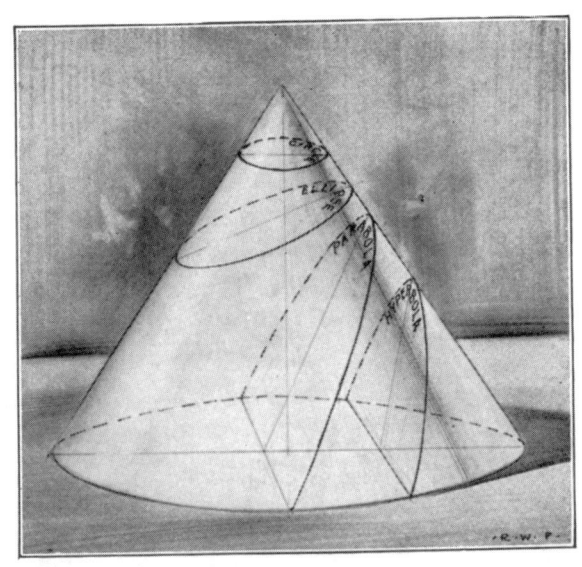

Fig. A.1.15: *The Various Conic Sections.* The curves that may arise as the mirror is worked may be expressed as sections taken at various parts through a cone. For purposes of instruction, an actual cone of wood may be cut across on each of these planes. It is well for the worker to become familiar with the nature of each type of curve or conic section. Any good encyclopedia describes them.

part as locally spherical; and then determine the degree of parabolization by ascertaining the difference in focal length of the respective spheres.

A.1.7 Silvering the Mirror

Silvering is now in order. It was some time before I produced a good, tough, silver coating, but if I had had access to the information in ATM3, Chapter C.1 there would have been no trouble.

Finally, if a lacquer diluted six times with amyl acetate, is poured over the mirror and allowed to dry with the glass on its edge, the lustre of the silver will be prolonged for years, without in any way impairing its optical properties.

I have intentionally selected the method of testing at the center of curvature as being the favorite among amateurs, notwithstanding the fact that it is not as rigorous as testing at the focus with an optical flat. Testing at the center of curvature is capable of yielding an acceptable glass in a reasonable time before the worker's patience has become quite exhausted. It is far better in the interest of the amateur astronomer that he go far enough

Fig. A.1.16: *Parabolizing by Overhang. The most common of the several possible methods. Light side pressure is being exerted by the left hand. The method of permitting the hands to touch the glass, as depicted above, especially during the later stages of polishing and during the figuring, has been criticized. The warmth of the hands will not be so likely to expand the glass and affect the figure adversely if an easily removable annular disk of corrugated pasteboard the diameter of the disk is slipped over the central handle. This is a good insulator.*

in the work to see and appreciate what a fair mirror will show him in the heavens, before he bogs down in the slough of despond and throws up the job as impossible.

A few years ago I tried out the method just mentioned with fifteen mechanics (taken at random from the industrial shops of Springfield, Vermont) and they all produced acceptable mirrors, and nearly all finished their mountings.

A reference to the method of testing at the focus will be interesting to one who has figured his mirror at the center of curvature. The setup is shown in Figure A.1.17, at the top. The optical parts are arranged just as they will be in the finished telescope, but with the addition of a flat, silvered mirror, placed as if it were just outside the telescope tube.

Light from the pinhole strikes the speculum via the prism, and if the mirror is correctly parabolized it goes out of the tube as parallel rays. It then returns by reflection from the flat, following back over its outward

A.1.7. Silvering the Mirror

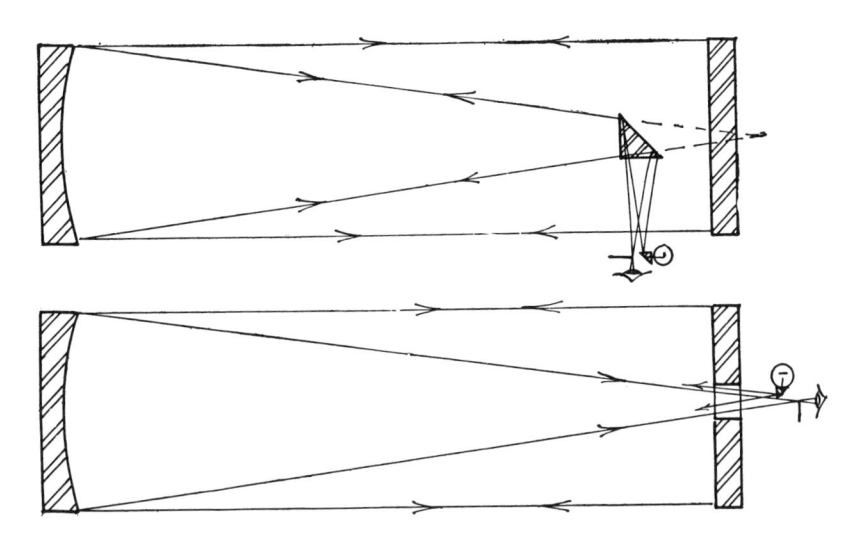

Fig. A.1.17: *Two Methods of Testing at the Focus.*

course and producing an image of the pinhole at the knife-edge. Thus, parallel rays are obtained, just as if they came from a distant star, and since they are parallel we may test at the focus instead of at double the focal length, as we formerly did.

In other words, we have manufactured parallel light in the laboratory. This is the more rigorous method of testing. The paraboloid will darken evenly all over, exactly like the sphere with the other test.

In testing at the focus the pinhole and knife-edge must be brought close together in order to avoid the necessity of providing a large diagonal. This is accomplished by placing a small ($1/4$-inch) prism over the pinhole, as shown in the upper figure. I parabolized about 100 6-inch mirrors by this method, modified as shown in the longer drawing. Here the flat had a central hole and the pinhole and knife-edge were located just back of it.

If these two arrangements are closely studied it will be seen that two reflections are avoided by the second. In the first, the light is reflected from the pinhole to the large diagonal, thence to the concave, thence to the flat, and returns over the same course in reverse order, to the knife-edge. In the second, however, the light goes from the pinhole to the concave, thence to the flat, and returns via the concave, back to the knife-edge.

Thus in the first method there are five reflections, against three in the second. Not only is this a great saving in light, but the ease with which the second arrangement may be put in adjustment and the image found shows at once the advantage of the second method.

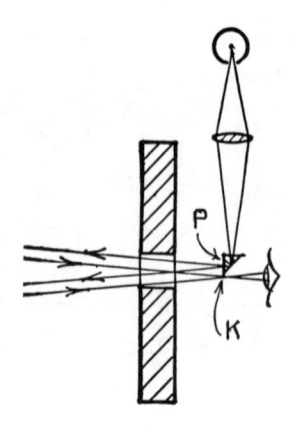

Fig. A.1.18: *A Modification of Figure A.1.17*

An additional refinement is the introduction of a condensing lens between the light source (I used a cylindrical acetylene flame) and the pinhole, and this is shown in Figure A.1.18. This allows the lamp, with its unavoidable heat, to be removed from the vicinity of the flat.

I covered the front of the small prism P with tinfoil in which the pinhole was perforated. Incidentally, the tinfoil was made to overlap the edge of the prism slightly at K, and thus it became the knife-edge itself. In this way I was enabled to keep both pinhole and knife-edge within only $1/8$ of an inch of each other, permitting the use of a flat which happened to have a central hole less than 1 inch in diameter.

The amateur's attention should be called to the fact that the setups shown in Figure A.1.17 are primarily intended for producing parallel light *artificially*. The returning light rays from the large flats are precisely like the light coming from a star, but with the great advantage that the rays are not affected by disturbances of the earth's atmosphere.

For testing at the focus, an ordinary engine lathe can be made into an excellent testing bench by removing the head and tail stocks, mounting the concave mirror on a suitable support at the head stock end of the lathe bed and fastening the flat and the lamp to the cross slide. Anyone who is familiar with the engine lathe will realize at once that the pinhole and knife-edge can thus be maintained in perfect control, both toward or from the concave, and at right angles to the axis of the mirror.

Nothing has been said here about scratches, effects of changed temperature on the glass, where best to work, testing with an eyepiece, the dreaded turned-down edge, sticking of the glass, the various strokes and altered laps, and so on. Ellison covers them all in Chapter A.4.

A.1.8. Biographical Note by the Editor

Sir Howard Grubb, the well-known English maker of telescopes, is credited with the remark that "when the mirror has been brought to a complete polish, the work is about one-quarter done." And while it is true that the long interval of figuring, with its interminable testing, tries the soul of the amateur, let him take pride in the fact that he is dealing with—and controlling—minute errors a thousand times smaller than those dealt with by a mechanic or machinist; and in the satisfaction of knowing that with this mirror made with his two hands he will be able to see the polar caps of Mars, Jupiter's bands, Saturn's rings, nebulae, clusters, and double stars—an instrument that would have excited the envy even of Galileo and Newton.

My experience has been this, that anyone who can use his hands, is possessed of moderate patience and sufficient reasoning ability to interpret the knife-edge shadows, can make a good mirror. Without these attributes he had better forego the venture.

Mirror making has many points to commend it. The tools are easy to make. The cost of materials is (compared to results) low. The work may be carried on at odd moments, day or night and in any available room of the home. In short, it contains the elements of a real indoor sport.

A.1.8 Biographical Note by the Editor

Russell Williams Porter, general advisor, source of inspiration, and collaborator with the editor in the preparation of this book—very largely, also, its illustrator—has had a romantic career; so much so that the writers of human interest articles for the magazines have not missed him (e.g., "One Happy Man," by Webb Waldron, *The American Magazine*, Nov. 1931). He was born in 1871. He studied architecture at the Massachusetts Institute of Technology and, while still an undergraduate, became enamored of arctic exploration, through the direct influence of Admiral Peary. He paid most of his college expenses from the proceeds of summer excursions to the Far North which he organized and on which he conducted other students eager for adventure. For many years thereafter he devoted his life to arctic exploration, making eight extended trips north of the Arctic Circle with Peary, Fiala, and others, acting on these expeditions in the capacity of artist, astronomer, surveyor, topographer, and museum collector in Greenland, Baffin Land, Alaska, and Franz Josef Land. Turning from these pursuits he became interested in telescope making, wholly as an amateur and very much as the general readers of the present volume have become interested. When the United States entered the World War he turned the skill thus gained to the advantage of the government, doing optical work throughout the war period at the National Bureau of Standards in Washington. Later

Fig. A.1.19: *Stellafane, or "Stellar Fane" (shrine to the stars) as it was formerly called, is the little clubhouse-observatory of the Telescope Makers of Springfield, perched on the summit of a fir-clad mountain three miles from that community in southeastern Vermont. The illustration gives only a meager impression of the equipment available. At the right, just out of the photograph, is the tower of a spectrohelioscope to be constructed. At the rear (south) is a polar Cassegrainian telescope of the type shown in Chapter D.1, Fig. D.1.2; also the Sun telescope shown in ATM3, Chapter F.1, Fig. F.1.1. Inside there is a star transit, a well-equipped optical shop, an equally well-equipped kitchen, sleeping quarters, and a collection of telescopes which may be taken outside for use. Inscribed on the facia of the roof, in front, is the verse, "The Heavens Declare the Glory of God." (Psalms 19:1)*

he returned to Springfield, Vermont, his birthplace, as Optical Associate with the Jones and Lamson Machine Co., well-known in the machine tool industry the world over, and there he developed the screw-thread optical comparator originally conceived by Governor James Hartness, head of that industry. As a side line he also built fifty "garden" telescopes, each with a 6-inch mirror, which embodied in pleasing form for the amateur user his invention of the split ring equatorial mounting. Discovering intelligent

A.1.9. Note Added in 1996

Fig. A.1.20: *Russell W. Porter by Dr. Clyde Fisher*

interest in amateur optics and astronomy among the expert mechanics in the Vermont community, in 1921 he organized the *Telescope Makers of Springfield*. Through this and numerous other activities in practical optics, he became known to the principals who built the world's largest telescope in California and was offered an opportunity to collaborate with them in the design and construction of that giant instrument. Amateur telescope makers in this country rightly regarded Russell W. Porter as the father of their hobby. He died February 22, 1949.—*The Editor.*

A.1.9 Note Added in 1996

In 1976, Berton C. Willard authored a wonderful book about Porter titled *Russell W. Porter—Arctic Explorer, Artist, Telescope Maker*[2] which

---
[2]The Bond Wheelwright Company Publishers, Freeport, ME (ISBN 0-87027-168-7)

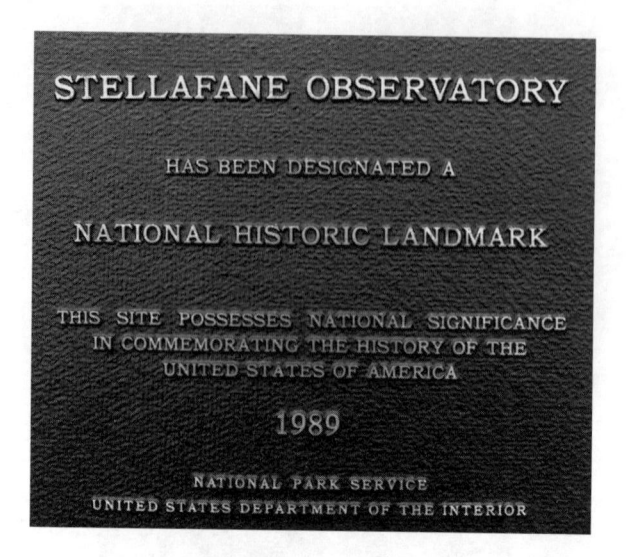

Fig. A.1.21.

provides much additional information about Porter and his work. Also, another book about a fellow Springfield Telescope Maker is Oscar Seth Marshall's autobiography titled *Journeyman Machinist en Route to the Stars—Stellafane to Palomar*,[3] edited and published posthumously in 1979 by his daughter Eva Marshall Douglas.

---

[3]Wm. S. Sullwold Publishing, Inc., Taunton, MA. (ISBN 0-88492-025-9)

Chapter A.2

The Beginner's First Telescope[†]

A.2.1 Editor's Note

This chapter originally appeared at the end of *Amateur Telescope Making*, (4th Edition, Eighteenth printing), starting at page 465. Ingalls noted then that

> Logically it belongs at the beginning because it describes the least complicated method of making a first simple telescope. The previous approach to the art was to shoot for a perfect mirror in an ideal mounting on the first try. Too often this worked out in one or the other of two unhappy ways. Believing the mirror must be almost perfect or it would not work, because the book says nothing to the contrary, many gave up when their mirrors were actually good enough to please them in a telescope. Or the beginner tried to design his final mounting without the practical experience that can be gained only by using some kind of telescope for a while. The result would be something too expensive to scrap but too faulty to please.
>
> For these reasons many old hands think it is better to start with a reasonably good mirror in a reasonably good first, or "pilot," mounting and later go in for the frills after gaining experience from actual use of the telescope. With this in view the method of mirror making described in the chapter has been cut to the bone. It omits the mirror handles, as many had long been doing to good advantage. It substitutes the softer, more easily worked plate glass for Pyrex. It omits the pitch polisher, which many beginners find cantankerous, and substitutes the easily manipulated honeycomb foundation. It omits the paraboloid and aims

---

[†]By James L. Russell.

at the sphere. Finally it entirely omits the Foucault test and substitutes careful control of the strokes. Many think that the average, isolated, uncoached beginner who tries to improve his first mirror by using this unfamiliar test, so easily performed wrong in any of 999 ways, and trying to make the many delicate mirror corrections it calls for, are as likely to downgrade a tolerable mirror, not as judged by the test but by the much less exacting test on the stars.

For telescopes after the first the author recommends the regular, orthodox, full-dress, Pyrex, pitch, Foucault test and paraboloid method. In classes led by him at the Cleveland Museum of Natural History hundreds of average people—chemists, cab drivers, physicists, truckers, engineers, school teachers—have made first or pilot mirrors and mountings. These invariably served to fire their enthusiasm for astronomy. Later, many made larger, finer telescopes of various types.

While the Russell approach should help those who find other parts of this book too mysterious, its use might well be considered by others as the most profitable approach. Here is one psychological reason. The perfectionistic methods described in this book and its companion call for long, deliberate painstaking work and mental calm, just when you are itchy to finish and put your first telescope to use. By the Russell method you arrive soonest. Then, with a telescope in use on the clear nights and your tension let down, you can calmly fuss on as many of the cloudy nights as it takes—probably months—to perfect a second mirror. By then you will be much better qualified to appreciate this telescope because of your practice in observing. It is a fact having nothing to do with the user's I.Q. that, to the inexperienced observer, a perfect mirror seems to perform no better than one considerably less than perfect.

A.2.2 Introduction

It is not the aim of this chapter to justify less than high-grade optical workmanship, nor to attempt to rewrite or debase the techniques of telescope making described elsewhere in this volume. What follows is a much simplified method for the immediate needs of a large number of would-be amateur telescope makers and astronomers who do not aim to become scientific wonders. Relatively few amateur stargazers will conduct astronomical research such as a few advanced sharks do, concentrating year after year on some barely visible, tiny detail of the Moon or a planet. Nor will many

A.2.2. Introduction

Fig. A.2.1: *Grinding mirrors at the Cleveland Museum of Natural History.*

go so far as to attempt celestial photography, which calls for a better-than-average telescope and no little fine machinery to go with it. Instead, the aim of the largest number of stargazers will be simply to look at the common celestial objects which they have read about, and perhaps to show their handiwork to admiring neighbors and friends.

Therefore let us pass by all theoretical things like the "Rayleigh limit" and "Wright's tolerance," as used in more advanced optical work. Some who skip these basic abstractions and underlying optical fundamentals when building a first simple telescope will look into them later, as they should when searching for ways to improve their workmanship, and this is well. Luckily the beginner does not need them.

For the purpose of this chapter neither an excess of gray matter nor more than a little money is necessary. I would show plain Joe Doakes how to make a plain telescope, though not a toy. Joe may be a college graduate or, due to circumstances beyond his control, not a graduate of anything except the school of hard knocks. If the latter, he will probably get the greater kick out of his handiwork. To complete a telescope Doakes does not even have to be a mathematician. He may barely be able to add up the grocery list on a Saturday afternoon marketing tour. He may, however, harbor a long-smoldering desire to own a telescope but just hasn't had the nerve to get it started. He says that the books read so complicated.

It is quite helpful, though not at all essential, if two or more people can get together and work in a group. Two heads are better than one, and it's more sociable. One of them may have a shop space just waiting for such a worthwhile project. It doesn't have to be a basement shop, as

Fig. A.2.2: *A six-inch reflector with Pyrex mirror, made by Prof. M. de K. Thompson of the Department of Electrochemistry, the Massachusetts Institute of Technology.*

in advanced telescope making where widely varying upstairs temperatures affect the more sensitive materials used.

The first decision you will probably make will be to construct a 500-power telescope right off the bat. You'd like to see the whiskers on the gentleman in the Moon. Get this idea out of your head right now. Through the telescope this celestial face disappears anyway and, furthermore, you could not make a 500-power job at first, no matter how hard you might try. That takes experience. Besides, it is not the power that counts so much as the resolution. In plain language resolution means clarity. To see an object clearly, with detail of the main features showing, is of greater importance. It is much more satisfying to see a planet, moon or nebula smaller but clearly than to behold them as huge but fuzzy objects with all detail gone. So you must be reasonable. A 500-power telescope would require a tube about 500 inches long. Imagine yourself going out of the back door with 41 feet of seven-inch furnace pipe on your shoulder.

A.2.3 The Poor Man's Telescope

Instead of the familiar spyglass type of telescope having a large lens, which is tedious to build, requiring a lathe and other expensive equipment

A.2.3. The Poor Man's Telescope

Fig. A.2.3: *Typical amateur activities. Members of the Amateur Telescope Makers of Los Angeles in their shop. In the background is their grinding and polishing machine. Also note pedestals of pipe cast into washtubs filled with concrete. This provides a pedestal heavy enough to "stay put," yet capable of being rolled about and moved when desirable.*

and expensive optical glass, the amateur usually chooses the reflecting type. Russell W. Porter, the telescope mentor, called this the "poor man's telescope" because it is less expensive to build. Yet it is as good as the type with a lens; in fact, the great $6,000,000 telescope on Palomar Mountain in California is the same type. This type has a circular mirror ground concave like a shaving mirror. The concave mirror reflects the light from a star in a conical, tapering beam to a focus some distance in front of it. An eyepiece containing a small magnifying lens is placed at this point of focus to magnify the image formed.

Here it will be well worth while, before reading on, to stop and study closely all details of a reflecting telescope in the diagram in Chapter A.1, Figure A.1.3. Note the main mirror, the focus at F, the diagonal mirror at D which prevents the rays from actually reaching F and, instead, reflect them to F' where the eyepiece lens magnifies them for the eye. We shall often return to this drawing, and it will help make things clear all along if you almost memorize it before going on.

The point of focus can be made to fall at almost any desired distance in front of the mirror, from a few inches to many, by choosing the depth to grind out the concavity. The deeper the concavity the shorter, not longer, the focal length (the length of the tapering beam). Speaking in terms most easily understood, the magnifying power of your telescope will work out roughly the length of this tapering beam in inches. Now, since you can make the beam as long as you wish, simply by putting less work on the concaving, you could therefore make a mirror that would magnify a celestial object as much as you might wish, if it were not for a number of practical limitations long known to professional and amateur telescope makers. Chief of these is that, beyond a practical point, the more you magnify the less clearly you see the object.

In the six-inch telescope we propose to make, because that size is probably the best for the beginner, the length of the tapering, reflected beam—in other words the focal length—will be about 48 inches and the magnifying power about 48 times. A six-inch mirror of this approximate focal length, well made, mounted and adjusted, should show four moons circling around Jupiter, Titan, the largest moon of the planet Saturn, the great nebula in Orion in the winter season, and the huge star cluster in summer, and other celestial wonders too numerous to mention here. Our Moon will be a most glorious spectacle, with its great mountain ranges, some of them almost as high as Mt. Everest, its craters many times larger than any on earth, a spectacle guaranteed to make all your visitors exclaim.

Of the parts of the telescope diagrammed in Figure A.2.8 we shall make the main mirror, cut the diagonal out of a piece of common plate glass, buy the eyepiece, and omit the tube by mounting the parts on a single board as shown in the illustrations in later pages. It is easy to make this mounting, and its cost is very low.

A.2.4 The Main Thing is the Mirror

The mirror is the thing that takes most of the work and time to make. Yet, most of this work is fun. The mirror is a thick round piece of glass with a spherical depression ground into one surface and polished incredibly smooth. This glass does not actually become a mirror until it has been sent away to have an extremely thin layer of aluminum or other reflecting metal placed upon the concave side (not on its back, like ordinary mirrors). Yet, for convenience and because everyone does, we will call it the mirror right from the start. However, it is not the glass but the shiny metal that reflects the light received. The glass is only a spherical form to hold the reflecting metal to the proper shape. As the light from the distant object observed does not pass through the glass it would make no difference if the glass was opaque. This is why the glass need not be a fancy variety, thick plate glass serving very well. Many substitutes for glass have been tried but no other has seemed to work so well. Glass was used in the world's

A.2.4. The Main Thing is the Mirror

Fig. A.2.4: *A homemade telescope made by E.L. Worbois and described in the* Scientific American, *May, 1928, page 448. The sights are like enlarged peep-sights on a rifle. This is almost as satisfactory as a finder.*

Fig. A.2.5: *The telescope made by E.L. Worbois and shown in Figure A.2.4 is shown folded up for transportation in a car. The mirror has an 8 inch diameter.*

largest telescopes after other materials were given up.

For your mirror you will need a circular piece of plate glass at least $3/4$ inch thick (hence windshield glass is much too thin), and a second piece having the same diameter and at least half an inch thick. (No harm, however, if this piece is as thick as the other.) This second glass is called the "tool" and will be discarded when the job is done. Instead of the often used Pyrex, plate glass is recommended mainly because it is softer and more easily ground and because it is a little less expensive. The chief advantage of Pyrex, its lower expansion with temperature change, need not worry us on this simple telescope because we shall not yet be expert enough to take advantage of its small extra value.

You will need abrasive grains and powders in a series of five or six finer and finer sizes for grinding, also sheet beeswax called honeycomb foundation and optical rouge or a similar substance for polishing. While all these materials could on a pinch be assembled separately, provided you could be sure of exactly the correct kinds of each, it is easier and safer and cheaper to buy them in prepared, packaged kits available from dealers in amateur telescope making supplies (see advertisements in amateur astronomical magazines, such as *Sky and Telescope*, Cambridge, Mass.).

As soon as you start grinding you will need an inexpensive steel rule, though any good metal-edged ruler should suffice provided it is not curved. In fact almost any rigid length of metal will suffice provided it is straight. If none is available you can buy a six-inch steel rule from a mail order house for a small sum.

You will need a thickness gauge to slip beneath the rule so that you can stop hollowing out the mirror depression when it is at about the correct depth. For this, a common dime may be used. From many dimes select one with the right-hand part of the date partly worn off, instead of a new or a recent one. Such a dime will have the most desirable thickness.

For the wax, order a ten-sheet package (the minimum sold) of "Medium Brood, $8\,1/16''$ by $16\,3/4''$ Comb Foundation H20" from the A.I. Root Company, Medina, Ohio. The price in 1955 was about \$2 including the postage. ("Plain unwired medium brood foundation" is also available from your mail order houses, in similar minimum packages—though not numbered H20.) Honeycomb foundation is pure beeswax in sheet form which bee keepers use for coaxing the bees to build straight combs, as it has the outlines of the cells embossed into it. Keep the flat sheets carefully covered and in their box in a cool place at all times when not in use and do not allow them to become bent or dented or nicked by handling them.

The beeswax foundation will take the place of the pitch that comes in the put-up kit. (The pitch may be more than welcome if you later decide to make a second telescope and have the common trouble learning to make a good pitch lap, and waste much pitch. Therefore, carefully save it.)

Before we start work we must provide a solid grinding table of some sort to work on. This may consist of a 12-inch or 14-inch square of plank mounted upon a length of two-inch pipe by means of a floor flange. The pipe may be from 30 to 36 inches long, depending upon the height of the operator. Fasten the bottom of the pipe to the floor with a second floor flange, screwed, bolted or concreted into it. Instead of fastening the pipe to the floor, some set it in a tub of concrete (see Figure A.2.3), so that it can be rolled out like a barrel and rolled back out of the way when not in use. Another suggestion is to mount the mirror atop a wine barrel filled with water, or on top of a flour barrel filled with stones or sand. There are many ways of providing a grinding table but the main object is for the operator to be able to work completely around it. Emphatically the stand must be sturdy and must not rock, since at times you will be pushing the mirror across the tool quite forcibly with both arms and really putting your back into it.

To mount the tool, place it either side down upon the center of the grinding table and space evenly around it three wooden cleats. These may be pieces of plywood carefully nailed to the table top with brads or something like those in the illustration (see Figure A.1.16). Leave space between one of them and the glass for the insertion of a paper match or a tiny soft wooden wedge, so that the glass can be removed after pulling out the tightener.

You should provide a bucket three-fourths full of water, in which to rinse the abrasive off your mirror from time to time. Abrasive washed down the sink will settle in the trap because it is heavy and will plug it solidly.

Provide a bottle of water with a sprinkler cork, a salt shaker filled with the coarsest abrasive grains (not coarser than No. 80 if you have bought them separately, No. 80 being roughly the size of table salt), otherwise the coarsest size in the kit.

Also provide a clipboard with ruled paper on which to keep accurate record or log of everything you do and the material and time it took to do it.

A.2.5 And So—

It is now our job to grind in our glass a spherical depression, meaning a shallow basin that a very large sphere would fit—and afterward polish this concave surface and have it coated with reflecting metal. It was discovered long ago that if two circular pieces of glass having the same diameter are placed one upon the other and rubbed together with loose abrasive grains between, the lower surface of the upper disk will gradually grind more and more concave and the upper surface of the lower glass more and more convex in like amount. See Figure A.1.2. The concave glass is the one we shall put in the telescope, discarding the other after it has served its purpose (or we could use its back later for making a second mirror).

The depth of curvature on both glasses and the speed of cutting will depend entirely upon how skillfully they are rubbed together. The curve is controlled by the strokes that we use. A long stroke back and forth will produce a curve too deep in the middle to be quite spherical. A spherical surface should result if we rub with a shorter stroke one third the diameter of the glass, or two inches long for a six-inch glass.

Let's get this matter of stroke lengths clear before we are any older, and have it in black and white to refer back to, otherwise we may later get into fatal trouble due to a misunderstanding. A two-inch stroke does not mean that the glass goes two inches beyond the tool at one end of the stroke, then two inches beyond it at the other. That would be a four-inch stroke. Instead it means a one-inch overhand at the farther end, and then a one-inch overhang on the end near you. Total, two inches of movement. On a six-inch mirror this would be a one-third stroke. The strokes shown in Figure A.1.2 are about one quarter strokes at A and one third at B.

Since no one could control his hands accurately enough to keep taking strokes just two inches long, you may now say, as many have said, "Why work by hand when a machine could do it more accurately and more easily?" The trouble is that the machine would do it too accurately. We don't even want each separate stroke to be exactly two inches long. What we want is a mixture of thousands of strokes, some a little but not much longer, others a little but not much shorter, all a little different but all averaging two inches long. Such a combination prevents the formation of rings of unevenness on the mirror surface, like those on a target. When you stroke freehand the mixture of different strokes is what you get. A machine might cause such rings because all strokes would be alike, unless you knew just how to keep changing its strokes. A few advanced amateur telescope makers do use machines. There are illustrations of some in this book. But the users could not get the hang of the correct use unless they had first got the sensitive feeling into their hands by making a number of mirrors by hand. The reason is that the machine can't feel and does not learn by experience.

Let us sum up and get it all clearly in mind. The evenness of the curve on the mirror is controlled by the average length of strokes which are never the same, but which vary between limits. The better you keep these limits controlled the better will be the result with our method, which depends upon careful attention to prescribed details and not on guidance by difficult tests.

There are two additional things which, combined with the varying strokes, are needed for obtaining an even curve.

1. Clear through to the end of the grinding job you must always turn the glass bit by bit in your hands as you push and pull it back and forth.
2. You must at the same time walk slowly around the grinding table, choosing whichever direction you like best.

A.2.6 Now We Start

To start off, sprinkle a little coarse abrasive on the upper surface of the tool and add a few drops of water. No, I do not mean a flood. Most people use many times too much water. About ten drops should suffice. Do not dump the abrasive in the middle of the tool; or, if you do, spread it and the water with your finger so that all the surface of the tool is wet.

Now place the mirror upon the tool, grasp it firmly between the hands and, pressing firmly, push the upper glass away from you across the lower glass with as long a stroke as possible without tipping it off the tool or chipping the edge of the tool from the pressure put on the edge by the overhang at the ends of the strokes. The mirror should overhang well over two inches at each end and almost three inches as soon as you catch onto the knack.

Thus by our rule this stroke, six inches long, is a full stroke. This is a far longer stroke than the one third strokes we have just been describing for a sphere, and we therefore seem to be disobeying our own rule right at the start. However, we shall use these very long strokes for the preliminary "hogging out" of the concavity and for that alone. The reason is that they remove the most glass from the central parts and accomplish this preliminary job much faster than the shorter strokes would accomplish it. Although they deepen these parts too much for a sphere, we shall alter the curve to a sphere by shortening the strokes to one third as soon as we approach the desired depth. In other words, the full strokes are just a trick for quicker hogging out.

You must now learn to turn the glass in your hands a little every few strokes or every few seconds. However, there is no need to count strokes or time them. Unless you keep turning the glass now and then it will not come out spherical but lopsided.

You must also learn to walk gradually around the grinding table as you work.

These three motions,

1. stroking back and forth,
2. turning the mirror,
3. walking around the table,

will at first seem confusing but will soon become as natural as the motions in driving a car. No permanent harm will result if you haven't picked up and fully combined all three before the curve is hogged out, but this is a good time to learn them.

After two or three minutes' work the loud grinding and tearing noise that you hear will change to a mere soft swishing sound and the mirror will push and pull hard and even tend to bind and stick. This is because the abrasive grains have broken down. They have done their work and, since they can cut no more no matter how long you continue the motions, it is time to stop and renew them. You have completed what is called one "wet."

So slide the mirror clear off the tool. Always *slide* it off, never lift it off or even lift it up when you have slid it almost off. To avoid a disaster make this an invariable rule. The reason is that if you lift it off instead of sliding it off, and the two glasses cling together only a little, as they often do, you will some time lift both of them, the tool will cling till you have it over the floor and then will let go and crash, and you never can put Humpty-Dumpty together again. So make this a discipline, always to slide off the mirror. On the other hand it is perfectly safe to replace the mirror on top of the tool to start grinding again. In fact, sliding it on would scrape off the new charge of abrasive.

While grinding, do not get into the habit of leaving the two glasses standing together without motion while stopping to talk or for any other reason. The reason is that as you progress from finer to finer abrasive grains the glasses will tend to stick together toward the end of the wet, more so when they are not in motion. This is because the water between the two surfaces of the glass evaporates and the air pressure from outside presses the disks tighter and tighter as the water and abrasive leave.

A.2.7 Keep Hogging

The process is now to repeat wet after wet until the desired depth of the curve has been reached.

Keep the stroke a good five or six inches long, or as long as you can without tipping the mirror at the ends of the strokes and flaking off little fragments of glass from the edges of the tool. If you have knocked off only a few, probably from actually teetering the glass over the edge, this will do no real harm. Hold the mirror tightly and try to keep it from tipping.

There is a second reason for the very long strokes in hogging out. Not only do they deepen the central parts of the mirror most rapidly, which is

A.2.7. Keep Hogging

Fig. A.2.6: *Mrs. Thomas A. Jenkins of Syracuse, N.Y. and "It," a 6-inch tubeless reflector made by herself. Up to 1932, three women were known to have completed telescopes.*

what we desire, but they at the same time remove the least glass from the outer parts, which we also desire. If instead we were to use short strokes for hogging out we would wear off perhaps as much as one fourth inch from the outer parts before the inner parts were enough deeper to reach the desired curve. This would leave the mirror glass pretty thin. It would also grind off the bevel that comes on the edge of the mirror, leaving a sharp edge that would easily chip. The chips would get between the two glasses and scratch.

The wear at the edge is not fatal, however, but just a little bit bad, or worse if excessive. If the edge is worn sharp, rebevel it under a running tap, using a stone. Keep the beveling strokes lengthwise of the glass, not crosswise. Then note in your log what happened, and go ahead. One purpose of this pilot telescope is to absorb as many of the beginner's mistakes as possible and to teach him what not to do.

The harder you press the faster you will excavate. If you don't perspire soon you are taking it too easy, so push, man, push—and pull too, always using pressure.

Now that you have learned some of the basic things it is time to attune your ear more closely to the music of the grinding. The noise made by the abrasive grains tearing into the glass should sound the same for the forward as for the backward strokes. An inexperienced worker usually bears down harder on the forward stroke and relaxes on the backward stroke, so that the sound is like that from a one-person crosscut saw, which gives a resounding buzz on the forward push but only a swish on the return stroke. On the mirror an experienced worker makes both strokes accomplish work and gets more done. Therefore, don't knead the work like a cook with a gob of dough.

A trick for squeezing a little more out of broken down abrasive while working out a wet is to slide off the mirror now and then and replace it at once, going on until the new tearing noise subsides.

The average adult, using six-inch strokes at the rate of about 120 times across and 120 times back a minute, can hog out the glass at the rate of about $1/64$ inch in each half-hour spell of work. At this rate our mirror should be hogged out in about an hour and a half. However, after the first hour of work let's call a halt and measure the depth to see what we have accomplished.

A.2.8 Checking Progress

Rinse the mirror thoroughly in the bucket near at hand, never in the sink unless you want to have to dynamite the trap to clean it. (Probably, however, it is safe to rinse the glass in the sink after practically all the heavy abrasive has been washed in the bucket.) Now dry the mirror and place it on a table with the newly ground face up. Place the rule on edge, never flat, spanning the center. Peer carefully under it. You will be able to note that it has already been hollowed out a little though not yet much.

Now place the selected dime at the exact center of the mirror. Place the rule on edge squarely across the dime's middle. Keeping the rule there, teeter it over the dime. You can hear the clicks and see the spaces between the rule and the edges of the mirror. The goal is to grind until there is no more space at either end of the rule and no more teeter or click. As the space gets small, test more often, thus feeling your way to the goal, and at least trying to hit it right on the nose. But if you don't there's no real need to weep unless your miss is quite wide. If you quit before the clicks disappear, thus leaving the curve too shallow, you will lengthen the telescope, making it harder to mount, carry, and use, also its field of view will not be quite so wide. On the other hand, the effect of deepening too much, with an open crack left between the rule and the dime, is of another sort and is not so good—images not quite too sharp, also less magnification. The compromise, about 48″ focal length, given by the worn dime, is based on the experience of thousands of telescope users. You will find some figures on this subject a few pages farther along.

A.2.8. Checking Progress

It may seem improbable that so shallow a dish can become a mirror in a telescope with the power of 48, yet it is true.

So now we have completed the hogging out of a hollow but we do not know its exact shape, and thus far neither have we cared. Though the curve looks spherical to the eye, it is actually steeper in the central parts. The difference is too small to be apparent, but is large enough to destroy the clarity of the telescope if left unchanged. From now on, therefore, as we smooth up the glass with finer and finer abrasives, we shall try to get it closely spherical at the same time.

To change the curve to a sphere we shall shorten the strokes, and to make it smooth we shall use finer and finer abrasives. At present the surface looks like rough frosted glass, which is just what it is. After the series of abrasives the surface before we begin polishing will resemble the finest satin.

So now, before changing to the next smaller grains we must clean house altogether more thoroughly than even the most fastidious would understand. Slip out the wedge from the tool and carefully pry up the tool with a putty knife. Dunk it in the pail of water to wash off all trace of the abrasive. Likewise wash the mirror and rinse both glasses under running water. Go so far as to scrub their edges with a brush. Scrub the surface of the grinding table and its supporting pipe. Remove every vestige of the coarse abrasive from any hand tools that you may have used. Scrub the water bottle. Clean under your fingernails.

The reason for this super-fastidious degree of fussiness is that in each successive stage of fineness of grinding one single grain of a coarser grit is a jagged chunk of destruction which can cause an ugly furrow across the face of the mirror. So clean every spot, corner and crevice where any grit may lurk and later hop out on your work and mar it. Then handcuff with their hands behind them all visitors, to prevent them from picking up grit-contaminated objects and thus transferring grit to your work, for they will fail to understand your fussiness and will simply continue to handle things. Grownups are the worst. If they still don't understand, read them this paragraph.

Do not, however, contaminate your abrasives yourself. A common way to do just that is to open the packages all at one time and, run a finger into each to sample its size. Instead, keep them apart and open each only when needed. There are 1001 other ways to contaminate abrasives, most of which you might never dream of. For example, one worker kept nervously running his hands through his hair, an unconscious habit. Coarser grains were left there, and when he later used finer grains he picked up the coarser ones in the same unconscious way. As soon as you get a little experience you will be able to avoid scratches, and if you use your head and think from the start you need never get any. Just don't tear along, but learn to move calmly. The born speedster usually has a hard time before optical work tames him.

A.2.9 No More Long Strokes

After the scrubbing operation fasten the tool again on the grinding table. Use a new salt shaker for the next finer grade of grains to prevent the mixing of sizes. Proceed as before but with one important exception. From now on to the end of the job the stroke must be shortened from the hogging-out full stroke to the normal one third stroke, or from about six inches clear down to about two inches. This means that the mirror will overhang the tool only an inch on the forward stroke and the same on the back stroke.

At first and for quite a time it will not be easy to shorten the strokes all the way, but shorten them you must—yes *all the way* to one third—and stick to it, if you want the mirror to become a sphere and define things. For we are depending upon this stroke-control method as a substitute for the Foucault test.

This does not mean that we must make every stroke exactly two inches long. It means that we want a mixture of thousands of strokes to average close to that length. Some will of a necessity be more or less half an inch longer or half an inch shorter. For, without an automatic stop arrangement no human hand can control the stroke more perfectly. And automatic stops we do not want, since it is the imperfect but not too imperfect stroke by hand that creates the perfect average that prevents the formation of an irregular zone. However, the harder you try to keep within limits the better will be the result, even though one stroke is never exactly like the one before it.

As a constant reminder keep a piece of wood cut one inch long on the grinding table close to the work. Otherwise the inch overhang at the ends may creep toward two inches.

Another tendency of some workers is to make the forward stroke too long, overhanging the tool more than an inch, and the backward stroke too short, overhanging the tool only half an inch or even less. This can be remedied by standing a few inches farther from the table, or a few inches closer if the tendency is to make the back stroke too long instead.

After half a dozen wets of fresh abrasive, substitute for the grains from the salt shaker the wet, gooey sludge that collects on the grinding table. This sludge is simply abrasive grains broken down finer and mixed with particles of glass. Dip two fingers into it, and a third finger into water (which should provide enough water), smear them evenly around the tool and grind. Add water only if the mirror binds and tends to stick, and then

A.2.9. No More Long Strokes

only a drop or two. If the mirror seems to float off the tool when first placed on it you are using too much water.

Fifteen or twenty wets with this stage of abrasive should suffice, provided they are followed by seven or eight with the sludge alone.

From now on this tapering off with sludge should be done for 15 or 20 minutes after each grade of abrasive. It will be well at each stage to re-examine the drawing in Figure C.1.10 and read the accompanying note and the "Last Word" note in Section A.4.12, provided you like to understand the reasons for what you are doing.

The finer and finer grinding with each successively finer grade of abrasive is done in the same way—15 or 20 wets of new abrasive followed by seven or eight of sludge, and with ever more thorough housecleaning at each stage, always including the water bucket. Always use a new, clean salt shaker for each step in grinding.

The pressures should be relaxed somewhat with the finer abrasives.

On the finer grades you will not hear such a loud tearing noise, yet the noise can be heard if you listen closely—less so, of course, with the diminishing sizes. It will pay dividends to slide off the mirror once in a while and replace it while grinding.

With the smaller sizes you must judge when the grains have broken down by the increased pull on the upper glass after the moisture has disappeared between the glasses.

The final grade of abrasive should be used much longer. Start with the grains from the shaker and, after 20 wets, use sludge conscientiously for a full hour.

Care should be taken not to put too much abrasive between the glasses, especially when using sludge, for there is but little grinding action with a thick blanket of wet abrasive. The finer the size of grains the thinner the spread, but never allow glass to reach glass. Most workers use too much sludge, hence this repeated caution. Using the middle finger, swab a little on and quickly spread it all over before its moisture evaporates.

Never for an instant at the stages where the surfaces are very near glass-to-glass contact should you stop the grinding motion, otherwise the glasses may stick together. Should the glasses tend to bind, slide the mirror off quickly and start afresh with a new charge of sludge.

Use water so sparingly as to keep the glass just off the sticking point.

A drop of castor oil, glycerin or turpentine on the tool to mix with the sludge will help prevent sticking.

If, in spite of all precaution, the glasses get stuck, never—*never*—try to pry them apart with a putty knife, chisel, or other tool. Instead, all attempts should be made with a view to sliding them apart. Sometimes they will separate quite easily if a few drops of penetrating oil are flowed around the edges; or if they are soaked overnight in water. Sometimes slowly warming one of the glasses over a hot plate or any source of heat will expand one glass enough to disengage it from the cooler one; but take care not to apply heat too fast as this will crack the glass beyond repair. As a final recourse lay a generous cushion of newspaper on the floor. On top of this lay a length of wood. Place the tool in contact with the wood and hold the wood down with your knee or lay it against a wall. The mirror, which is stuck to the tool, should overhang the wood. Now rest a second length of wood against the mirror and hit this length with a hammer. Or the two glasses may be stood on edge on a block of wood and the higher one hit with a rubber hammer, though this could be rough treatment.

At the finish of fine grinding, the surface of the dry mirror should have a smooth, gray, satiny appearance and be like the finest velvet to the touch. Ordinary newsprint can be read through it when it is dry.

A.2.10 Scratches

The surface may be inspected with a reading glass. Very tiny scratches, invisible to the unaided eye, may be neglected. Deeper scratches ought to be removed by continuing to grind with the finest abrasive until they disappear. Scratches and imperfections appearing near the center of the mirror are not as difficult to grind out as those near the edge. This is because the center naturally grinds fastest. Very fine scratches near the center generally disappear during the polishing operation, but those near the edge should be removed by the finishing grit beforehand.

Here is one cause of scratches. As the fine grinding work progresses the abrasive between the glasses tends to spill over the side or is scraped off by the motion of the upper glass. This gives an opportunity for glass to rub upon glass, resulting in numerous fine, hairline scratches around the edge of the mirror. These can be largely prevented by using great care, when removing the upper glass, to slide it to the *very edge* of the tool, and by keeping the edge of the tool always moist. For the latter it is well to get a helper frequently to sprinkle a few drops of water on the job while you are actually doing the grinding and the upper glass is in motion.

If many scratches and pits are large enough to be plainly seen with the unaided eye, perhaps going back a size or two in abrasive will clear them. If they can be seen only with a magnifying glass and are only near the edge I would try to eliminate them with the finest grade of abrasive, without going back farther. A few scratches of this kind near the center will probably polish out later. The nearer the edge the slower they polish out.

Note often during fine grinding whether the edge of the mirror has again

A.2.11. Polishing

been ground to a sharp edge. If left so, tiny chips started by the grinding may break off later, get under the glass and scratch it. Therefore, hone the edge as already described, always stroking lengthwise, for crosswise strokes are likely to press off little flakes from the face of the glass all around the mirror, and heavy stroking pressures even lengthwise may leave invisible cracks and microscopic fissures in the glass. Then, later on, particles break off, get between the glasses and cause deep scratches.

Before the grand final housecleaning store the sludge from the finest abrasive that is left around the top of the tool in a small, wide-necked, tightly closed bottle with a little water added. There is a grim little joke about this. You'll come to it later.

A.2.11 Polishing

Now wash everything within an inch of its life, for we are going to polish, and a scratch on a polished surface really looks bad and will make you cuss. True, the telescope may work about as well, but will not feel proud of your job, especially when it is finished and coated with reflecting metal, for the metal will not cover up the scratches but will make them a great deal more noticeable.

The bucket will not be needed again, so take it a mile or two away.

Now, with clean hands, cut a sheet of the honeycomb foundation crosswise in two equal parts, using an old razor blade. Take care not to bend or kink it. Avoid crushing down any of its tiny pinnacles by gripping it between the fingers. Instead, hold the sheet near a corner, as the corners will be cut off and discarded. Each battered tip will tend to cause uneven polishing.

We will now attempt to stick one of the square sheets upon the working face of the tool. Scrub the bone-dry upper face of the tool lightly with absorbent cotton dampened with turpentine. Then span the tool with your thumb and fingers and, keeping them there, expose the convex side to a source of warmth such as a hot air duct or hot plate. Do not warm the mirror in hot water. Heat it up slowly and evenly by swinging it back and forth across the source of heat. Your fingers will inform you if you are heating it too fast. If heated too fast the glass may crack and thereafter be entirely worthless, so don't rush it.

From time to time apply the heated surface to the bare underside of your arm, just as a parent does in feeling the temperature of a baby's milk. Our glass should be heated only to about this pleasantly warm temperature, because if it is too warm the beeswax covering we are about to place upon it will melt.

On the other hand, if the glass is too cool the beeswax will not stick.

The idea is to have the glass just warm enough to soften the tips of the little pinnacles on the underside of the sheet next to it so that they will melt and very lightly attach themselves to it, yet not warm enough to soften the rest of the sheet. This is easy after a little practice. You might have to try it twice.

When you are sure that the warmed glass would not burn a baby's skin and is dry, place it on a table, convex face up, and quickly place on it the half sheet of the beeswax. With no delay dip the mirror in water to wet it, so the wax will not cling to it, and lower it on the beeswax sheet—gently so as not to injure any of the tiny tips—and leave it there a few minutes.

You now have a glass sandwich with a beeswax filler. Bear down lightly upon the mirror with one hand while you trim the excess wax from around the edges with a razor blade, like trimming off dough around a pie pan. The beeswax pad between the glasses will then be round.

Keeping the sandwich together, place the whole thing upon the grinding table and wedge the tool in. The cooled mirror may now be slid off the wax.

If the sticking job was done properly the round wax sheet should cling to the tool for quite a length of time, at least until enough water works underneath it to release it, and will absorb quite a little punishment. When and if it comes off, do not try to attach the same piece again, since this cannot be done. Instead, clean and dry the tool again thoroughly and make a brand new lap as you made the first.

From this point on we shall drop the word "tool," for the glass no longer does any work. It is now only a rigid support for the "lap," as the polisher is correctly called.

Before proceeding, mark the back of the mirror with a couple of little strips of adhesive tape. Otherwise, some time when you are weary or absent-minded you may lay the flat side of the mirror upon the lap, as some have done. This will immediately ruin the lap by crushing some of the tiny tips ever so slightly. If you ever do this there is only one thing to do: make a new lap, for the old one will polish an uneven curve.

A.2.12 All Set to Begin Polishing

The glass, with the lap on it, is now cleated to the grinding table ready for business.

Place about two heaping tablespoons of rouge or whatever polishing material your kit contains in a pint Mason jar not quite full of water so that you can mix it thoroughly by shaking it with the cap on.

Allow it to settle a minute and pour some of it into a clean new oil can having a spout, or half fill an eight-ounce medicine bottle with a tiny

A.2.12. All Set to Begin Polishing

hole bored in the cap. Squirt a generous amount on the surface of the lap. Shake the container vigorously each time you do this, because the polishing material is quite heavy and quickly settles to the bottom. Place the mirror upon the lap and commence to stroke with the greatest care you can muster, always using a two-inch stroke and walking slowly around the table.

During polishing it will not be necessary and may even be harmful to rotate the mirror regularly in your hands during the stroking motion. Rotating it more than a little may cause the outer part of the mirror to polish faster than the central parts. The result is called "turned-down edge" or "rolled-over edge," or a "toadstool mirror." The edge part of the curve is ever so slightly less curved than the rest and the mirror will not focus well.

Do not, however, stop walking around the working table.

Not as much pressure is needed upon the mirror in polishing as in grinding, though just the weight of the mirror and hands is not enough.

From now on do not lay your hands long in one place upon the back of the mirror while you stroke. Also hold the mirror with the finger tips only, when you pick it up. The reason is that the heat from the fingers will expand the areas of the glass with which they are in contact. These expanded areas will extend down through the glass and be polished off because they stick out. Then, after the glass has cooled, these same areas will collapse and come out as depressions, or just the reverse of the way they started. The resulting unevenness of surface would cause unevenness of focusing. This trouble may be avoided by changing the position of the hands upon the mirror from minute to minute.

At first, after the polishing material is added, the wax of the lap will be covered with bubbles with dry areas between. This will not last long if you remove the mirror and squirt more rouge water around the dry spots and about the lap in general.

After the lap has become one color all over and the little facets in the wax seem filled, stop being too generous with the rouge water, since a little rouge goes a long way and, believe it or not, too much actually slows the polishing. If a bubbly mess of water and rouge collects around the edges of the lap, you are keeping it too wet with rouge water. On the other hand, do not allow the lap to become dry.

After polishing for half an hour it is time for a look-see to determine what may be happening. Remove the mirror and cover the exposed lap with a clean paper to keep dust from settling upon it. Rinse the mirror under the cold water tap. The polishing water will not stop up the plumbing. Pat the mirror clean with absorbent cotton. Do not rub it, as this may scratch it if grit has gotten into the cotton.

When the surface of the mirror is dry you will see that it has begun to polish. This appearance has fooled many into thinking the job will be done in only a few more minutes. No such luck. The parts that are polishing so rapidly are the flat areas shown in Figure C.1.10, between the millions of tiny pits made by the fine abrasive grains. The real job is to polish the surface down below the level of the bottoms of the pits, as shown in the same drawing.

Perhaps the central part of the mirror is smoother than the edge part, which may not have polished out at all. This common condition results from one or more of several common causes, one of which is generally poor contact between the mirror and lap. Or the beeswax may be making firmer contact with the central part of the mirror than the outer parts. A way to test this is to rest the mirror upon the moist lap and look down through the glass while trying to rock it on the lap by pressing gently with a finger on one side, then on the other, alternating. If the liquid squeegees out at the edges they are a little apart. Try turning the mirror slowly for a minute or so. Or try applying great pressure evenly on it, which may squash the central parts of the lap to even bearing with the outer parts. Or try warming it very slightly in lukewarm water and resuming polishing while it is warm. The mirror may melt the lap just enough to bring back perfect contact.

Now recharge the lap with a few more squirts of rouge water, polish carefully for another half hour, and look again.

Keep this up hour after hour if need be until the mirror is completely polished clear to the edge.

In polishing, we can now keep count in hours instead of in wets, because the polisher does not wear out as the abrasive grains did.

Throughout the entire time it is important and vital not to allow your mind to wander and your strokes to lengthen beyond one third. It is better to quit for the day than to go on with lengthened or careless strokes.

If you must rest for long, or put the work away "come another day," lay a paper napkin on the lap before adding the mirror. The napkin will get wet but no harm will result. Do not use wax paper or anything stiffer, since this will injure the little wax pinnacles. Then wrap the whole sandwich in an outer newspaper and put it away; or, if you leave it on the polishing table, cover it well and securely.

When you resume work you will probably find the lap and the napkin dry. No harm has been done. Just remove the napkin, shake up the oil can, give the lap a few squirts of rouge water and go to work.

As you polish it is well to look often to see what is happening to the mirror surface. All is normal if the edge, though somewhat behind the central zone, is polishing too. Possibly the final half inch will be stubborn.

A.2.12. All Set to Begin Polishing

Fig. A.2.7: As you read "ATM" you naturally will be choosing some design for your maiden effort. Some plan too ambitiously. Consider whether it is not better to make a simple mounting and use it a few weeks before expending time and materials on a really fine instrument. Inevitably, if you do the former, you will discover practical points that will modify your design to the good. Above mounting is neat, trim, easy to make, inexpensive. Tube is square, built up of eight slats of wood, edge-nailed in pairs. Very rigid. If near lights build lower 12 inches in solid to cut off reflection. Wooden tripod has long leg parallel to earth's axis. Mounting is made of pipe fittings. Better to make them heavier than lighter—unless you wish object lesson in shaky telescopes (beginner's besetting sin). Through hole in the wood put long nipple with "T" on top, pipe-nut on lower end. Bore out "T" internally, insert another long nip screwed to another "T" similarly bored out to take declination axis made of pipe. Above design was by Dr. and Mrs. R.M. Watrous, Evanston, Ill. They cast a lead counterweight around a sleeve of larger pipe and anchored it at balance point with set-screw. Modify details to suit your taste, whims, available materials, pocketbook. Almost none of the thousands of telescopes made by amateurs have been duplicates. When he duplicates another's telescope the maker robs himself of the fun of putting some of himself into his design and execution. His chest will never protrude so far out in front when he shows the job to friends and visitors. This gives answer to occasional inquirers for blueprints: make your own "blueprints" with a pencil stub on an old envelope or in the pages of your mind. Your telescope will then be uniquely yours.

I have seen many mirrors refuse to polish clear to the edge at all. This is probably traceable back to fine grinding and is due to my failure to scold you enough at that time, so it is all my fault after all.

If the edge is still cloudy half an inch or so in, you may have to abandon the polishing operation and go back to fine grinding. However, since the cause might even yet be some unobserved condition in the lap, before committing suicide, try putting on a new lap, polishing on it an hour or so, and seeing whether there is progress. Make sure often that the contact is even, as already described, also that no sharp mirror edge remains to shave off the pinnacles on the lap. If this last resort fails, the sad word is to peel off the lap and fine grind for an hour or two more with the joker sludge that you saved up in the bottle from the finest stage of grinding, then polish all over again.

When the mirror is polished clear out to the edge and not before, it is time to call it finished and celebrate.

A.2.13 Mounting the Poor Man's Telescope

Since we are making a pilot mirror, why not also make a pilot mounting. This is the cheapest, most practical way to learn certain facts about mountings that books somehow can't make nearly so real. About the most inexpensive mounting to build is the tubeless or "spinal column" type of which those in Figures A.2.2 and A.2.4 are simplifications of the original designed by Russell W. Porter who later helped design the great 200-inch telescope on Palomar Mountain in California. I shall describe another simplification of the same mounting, which has been successful among amateurs in Cleveland, Ohio.

Really, the reflecting type of telescope need have no tube. Under average conditions, away from brilliant street or back-yard lights, a tube is not necessary. (But if you do have these lights the type shown in Figure A.2.7 is perhaps the next simplest, with the lower part of the tube enclosed as suggested under the illustration.) The big advantage of tubeless telescopes is that the parts are easier to mount and adjust. Nothing is in the way of the hands while adjusting. There is no reaching down inside a narrow tube to make an adjustment and spoiling another adjustment by catching it with the sleeve. Best of all, an open model can be adjusted in the dark and while looking into the eyepiece, which is a very comforting advantage. Finally, the simple wooden parts are so inexpensive that they may be discarded painlessly for a tube telescope after the lessons of mounting and adjustment have been learned. An example of such a lesson learned without grief on a tubeless telescope is the common headache of locating the exact place on the tube of a tube telescope to make a big round opening for the eyepiece. In many cases a second opening afterward has to be made, leaving an ugly gap in the tube.

A good drawing, such as Figure A.2.8, itself explains so much that I will agree not to repeat in words what it tells graphically, if you will agree to pore over it a long time and get it clearly in mind.

A.2.13. Mounting the Poor Man's Telescope

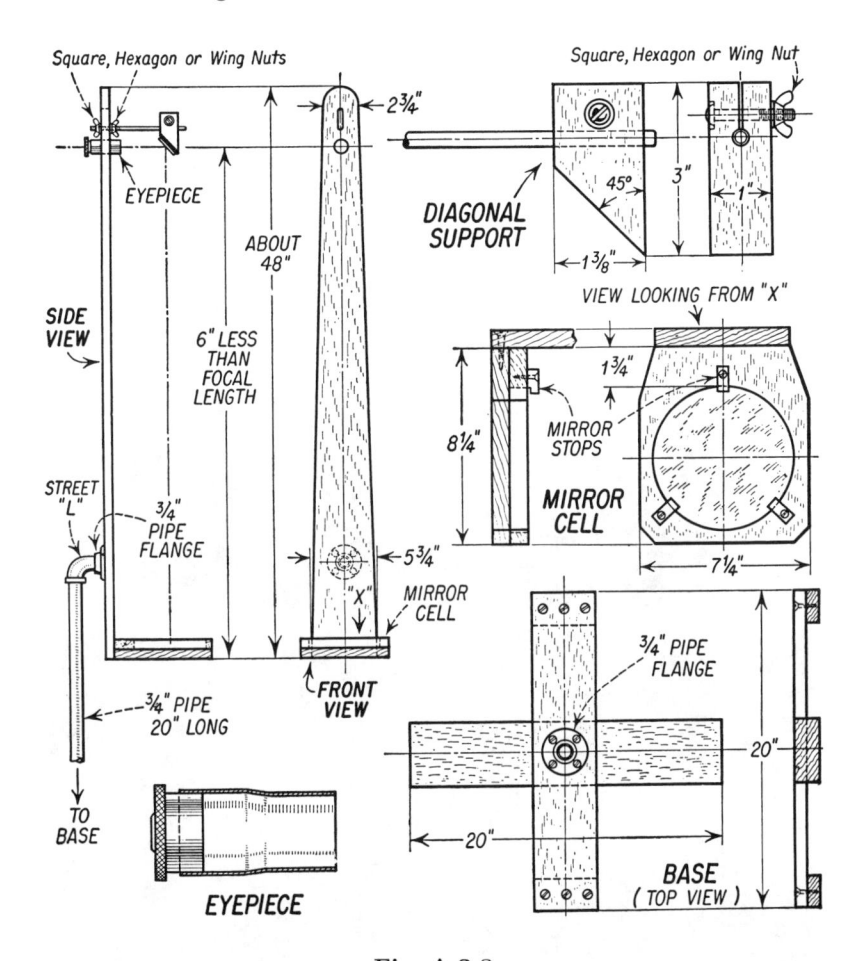

Fig. A.2.8.

There is very little on this telescope that cannot be obtained almost anywhere you live.

By planning a sequence we can hasten results. First thing is to put in an order for the eyepiece, standard outside diameter (1 1/4″), focal length one inch (25 mm), inexpensive type, from one of the telescope supply dealers who advertise in astronomical magazines for amateurs. Do not be tempted to buy a "high-power" eyepiece of half-inch or quarter-inch focal length right now. It will only add to your troubles. Refinements such as this may be added after you have become familiar with the use of eyepieces on fancy telescopes.

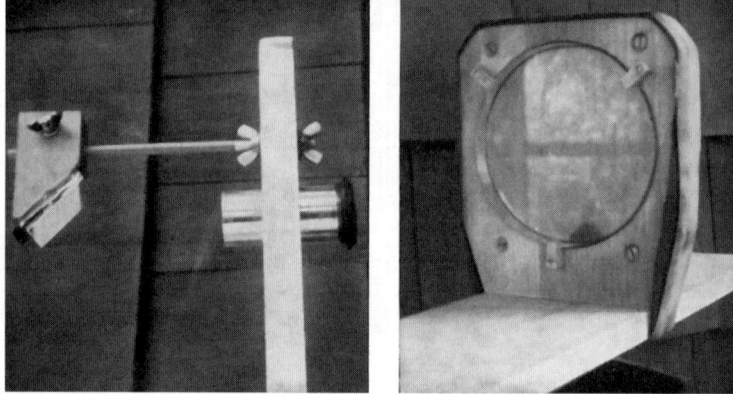

Fig. A.2.9: *One of the Cleveland versions of the spinal column mounting.*

While the eyepiece is coming, let's make the diagonal mirror. Closely inspect a piece of plate glass to avoid scratches, and cut out or have a hardware man cut out a rectangle $1\,3/8''$ by $1\,3/4''$ within about $1/16''$ (it is not easy to cut plate glass). Put a dab of medium-sized abrasive and water on an old piece of glass and grind down the sharp edges without scratching the face, or hone its edges with the stone you used for honing the sharp edge of your mirror while grinding.

A.2.13. Mounting the Poor Man's Telescope

Accurately measure and record in your log the greatest diameter, the thickness and focal length of the main mirror. To measure the focal length prop the mirror up on a pillow facing the Sun. With a strip of cardboard about an inch wide explore the space about four feet in front of it. When you catch the reflected round image of the Sun on the cardboard keep adjusting the pillow until the shadow of the cardboard is on the mirror (anywhere on it will do) at the same time that the image of the Sun is on the cardboard. Now move the cardboard toward and away from the mirror and locate the place at which the Sun's image is sharpest and smallest on the card. Mark the distance on a long stick held against the mirror.

(The bright spot will not burn your hand or set fire to anything, because the mirror has not yet been coated with metal, and most of the sunlight passes through the glass. However, even the small fraction that is reflected by the uncoated glass is enough to cause permanent blindness if it is shined into anyone's eyes.)

Reverse the stick, end for end, so that the first marking will not influence your judgment, and do it again. The average of the two measured distances should not be far from the focal length of the mirror. Within an inch or so is close enough for the present although we were aiming at 48 inches. No big harm will result—don't throw the mirror away—if this measurement is as high as 54 inches or as low as 42 inches. Even beyond these the telescope will work but not quite so well, especially on the lower side of the 42″.

Send the mirror and diagonal to one of the laboratories that advertise to coat telescope mirrors with reflecting metal. Have them coat both mirrors.

Now, because we recorded the diameter and focal length in the log, we can be preparing the mounting while awaiting the mirrors.

First thing to obtain is the eyepiece holder. This calls for a $2\,1/2$-inch length of tubing with $1\,1/4$-inch inside diameter, so exactly $1\,1/4$-inch that the eyepiece will slide smoothly in it for focusing, neither binding nor sloppy but what mechanics call "slip." Until I happened to find a standard plumbing fitting properly called a "brass trap extension, slip joint pattern, $1\,1/4$-inch," and available inexpensively from plumbers wherever plumbing exists and sometimes at hardware stores, it was often quite difficult for the average amateur to scout out a suitable eyepiece holder. The trap extension comes several inches longer than we need on our telescope, which will not be equipped with plumbing. So cut off for use about $2\,1/2$ inches from the larger end, using a file and then a hacksaw or the file alone, and avoiding the use of a vise. Near the same end is a slight reduction in diameter having usefulness to the plumber and none to us, except that it stiffens the tube a little.

Before cutting the fitting, make sure that the eyepiece fits it nicely so that you can adjust the focus without annoyance. If it is loose you can crimp the edge a little by holding it in the hand (a vise would deform it) and pressing a metal object diagonally against the end while you rotate it.

Unless you paint the outside of the eyepiece tube with black stovepipe paint the reflections from it at night will bother you when observing. Another way to prevent this is to cover the ends of the tube with black friction tape.

A $1/4$ by $7 1/2$-inch machine bolt with the head sawed off will serve for the diagonal support. Ordinary nuts with washers may take the place of the wingnuts, though the wingnuts considerably facilitate adjustment in the dark.

Unless the wooden block for the diagonal is cut accurately to a 45-degree angle and the hole for the bolt is bored square, there will be difficulty in adjusting the telescope accurately. Hence, use a drill press if possible, likewise use a miter saw to cut the angle of the block.

If we could depend upon closer precision than a quarter inch or less from our determination of the mirror's focal length we could safely cut the wooden spinal column to exact length and attach the wooden cell that holds the mirror. We would cut it to three inches less than the focal length and it would come out correct. However, our determination of focal length is not that dependable but is only an approximation. Therefore we might need as much as three inches more than we think we need—say the same length of wooden spinal column as the focal length of the mirror. Even this could be too short and it therefore will do no harm, provided the lumber is available, to cut the piece a whole foot longer than the focal length. The reason is elementary: It is easier to saw off and discard a surplus afterward if we find one—though as a Scotsman this pains me—than to fasten a piece on if, after all, we discover a shortage.

Bore the big hole in the eyepiece holder about three inches from the end of the board. Do not yet taper the board.

Unfortunately the hole diameter will not be that of a standard bit and therefore the hole must be bored with an expansion bit. Start on scrap wood with a hole too small and sneak up on the correct diameter for a snug fit for the tube by boring one hole after another until the tube fits snugly in the last one. Hold the bit square in both directions—easier said than done—and, as soon as the tip first breaks through, turn the wood over and finish on the other side. Unless the fit is pretty snug the eyepiece tube will soon be loose. Therefore, since you can't unbore a hole if it proves to be oversize, it is better to practice on scrap wood. By the time we finish the diagonal support and mirror cell the coated mirrors should be back. Cement the diagonal to its wooden block with comb foundation. It is best to rehearse this with scrap wood and an old fragment of glass. Melt up some comb in a kitchen spoon without boiling it, pour it directly onto the

A.2.14. Collimation 53

wood and apply the glass. You will probably find that the wax sets too quickly to permit careful positioning of the wood, so repeat after warming the glass and the wood. When you do it on the diagonal mirror, center the block neatly. The penalty for putting the coated side of the diagonal to the wood instead of toward the main mirror, as some have done (just the opposite of a looking glass) is to buy a new diagonal and have it coated.

A.2.14 Collimation

To collimate your telescope means to line up the optical parts so that the conical beam of light from the mirror will be reflected off the diagonal mirror straight through the center of the eyepiece holder to the eyepiece itself.

The first part of collimation should be done in a well-lighted room or out-of-doors in the daytime with the spinal column upright. It may be clamped to the back of a chair so that the eyepiece hole is about at your eye level.

There are three optical parts in a telescope, the mirror, diagonal, and eyepiece.

We will start to adjust at the eyepiece, next adjust the diagonal, and finally the mirror and cell. The eyepiece will not yet be needed.

First, make sure that the eyepiece holder fits snugly in the hole you have bored. Great care should have been taken to bore this hole square but, if the eyepiece holder is out of square as determined by a small square or square object, it can be wedged straight with toothpicks.

Saw off the head of the $7\frac{1}{2}$-inch bolt and smooth the burrs with a file. Insert the threaded end of the bolt in the slot in the spine and snug it up with the two nuts. Because the nuts bear on the washers, which bear on the surface of the spine, the bolt should automatically square itself in both directions and the diagonal should come to an angle of 45 degrees with the spine.

Adjust the diagonal so that the center of the slope is $4\frac{3}{4}$ inches from the spine, the same distance as between the center of the mirror and the spine.

Test the alignment of the diagonal by running a long thin stick or a pencil centrally through the eyepiece holder and judging whether it would touch the center of the diagonal. If not, loosen the nuts and readjust, not by changing the $4\frac{3}{4}$-inch distance but by sliding the bolt endwise in the slot. Check the adjustment by looking through the eyepiece holder from a distance of about 8 inches.

Next, prepare the mirror cell for the mirror. If at all possible, cut the side to be attached to the spine with a mitre saw or a power saw, and then the cell will be accurately square with the spine.

If the opening for the mirror is too snug and the mirror is forced into it, or if the mirror is cemented into it, the pressures will deform the glass, thick as it is, too much for the mirror to focus correctly and the telescope may not work at all. If, on the other hand, the opening is more than about $1/32$ inch larger than the mirror this may affect the collimation a little. The space may therefore be filled in with a circular band of blotting paper or a light rubber band. But beware of a tight pinch! Best of all is a very slight but audible sidewise shake and no fillers.

Our next need is a way to slide the cell up and down along the spine at will, yet hold it tightly at any one place while adjusting it to the correct distance from the eyepiece via the diagonal, also square with the center line of the spine. A powerful, moveable, easily obtained "clamp" of this temporary kind is shown in the halftone illustration (A.2.9, lower right): two or three rubber bands about two inches wide, cut from an old inner tube, are snapped over the cell and around the spine. Before inserting the mirror try shifting the cell along the spine, using a tapered wooden lever beneath, also moving it by tapping, so that you will know how to control it. As a gauge for the eye to keep the cell square on the spine in the crosswise direction draw on it several parallel lines, carefully using a square. This use of the square is why we did not taper the spine at the start.

We shall need an additional way to make a closer and final collimation adjustment of the mirror within the cell itself, after the cell is permanently attached to the spine. Here in Cleveland, on the pilot telescope, we glue three or four strips of sponge rubber about a quarter inch thick and an inch wide to the bottom of the cell with shellac, as shown in another halftone illustration (A.2.9, top). With these cushions under it, the top face of the mirror stands about one eighth inch above the cell. We finally adjust the mirror with the three screw stops shown in the line drawing, also by trimming the stops. The amount of pressure put on the mirror is then no greater than that of the sponge rubber pushing lightly upward against the stops. On these simple telescopes this very light pressure, not recommended for a second and larger telescope, has never given us trouble.

With the mirror mounted in its cell you should now see three things through the center of the empty eyepiece tube:

1. the round mirror as reflected to your eye by the diagonal,

2. the image of the diagonal reflected by the mirror,

3. the image of your eye reflected to it by the mirror.

When the parts are properly adjusted you should be able to see, while looking straight, not crookedly, through the same eyepiece tube, the following:

A.2.14. Collimation

1. the image of the whole mirror and not just a part of it,
2. the diagonal should appear centered in the mirror, not off to one side or the other,
3. the image of your eye should also be centered on the diagonal.

In other words, if collimation is correct, the diagonal should look as if in the center of the mirror and your eye should look as if in the center of the diagonal. These are essentials if the telescope is to work efficiently.

If the diagonal does not appear to be at the center of the mirror, put your hand behind the spine while still peering into the center of the eyepiece holder and move it up or down slightly on the bolt and note which way it moves. If it moves still farther out of line, adjust it in the other direction until it appears centered with the mirror. Or twist it a bit to left or right on the bolt and see which way it seems to move while peering into the eyepiece holder. If it moves still farther out of line adjust it in the opposite direction.

Keep at these two adjustments until you get the image of the diagonal centered with the mirror.

However, before refining the collimation further, we must stop and set the cell at the correct position along the spine so that the eyepiece can most conveniently reach the focus of the mirror. Then we shall return to closer adjustments for the collimating job. Final adjustment of a telescope is seldom gained on the first time over. Instead, we come as near as we can and then go over it again, and keep improving the job by closer and closer refinement. Some call it finished as soon as they can see anything at all, while others who want to get the utmost out of the telescope keep improving the collimation even after the performance seems fairly good, until they are sure no further gain is possible, sometimes making small additional gains for weeks. So it isn't a case of either not seeing at all or seeing perfectly the first time.

The next stage is done on a bright celestial object, with the eyepiece in its holder. The most brilliant object in the sky, and the easiest for collimating use, is the Moon. It is hard to see details on a full Moon, but the Moon at first quarter is in the most convenient position in the sky.

Insert the eyepiece and, while peering into it, have someone squint over the edge of the spine and point the telescope toward the Moon. When it is aimed correctly you will see a brilliant flash of light in the eyepiece, undoubtedly fuzzy because the eyepiece is not yet in focus due to wrong distance. Push the eyepiece in while looking through it. Is the fuzziness resolved into a moon on the way? If not, pull it out as far as it will go. Still fuzzy? If so, the mirror cell is either too near or too far. Shift it an inch or so along the spine and repeat the performance. Explore along the spine in this way with the cell until you can focus the Moon clearly with the eyepiece at some position in its holder. You will have to readjust the squareness of the cell each time; but it gets much easier after you have done it once.

You should keep pinning it down closer by adjustment until the clearest focus is had when the eyepiece is about halfway in the holder, or about an inch from the nearer face of the spine.

Now go back and check all the work you did before and, when you are sure you have the cell in the position where you will continue to like it best, attach it permanently with three screws and saw off the excess of the spine board beyond the cell.

Then taper the spine with a ripsaw and plane and paint it dead black.

A.2.15 Albert G. Ingalls Editor's Note

The following data have been adapted from articles in *Scientific American*, 1954, August and September:

Each August for many years the Cleveland Astronomical Society has entertained the city of Cleveland with a public telescope and star party. James L. Russell, their organizer, writes:

> We have had as many as 10,000 people in the public park where we hold them. The city turns out the park lights; we line up 35 of our homemade telescopes and, while a long line moves slowly past each telescope, a professional astronomer addresses the crowd from a sound truck. After looking at the Moon or a planet the people sit on the grass and watch the astronomical movies that we show, or study the celestial objects that we point out in the sky with a huge searchlight. We have so many people that it takes three evenings to run them all through. You would have to attend one of these parties to realize their magnitude.

During the rest of the year Russell leads Tuesday evening classes in telescope making at the Cleveland Museum of Natural History, which has equipped a large laboratory for the amateurs. He says:

> We have 40 people working at a time, and in the first five years, up to 1954, they have made 350 mirrors. We have 16 mirror-grinding tables, places for 40 people to polish mirrors at one time, hand tools and machinery for making mountings. Some grind or polish while others saw. It is sociable as well as astronomical. The din and dust are terrific, with 50 people in the huge room grinding, polishing, hammering, hollering and yap-

A.2.15. Albert G. Ingalls Editor's Note

ping their heads off, all at the same time, from five until nine every Tuesday.

For each one who finishes the grinding stage we have 10 waiting to begin. Those who have made two or more mirrors, including myself, act as instructors five at a time. No one is paid, for us it is recreation and we just love it. I am a lawyer, not a mechanic, and definitely not one of those who can turn out a perfect mirror, though as a beginner many years ago, using *Amateur Telescope Making*, I made 6-inch, 8-inch and 10-inch Newtonian telescopes and a Cassegrainian.

For our group the trouble is that *ATM* wants the beginner to reach perfection on his first mirror. Now what does a trucker—or a lawyer—know about the mathematics of paraboloids? So the struggler asks, 'Must I become a mathematician or am I making a telescope?' The mirror-making manuals, simply by saying nothing to the contrary, leave the novice to suppose that he must achieve perfection or else the telescope won't work. Yet the fact is that a considerably less than perfect mirror will work so well that the average beginning observer will not be able to distinguish the best from the worst. The eye learns only with practice, and the learning goes on for months, years. True, under the Foucault test the mirror may not look good to an expert but, as Ellison says in Section A.4.1, paragraph *Perfection*, that test is actually more delicate than is necessary. A few do make fine mirrors but ironically the payoff comes when they discover that, for the inexperienced observer, these show no more than the one made by Joe, who is no optical wizard.

For the first mirror we had to abandon Pyrex, the mirror handle, the pitch polishing lap, and the paraboloid. We teach the beginner some of the rudiments of the Foucault test but only because we are present to coach the worker in its use. Although pitch laps give finer polish and fewer zones, they are so difficult for the average beginner to make and alter that we regard them as the principal bottleneck in mirror making. They have discouraged more beginners than any one thing, or any ten things. Although we get fewer fine mirrors with honeycomb foundation, it best suits our purpose, which is to finish the first mirror while the maker's enthusiasm lasts.

Our objectives on the first mirror are only to teach a simple technique and to produce a mirror of tolerable quality.

A few of our candidates have failed mainly for two reasons.

Some, in their haste, refused to give attention to each working detail. Others failed because they insisted upon being too technical on the first mirror and became mired in the quicksands of complication. Ironically, a few hit better than average beginner's curves the first time, perhaps not without a grain or two of good luck.

Because we consider the human angle and help the beginner to bridge the gap between his capabilities and what he may regard as the wizards' and bookworms' book, 80 percent finish the first telescope. Most of our workers don't want to be experts; they just want a telescope. After climbing aboard by this preliminary pilot or gangplank job, some go on to build fine telescopes in our laboratory, up to 18 inches in diameter, Cassegrainians, Schmidts and other advanced kinds, using the entire orthodox Pyrex-pitch-paraboloid-Foucault method.

The rest of this note is for the "wizards and bookworms" whose activities mystify Joe Doakes. It offers, instead of the same Rayleigh limit standard for all, a different standard for each worker's needs—easier for many, the same for some, and tighter and tougher than ever for the advanced.

First, on the easier side of the Rayleigh limit.

It has been said many times that in any reflector the Rayleigh limit calls for precision of the surface to one eighth of a wavelength, or $1/400{,}000$ inch. So often has the term "good, honest eighth-wave optics" been used in speech and writing that many have supposed it an inexorable standard. Actually, however, the tolerance is relaxed at the outset to $1/200{,}000$ inch by the fact that the eyepiece may be adjusted to an average focus. This has been explained by Franklin B. Wright in Chapter B.3, and by John R. Haviland in ATM2, Section A.3.2. Alan E. Gee now explains it graphically.

In the Figure A.2.10 (Left), the curve being hyperboloidal, the observer automatically focuses the eyepiece at the best focus, b, where the circle of confusion is smallest, ignoring a. In the second the curve is spherical (undercorrected mirror) and he again selects b instead of a. For regular undercorrection or overcorrection this ability to focus results in a reduction of the *effect of* the spherical aberration by a factor of four. For the effect of zones no such statement can be made. The observer will still work at best apparent focus; however, where this will be, and how good it will be, depend upon such factors as the relative area or areas of the zones producing the errant rays, their inter-

A.2.15. Albert G. Ingalls Editor's Note

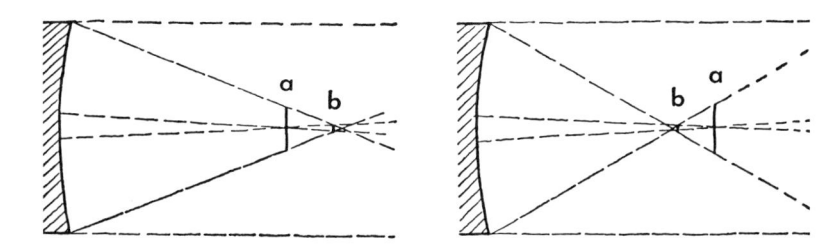

Fig. A.2.10.

cept on the optical axis, and their angular subtense. Based on the Rayleigh limit, Wright's tolerances in shown in Chapter B.3, Table B.3.1 are all too stiff by a factor of two. A six-inch mirror is almost at the Rayleigh limit if left spherical at $f/8$. I suspect more of these 6-inch, $f/8$ mirrors are worsened by attempted parabolization than are improved.

In this basic matter, governing all amateur telescope makers' exertions, Wright now assents with Gee and others who have thought his standard too exacting. He says, "I based my limits on smooth curves of surface viewed from the position of best average focus, then made the limits twice as strict as this calls for to take care of imperfect measurements or imperfect focusing, and from just plain conservatism." In accordance with this revision Wright has now added an amendatory note to his chapter (see Section B.3.1).

Other reasons why considerably less than perfect mirrors work well enough to please inexpert users are:

1. A novice observer is likely to look most often at the Moon, because it is such a spectacle, and, as Horace H. Selby has pointed out in ATM3, Section B.3.2, with a low-powered eyepiece almost any telescope will give a good general impression of the Moon.

2. On many nights the turbulence of the earth's atmosphere tends to level the observational difference between a fine mirror and a poor one. (But the better the start, the better the outcome.)

These reasons may seem to provide an alibi for sellers of poor telescopes. As a matter of fact, the complaint against such merchants is not that their instruments are too poor for use by the average beginner but mainly that their claims of precision are sometimes too strong.

Having dealt with the soft side of the Rayleigh limit as generously as the facts of physical optics permit, let us now climb over the wall into the rarefied realm of the perfectionists, who naturally do not want to be robbed of their enjoyment of high precision. Nor do they want their near-religious esteem for it pulled down. Perhaps Doakes may not completely realize the mainsprings that make the typical amateur telescope maker tick. Most of these "wizards," as they seem to him, are bookworms at that. They are also as interested in basic physical optics as in its poor but respectable humble relative, precision optics. Always curious about the underlying reasons for the things they see, the wizards willingly stop to nose into causes. Since curiosity is the scientist's prime mover, this makes them scientists. Nor is the typical amateur telescope maker in quest of an easy job. One of them says, "If an optical job isn't tough and temperamental I'm bored stiff with it. I want real opposition." Joe, who just wants a telescope, need not worry the wizards—for whom, incidentally, there are higher mountains than ever to climb, as will shortly be seen.

It is true that only a minority of observers are expert enough and only a minority of nights have good enough seeing to exploit the superiorities of a mirror made to a tolerance of half the Rayleigh limit, that is, twice as precise, or about $1/800,000$ inch (provided always that the diagonal and eyepiece do not wipe out the mirror superiority). It is also true that the law of diminishing returns sets in rapidly after the Rayleigh limit is surpassed. However, there is something in Rayleigh's original writings that provides an "out" for the worker who loves to surpass the Rayleigh limit. Few have bothered to dig out and read just what Rayleigh wrote.

The classic article in which he discussed the subject appeared in the *Philosophical Magazine*, Fifth Series, Volume 8 (1879) pages 261–274, also 403–416, and 477–486, and was concluded in Volume 9 (1880), pages 40–55. The thing now known as the Rayleigh limit was an incidental part of the article, which was largely mathematical, titled "Investigations in Optics, with Special Reference to the Spectroscope." The entire series is reprinted in Volume 1, pages 415–459, of Rayleigh, J.W.S., *Scientific Papers*, six volumes. The language Rayleigh used should interest the advanced amateur, especially if he examines it closely. He wrote:

> Foucault was, I believe, the first to show that the errors of optical surfaces should not exceed a moderate fraction of the wavelength of light. In the case of perpendicular reflection from mirrors the results of §4 lead to the conclusion that no considerable area of the surface should deviate from truth by more than one eighth of the wavelength.

Thus he mentions surface error, not optical path difference.

Rayleigh then says that

> the aberration *begins to be* (not his italics) decidedly prejudicial when the wave surface deviates from its proper place by about

A.2.15. Albert G. Ingalls Editor's Note

a quarter of a wavelength" [i.e., half that for a mirror]; and elsewhere, it begins to be decidedly mischievous.

If the definition of the Rayleigh limit is similarly examined closely in modern treatises on physical optics, inconspicuous qualifying words to the same effect will be found. Thus Conrady says that an optical instrument "would not fall seriously short" of the performance of a perfect system if the difference between the longest and shortest optical paths was not greater than one fourth wavelength.

Conrady adds that Rayleigh's deduction remains probably as good a single statement as could be desired today, and that the large amount of research that has been done on it has not shaken its validity. He too points out the advisability of departures from it in some circumstances—downward for best apparent focus, as per Gee above, also upward. Rayleigh gave as an example that a spherical mirror of 3-foot focus might have an aperture of 2 1/2 inches "and the image would not suffer materially from aberration."

The one kind of observing for which it is profitable to surpass the Rayleigh limit is the close observation of fine detail on the Moon and planets. However, this is just what most interests the average advanced amateur. The detail is made visible by contrasts between adjacent areas. Without the contrasts no detail shows and all is flat. These contrasts are heightened in proportion to the quality of the telescope optics, as can be proved by physical optics or empirically by observing. Here Haviland's statement in ATM2, Section A.3.2, that perfection beyond the Rayleigh limit will not noticeably improve the image is inadequate. It has been shown in France by Albert Arnulf[1] that, for the faintest perceptible contrasts, the efficiency of a mirror rises from 64 percent when corrected to the Rayleigh limit, to 92 percent when the correction is carried to one fourth of that tolerance.

The optical designer James G. Baker says,

> At very low contrast levels, such as obtain on the planetary disk, a mirror made as poorly as the Rayleigh limit will not perform well and a much better mirror should be the goal. French observations in the laboratory indicate that there is no real lower limit to the accuracy requirements for the observation of maximum contrast of faint details. For example, if the contrast level is as low as 1.01 to 1, it may be necessary to have the optical

---

[1] See Arnulf in *comptes rendus,* Académie des Sciences, Institut de France, Vol. 200 (1935); also Maurice Françon in *Cahiers de Physique,* 944, No 26 and in *Révue d'Optique,* 1947 and 1948. Also see *Optical Image Evaluation,* National Bureau of Standards Circular 526 (289 pages). Most of this literature compounds foreign language with physical optics and is not light summer hammock reading.

system perfect to within one fiftieth of a wavelength. Any amateur sincerely interested in a high-quality mirror is likely to continue, as at present, seeking the best curve he can obtain.

A.2.16 Notes Added in 1996

1. Today, the use of a worn dime to measure the depth of the mirror is not valid (if it ever was) since dimes are no longer made from silver alloy and will not wear the same. Temporarily "pseudo-polish"' your mirror by working with spent sludge until the surface takes a superficial fine-ground appearance. Then use the method of section A.5.1.3 to judge radius of curvature. A 100- to 110-inch separation between the mirror and the eye-flashlight combination should be your goal (the focal length is half that distance). Focal length will diminish further as you fine-grind.

2. Glass making has changed and cheaper methods are sometimes used to make plate glass. You can no longer rely on a stray piece for the diagonal. Already-coated diagonals are sold quite inexpensively today. Obtain one of these, and use silicone cement to attach it rather than beeswax.

3. The tolerance on focal length from 42 inches to 54 inches is probably based on balancing the mounting. Forty-eight inches is already nudging the edge of the optical tolerance for spherical mirrors. The mirror will perform better as the focal length is increased. Sixty inches is a workable upper limit that allows increased image quality while not extending the tube too far. *H.R.S.*

Chapter A.3

The HCF Lap[†]

HCF is Uncle Ephram's abbreviation for honeycomb foundation, a material universally used by beekeepers to encourage the bees to build their combs straight in the hives. For the polishing lap, it should be the pure beeswax variety, unwired, sold by dealers in beekeeping supplies under the trade name "Medium Brood Foundation." This material runs seven or eight 8-by-17 inch sheets per pound and costs about $1.00 for that amount. It is also available in the "Jumbo" size, 2 inches wider than the above.

HCF is made by running thin sheets of beeswax through embossing rolls which fill it with small tetrahedral depressions called "cells," and it is the delicate walls between adjacent cells that form the facets of the lap. The stroke of the mirror causes a whirlpool action of the rouge and water mixture in the cells, agitating the rouge particles upward, so that they wedge against the edges of the facets in the direction of the stroke. When the stroke is reversed, these particles break away and are replaced by others on the opposite sides of the cells—and so on, back and forth. Apparently it is these tight wedges of rouge that do most of the work, practically no rouge becoming embedded in the surface of the wax facets.

A.3.1 Advantages of HCF

The advantages of HCF are: comparative freedom from scratching, sleeks and fog; retention of rouge charge for long spells of polishing; ease of making laps, either full-size or special shapes for zonal correction; and speed of polishing—although the latter is not so noticeable with the new Pyrex disks now in general use. On these, harder pitch tools and more pressure may be used. The main drawback attached to HCF is the in-

---

[†]By A.W. Everest.

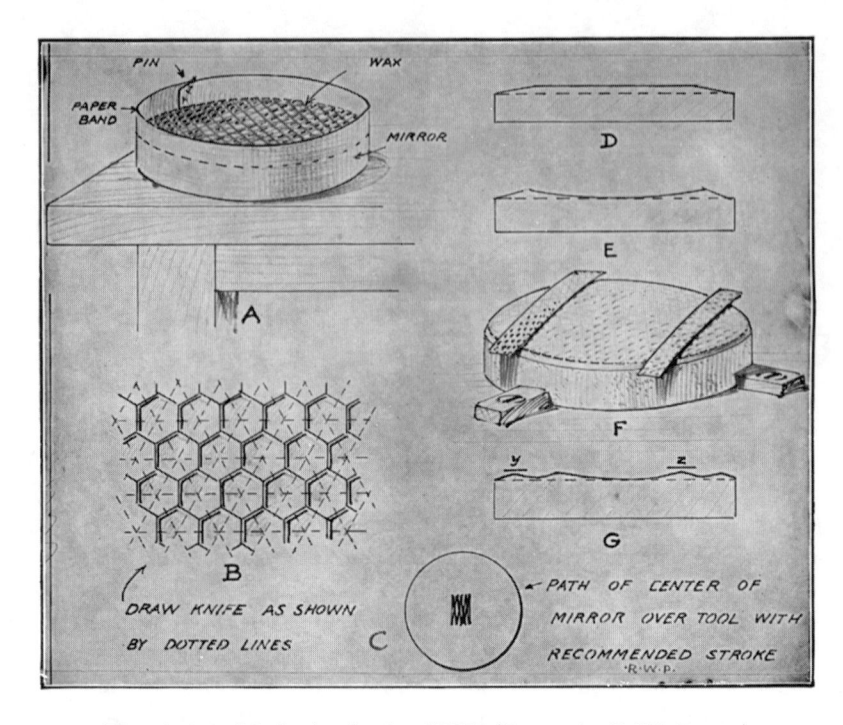

Fig. A.3.1: *Methods of using HCF. (Drawn by R.W. Porter)*

ferior optical polish produced. Its action is so drastic that the marks of facets show plainly under the knife-edge test. For small optical surfaces, this tendency can be reduced to a satisfactory extent by tapering off with fine washed rouge and reduced pressure. But on larger work, enough of this uneven surface texture will generally remain to be injurious to fine star images reflected from an otherwise perfect surface. Therefore, HCF can be recommended only as a medium for bring the surface to a complete polish and reducing zonal irregularities that may be found at the end of this operation. From then on it becomes, at the most, an accessory of the pitch lap.

A.3.2 Making and Using an HCF Lap

Various methods of using HCF will suggest themselves to the mirror maker, a suggested routine being as follows.

For the full-sized lap, warm the glass tool to about 110°F, dry it thoroughly and lay it, face up, on the work bench. Pour on pitch hot enough to spread out evenly to a thickness of not over $1/16$ inch and to hold the HCF

A.3.2. Making and Using an HCF Lap

to the tool. While the pitch is still warm, lay on a sheet of HCF. Wet the mirror with soapy water and press the HCF through the pitch, right down to the surface of the glass. Trim off the surplus HCF around the edge, and the lap is made.

This produces a lap which does not yield to cold pressing, and contact is to be obtained and retained through the tendency of the facets to wear slowly away. If the pitch is accidentally applied too thick the lap may still be used, but it must be channeled similar to the pitch lap. Otherwise it will polish irregularly, mostly at the center, since the pitch in the center has no place to flow.

Use soap water for the rouge mixture. To get the proper degree of soapiness, splash a bar of soap around in a pan of water until a sample of HCF immersed in it for a moment will retain a film when removed. The function of the soap is to make the mixture stay on the lap and flow in the proper manner. The worker was advised to draw a blunt knife through each row of facets in all three directions, as shown at B, but this is unnecessary when using soap. (See Figure A.3.1.)

To get contact, dip the mirror into the soap water, hold it on edge for a moment to allow the surplus to run off, and then dust on as much rouge as will stick. This will produce a grinding action when applied to the lap and, since wax is softer than glass, the lap will receive the grinding. Use short, straight strokes, as shown at C, for a few moments or until you can see, upon removing the mirror, that every facet on the lap has been touched, indicating complete and uniform contact. As a further refinement, rub a wet cake of soap lightly over the surface of the lap, wet the mirror, and then use the same strokes to work up a good suds. At this point the facets can be seen and the condition of contact determined by looking through the mirror.

For polishing, use a thin rouge mixture. A heaping teaspoonful of rouge in half a glass of soap water is about right. With this thin mixture there is a pronounced drag, that is, the friction between mirror and lap requires considerable muscular effort to overcome. The facets are visible through the mirror, and the hair-line wedges of rouge on the working sides of the facets can be seen as they form and break away. This is the condition necessary for rapid polishing.

After the first few minutes of work, wipe up the mirror to see whether the polish is coming up evenly. If either the center or the edge starts to polish first, a wide departure from a spherical surface will result. Any such tendency must be corrected at the start by further "grinding" of the lap, as noted two paragraphs above. As a rule, the HCF tool with solid backing, as recommended, will start polishing slightly faster at the edge than at the center, but as long as some polish can be seen all over, this may be ignored, since the lap will then soon straighten out of its own accord.

After replacing the mirror on the lap, following an inspection, the rate of polishing may prove to be reduced, due to the rouge having settled in the bottom of the cells. This should be agitated back into suspension with a camel's hair brush. Slight additions of rouge with an eyedropper at five-minute intervals will also help; or maybe just water to keep the cells full, thus "building up" gradually to the correct dilution for maximum drag. This may be done by sliding the mirror nearly halfway off the tool, adding rouge to the exposed area, and then repeating the procedure for the other side.

After an hour's polishing, the rouge mixture will become more or less contaminated with ground wax and glass, losing some of its drag and acting more and more like a lubricant. When this becomes very noticeable the tool should be flushed off and recharged with fresh material.

As an indication of what to expect, a properly ground Pyrex disk, say 10 inches in diameter, should be brought to a complete polish in about five hours, and should prove to be very nearly spherical if stroke C of the illustration in Figure A.3.1 has been used throughout.

Caution: Be careful when removing or replacing the mirror. A single slip may result in shaving off an area of the facets, which will not come back to contact until the rest of the surface wears down. Do not leave the mirror on the lap from one day to the next as a protection against dust, for beeswax contains a trace of acid put there as a preservative by the bees themselves, and after a prolonged period of contact this acid will attack the glass, leaving an imprint of the honeycomb pattern which can be seen both visually and under the knife-edge test.

A.3.3 Using HCF for Zonal Correction

For zonal correction, strips, arcs, or special shapes of HCF are used on a lap especially prepared to hold them, since it would be unsafe to lay these on the good lap. To make this, lay a sheet of HCF, trimmed to size, on the face of the mirror. Wrap a paper band around, as shown at A, and pour in plaster of Paris to a depth of about one-sixth the diameter of the lap. As the plaster sets, press it down with the hands, to insure contact; and before it is too hard, level off by scraping across in several directions with a straightedge.

The HCF strips for zonal correction, as at F, are cut to shape with shears and dipped in soap water, laid on the special lap where required, slightly crushed into contact with the mirror, and then given a rouge charge from the eyedropper. Strokes should be very short, and the mirror should be rotated slowly but uniformly for one or more *complete* revolutions. Test

A.3.3. Using HCF for Zonal Correction

Fig. A.3.2: *The Discoverer of the HCF Lap. From a pencil sketch by R.W. Porter.*

often, as the action is fast. Ten revolutions should be about the limit per spell.

As a typical example of zonal correction, suppose we had the turned-down edge shown in apparent section at D. It would require the removal of a considerable amount of glass to wear the mirror down to the level shown by the dotted line. But by drawing the knife-edge back slightly to get the appearance shown at E, the material to be removed will be found located mostly in one zone. At this position the center appears to have the same depth as the edge. A similar setting of the knife-edge should be selected for all zonal irregularities when any sign of turned edge is present—that is, the depressed zones should be made to appear at the same level as the edge of the mirror.

Returning to the example of correction, the problem is to remove the material above the dotted line. Strips of HCF may be laid on the lap, as shown at F, to wear down the crest, although arcs would be somewhat better for bringing the action where it is wanted. Strokes parallel to the strips will concentrate the action to about the width of the strips, while strokes at right angles will ease off the effect. This will give something like G. The next step is to peel off the strips and relocate them, one just inside and the other just outside the first position, to remove the new crests formed by the previous operation. These maneuvers are continued until the surface appears practically flat, when a final blending should be given on a pitch lap of the conventional type.

Sometimes it is necessary to make a correction in some area where it is awkward to balance the mirror on the small strips required. Take, for example, a bump at the center, or a raised zone very near the center. In such cases a useful stratagem is to add near the edge of the lap three small balancers of HCF which have been dipped in soap water but no rouge applied. The mirror slides on them easily, and in perfect balance, but no polishing takes place.

For accurate location of zones when the mirror is on the testing stand, drive in a row of pins at half-inch intervals along a straight piece of wood $1/4$ inch or so square and about 2 inches longer than the diameter of the mirror. Hang this across the horizontal diameter of the mirror, with a supporting loop or semicircle of wire up around the upper half. When the mirror is wholly illuminated, the pins look black, though when the knife-edge shadows appear they become brilliantly illuminated by diffraction. In other words, they are easily seen at all times.

When measuring zones, measure them on both sides of the mirror. The shadows travel over the crests of the zones in much the same manner as shadows 1 and 3 of the true paraboloidal figure, as explained in "A Study in Shadows" (ATM1, Chapter B.2), and an inspection at both sides is advisable in order to make sure of the location of the highest point.

For the final blending on the pitch lap, use short strokes and stop to press often, the idea being to keep the surface of the lap complementary to the surface of the mirror and then to make the high spots on the one ride the high spots of the other.

The method of parabolizing on the full-sized pitch lap is covered elsewhere in the book. If trouble is experienced when changing from spheroid to paraboloid in a uniform manner with the strokes recommended, careful use of the HCF strips before each spell on the pitch lap will help to remove the surplus material.

Chapter A.4

The Amateur's Telescope[†]

A.4.1 Introductory

Nearly 150 years have passed since William Herschel, then the unknown organist of Bath, began those experiments in telescope construction which in a few years made him the greatest astronomer and greatest master of the telescope that the world has yet seen. The Newtonian reflector was his chosen form of instrument almost of necessity. The achromatic lens was as yet in its infancy, and the Gregorian construction was too complicated for an experimenting amateur, and was, besides, only suitable for small sizes, and consequently limited in its possibilities.

A distinguished band of workers have, during the past century and a quarter, followed Herschel's lead, and nearly all have imitated his preference for the Newtonian, although the achromatic object glass has become such a formidable rival to the concave mirror. The reason for the popularity of the Newtonian is still the same as in Herschel's day. It is the amateur's telescope, because, now, as then, it is the easiest to make and the easiest to mount, and far the cheapest either to make or to buy of any class of telescope. And, though cheap, it is not "cheap and nasty."

A good mirror is to the astronomer a thing of beauty and a joy forever, capable of unfolding all the myriad glories of the heavens and of holding its own with the most costly products of the optician's art in definition and power. The names of those who have worked on it and helped to perfect it include many of the very brightest lights of the world of science in the nineteenth century. Lord Rosse, Lassell, With, Calver, Draper, Common, Foucault, von Liebig, all gave of their best brains to the problems—most

---

[†]By the Rev. William F. A. Ellison. M.A., B.D., F.R.A.S., F.R. Met. Soc., Director of Armagh Observatory, Member of the British Astronomical Association, and of the Société Astronomique de France.

fascinating and alluring problems they are—set for solution by that beautiful sphinx, the paraboloidal mirror. That simple-looking disk of glass, with its almost imperceptible curve, shows little indication of the magical and mysterious powers which lie latent in it, powers to open to the astonished gaze a universe of glory and wonder; or of the amount of thought and anxious endeavor which have gone to the perfecting of its subtle curve and the depositing of its shining skin of silver.

Briefly, the advances since Herschel's time have been:

1. The invention of the method of depositing a film of silver on glass by von Liebig, improved later by Brashear and others, which made glass possible as a material for optical specula; and

2. Foucault's lamp and knife-edge, which made it possible for the workman to see exactly what he was doing while figuring the reflecting surface.

It was unfortunate that Foucault's test was not known to With. Had it been, it would have immensely increased that great master's output, and also improved the quality of his work. As it was, Calver, though little if at all With's superior in skill, reaped the harvest of accuracy which With just failed to gather in. Many of With's specula have been retouched, to their great benefit, both by him and by the present writer, though one always handles a With mirror with reverence and wonders at the skill which could come so near perfection working in the dark.

Perfection—that is what the concave speculum has attained of late years. It is possible for a worker of sufficient skill and experience now to set about making a mirror, secure in the knowledge that he can produce a surface in which the most extraordinarily delicate test the human brain has ever devised can detect no flaw. Nay, more, he can produce a mirror whose accuracy is really *beyond* the requirements of telescopic vision. A mirror which has faults quite visible to the expert using Foucault's test will often perform in the telescope just as well as one which has none, because the faults of the faulty one are too small to affect the visual image. The test is, in fact, *unnecessarily* delicate. But then nothing is ever "good enough" so long as it can be improved, and no good speculum maker will let a mirror out of his hands which has a defect which he can see and can remove, however small that defect.

The earnest and industrious speculum worker never really knows to what his efforts may lead him. In most trades the amateur must be content to follow humbly, and at a distance, the steps of the trained professional man. In telescope making, and especially in the making of the essential parts of the reflector, it is the other way about. The amateur has shown the way to the professional, and forced the pace for him, ever since Herschel's

A.4.2. Literature

time. Herschel himself was an amateur, so was Lord Rosse, so was With, so were Draper, Common, Calver, Wassell, and Alvan Clark. That many of these *became* professionals only emphasizes the fact that they began work as amateurs and ended by beating the professionals at their own trade. That they did so is largely due to their recognition of the principle expressed in the phrase I have just now used. "Nothing is 'good enough' so long as it can be improved."

The chapters which follow are dedicated to the amateur telescope makers of the world in the hope that some at least of them may be thereby helped on the road which led Herschel, With, Calver, and Clark from humble amateurism to the headship of the world's professional makers. The writer's first telescope was constructed when he was aged ten years, and consisted of a spectacle lens, a sixpenny microscope, and a pasteboard tube, with which humble instrument, innocent of achromatism, he first viewed Jupiter's satellites, the phases of Venus, and the lunar mountains. This was the beginning of the ladder which has already reached more than 140 mirrors of apertures from 6 inches to 12 inches, and object glasses of 4, $4\,1/2$, 5, and $5\,1/4$-inch aperture.

A.4.2 Literature

The beginner who seeks for literature to direct his efforts will meet with the difficulty that most of the works on the subject of speculum making are out of print, or buried in back numbers of magazines, especially those of the *English Mechanic*. Draper's papers are only to be got at in the records of the Smithsonian Institution. The very excellent articles of Francis may possibly be obtainable in a public library in Vol. VII of *Amateur Work*. Wassell wrote in the *English Mechanic,* 1881–83. Browning's *Plea for Reflectors* and Horne and Thornthwaite's *Hints on Reflectors* are both out of print, as is also a useful little book by W. Banks, F.R.A.S. The only thing of the kind still in print is a helpful chapter on the subject in Hasluck's *Glass Working by Heat and Abrasion,* published at 1s. 6d. by Crosby Lockwood & Co.

Even if all these were obtainable they have one defect in common. They are more or less out of date. The most recent of them represents the state of progress in the art of mirror making existing in 1890–95 or thereabouts. And if nothing else had happened since then, the invention of Carborundum in 1898 was sufficient to revolutionize the whole process of grinding, and to place emery out of court as an abrasive. This material, a carbide of silicon, and manufactured much in the same manner as carbide of calcium, was first made at the Niagara Falls Electric Works, and began to come into use for glass working about 1900. In that year the writer obtained a sample in

Dublin, and since then has used no other grinding material, except for the very last stage of fine grinding. Carborundum cuts about six times as fast as emery, and with No. 80 a 6-inch mirror can easily be rough-ground to curve in less than half an hour, and the whole process of grinding can be done in two and a half to three hours.

A.4.3 Tools and Materials

For the amateur, speculum making has one great advantage; it does not require an extensive or expensive outfit of tools. Indeed the essential ones need not cost more than a very few shillings. Opticians, it is true, use cast iron or brass tools, carefully made to gauge and ground true, for forming the curves both of lenses and specula. And where a large number of curved surfaces have to be produced, all of the same radius, these are indispensable. But for the purposes of the amateur mirror or object glass maker glass tools are preferable, and both their cost and the trouble of making them are negligible. Here, then, is a list of the things necessary to be provided before we begin. It is neither long nor costly:

1. A pair of equal glass disks—one for the mirror, the other for the tool. The mirror-disk should have a thickness not less than one-eighth of its diameter—perhaps one-sixth is better still. The tool may be of lighter stuff.
2. A barrel—or, better still, two barrels.
3. A pound or two of pitch.—There are three kinds usually on the market—English, Russian or Archangel, and Swedish or Stockholm. Any of these will do, but Swedish is commonly preferred. As now sold, no cleaning or straining is needed. Purchasable of chemists or oil and color shops.
4. Carborundum, No. 80, 220, FFF—15M, 30M, and 100M powders are suitable and sufficient. Sold in 1-lb. and 2-lb. tins. Buy from the makers, on no account from the "local shop," or grades may be found disastrously mixed.
5. Jeweler's rouge.—One pound will polish quite a number of mirrors.

The above are indispensable. To these may be added a list of articles not indispensable, but very desirable:

1. A lathe.—The possessor of a lathe can turn up the blocks and handles which he requires for mounting and holding disks during working, and can do a lot of neat jobs when he comes to mount his mirror and flat. To make a flat a lathe is indispensable.

A.4.3. Tools and Materials

It is very desirable that the grinding and polishing should be done in different rooms. If the worker is lucky enough, or rich enough, to possess the house-room for this, the polishing-room should be provided with its own barrel, and nothing contaminated with Carborundum should on any account be allowed to enter it.

2. An oil stove or gas ring is almost indispensable for melting pitch, warming glass disks before cementing, etc. (Pitch will not stick to cold glass.)—If the worker resides out of reach of gas and uses an oil stove, it should be of the central-draft variety, with a circular burner, and its upper works should be stout enough to support a fairly heavy weight.

3. An assortment of enameled iron dishes and basins will be found useful for quite a number of purposes, such as holding water and rouge, and covering tools and mirrors to keep off dust during the intervals of working. Porcelain or earthenware ones are objectionable, owing to the danger of damaging a mirror by an accidental blow against them.

Fig. A.4.1: *Two Ways of Attaching the Tool to the Workstand.* **Left:** *A is the tool; B is the beveled disk of wood. Three wood screws hold it in place.* **Right:** *A is the tool; B is a thick, wooden disk; C is a disk of sheet iron; S is one of three screws, with washers. Drip from the edge of A may be caught by sheets of thin tin, cut to fit close to B.*

If the worker possesses a lathe, his first job will be to turn up handles and supports for tools and mirrors out of hard wood (oak, box, or mahogany for choice). The usual way of mounting the glass tool is to cement it to a disk of wood an inch or two larger in diameter than the glass. The disk has a wide bevel, and is gripped by three large countersunk screws, the heads of which hold the bevel, as shown in Figure A.4.1, left.

A preferable plan, however, is to have a thick wooden disk, considerably smaller than the glass, screwed securely and concentrically to a stout sheet-iron disk somewhat larger than the glass. The edge of this is gripped by three screws, with a small washer on the head of each. The glass tool cemented to the wood forms a sort of mushroom top. The advantage of this is that it enables us to keep the whole arrangement clean.

Get a sheet of thin zinc, and cut a circular hole in it the size of the wooden disk. Then cut it across, dividing the hole in two. This is placed under the tool, while grinding, with the two semicircular openings closely embracing the wood disk and catches the mud and water which drips from the tool. It is removed and washed as often as may be required, and saves a lot of mess and much *risk* of scratches in the fine grinding.

It is well to have a piece of soft deal plank, nicely planed up, and cut to fit on the top of the barrel, to which it is secured with screws. To this the various disks carrying tools are screwed, and when frequent screwing and unscrewing has damaged the wood beyond repair it can be easily replaced. Another disk, smaller than the glasses, must also be turned-up. It should be about 3 inches diameter, and is for the handle to hold the mirror. A socket is turned in the center of it, and a cylindrical piece fitted and glued or pinned in place. Now attach the handle to the mirror and cement the tool to its wooden disk, and screw the latter to the barrel top, and we are ready to begin.

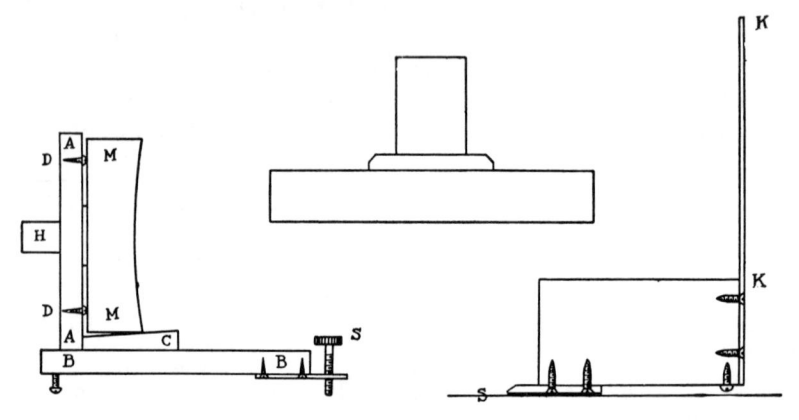

Fig. A.4.2: *Three Necessary Accessories.* **Left:** *The adjustable easel for carrying mirror when testing.* **Center:** *The glass disk with wooden handle attached, ready for work.* **Right:** *The knife-edge on its block—Ed.*

A.4.4 Rough Grinding

The next stage is roughing out the curve. The tool is warmed, slightly smeared with spirits of turpentine, melted pitch is poured on the wooden beveled disk, and the smeared side of the glass pressed down on it, squeezing out the excess all round. When cold, the disk is attached to the barrel top. We next warm the back of the mirror, smear a little turps in the center, pour sufficient melted pitch on, and press the handle firmly on. When cemented on it looks like Figure A.4.2, center.

Ordinarily 8 oz. cocoa tins are convenient for melting pitch in. Of this more later on, when we come to polishing. We place a basin of water and a handful of absorbent cotton handy on the work bench and a tin of No. 80 Carborundum, strew a little Carborundum on the top of the tool, dip the

A.4.4. Rough Grinding

face of the mirror in the water and place it on the tool.

We must now make acquaintance with the mirror maker's "three motions." In order that the desired curve may be produced, the upper disk must

1. travel to and fro across the lower,
2. rotate about its own center, and
3. the worker must walk slowly round the barrel.

It is obvious that these three motions could easily be produced by machinery, and for grinding, such a machine would work admirably. But it would fail in polishing and figuring. The labor of grinding a mirror of moderate size is not great, and hand work is all that we shall need. Motion (1) is known as the "stroke," and by lengthening or shortening it we can produce most useful modifications of the curve. In roughing out, a long stroke is useful—i.e., one as long as possible when the center of the mirror reaches the edge of the tool at the end of each traverse. It produces an irregular curve, of greater depth in the center; but at present this does not matter, as we are more concerned about excavating a hollow as quickly as possible than about its shape. The shape will quickly come right when we begin fine grinding and shorten the stroke. An elliptical stroke, or "side" stroke, is useful for some purposes, but it slows the cutting.

After working for a few minutes the charge of Carborundum begins to get dry; so we lift the mirror and add a few drops of water, and so go on till the feel of the grinding tells that the abrasive is ceasing to cut. We then lift off the mirror and renew the charge. The older authorities recommended washing the tool and mirror to remove the mud which has accumulated. But this is not necessary at the present stage, and wastes time and labor.

When it is deemed advisable to try how deep the curve is, then we must wash the mirror free from all traces of abrasive, using for this purpose the cotton before mentioned. Absorbent cotton is used instead of the sponge once recommended, because when each stage of grinding is completed the cotton can be thrown away and a fresh, clean bunch taken for the next stage, thus avoiding all possibility of carrying over grit from one operation to another.

To test the curve, the mirror is stood on its edge on a table or shelf, with its surface well swilled with water to make it reflective. The worker stands before it, with his eye on a level with its center, and holding in one hand a candle or small lamp level with his eye. He moves the light to and fro at right angles to the axis of the mirror. If the eye is nearer the mirror than its center of curvature, the reflection will move the same way as the light. If further off than the center, the reflection will move the opposite way. In this way, even with a rough-ground surface, the radius of curve can be ascertained within a few inches. The further the fine grinding proceeds the more accurate this test becomes.

It is not usually necessary to be particular to a few inches about the focal length of a mirror. But sometimes it is required to work as close as possible to a given focus. In such a case the rough grinding should stop whenthe radius of curve is still about 8 inches to 1 foot longer than is required, leaving this overplus to be brought down in the next stage. Being more finely ground, the wet surface is then capable of producing a fairly definite image of a light, and we can supplement the method above given of testing the curve as follows: Taking the testing lamp to be described in the chapter on Foucault's test, we substitute for the brass tube carrying the pinholes a tube of perforated zinc. For the knife-edge we substitute a vertical piece of ground glass, and mount this and the lamp abreast on one base. By placing these in the center of curvature of the mirror, we get an image of the perforated zinc thrown on the ground glass. When it is focused as sharply as possible, the distance from mirror to ground glass is the radius of the mirror's curve. A long lath marked in feet and inches is useful for measuring.

It is well to remember that the radius of curve usually shortens about $1/2$ inch in the process of polishing, and this should be allowed for if an exact focus is to be worked to. With care, it is possible to get within $1/4$ inch of a given focus.

Having roughed out the curve to within a little of the required depth, the next step is to cleanse away most thoroughly all traces of the coarse abrasive. The mirror is well sluiced with water, the tool and its block are detached from the barrel and sluiced under the tap, and crevices scrubbed out with a brush (an old tooth brush is excellent), and the top of the barrel is also well washed with plenty of water.

Now we replace the tool in position, throw out the water in the basin and the old cotton, get a fresh handful and refill the basin with clean water. We now proceed, using 220 Carborundum.

If the curve is very near the required depth, we shorten the stroke to one-third diameter; but if we are still several inches off the required focus it must be kept long for the present. If the curve requires no further deepening, six "wets" of 220 Carborundum will suffice for this stage. (Each time a charge of Carborundum is ground down and a fresh lot applied with water is called a "wet.") Five minutes is an average time for a wet; so each grade of fine grinding will last roughly half an hour.

We may here add that if a very long stroke has been used in roughing the first effect of changing to short stroke may be to lengthen the radius of curve a couple of inches. It will, however, shorten again, though more slowly, and the effect of the short stroke will be to bring the curve of the

A.4.4. Rough Grinding

mirror approximately to a part of a sphere which is what we want at present. We must persevere with 220 Carborundum till the desired radius is quite reached, for the effect of subsequent grades in deepening the curve will be almost nil.

When at length ready to proceed, the washing up process is carefully repeated, and we change to FFF. Six wets each of this, of 15M, 30M, and 100M, are given, washing up with care after each grade of powder. There is no need to elutriate Carborundum, as was formerly done with emery, to obtain the finest grades of all. No finer can be produced than the 100M[1] supplied by the makers. Indeed, it may be doubted whether this gives any finer surface than 30M or 60M Polishing may be commenced on a surface fined with any of these three.

But we may with advantage use a sixty-minute settling from finest washed flour emery to finish with. A pound of this emery is placed in a large glass jar, the jar is filled with water and well stirred, and then let stand for an hour, after which the liquid is drawn off with a siphon of rubber tube into another jar, care being taken not to disturb the emery at the bottom. The siphoned liquid is let stand till it deposits all the solid matter suspended in it, and this sediment is used to give the final fining to our mirror.

Great care is necessary in this final state, and also with 60M Carborundum. The quantity of abrasive between the close-fitting glass surfaces is so small that they sometimes seize each other and cling so fast together that it is difficult to separate them without a serious scratch. If the disks begin to cling, they should be slid apart at once, lest worse happen.

When the mirror is properly fine ground it will be possible to read large print through it at some inches distance, or to obtain sharp vision of the sashes of a window at several feet. But the beginner should beware of accepting this as an infallible test of fitness for polishing, as it may consist with the presence of large pits and scratches surviving from the coarser abrasive. The best safeguard against these is to be thorough in the earlier stages of the fining.

The six wets of each grade above described will be found sufficient, if care is taken to grind each down completely. Pressure on the glass is a help to the thoroughness of the grinding, and can have no ill-effect provided the thickness of the mirror is not less than the $1/6$ diameter before prescribed. Some workers wash the surface of both mirror and tool with the cotton after each wet, but this is not necessary, though perhaps advisable in the 30M and 60M stages.

The writer has always made a practice of mixing the finer Carborundum

---

[1] This size is no longer manufactured in the United States.—Ed.

grades (15M to 60M) with alcohol, and keeping them in corked bottles. The bottle is shaken and a few drops poured out on the tool for each wet, a few drops of water being added. These powders are so clinging that it is difficult to distribute them over the tool dry, and they do not readily mix with water. Moreover, a certain gradation of fineness can be obtained by shaking the bottle vigorously for the first wet, less vigorously for the second, and so on, and very slightly for the last.

Another way of obtaining a final fine grinding is by making a pitch-tool out of hard pitch. After 60M is finished with, we carefully wash all up, and replace the tool on the barrel. Dry the surface with a cloth, and smear a little spirit of turpentine over it with the finger tips. Now melt some hard pitch in a tin over the oil stove. It must be quite thoroughly melted. When completely liquefied, smear the face of the mirror thoroughly with a lather of soap and water, pour the melted pitch all over the tool, take mirror at once and press it down on the soft pitch, moving and twisting it about till the pitch is judged hard enough to retain its shape of itself. Now slide the mirror off, and you have a pitch surface of the same curve as the mirror itself. Let this cool completely, and then apply a thin layer of the 60M Carborundum, mixed with alcohol, all over it. Work the mirror on this in short strokes for ten to twenty minutes, and you get a semipolished surface, which, if thoroughly done will polish on the rouge tool in about three hours' work instead of the usual six. This method of doing the last state of fining has the merit of being perfectly safe, both from seizing and scratches.

It is in the last stages of fine grinding that scratches usually originate. A scratch during polishing is rare, and can only result from carelessness, such as approaching the polishing tool with garments or person contaminated with grit, leaving the tool exposed to dust, "spring cleaning" the room when polishing is in progress, etc. (N.B.—The polishing room should *never* be swept, and no one should ever be allowed, on any pretext whatever, to "tidy" it. Total destruction of a polished surface may be the result of neglect of this rule. "Let sleeping dust lie" should ever be the mirror maker's maxim. It will do less harm on the floor than on his tools or on optical surfaces. A lock on the door and the key in his pocket is the best safeguard.) But all endeavors should be directed to securing immunity from scratches in the fine grinding. Extra care in the washing of tool and barrel and mirror, the provision of a separate basin for the final operations, and covering the work bench with sheets of clean paper will usually secure a clean surface on which to begin the polishing. Very slight scratches need not be regarded, as they will polish out. But even serious ones will often be invisible on the fine-ground surface, to start into conspicuous and ugly visibility when polishing begins.

Many workers recommend the use of a pocket lens to examine the surface

A.4.5. Testing; Foucault's Shadow Test

after each state of the grinding, in order to ascertain whether all marks left by previous grades have been ground out. The experience of the writer is not favorable to this as a test. Even the microscope is not always able to tell whether a surface is well or ill-prepared before polishing has been begun. The large deep pits and scratches are so disguised by the presence of the mass of small ones that the most experienced eye may fail to detect their presence. It is only after about half an hour's polishing that they start into disastrous prominence. The only real safeguard is to be very thorough with the last, or last two, stages of the fine grinding. Carborundum cuts so fast that 30M or 60M will quickly remove even quite deep pits and scratches, and if any doubt remains as to the existence of these, a double dose of 60M will usually make sure of them. A plan which the writer has sometimes practiced is to make a scratch with a diamond somewhere near the middle of the tool, deep enough to be quite certainly deeper than any possible abrasive pit. The edge of the mirror and middle of the tool are the parts which grind most slowly. Therefore, when this scratch grinds out, it may be confidently assumed that all lesser pits are gone.

For the reason just mentioned, that the edge of the mirror gets the least grinding, it is a very good plan to do the last stages of the fine grinding with the mirror face up, reversing the relative positions of mirror and tool. The "mushroom" form of support already described (Figure A.4.1, right) will facilitate this change. In this way that part of the mirror which polishes most slowly (viz., the edge) will get the most fine grinding, and the benefit will quickly become manifest when polishing begins.

A.4.5 Testing; Foucault's Shadow Test

As soon as we begin to polish we immediately require some means of ascertaining what is happening on the concave surface of the mirror, though this knowledge is not of the greatest importance until later on, when we begin the most difficult process of all, viz., the figuring. We might, indeed, go ahead with the polishing, confident that no irremediable error would develop in the curve, so long as the tool was made and used as we shall describe by and by.

Once upon a time it was believed that if a figure of a mirror once became hyperbolic it could not be remedied, and all sorts of devices to avoid the "fatal hyperbola" were resorted to. As a matter of fact, a hyperbolic mirror, though less easy to remedy than some other faulty curves, presents no difficulty to a moderately expert hand, and we could, if we desired, polish away merrily till all marks of abrasion had disappeared, without any testing at all, careless which of the list of possible regular curves turned-up at the end of the process.

There are only two limits to this possibility: (a) the limit of human endurance, which prescribes that half an hour's polishing at one spell is enough for the worker's patience; and (b) the limit of endurance of the pitch tool, which begins to soften from the heat produced in polishing, if one goes on much longer without letting the tool cool down. To this we may add human curiosity, which naturally desires to see how the figure is shaping.

Up to the time of With, speculum makers had to work in the dark, except for tests upon stars, for which one had to wait till the sky chose to clear, and also to dismount the mirror from its handle and mount it in a telescope tube. Naturally, mirror making under these circumstances was a slow and uncertain process. It is owing to the genius of the great French Scientist Foucault that we now have a simple and easy method by which the figure of a mirror is made actually visible to the eye, and so delicately visible that the expansion due to the heat communicated to the glass by the touch of a finger can be clearly seen.

Foucault's method consists essentially in the provision of an artificial star, in the form of an illuminated pinhole. This, of course, cannot be placed at a great distance, so that the light from it will be sensibly parallel, like that of the real stars. It is therefore placed at the center of curvature, and although the resulting appearances differ from those seen when parallel light is converged by a mirror to its principal focus, the two sets of phenomena can be connected by a simple formula.

If the pinhole is placed at the center of curvature of a spherical mirror, it follows from optical laws that all light from the pinhole falling on the mirror will be reflected back exactly to the pinhole again, and will form an image of the pinhole, the same size as the pinhole, on the pinhole. In this position it could not be examined. We therefore slide the lamp a few inches to the left, causing the image to move the same distance to the right, where it can either be examined by an eyepiece or received direct into the eye. Both methods are useful. If the image is allowed to enter the eye, the mirror is seen full of light, like a full Moon. Now comes in the second part of Foucault's ingenious plan.

A vertical knife-edge is mounted so that it can be made to slide laterally across the path of the pencil of light, close to the point where it focuses just before entering the eye. When this is done the eye sees the shadow of the knife-edge cross the mirror. Hence the name "shadow test." But the manner of its crossing differs according to the position of the knife-edge with respect to the focus. If it cuts the beam within (nearer the mirror than) the focus a vertical shadow crosses the bright face of the mirror the same way that the knife-edge moved. If it is *outside* the focus, the shadow crosses the opposite way. If it is exactly *at* the focus, the surface of the

A.4.5. Testing; Foucault's Shadow Test

mirror darkens all over evenly, and looks flat, no moving shadow being seen either way. These are the appearances characteristic of a sphere.

But if the mirror is not spherical, but has some parts of a greater and some of a lesser radius, however small the difference, what will happen? Obviously we shall see the shadow broken up into parts crossing the mirror opposite ways, if the knife-edge is so placed as to be within the focus of some parts and outside that of others—and that however minute the difference may be.

In practice the mirror is mounted on a kind of easel having one screw-foot in front, as a fine adjustment for raising and lowering (Figure A.4.2, left). The lamp should be as small as possible, and it is convenient if it can be made to sit in one of the rings of a retort stand, so that it can easily be clamped at any height. A brass tube, with two or three pinholes of different sizes, drops over the chimney. The knife-edge is a vertical strip of steel with a sharp edge, mounted in a wooden block weighted for steadiness and having its forward edge provided with a smooth metal straightedge (Figure A.4.2, right). The object of this will be seen later on.

The whole apparatus, with its table, should rest on a stone, tiled, or concrete floor—not boarded. If this is not attainable, then let the mirror easel rest on a bracket bolted to the wall, and the table carrying the lamp be of the stoutest make, and stand on a hearthstone or some similar spot.

The quantities which the apparatus is designated to measure being of the order of millionths of an inch, no movement of the apparatus itself is tolerable.

A convenient form of support for the lamp and knife-edge is a very stout tripod table, with a small top made of soft pine board, planed nicely smooth. Its height should be such that a man kneeling at it can easily rest his elbows on it. It is advisable for the worker to make his own.

We have already seen what are the appearances presented by a spherical mirror when put to the question by the pinhole and knife-edge. But we shall rarely see a truly spherical one. Nearly always we shall have curves with a radius differing more or less in different parts. They will be either

1. radius longer in center than at edge (oblate spheroid),
2. radius very slightly shorter in center (ellipse or prolate spheroid),
3. radius a little shorter still in center (paraboloid), or
4. radius very much shorter in center (hyperboloid).

Of these, 1 and 2, as well as the sphere, are undercorrected; 4 is overcorrected, and 3, the paraboloid, is truly corrected, and this is the curve which we desire to produce.

There are, of course, besides these an infinite variety of irregular figures, which may be combinations of any two or more of the above, or figures not symmetrical, such as astigmatic curves produced by "flexure" (strain, or bending of the glass). The last is not likely to be met with if good glass of proper thickness is used and the directions already given as to mounting and holding it are followed. But if a disk of glass is cemented to a stout wooden block covering the whole of its back, as we have known some beginners do, it is pretty certain to be flexured; and a flexured mirror is rarely curable. The most usual irregularities are hills or hollows in the center, rings, and "turned-down edge." For the present it will only be necessary to be able to recognize the appearances presented by the principal types of regular figure, viz., oblate spheroid, sphere, paraboloid, and hyperboloid.

We have put the lamp on the left and knife-edge on the right, reversing the order of Francis, Wassell, and Draper, as a matter of convenience. It is most important that the knife-edge should be next the observer's right hand, as very delicate movements of it have to be controlled. The lamp is never moved when testing. A little practice and thought on the cause of these phenomena will enable the beginner to distinguish a hill on his mirror from a hollow, even if the nature of the irregularity is not obvious at first sight.

The lamp used in these tests may be any small-flame one. A small Argand burner, with a narrow chimney, is perhaps the best. The smaller the source of light, the nearer the knife-edge can approach the pinhole, and the less will be the offset between the pinhole and knife-edge from the actual optical axis of the mirror. They should not be too far from the axis, or distortion of the shadows will result; and, contrary to the opinion of many workers, the pinholes should not be too small. It is convenient to have two—one of liberal size, pierced with an ordinary sewing needle, for rough testing, and a small one, made with the point of a very fine needle, for fine work, eyepiece testing, and testing a mirror after silvering. We have known grotesque errors to result from using too fine a hole. Indeed, on one occasion an absolutely perfect mirror was sent to the writer to correct for a "turned-down edge," which was entirely nonexistent except in the owner's testing apparatus. He used an acetylene flame and an excessively tiny pinhole, with the result that he saw a series of diffraction bands inside the margin of his mirror, and took them for a turned-down edge. An eyepiece test will instantly detect a turned-down edge if any is present, even if too slight to notice with the knife-edge. But more of this later.

It is convenient to cut two openings in our brass tube, opposite each other, cover the whole with a sliding collar of very thin sheet zinc, and pierce the holes in this, one in each opening. Lest the arrangement should collapse if the chimney should overheat, it is perhaps well to avoid the use of a solder, and to secure the zinc collar to the brass by means of a spring clip. A piece cut from a clock spring, of the requisite curve to encircle the

A.4.6. Polishing 83

tube, answers well.

Another way is to file the brass tube nearly through, and then pierce the thin remaining metal with a needle. But there is considerable advantage in being able to adjust the pinholes higher or lower in the tube, which, of course, is impossible if they are pierced in the metal of the tube itself.

A.4.6 Polishing

Having provided ourselves with a Foucault's testing apparatus, we may now proceed to polish. The glass tool on which the mirror was ground is, with its wooden block, removed from the grinding barrel, thoroughly scrubbed and sluiced with water, to clean away all trace of abrasive, and transferred to the polishing barrel (and polishing room if we have one). Dry its fine-ground surface, and smear a little turpentine over it. Meanwhile a tin of pitch will be melting over the oil stove. This should be carefully watched, and stirred frequently with a short stick.

It is a very great improvement to add to the pitch about 5% to 10% of beeswax. The effect of this will be appreciated when we come to cutting out the facets on the tool. Pure pitch is abominably sticky, and also is liable to fly into tiny chips when cut. These adhere to the skin, hair, and clothing, and are more than likely to conduce to profanity, being very difficult to get rid of. The wax-pitch mixture does not chip, and its stickiness is so much reduced that it can be molded with the fingers when soft without adhering. Still more important is what may be called its "flexibility," using the word in the motor-engineer's sense. Pure pitch, to work well, is required to be very close to the ideal degree of hardness. If it is too soft it very rapidly produces a deep hyperbola, and also a "turned edge," the dread of all the old mirror makers. If too hard it scratches. But the wax-pitch mixture will work well, within much wider limits of hardness, and seldom produces a turned edge. It should be just possible to mark it with the thumbnail when cold.

Watch that the pitch does not boil, and if it shows signs of doing so, lower the lamp. Prolonged heating hardens it. To soften, add carefully a little spirits of turpentine. (Do not spill any over the lamp, or there may be an explosion.) Not more than a teaspoonful should be added at once, as its effect is quite disproportionate to its quantity.

While the pitch is melting we get the mirror ready by standing it face up and painting it all over with a thin paste of rouge and water. We will require two or three glass jars (such as 1-lb. jam jars) and a plate of glass to cover each to exclude dust, a camel's hair brush, large and flat, and a knife, or, better still, an old razor. We also need a stamping tool. To make this, take a piece of nice, cleanly-planed wood with straight faces about 9

× 1 1/4 × 2 inches. Cut two pieces of thin hoop iron 9 inches long, and file one edge of each to a wedge-shape. Clean both pieces up well with a file, drill three or four holes in each, and screw them to the sides of the piece of wood, so that their sharpened edges are parallel and 1 1/4 inches apart. This is our stamp for marking out the facets of the tool.

Mix in one of the jam jars a tablespoonful of rouge with water to a thin cream, and paint some of it over the face of the mirror with the camel's hair brush. When the pitch is quite liquid all through, grasp the tin with a cloth, and pour it out rapidly on the tool, beginning at the edge, going inwards with a spiral motion and ending at the center, where a considerable excess may be poured.

Lay the tin aside and at once take the mirror and press it face down on the semifluid mass, twisting it around and moving it to and fro for several minutes, or till the pitch is cool enough to retain its shape. It will overflow all around the tool. Let it stay there. It will do no harm and will safely imprison any bits of loose grit that may be present.

The layer of pitch on the face of the tool should be about 1/8 inch deep, not more. When it is judged time, slide the mirror off, lay it aside, and take your stamp and press it lightly on the still soft tool.[2] It leaves the impression of its two parallel edges. Lift it and place the second edge in the furrow of the first and press again. When you have gone across from side to side repeat the operation at right angles. The tool is now covered with systems of parallel lines, the two sets being at right angles to each other and dividing the surface into squares of 1 1/4-inch sides.

Now we take the old razor and proceed to cut out V-shaped channels about 1/8 inch wide along these lines. Cutting down to the glass, the strips cut out can be lifted clean out, leaving clear V-shaped furrows. We could not do this with pure pitch. We also get a criterion of the hardness. If the strips cut and lift out without either elongating or breaking up into bits, the hardness is just about right. A strip, when cold, should bend just a little before breaking. If it will bend nearly double without breaking it is too soft.

But there is one point about the facets which we have not mentioned, and it is an important one. The center of the tool must not be near the middle of a facet, nor must it be in a channel. It should be *in the corner* of a facet. If not, the mirror will polish in rings. It is a good plan to mark the center with a pair of compasses before using the stamp and then to stamp the lines embracing the center first.

The older workers, and especially Wassell, were very particular about the *shape* of the V-grooves, that the slope of the sides should be exactly

---

[2] See two paragraphs below about decentering the central facet.

A.4.6. Polishing

45°. This, like too many of the older refinements, is pure bosh. It does not matter in the very least what the shape of the grooves is, provided they are hollow enough to give clear air channels under the face of the mirror. Nor does it matter what the shape of the facets is either. We only make them square because that is easiest. The $1\,1/4$-inch squares will do for all sizes of mirror from 6 inches to 10 inches. Above 10 inches we may double the size and make them $2\,1/2$ inches. Below 6 inches they may be dispensed with altogether.

The cutting of the facets will somewhat disturb the curve. As soon as it is finished, therefore, while the pitch is still a little soft, we must paint the tool over with the rouge and water, and place the mirror on and work it a bit, say for five to ten minutes, meanwhile observing through the glass what is happening. Probably at first the central facets will not be entirely in contact with the glass. We must work, if necessary, with pressure, till all air bubbles disappear and all facets are in contact. And here we find the advantage of having the back of the mirror transparent. We will find it again when silvering. Many makers grind the back; a foolish and a totally unnecessary proceeding. The only possible object of this is to prevent light which passes through the silver film from being reflected from the back to illuminate the field of view. I have elsewhere shown that the maximum possible amount of light so returned could not exceed that of a 12th mag. star distributed over the entire field. So do not grind the back, and both figuring and silvering will benefit.

Now, having the tool in order, we proceed to polish. The motions are just the same as in fine grinding. But now we must time ourselves. It is by time that we judge how much polishing the mirror has had. So a clock forms part of the furniture of the polishing room. Another useful article is a thermometer, which should be hung not against the wall, but freely out in the room. A rise or fall of even 5° in temperature will greatly affect the behavior of the tool. A rise softens the pitch and a fall hardens it; consequently, a change of temperature may quite alter the character and effect of our tool. For this reason, as well as for another (to be explained later on), a light building of wood or iron is totally unsuited for a polishing place. The best place of all is a cellar or basement below ground, where the temperature will remain reasonably constant. In any case it must be a building with substantial stone walls. And the Sun should not be allowed to shine in if it has a large window. Especially it must not, on any account, shine on the tool, or we may have to remake the latter.

Having fully taken in these *caveats,* we may begin our first spell of polishing and go on for half an hour by the clock, using short strokes, and only stopping occasionally to renew the rouge and water. A dish of clean water, with a short stick having a large handful of absorbent cotton tied round one end in it, will be found useful on the workbench. This is to wash the rouge off the mirror when we stop to test, and keep our hands out of the mess. We also will need a few pieces of old linen, quite clean, for wiping the optical surface dry.

It is well to keep a written record of the spells of polishing and of the figure found at the end of each. At the end of the first spell the mirror should be semipolished all over, rather more in the center. We may possibly find a very eccentric figure at this stage; perhaps a very exaggerated hyperbola. But do not mind. Go on, and it will come to reason later on. And if it does not, but gets worse, which is not likely, remember that any figure made by polishing can be unmade by polishing. It can never be necessary to regrind, no matter how eccentric the figure.

But it is more than likely that it will be found to be somewhere near a sphere. In any case go on. We are only polishing, not figuring, and have at least three hours of polishing before us before the figure matters at all.

The mirror should work easily and smoothly on the tool. If it does not, but sticks and clings, the curves cannot be truly coincident. In this case it is useful to leave the mirror on the tool for several hours, with plenty of rouge and water between, to prevent sticking together, and three blocks of wood round to prevent sliding off.

After the first half-hour of polishing we may use the microscope to ascertain the prospects of quick polishing. A 1-inch objective is suitable. With this all pits and scratches are visible, and we can see if any deeper than the average are present. If none but the finest are visible we may expect to polish in about three hours, or even less. The outer $1/2$ inch of the mirror is all that need be examined, as this zone is always the slowest in polishing.

If the successive stages of fine grinding have not been done with sufficient thoroughness it will be quite easy with the microscope to identify the pits due to the successive grades of abrasive. Moreover, the appearance of emery pits is quite different from those due to Carborundum. The latter are sharp-edged tiny holes, deeper than they are wide; those from emery are much more diffused, and naturally polish out quicker. Hence the advantage of an emery finish to the fine grinding.

We shadow test just for curiosity before going on, for it is little likely that any errors gross enough to require a change of tool will be seen. Then we proceed to another half-hour's spell, and so on till the surface begins to look well-polished to the naked eye.

During the preliminary polishing, the water in the dish will be getting more and more stained with rouge. When it is no longer able to cleanse the mirror, set the dish aside and take a clean one and a fresh lot of water. After the first dish has stood for twenty four hours its charge of rouge will have settled. Pour off most of the water and with what is left rinse the dish round well, pour off all but the last few drops into a clean 1-lb. glass jam jar, and put it aside covered with a glass plate. This will give us a reserve of extra-fine rouge for the final stages of figuring. This may be repeated as often as the water in the dish becomes too deeply stained to cleanse, and the fine rouge settled from the rinsings is most important in obtaining a surface clear of the tiny scratches rouge is apt to make.

Some samples of rouge are very scratchy, and cannot be used at all without treatment. They may be stirred up in a jar of water and poured off after $1/4$ to $1/2$ minute's settling. But a better plan is to make the rouge into a paste with water and then ladle the paste by spoonfuls into a flannel bag. Place this bag when full in a jar of water and knead it well under water. The fine powder comes through the bag and all coarse particles are left inside. It is a messy job, but worthwhile.

When we can no longer see any defect of polish with the naked eye we once more inquire of the microscope, and if no abrasive pits can be seen in the marginal zone of the mirror all is well so far as the polishing is concerned.

A.4.7   Figuring

We now come to the *crux* of the whole process. Grinding and polishing are purely mechanical processes, which any handy man should be capable of learning in a few lessons. But the man who can produce a perfectly true paraboloidal curve right up to the edge of a mirror is not a mechanic, but an artist; and the artist is born, not made. Volumes might be written on the art of figuring, and the reader of them would be no nearer being able to produce a true curve after reading them than before, if the talent were not born in him.

Much depends on the figure which we find present at the completion of polishing. If it is a sphere, or an oblate spheroid, of a moderate amount of oblateness, and with no complications, such as turned-down edge, we may go ahead with the tool we have been using, taking to the fine rouge collected as described already. We have merely to deepen the curve very slightly towards the center. There are several ways of doing this. We may classify them:

1. Parabolizing by long stroke
2. Parabolizing by graduating facets
3. The small polisher system
4. Parabolizing by overhang.

The first two are old methods, and are described in all articles on the subject, from Herschel to Wassell, and later.

A.4.7.1 Parabolizing by Long Stroke

We have been polishing in short strokes of about one-third diameter, and straight (center over center). If we now increase the *length* of stroke to one-half, two-thirds, or whole diameter, we shall get a more or less rapid hollowing of the central region of the mirror.

If there were no complications to this method, parabolizing would be easy; but, unfortunately, using a long stroke very often means producing a turned edge. It must, therefore, be used sparingly, and with discrimination. Supposing, for example, that the figure we have to parabolize is a sphere with a turned-*up* edge, we may use long strokes for a while, being careful not to overdo it. Dealing with an oblate spheroid in the same condition, we might use it a bit more freely. Resort must be had to Foucault's test at frequent intervals. But where there is a question of a turned-*down* edge we require a control on Foucault's test.

Very few mirror workers are capable of detecting a turned edge by Foucault's test alone unless the turn-over is very gross. But an *eyepiece* test will decide the matter at once. For this we require an eyepiece of about one inch equivalent focus and a fitment to hold it which will stand in another ring of the retort-stand which carries our pinhole lamp. To use it we simply remove the knife-edge out of the way, and slide the ring carrying the eyepiece down into its place till we can see in the latter the image of the pinhole. If the figure of the mirror is near a sphere the image will be nicely sharp. Now slide the whole stand, lamp, eyepiece and all, alternately nearer to, and further from, the mirror, so that we get the image out of focus. If the image, when some distance inside focus, is a circular disk, with a clean, well-defined edge, the curve of the mirror is true to the very edge. But if the disk has a hairy, fuzzy, and ill-defined outline, the edge of the mirror has a flatter curve than the rest; in fact, it is "turned-down." This test is infallible and extremely delicate for this one defect.

The expanded disk, outside focus, is *always* sharp in outline. Even when the mirror has a turned-up edge the appearance is not reversed, as we might suppose. In this case the outside focus disk will have a slightly softer outline than usual, while the inside focus one will be very sharp. In using this test, however, we must remember that the paraboloid which we are trying to produce is a figure which becomes flatter towards the edge. Therefore, we must not aim at getting the inside focus disk too sharp. It should be just a little softer in outline than the outside focus one when the mirror is finished. But a hairy edge to it cannot be tolerated.

A.4.7. Figuring

The turned-down edge is the mirror maker's *bête noire*. Formerly it was considered impossible to escape it, and older makers, as a matter of course, used to grind off the outer half-inch, or advise stopping the mirror down. But we know better now. As the "fatal hyperboloid" is now no longer incurable, neither is the almost equally dreaded turned-down edge.

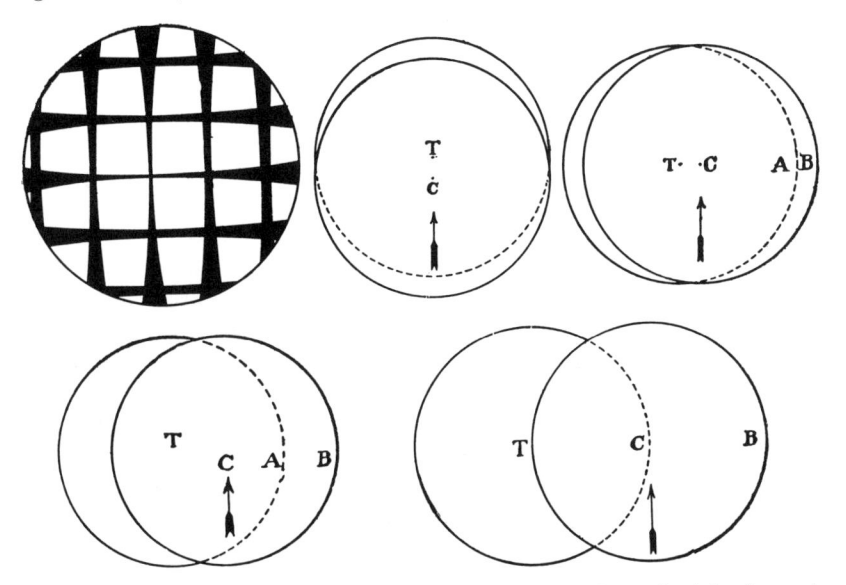

Fig. A.4.3: Upper left: Tool graduated for parabolizing. In each of the four other diagrams the complete circle represents the mirror and the arrow, the direction in which its center, C, is moving to and fro. T is the center of the tool. In the first drawing, C passes over T at each stroke. In the second, third, and fourth, C passes at greater and greater distances to the right of T, according to the effect required. In the last drawing the center of the mirror works on the edge of the tool. Not to be recommended unless central hill is to be removed, and then very cautiously.

A.4.7.2 Parabolizing by Graduating Facets

This is the standard method of bringing an oblate spheroid or a sphere to a paraboloid. It seems obvious that if we reduce the size of the facets of our tool, progressively from center to edge, by widening the channels between them, we shall increase the amount of abrasion in the central parts relative to that near the edge, and thus will obtain the desired result of a deepened curve near the center, without increasing the stroke or imperiling the curve of the edge. The alteration will be made as shown in Figure A.4.3, upper row, left.

This is, perhaps, the easiest method of graduation. It might also be done by leaving the middle facet as it is, trimming off the corners of the three nearest to it, and cutting those next in order down to circles, and yet smaller circles. This will often produce the desired effect, but not always. It is well to use a broad chisel and light wooden mallet for trimming facets and keep a large, soft brush handy for sweeping the tool free of chips. This should be kept in a dustproof receptacle.

The above two methods of parabolizing are standard methods of all the old masters of the mirror working art. Those that follow are the result of experiments of the present writer.

A.4.7.3 Parabolizing by the Small Polisher System

It has always been laid down as axiomatic that mirror and tool must be the same diameter; but, like many axioms of the old workers, this principle has no other foundation than their fear of attempting new methods. It is quite easy to both grind and polish on tools considerably smaller than the mirror. And what we have called the "Small Polisher System" is often useful as a remedy for the great enemy, "turned edge." A polisher a little less in diameter than the mirror, of hard pitch, used with short, straight strokes, will often remove a turned edge when everything else has failed. A polisher considerably less than the mirror, even so far as only two-thirds diameter, will rapidly hollow an obstinate oblate spheroid. Naturally, this must be used with caution, stopping every few minutes to test, lest a deep hyperbola result.

A.4.7.4 Parabolizing by Overhang

Cutting and trimming the pitch tool is always more or less objectionable, if for no other reason than that it cannot be undone, once done, without the trouble of destroying the tool and making a fresh one. But without making any alteration in a satisfactory tool it can be made to cut the central region of the mirror faster, as follows. Observe the four positions in Figure A.4.3 (the figure at top left being excluded).

The first one shows the ordinary center over center stroke, with which we have been working. When the centers coincide, of mirror and tool, the weight of the mirror is equally supported all over by the tool.

But if instead of letting the center of the mirror pass over the center of the tool at each stroke, we bring the mirror a little to one side (as in the second) so that the crescent AB of the mirror is always off the tool, obviously the region of the mirror between C and A will be supporting the whole weight of the mirror, and will be pressed against the tool more forcibly than the rest of the surface. If, therefore, we polish with the mirror in this position, the mirror, as before, rotating about its center and the

A.4.7. Figuring

worker walking round the barrel, we will get a greater abrasive effect in that part of the mirror, whose radius is CA.

And if we increase the overhang, as in the lower left, we further narrow the part of the mirror supporting its weight, and further concentrate the abrasion in the central region.

While if we place the mirror as in the last drawing of the group of four, with its center on the circumference of the tool, we get the whole pressure, due to the weight of the mirror, supported on its center alone. To work in this position would rapidly produce a deep hollow just in the center. It is therefore useful for removing a central hill, but it is not to be recommended except in such a case, and then very cautiously.

We have, therefore, in this method of "overhang" (so called because the mirror always overhangs the tool by a certain amount laterally during the entire stroke) a most valuable means of controlling the figure. It is obvious that we can work the overhang method with an elliptical or circular as well as with a straight stroke, and also that we can alternately increase and diminish the amount of overhang as we work. In this way we can distribute the extra abrasion due to overhang freely over any required area about the center of the mirror, and therefore can probably produce a paraboloidal curve more easily by this method than by any other.

In all these operations it is well to have the back of the mirror protected from the heat of the hands by some nonconducting material. Otherwise a good deal of trouble may result, and the success of figuring may be considerably retarded. A piece of thick pasteboard cut to a circle the size of the mirror with the center cut out to admit the handle will be found useful.

A.4.7.5 Working Uphill

So far we have been trying to bring the figure of our mirror from sphere to parabola, or from oblate spheroid to parabola, via sphere. Now this is the way that the figure of a mirror, if left to itself during polishing, will travel nine times out of ten. We therefore call it working "downhill."

But, supposing we have, to begin with, a more or less hyperboloidal figure, or that we have, in bringing the figure downhill by any of the above methods, overshot the mark, as very often happens, and obtained a hyperboloid, we will have to find means to make the curve retrace its steps. As this is considerably more difficult (it used to be counted as impossible, hence the term "fatal hyperboloid"), we may call it "working uphill."

To begin with, we may lay down the principle that a hard polisher pulls the figure uphill, whereas a soft one lets it downhill. But we may as well say here at once that although various types of pitch polishers are calculated to produce certain definite effects, the pitch tool is most

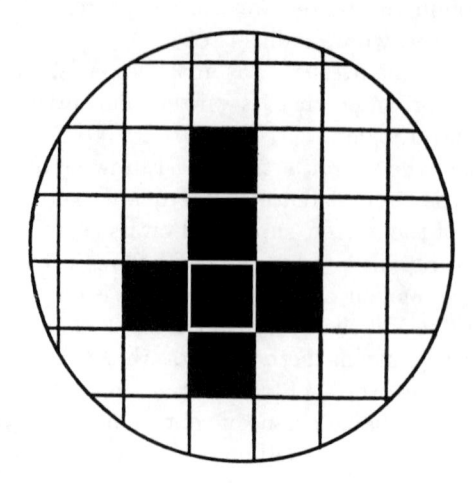

Fig. A.4.4: *Facets shaved from the tool to reduce a hyperbolic figure.*

delightfully inconsequential in its behavior, and one never knows for certain what any given tool will do till it is tried. It is as variable and changeable as the weather, and a speculum maker may fancy that after some years of experience he has fathomed all that pitch can do when it will suddenly surprise him by some totally fresh whim.

However, the first thing to try for a hyperbola, as well as for a turned edge, is a hard tool and short strokes, and three times out of four this will be successful. If, however, it is not, we must try graduating the facets in the opposite way to that shown in Figure A.4.3 (upper left) cutting down the central ones and leaving the marginal ones full size. In extreme cases of a very deep or obstinate hyperbola, the whole center of the tool may be removed bodily, leaving only a ring of pitch. This will probably result in an irregular oblate spheroid, whereupon we make a fresh tool, and proceed to work downhill again.

A shape which the writer often finds useful to bring a hyperbola to reason is an ordinary tool having a wide, rectangular strip the breadth of one, or even two, facets, removed right across the center, or having two such strips, one a facet longer than the other, crossing each other, as shown in Figure A.4.4. It must not be supposed, of course, that this arrangement will produce a regular figure. It will probably result in a large hump in the center. When the center is sufficiently raised we can then make an ordinary "downhill" tool, and work towards a paraboloid again as before. If we have, as sometimes happens, a figure which is not exactly a hyperbola but has a big hollow in the center, the above tool, carefully used, may (with

A.4.7. Figuring

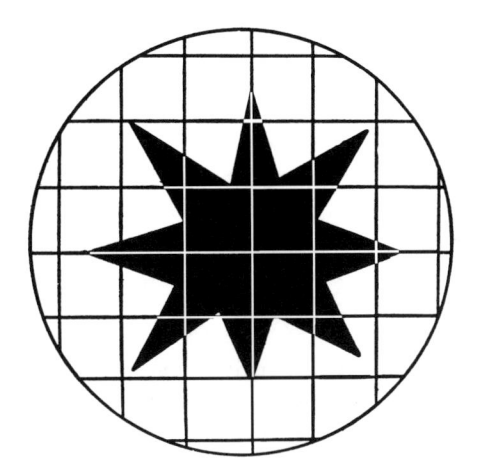

Fig. A.4.5: *Another way to reduce a hyperbolic figure. This is more suitable than Figure A.4.4 for small sizes.*

luck) bring it just right.

It sometimes happens that a ring, either raised or depressed, appears on the mirror. A polisher which rings the mirror probably has its center in the wrong place with respect to the facets. The ringing action may be reduced, or even entirely suppressed, by using an elliptical stroke, or "side," as it is sometimes called. "Side" tends to turn the edge, though not to the same extent as long strokes, and therefore should be used sparingly. A single depressed ring on an otherwise promising figure may be eliminated by cutting out two or three facets along the path of the ring, reducing the abrasion of that particular zone. A raised ring is more difficult to deal with. If not far from the circumference it may be removed by the "overhang" method, working so that the edge of the tool just traverses along the raised zone. It may also be dealt with by very cautious use of a polisher all cut away except a ring of the same size. A very few strokes at a time should be attempted on this, or we may get a depressed ring instead.

Having overcome these difficulties we will suppose we are now approaching the desired paraboloid. There are several difficulties of another kind to be met before we "get there."

If the worker is at all observant he will have noticed before this that the figure seen on testing immediately after removing the mirror from the tool is quite different from what appears some time later, when the mirror is allowed to rest awhile. The cause of this is, briefly, temperature. The friction of mirror on tool produces quite a considerable amount of heat, and we shall find that there is a very definite law connecting the figure of a mirror with its condition of temperature relative to the surrounding atmosphere. Briefly stated, the law is as follows:

When a mirror is cooling, its figure is temporarily pulled "downhill," or in the direction of the hyperbola. When a mirror is warming, its figure is temporarily pulled "uphill," or towards the sphere. (This, by the way, is the real reason why a slight undercorrection is always given to specula by the best makers. A mirror, when in use on the heavens, is nearly always cooling slowly, with the atmosphere of a clear night. Consequently, if its figure were a full paraboloid it would always, under working conditions, be a little overcorrected. To avoid this it is left a little undercorrected in the workshop.)

The result of this law is that we can never see the real figure of a mirror immediately after taking it off the tool. In fact, if we see a parabolic figure then, and leave it awhile to cool, we shall find that it has changed to a sphere, or even an oblate spheroid. To obtain a paraboloid when cool we must have a very pronounced hyperboloid on taking it off the tool. At least half-an-hour must elapse before the true figure can be seen. Ignorance of this has caused endless trouble to beginners, because in order for the figure of a mirror to be all right, it must look all wrong when just taken off the tool. Consequently, when approaching completion of our mirror, we have to pass through a period of working for a few minutes and then waiting half-an-hour to see the result.

A.4.8 Editor's Note

Many old-timer, advanced amateurs, vividly recalling the misgivings caused when they were novices by the opening paragraph of the A.4.7, have urged with feeling that it be struck out of the book. It also must have discouraged many from even starting.

It is probable that many—perhaps most—novices run away with the belief that a telescope either has a perfect mirror or it will not perform. At the risk of seeming to encourage slovenly standards it ought to be explained by the Society for the Prevention of Cruelty to Beginners that a mirror has to be pretty bad before it won't perform; also that a mediocre mirror will usually seem, at least for some time till your observing acuity has been educated, to be about as nearly perfect as your first grandchild; and that the reasonably good mirror you should aim to make on your first try will probably seem about perfect for a while. Thus Ellison's paragraph misleads by omission.

Many, before beginning, think of making one telescope and one mirror, which must of course be tops or else, and thereafter using it indefinitely. More common is something like this: On trying out your first mirror you exclaim, "I did better than I thought I could." After a few weeks you add, "but I know its faults and can do better, and while I'm at it I'll make a bigger telescope." A year or so later you build a third, again larger, and your third mirror will be on nodding terms with Ellison's described ideal. All along, as your observing eye has become more exacting, your manual skill has improved to keep pace. (About this time you go back and, out of curiosity, retest your first mirrors—the ones you once thought good—and maybe grin a little.)

As for "talent," Ellison doesn't specify how much talent is needed but everybody has talent, which is mainly perspiration. The words "artist ... born, not made" have been a real bogey to some, but read Ellison closely; he applies those terms to perfect mirrors alone.

All sorts of plain average folks, thousands of them, have made successful mirrors, even editors. Cheer up.

A.4.9 How to Recognize the Paraboloid, Zonal Testing

The "Temperature Effect" is not the only difficulty to be encountered at the critical stage when we are nearly "there," but not quite. The question arises, "How are we to recognize the paraboloidal curve when we get it?" This is a very real difficulty. The only difference visible in testing between a true curve, one a little under, and one a little overcorrected, is a difference of *depth of shading.* We cannot leave this to mere personal judgment. We must have a means of distinguishing accurately. And this is the more necessary as the depth of shade in the parabola *varies enormously with the ratio of focal length to aperture.* The same shadows which would indicate a beautiful parabola in a mirror whose ratio was $f/8$ (as the photographers conveniently call it) (i.e., whose focal length was eight times its aperture) would mean a deep hyperbola in a mirror of $f/9$. And when we get much beyond $f/10$ or $f/11$ we can *see no parabolic shadows at all.* If shadows can be seen in such a mirror it is hyperbolic. In other words, the difference between a sphere and a paraboloid in such a mirror is too small to be seen; it can, however, be measured.

We have already learned that when testing a spherical mirror, with the knife-edge just cutting the converging beam exactly at the center of curvature, the mirror is seen to darken evenly all over. In other words, the difference of radius of curve for center and all zones distant from the center is zero. In the paraboloid and hyperboloid the radius of curvature increases outwards from the center. In the oblate spheroid it diminishes outwards.

Therefore, if we place over the mirror a stop which covers all its surface except a small area about the center, and adjust the knife-edge till this area darkens evenly, and then substitute for the first stop another which covers all except a strip of margin, if the mirror be a paraboloid we shall have to move the knife-edge a little way *further from* the mirror to get this marginal strip to darken evenly; and if it be a hyperboloid, a little further still.

And if we know *how much* the knife-edge must be moved back for a paraboloid we shall have an accurate means of recognizing the desired curve when we see it.

Now, the mathematicians have given us this information, and it is enshrined in the simple and easily remembered formula r^2/R where r is the radius of the mirror, or of any zone thereof, which we desire to test, and R is its radius of curvature. For example, if we are testing a $6\,1/4$-inch mirror of 4-foot focus (i.e., 8-feet, or 96-inch radius), $r^2/R = 3^2/96 = 9/96 = 0.093$in.

Therefore, if we find that, after adjusting the knife-edge to the radius of the center we have to move it back 0.093 of an inch (nearly one-tenth of an inch) to get the marginal zone to darken evenly, the mirror is a paraboloid, always provided that the mirror, tested without a stop, shows a *regular* curve, i.e., it is free from rings, hills, hollows, or a turned edge.

Some workers prefer to test a whole series of zones, and tables are given of the values of r^2/R for all diameters of zone and all focal lengths. It is, however, very trying for the eyes to test a long series of zones, and it is not necessary, for the open shadow test tells very clearly and unmistakably if the *general curve is regular*. And then, it is only necessary to test the difference between the center and marginal zone, for if this is right and the curve is regular all the other zones *must* be right.

Now we can understand the object of the metal straightedge attached to the stand of the knife-edge (see Figure A.4.2, right). We pin a white card to the top of the testing table with a couple of thumbtacks, stand the knife-edge on it, and when we get it adjusted for the center of the mirror, hold it firmly down, and draw a sharp-pointed pencil along the straightedge.

Next we adjust for the marginal zone, and repeat the process, taking care to keep the block in which the knife-edge stands parallel to the edges of the card.

Now we have a pair of parallel lines on the card, and the distance between them will be r^2/R *if our mirror is parabolic*.

Many and elaborate are the arrangements devised by Wassell and others for making this simple measurement: micrometer-screw movements in two directions, springs to take up slack, scales, and microscopes to read them. The writer has tried them all, and ended by "chucking" them in favor of the simple method above described. The straightedge, pencil, parallel lines on the card, and a pocket lens and scale of 50ths of an inch, to measure the distance apart of the lines, are sufficient. The simpler a method is the better, provided it gives the desired result. And in all this little book, the

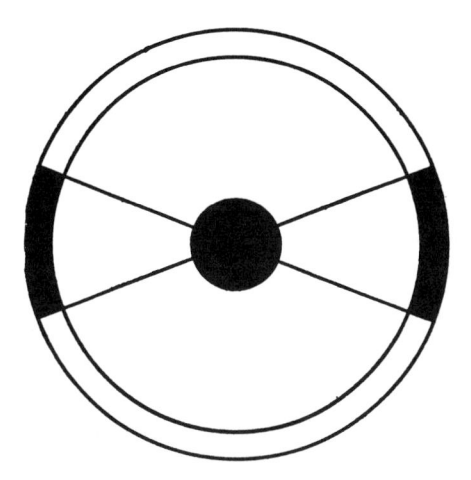

A.4.9. How to Recognize the Paraboloid, Zonal Testing

Fig. A.4.6: *A cardboard stop for use in measuring the depth of parabolization. This style is termed Stop I.—Ed.*

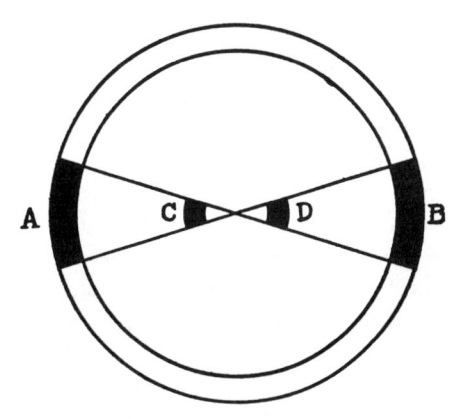

Fig. A.4.7: *Another kind of stop, termed Stop II. The bottom of the stops may be shaped so that they will stand steadily.—Ed.*

object before the writer has been to reduce everything to the maximum of simplicity consistent with efficiency.

The stops referred to above may conveniently be made of pasteboard. For each size of mirror two circular pieces may be cut, the exact size of the mirror. One has a circular hole cut in the center, a little larger than the minor axis of the flat which is to be used with the mirror, e.g., for a $6\,1/4$-inch mirror a hole $1\,1/2$ inches in diameter will be suitable. From opposite sides of the margin two pieces are also cut, as shown in Figure A.4.6. The shaded portions are those cut out. The width of the marginal zone may be 1/2 inch for mirrors of 4 to 5-foot focus, but must be wider for ones of longer focus, owing to the greater distance from which it must be read. The other disk may be cut thus (Figure A.4.7). The cross lines and the inner circle are only for guidance in cutting.

The stops having been prepared, the mirror is placed on the easel, a card is pinned to the top of the testing table, and the knife-edge put in position and adjusted to the proper position for shadow testing. Care should be taken to place the straightedge parallel to the sides of the card, and the sides of the latter parallel to those of the table.

The stop is now placed on the mirror. (The writer prefers stop I, but II is the form most frequently used.) Let us suppose we are using No. II. What shall we see? Let us consider first zones A and B. As the knife-edge is advanced across the beam of light, one of three things will happen. Either zone A darkens while B remains bright, or vice versa, or they both darken equally.

The observer must practice his judgment in deciding which of the two is the darker, as in reading a Bunsen's shadow photometer. If both are equally dark, the knife-edge is *at the focus of this zone,* and the pencil may be used to mark the position of the straightedge.

Now *before moving the knife-edge,* observe the appearance of the zones C and D. If the mirror is parabolic or hyperbolic, the knife-edge when at the focus (center of curvature) of the zone A B will be *outside* the focus of zone C D, because in these curves the center has the shortest radius. Therefore, when A and B are *equally* dark, C will be *darker than* D. The knife-edge must therefore be moved *towards the mirror* a little to make C and D become equally dark. When they are judged equal, the pencil is used again. If the distance between the pencil lines is equal to r^2/R, the mirror is exactly corrected.

In practice it is always advisable, for the reason already stated, to leave the correction a little less than r^2/R. Thus in a mirror where $r^2/R = 0.10$ of an inch, 0.08 of an inch would be a good correction. Even 0.06 of an inch might pass as sufficiently corrected. In any case a number of readings should be taken, and the mean of them all adopted as the true reading.

The chief difficulty the tester will meet with is the difficulty of deciding the exact point where C and D are equally dark. It is far easier to decide between A and B, owing to the more obtuse angle subtended by these at the center of curvature. Therefore, it is well to take a number of readings of C and D, sometimes taking the knife-edge well inside their focus and working outwards, and at other times well outside and working inwards. It is for this reason also that we recommend the stop of the form No. I, as it is somewhat easier to decide when the shadow in the central hole *travels*

A.4.9. How to Recognize the Paraboloid, Zonal Testing

neither way, but comes on equally all over the hole, than to decide between the relative shades of the two slots C and D. If these slots are too narrow, the eye is very apt to be misled.

In the case of a long focus mirror, the decision is naturally very difficult. Take for example, a 6-inch of 6-foot focus. In this case r^2/R =7.29/144=0.05 of an inch.[3] Therefore 0.04 of an inch will be a good correction. The observer will find that when A and B are equally dark it will be difficult to decide that C and D are not equally dark also. In fact, 0.04 of an inch is a very usual margin of error between two readings of C and D, so that the real difference between the two pairs of zones is not greater than the possible error of observation.

Everyone who is accustomed to making delicate instrumental measurements knows how difficult it is to obtain a reliable reading under such circumstances. Therefore the mirror maker should beware of extreme focal lengths; $f/8$ to $f/9$ is the easiest for purposes of accurate correction. We have never yet seen a mirror of $f/10$ and upwards, even by well-known makers, that was not overcorrected. Foci of *below* $f/8$ are objectionable for another reason, though when we get to apertures of 15 inches and above, considerations of space, weight of mountings, etc., make it necessary to adopt short foci. The trouble with these is that the wide-angle cone of rays upsets most eyepieces, unless they are specially calculated for the purpose. Ordinary negative eyepieces give quite an unpleasant amount of false color on mirrors of $f/7$ and $f/6$, and really spoil the perfect achromatism of the reflector. Achromatic eyepieces should be used with all mirrors of focus below $f/8$ if the finest definition is to be obtained.

In conclusion of this part of the subject, let us again repeat the warning given before with respect to temperature. It is absolutely necessary, the *sine qua non* of success, that testing shall be carried out in a place in which equilibrium of temperature can be maintained. The best of all places is a cellar completely underground, and free from drafts. If this is not available, the next best place is a basement room, partly underground. And if it must be above ground, it should be done in a substantial stone building, with as little window space as possible. It is *quite impossible* to get true readings in a living room, or any place artificially heated. It is equally impossible in a lightly-built workshop of timber or galvanized iron. In fact, such a building is hopeless for the mirror maker's work, except for the roughest parts of it.

Polishing, not to mention figuring or testing, is impossible in a place where the temperature may vary 20° or 30°, according as the day is sunny or cloudy. Testing *in the open air* is equally impossible. And this last is

---

[3]The fraction 7.29/144, *appears* to be in error but is not. The square root of 7.29″ is 2.7″, the radius of the *center* of the zone, which is thus seen to be 0.6″ wide in this case.—Ed.

tantamount to saying that *testing on a star* is useless, for a star test is necessarily in the open air. No doubt, it sounds plausible to say that as a telescope is made for observing stars, its performance on a star must be the best criterion of its quality. But it is to be noted that this plea is used by those with the most limited experience of up-to-date methods of testing. And when we find that a star test is capable of pronouncing *one and the same mirror on the same night by turns undercorrected, truly corrected, and overcorrected,* we see how little reliance is to be placed thereon. The writer has often had this experience when using various mirrors in the open air.

It is easy enough to see, by the out-of-focus images of a star, what is the state of correction of the mirror. A truly corrected mirror, out of focus, will give an expanded disk, uniformly illuminated except for faint traces of diffraction rings, having a clean, sharply-defined edge, and a round black spot in the center. This black spot is the shadow of the flat, and it should be *the same size at equal distances inside and outside focus.*

If it is larger *inside* focus, the mirror is undercorrected. If it is larger *outside* focus, it is overcorrected. And many a time on a night when temperature was variable the writer has watched a mirror *change through all these phases* within not very many minutes, the changes of the black spot answering faithfully to those of the thermometer in the screen close by. And the changes are *not small.* A rise of 4° or 5° in the air temperature will instantly reduce the figure of the mirror from a true curve to a very much undercorrected one.

And such changes are frequent, especially on mild nights in autumn. The passage of a light cloud, the springing up of a breeze, or the formation of a fog will raise air temperature suddenly by several degrees. And the mirror instantly responds by lowering its correction. And in the opposite event, of temperature falling sharply, the mirror goes the opposite way, the correction being raised. But this is less serious, because, as already pointed out, most mirrors are slightly undercorrected.

The change, of course, is temporary, and is only due to the fact that, a mirror being a thick and massive piece of glass, and glass having a high heat capacity, its warming and cooling cannot keep pace with that of the air, but *lags behind it.* In point of fact, makers of specula, without knowing it, have been in the habit of *correcting them for a falling temperature.* If it were desired to use a speculum for daylight work, e.g., for solar observation, it would be advisable to considerably overcorrect, or in other words, to correct for a rising temperature. And in making a mirror for use in a climate such as that of Mexico, where the temperature drops very rapidly after sunset, a figure in the neighborhood of a sphere might perform better than a paraboloid.

A striking illustration of this propensity of mirrors may be mentioned

A.4.10. Silvering

here. A 9-inch mirror made by Mr. Maurice A. Ainslie, a well-known and expert amateur speculum maker, was sent to the present writer for examination, and was found to be a little overcorrected. At the owner's request it was retouched, and the correction lowered. Some time afterwards the author was introduced to Mr. Ainslie at a meeting of the British Astronomical Association, and mentioned this mirror and its defective correction. It proved to be one which Mr. Ainslie had specially made for observing planets *in the morning twilight,* when temperature would be beginning to rise after the night, and was overcorrected because it was found to perform best in this condition.

It sometimes happens that a mirror is made of inferior glass or glass of unusual quality, and such mirrors will often have a whole set of peculiarities of their own. A thin mirror will sometimes perform quite well with one particular diameter vertical, but in all other positions give double images or reveal other signs of flexure. And twice within the writer's experience a mirror was submitted to him which *would not keep a figure.* A 9-inch mirror was on one occasion refigured *three times,* and on each occasion after a week or two was found very much undercorrected. At last the plan was adopted of strongly *overcorrecting* it, and then the figure after a time came back to a paraboloid, *and stayed there.*

In concluding this part of the subject, let me give one caution to the beginner: *Do not be too ambitious.* A 6-inch mirror is quite large enough for a first essay. If not a success, as is more than probable, the loss is only a few shillings, and some time and labor. The difficulty of working a mirror increases by leaps and bounds with its aperture. Six inches to 8 inches soon become fairly easy with some experience. But 12 inches is a tough proposition even for a skilled hand.

A.4.10 Silvering

The mirror completed, it now only remains to provide its surface with its reflecting film of silver.[4]

A.4.10.1 To Polish the Film

If we have hit off the right moment to remove the mirror from the silvering bath, very little polishing will be required. This is a great advantage,

---

[4] See the directions for silvering issued by the U.S. Bureau of Standards, (ATM3, C.1). These are in some ways better suited to the American worker. The method chosen is especially recommended by Mr. R.W. Porter, author of Chapter A.1. After the omission of Ellison's directions for the actual silvering process, and the substitution of ATM3, C.1 for them, the text of the Ellison book continues below.—*Ed.*

as there is always more or less risk of injuring the fragile skin of silver in polishing it. Make two pads of fine, soft chamois or washleather, stuffed tightly with absorbent cotton. Keep them in a clean jar or tin covered to exclude dust. One is to be used plain, the other covered with the very finest possible rouge. A regular speculum maker has no difficulty in obtaining the right stuff. With constant rubbing down on the tool, a certain amount of rouge becomes so fine that it takes several days to settle out of suspension in water. When this is observed to be the case some of the red-stained water may be poured off into a glass dish and set aside in a warm, dry place, covered with a plate of glass. When observed to be settled clear, the water is poured off and some more of the same stuff added, and let settle in its turn. When enough sediment is seen to have collected let it remain till dry, always, of course, covered. The polishing pad is dipped in this, and takes a charge which will last almost *ad infinitum* without needing renewal.

If this method is not available we must put some dry rouge into a jar, fill up with water, and stir well. After settling, take on the finger tip some of the red scum and froth which floats and smear it on a clean piece of glass. Dry carefully, and rub the polishing pad on it.

First rub the film carefully all over with the plain pad in small circular strokes. Then follow with the rouge pad. Very little of the latter will be needed if the mirror was not overimmersed. If it was, there will be a whitish film on it which will require a good deal of polishing to get rid of it. But even many hours of overimmersion will do no harm beyond a little extra trouble in polishing.

Once polished, leave the film quite alone. *Never attempt to get rid of tarnish by repolishing.* You will lose far more light by *thinning* the film than you will gain by polishing it. Even when appearing very badly tarnished, a silvered mirror retains nearly all its original light grasp. Silvered glass, even at its best, never *looks* as bright as speculum metal. But appearances are deceptive. Silvered glass *looks dull* and *is bright*; speculum metal *looks bright* and *is dull*. When equally well-polished a silver film reflects *about double* the light of a speculum metal surface the same size. Even when very badly tarnished, the silver film is still vastly superior to the metal at its best. This is a fact not always realized even by professional astronomers.

A.4.10.2   A Few Hints on Silvering

Distilled Water. It may not be easy to procure this in quantities sufficient for operating on a mirror of any size, but, unless one resides in a large town or near a manufacturing district, *clean rainwater* will answer all purposes if certain precautions are observed in collecting it. In the country, rainwater off any clean roof will do. A glass roof is best; next best is a

A.4.10. Silvering

galvanized iron roof, and next a slated one without chimneys and free from moss and lichen. The best time to collect is on a very wet day after it has been raining for several hours. Do not dip the supply from a barrel or tank, but let it run from the eave-gutter directly into a clean vessel. Store a few gallons in clean corked bottles, replenishing the supply as occasion serves. Quart whiskey bottles answer well, as a trace of alcohol does no harm. If you are near the sea collect water only when the wind is *off the land*. Traces of chlorides are the most injurious impurities likely to be present, and they are sure to be there "when the wind bloweth in from the sea." Test a sample with a drop of silver nitrate. If no cloudiness forms in it the water is all right.

A.4.10.3 Care of the Film

Next in importance to success in producing a film is success in keeping it. Two or three years are generally considered to be a good life for a film. But this can be very greatly extended, at least in the country, by suitable precautions in protecting the silvered surface. Its two enemies are *sulphur* and *moisture*. The first is only troublesome in a town atmosphere. A close-fitting cover, always on when the telescope is not in use, is the best remedy.

Moisture (the dewing of the mirror) cannot always be prevented unless the telescope is housed in a covered observatory. In the open air a rise in temperature of a few degrees, such as often happens on a fine night in autumn, will almost certainly dew the mirror. The only thing which can then be done is to close up the telescope and retire, especially as the rise of temperature will also upset the correction of the mirror for a time and play havoc with definition. But, short of a sudden rise of temperature, a mirror at the bottom of its tube, especially a wooden tube, is almost beyond the reach of dew. The flat *always* dews first.

The close-fitting cover, which every mirror should have, ought to be provided with an absorbent pad inside to take up moisture. I find the following plan so effective that I believe a film thus protected will last almost *ad infinitum*, certainly for very many years. Cut from stout pasteboard a circle large enough to fit loosely inside the mirror cell. Cut also similar circles from white blotting paper and clean absorbent cotton. Make a sandwich of them, cardboard one side, blotting paper the other, and absorbent cotton between, and stitch them together. Attach a loop of string to the pasteboard side. Before putting the cover on the mirror lay this pad on the face of the mirror, blotting paper next to the surface of the glass, and put a cover on over all. The string is for lifting off. If this pad is now and then well-toasted before the fire or exposed to hot sunshine for an hour or two,

Fig. A.4.8: *Armagh Observatory, at Armagh, Northern Ireland, of which the Rev. Ellison is director. On the left is the fine old stone residence, built about 1790, with a dome on top which houses an ancient unused telescope. The low wing on the right is Ellison's library and study. Next to the right, connecting through a passageway, is a square-roofed transit house, and beyond is a high old stone tower with dome which now contains a 6-inch refractor made by Ellison. At the extreme right is the low roof of Ellison's compact stone-walled optical shop. The two modern telescopes he uses, a 10-inch refractor and an 18-inch reflector, are in domes at ground level but these do not show, being behind the reader's point of view. The observatory is romantically situated on a low knoll and is surrounded by a grove of fine old trees.*

it will be dry enough to absorb all moisture within the cover and permit none to remain on the mirror. If the mirror ever gets dewed when open, the pad placed on it for a few minutes will cause the dew to vanish completely. A similar pad is not without its uses inside the cover of an object glass.

Protected in this way, I have had films in use for years, just as bright as the day they were deposited.

A.4.11 Mounting the Mirror

If we are to be particular, and want our mirror really finely mounted, the making of a suitable cell will be an engineering job, calling for a heavy iron casting and a powerful lathe to machine it—a job for a machine shop. But the Newtonian is a very tolerant instrument, and will perform well in a mount in which the most unpretentious object glass would disdain to be seen.

Draper and With used to sling their mirrors in a simple leather strap, with a stout board for a backing, at the bottom of a square wooden tube. And though it is desirable to have something less rough-and-ready than this, we can provide a very excellent cell which will hold the mirror quite securely in adjustment. We can also protect it from moisture and tarnish

A.4.11. Mounting the Mirror

when not in use with no more elaborate tool than the plain lathe aforesaid, a set of stocks and dies for screwing, and a few hand tools.

The first item required will be a thick disk of hard wood (oak is excellent), very slightly larger than the mirror and at least 1 inch thick, more if the mirror is 8-inch aperture or over. If a sufficiently thick piece cannot be obtained, get two, and screw and glue them together, with the grain crossed. The side on which the mirror is to rest should be faced up truly flat, or, better still, have a raised ring turned on it about three-quarters the diameter of the whole thing.

We next require a strip of sheet brass about 1 inch wider than the combined thickness of mirror and wooden disk; $1/16$-inch thick will do for a 6-inch to 8-inch mirror. For larger sizes it should be thicker. This is cut just long enough to go round the wooden disk and let the ends meet, but not overlap. Cut a strip about 1 inch wide with the length equal to the width of the large strip, and sweat it onto the joined ends. The strip and ends should be tinned and well-smeared with soldering flux. Then hold in a hand vise, ends meeting, and drill and put a couple of rivets through and tap up. Reverse the vise, and repeat the riveting at the other end. Now solder the strip and ends. Now you have a joint that will not give way for a trifle. Slip the collar so formed over the wooden disk, drill half a dozen holes round into the wood, and insert screws; and there is your cell. It requires nothing more but means of attaching to the telescope tube, something to keep the mirror from falling out, and a cover cap.

To keep the mirror in, we need not go to the trouble of fitting a rim. Three small blocks, cut out of $1/4$-inch sheet brass, will do. They should be spaced at intervals of 120° round the inside of the cell, and either soldered to it, or, better, attached by drilling and tapping and putting a small countersunk screw through from outside. When thus attached, the mirror can be removed and replaced by merely unscrewing the blocks. If they are soldered, the screws holding the brass to the wood backing must be removed and the whole metal ring lifted off. As these are wood screws, frequent removal is apt to make them loose in their holes. The method of attaching the blocks with screws is therefore to be preferred.

A simple method of attaching the cell to the tube is shown in Figure A.4.10. AB, CB, and DB are pieces of flat iron or brass bar attached to the wooden back of the cell by screws as shown, making angles of 120° with each other. Each is long enough to extend considerably beyond the circumference of the cell, and the outer ends carry holes, the centers of which are at the angles of an equilateral triangle. These three holes fit over three screwed brass or iron rods, rigidly attached to the telescope tube. Each rod carries two nuts. One nut is placed on each rod and screwed up

Fig. A.4.9: *An 8½-inch telescope built in 1930 by Rev. Ellison, for José Fernandez of Argentina.*

about ³⁄₄ of an inch. Then the cell is slipped over the rods, the screwed ends entering the holes until stopped by the nuts. The second nut is then screwed onto each rod and screwed home. Obviously, any desired adjustment can be obtained by slacking off one nut and tightening the other where required. The cover cap merely requires a piece of sheet metal cut to a circle a little larger than the cell, and a strip of length equal to the circumference of the latter. The ends of this are joined, as described for the body of the cell, and it is soldered to the circular piece. It should be an easy fit for the cell, and to secure that it is a circle it should be slipped partly onto the cell, the latter inverted onto the piece of sheet and the strip soldered while in this position, taking care not to solder it to the cell itself. A piece of the same strip, bent to a suitable shape and soldered to the middle of the cover outside, makes a handle for lifting off. The whole thing is much like the lid of a saucepan.

For a large cell (anything over 8 inches) it is desirable to have the cover cap convex. It can easily be made so by laying the circular sheet of metal on an anvil (failing one, the lathe bed will do) and tapping lightly all over with a hammer with a convex face. Begin in the center and go round in increasing

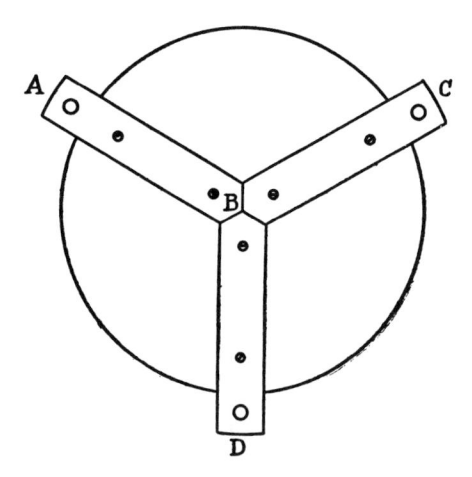

Fig. A.4.10: *A simple method of attaching the cell to the tube.—Ed.*

circles to the circumference, repeating the operation till the desired degree of convexity is attained. Any spot which remains too flat should get a few extra taps of the hammer. If the sheet is brass this operation may make it brittle. To prevent cracking, it should be heated dull red and then plunged in cold water, when it will be as soft as before hammering.

A.4.12 A Last Word to Beginners on Insufficient Grinding[†]

The beginner's "Public Enemy No. 1," the evident cause of more grief and disappointment than any other factor in his endeavors, has been insufficient grinding, mainly at the middle stages, despite the warning first published in an earlier edition. Literally hundreds have written appeals about like the following:

> I have now been polishing my mirror for more than 20 hours, and am at last beginning to wonder whether it will be worthwhile to go on. There remain two or three big pits per square inch, which I could not seem to see until I had polished for several hours and cleared away the many smaller pits that for a time must have camouflaged them. Only then did I realize that the pits I am now trying to polish out are very considerably deeper than the others. The longer I polish, the more reluctant I feel about making a return to grinding. On the other hand, if I

[†]By the Editor.

don't return to grinding, there appears to be little prospect of finishing short of another 20, 30 or 40 hours. Must I give up and go back?

If the extra and unnecessary time spent could be totalled, represented by all such cases that have occurred up to 1935 when the present note was inserted in a later edition of this book, the sum would reach a probably 50,000 man-hours. Many have polished 10, 20, 30, or even 50, extra hours on isolated deep pits. It is in the hope of enabling the beginner from the start to avoid falling into such a trap, that the present note is now (fourth edition) added to this book as the last word.[5]

It is not the very beginner alone, working on his first job, who finds himself in a dilemma like this. Until the lesson has soaked in well, through repetition of the same tragedy perhaps twice or even three times, a worker is likely to find himself in almost as bad a fix right over again. On the second occasion he may even have taken previous pains to make a prolonged search, and studied the mirror with a microscope under both low powers and high, after each stage of grinding, and concluded that "this time, anyway, there are no big pits—they absolutely aren't there." So he begins polishing the mirror. After about four hours, the finer pits being by then cleared away, there in front of his eyes stand the same damned pits again, thumbing their noses at him. Where did they come from?

The catch is this: It is difficult or impossible to decide positively, after examining a surface in strong light or under a microscope, whether large pits are hiding out in the confusing complex then seen. Such a surface is most deceptive. The pits do not stand out in the form of neat, round, isolated craters of different sizes, as they will after the rounded spaces between them have been leveled off by polishing for an hour or so. Instead, what is seen is a tangled microscopic landscape—folds, creases, ridges, knolls, and whatnot, all jumbled together, like the relief map of a New England state. Therefore most old hands, having "been there" themselves in their own time, no longer try to judge when the pits of a previous stage of grinding have been wiped off entirely and replaced by those of the next finer stage. Instead, they simply give each stage one full hour of grinding—for Pyrex it is better to double it. (The first stage, of course, is roughing out and need not be timed at all; it will come to whatever it comes to.) Particularly should this be done on Pyrex, which is much harder than plate glass—maybe two hours.

This doubling of the half-hour (six five-minute wets) (see Section A.4.4) is definitely not an expedient to offset the very beginner's inexperience. It

---

[5]This problem continues to the present day and consequently, in this edition of the book, it has been moved to a more prominent location.

A.4.12. A Last Word to Beginners on Insufficient Grinding

Fig. A.4.11: *The Rev. Wm. F.A. Ellison.*

is true that Ellison, a very experienced hand, could reduce the hour's time; though Everest, who had made and mothered some 150 optical surfaces up to 1935, does not find it worthwhile to try.

Six wets of the finer sizes of Carbo will come to about half-an-hour, as already stated, but the coarser sizes break down so rapidly that six wets of these would equal hardly more than 15 minutes, unless each wet were dragged out past the point where the reduced abrasion from the broken up grains was worth the abrasive saved. So it may be best to measure the work by the clock.

This recommendation—one hour per stage—apparently means more work. But does it? Are not six hours of grinding and ten hours of polishing (on pitch without prepolishing, otherwise less) a whole lot better than three hours of grinding and 20 or 30 hours of polishing, or even more?

If in spite of precaution the beginner does find himself in such a jam— that is, deep pits after about 10 hours' polishing—the ideal thing is, of course, to go back to grinding. The return in most cases should not be merely to the final stage of grinding but to about the third stage. This is because the offending pits are not last stage pits; if they were, the long period of polishing already given the glass would have removed them. Instead, they are probably the bottom halves or two-third portions of first or second stage pits, and even quite a long time spent with No. 600 on these coarser pits will hardly whip them, at least as economically of time as a succession of about the three final stages. This may be looked at in a quantitative sense, thus: Pits are about proportional in depth to the size of the grains that make them. The grain diameters of the Carbo series 60, 120, 220, 280, 400 and 600, as given by the manufacturer, are in the following proportion 1170, 490, 290, 156, 81, 42 (stated in 100,000ths of an inch); M 303 $1/2$ on the same scale is 16, Levigated Alumina 4, rouge 2—these as measured by S.H. Sheib of Richmond. These figures tell the whole story and should be studied closely. If whipping the 1170- or 490-sized pits with a 2-sized abrasive requires 20 or 30 hours' work, will even a 42-sized abrasive, used alone, be likely to whip them as economically of time as 290-, 156-, 81- and 42-sized abrasives used in logical sequence?

The above argument also shows the inadvisability of believing that "a few extra wets on the final stage will clear the slate," as has sometimes been advised in cases where there has not been good grinding at earlier stages. It most likely won't. Instead, this bit of bad advice may cause the beginner to be less careful in the earlier stages than he otherwise would, serene in the belief that everything will come out all right in the end if he only gives the job some extra licks on the last stage. Or he may believe that, as the everoptimistic Mr. Micawber said, "something will turn up." Something will. Pits. Yet if ten hours' hard labor doesn't exterminate these optical bedbugs, use the mirror in disgrace—like sweeping dirt under the rug.

Chapter A.5

An Amateur's View of Mirror Making[†]

A.5.1 From One TN to Another

A.5.1.1 Pitch

This, as far as the optical worker is concerned, is any material which will gradually yield or flow, and take on permanent deformation under pressure. If it fulfills these requirements, the main thing we are interested in is its "temper," which, for clarity, we may define as the number of seconds required to produce a quarter-inch dent in its surface with one pound pressure of the thumbnail. Thus we may speak of 5-second pitch, 20-second pitch, etc. The proper temper is secured by boiling to make harder, or melting and thoroughly stirring in turpentine to make softer, and then giving the thumbnail test on a sample teaspoonful which has been submerged in water at the working temperature for at least ten minutes. Optical workers use everything from the highly refined wood pitches of various trade names sold by the dealers, down to ordinary road tar. It is said that the mineral pitches have a much wider latitude in useful working temperature, and this is certainly desirable. *We* were brought up on common hardware store rosin and turpentine, probably the most cussed of the cussed, but we are still thriving on it, and are in no position to recommend anything else.

A.5.1.2 Abrasive Action, Rolling and Stationary

Most tyros start out with the idea that grinding will result from the use of Carborundum, and polishing will result from the use of rouge. That's wrong. Both are excellent abrasives, and either may be made to grind or

---

[†]By A.W. Everest, Pittsfield, Massachusetts.

polish at will, depending upon how it is used. Two pieces of flat plate glass will soon become fine ground if rubbed together with rouge mixture between. On the other hand, a pitch tool, properly charged with finest Carborundum, will produce quite a respectable polish, compared with what might be expected. So here's the rule. Rolling abrasives grind, due to a chipping action on the brittle surfaces between which they roll. Stationary abrasives on one surface polish the other, the action this time being a fine scraping or smooth wearing away.

A.5.1.3 The Spit Test for Radius of Curvature

This rough-and-ready procedure for keeping track of the radius of curvature during the roughing out has stood the test of time. The routine is as follows: Provide a mirror rack in some corner of the cellar, to be used only for this purpose. Place the mirror in position and drop a plumb line from its face to the floor. Make a chalk mark on the floor at this point, measure back to where the desired center of curvature should be, and mark this point also on the floor. To test the mirror, wash off the grit and, while the surface is still wet, place the mirror on the rack, hold a lighted candle close to the side of the observing eye, find the image in the wet surface, walk back till the mirror fills with light, and drop a gob of spit. Compare this with the chalk mark and there's the answer. One TN's wife added a can of chloride of lime to the technique.[1]

Nonevaporating liquids such as thin oil or glycerine are unnecessary for thus keeping track of the radius of curvature since, with a few trials, this method may be performed so quickly that the plain water film will last long enough for the purpose. When the correct radius of curvature has been about reached and more sensitivity is desired, bobbing the head will cause the characteristic knife-edge shadow to dart across the mirror. During the rough grinding, the water will generally form in vertical streaks on the mirror, making it difficult to see the shadow go crosswise. The head should therefore be bobbed *up and down,* causing the shadow to move vertically between the streaks, and making it easier to read.

After the first stage of fine grinding, the surface will be smooth enough to permit the actual figure on the mirror to be seen, the iris of the eye acting as a knife-edge. This effect, of course, will show up only when the eye is at a very critical point very near the exact center of curvature, making the test at this stage accurate to a fraction of an inch. Needless to say, the spit must be allowed to drop straight down, and not be ejected in a parabolic path.

---

[1] Today, replace the candle with a flashlight and renew the film of water with a spray bottle.

A.5.1.4 Tool Effect

This term refers to the more rapid grinding or polishing which always occurs in that zone of the mirror which is at the edge of the tool at the end of the stroke, provided the edge of the tool is in contact. The cause is obvious. At the end of a stroke, the unsupported area of the mirror partially counterbalances the opposite side, relieving the pressure there, so that the greatest pressure is at the edge of the tool nearest the unsupported area. The net effect is, to a minor degree, the same as with the overhang stroke made parallel to this edge of the tool, except blended out more into the adjacent zones. If the edge of the tool has been ground down excessively, or the edge of the pitch lap has been pushed down, the cause in either case having been long strokes or overhang, the tool effect will, with shorter strokes, be somewhat farther in where actual contact ends. A related effect is present when grinding or polishing face up with *circular* subdiameter tools, unless there is a frequent change in the chord of the mirror over which the stroke is directed.

A.5.1.5 Thermal Effect

This is distortion of the mirror's surface due to localized heat or cold, caused by the heat of polishing friction, or the temperature drop of evaporation from areas of the tool and mirror exposed during the stroke; and since glass has a coefficient of expansion running into significant figures, the hot spots will tend to swell out and the chilled areas shrink away. If the ball of the thumb is held against the face of the mirror for a moment, and the mirror is then placed on the testing stand, a pronounced thermal bump will be seen. In fact, with a little experience and our knowledge of the amount of magnification of the test, we may estimate approximately how many millionths of an inch the bump protrudes.

If polishing is resumed before the bump has receded, the first effect will be a more pronounced swelling of the bump, since it will receive more friction than the surrounding area. But, due to this additional friction, the bump will also be polishing faster up to the point where its excess heat is dissipated as fast as produced, when the bump will suddenly shrink back and be replaced by a hole. This experiment may be readily performed with ordinary glass, and seems to explain the cause of the dog-biscuit surface such as shown in Figure A.5.1. The protruding areas will behave in exactly the same manner as the thumb mark; after a little polishing they will recede and the valleys will become the bumps. This would indicate that the whole surface of a mirror is in a constant state of slow, irregular oscillation during polishing, if the polishing speed exceeds a certain limit. Therefore, there is a definite limit to the speed with which fine optical work may be performed,

Fig. A.5.1: *Focogram showing "dog-biscuit."*

this limit becoming a matter of instinct if the worker is alert to what is going on. The greater the polishing drag, the slower the stroke must be, so that the result of the two will not exceed the allowable heat of friction per square inch. Here the hand worker has the advantage over those working with machines, since he can tell the amount of drag much better by the "feel" than he can by the groans of a machine. This is probably offset, though, by the fact that the machine worker is generally content to take several times as long to polish as by the hand method, since he doesn't have to work. Assuming that he keeps out of this dog-biscuit mess, as is quite likely if Pyrex is used, there are still two general thermal effects which must always be considered, since they are present in all polishing operations, and particularly with the usual amateur practice of polishing face down on a full-sized tool. One of them is also present during the grinding operation, if water is used as the abrasive vehicle. These will be given separate headings for easy reference later on.

A.5.1.6 Evaporation Effect

The relative humidity of the average cellar with the furnace running is quite low, a difference of 10° F between wet and dry thermometers being common. From the thermal coefficient of Pyrex, 0.036×10^{-4}, we may calculate that a standard 10″ disk of this material will shrink over 30 millionths of an inch in thickness if kept wet and exposed all over long enough to

A.5.1. From One TN to Another

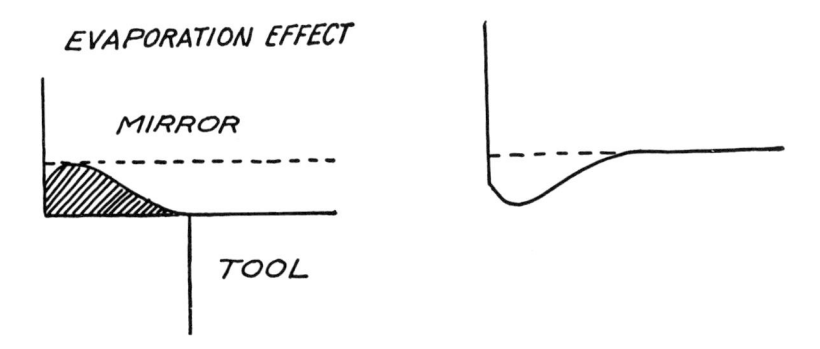

Fig. A.5.2: *Evaporation effect.*

assume a uniform temperature throughout. Under actual working conditions, when only the under side is wet, and only partly exposed part of the time, what really happens would defy analysis. But we may reason that the net effect would be a shrinkage, about as shown in the shaded portion of Figure A.5.2, left; deepest near the edge where it might be five or six millionths, and tapering down to zero at the edge of the tool for the position shown. With plate glass the effect would be three times this amount—certainly something to be considered. If the edge of the mirror is *dry*, the shrinkage will be held back one-half at the very edge, resulting in the little downward hook shown. Grinding the mirror in this distorted condition will bring it spherical for the moment, as shown by the dotted line. But when it comes to equilibrium, after removal from the tool, the zones affected by shrinkage will swell out again as shown in Figure A.5.2, right, resulting in a final figure exactly the opposite of that caused by evaporation—oblate spheroid with *turned-down edge*. With Pyrex, this is not a gross affair, but with ordinary glass it results in a grinding hangover with turned edge as deep as the final overall correction. The remedy is obvious. Work in an atmosphere of high humidity, keep the mirror and equipment wrapped in wet packs while working with them, or use nonevaporating mixtures. For the amateur's purpose, kerosene[2] is recommended for the final states of grinding. As an interesting experiment, wrap a bit of kerosene-soaked cotton around a thermometer bulb and hold it in front of a fan, to get quick action. Nothing happens. Try the same thing with water and watch the temperature go down. Of course, kerosene cannot be used for polishing because of its softening effect on the materials of which polishing laps are made. But here we shall find an offsetting factor, as far as the edge is concerned.

---

[2]Kerosene is combustible. Be careful, particularly with discarded kerosene soaked rags or paper towels.

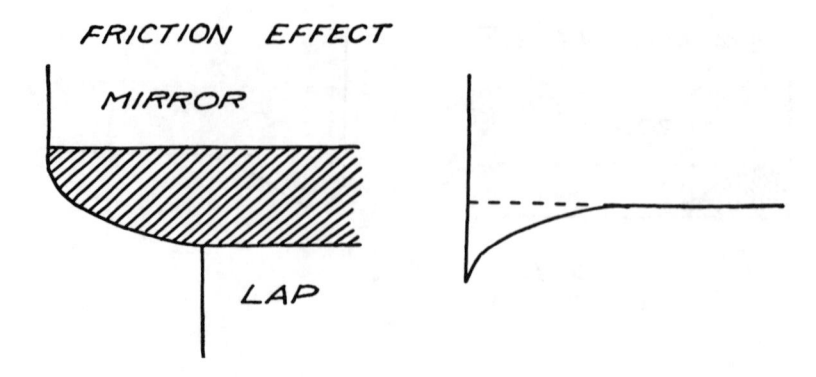

Fig. A.5.3: *Friction Effect*

A.5.1.7 Friction Effect

From our consideration of the tool effect, it is evident that the heat generated by polishing friction from one-third strokes will swell the surface about as shown in Figure A.5.3, left, the effect at the edge being held back again due to the more rapid dissipation of heat at this point. The final result will be about as shown in Figure A.5.3, right—oblate spheroid again, but this time with *turned-up edge*. The overall effect on the surface accounts for the reduction in the figure of a mirror which is cooling on the testing stand after polishing with normal one-third strokes. But, lest we fall into the error of accepting this as an infallible rule, let's consider another case—polishing off a central bump. The first effect will be as in the thumb experiment above. The bump will swell up, and even though we polished off much of its surface, we cannot see this immediately after its return to the testing stand. Assuming that nothing has happened, we may place it back on the lap and scrub a little harder at that central area. Perhaps, again, no results are seen. But after several such attempts, the bump will suddenly start to shrink back as we watch, winding up in a decided hole. Most amateurs do this at least once in their experience. And here we have a case where the figure of a cooling mirror changes in the direction of overcorrection. In other cases, where the friction effect has been somewhere near an average of the two above conditions, very little change will be seen. The average worker doesn't realize the amount of heat generated. Almost all polishing work is dissipated in the form of heat, in contrast to grinding where most of the work performed goes into the actual removal of glass. The specific heat of glass is about 0.16, and assuming a polishing drag of 2 or 3 ounces per square inch—nothing unusual—and wading through the calculation of Btu's, calories, foot-pounds, etc., a 10° to 15° F rise in temperature will be found to be a normal occurrence after a spell of polishing. This was actually

A.5.1. From One TN to Another

checked by a thermo-couple sealed in the bottom of a hole drilled nearly through a mirror. All of which means that, as far as the edge is concerned, the friction effect may, with luck, just offset the evaporation effect, so that the extreme edge stays clean-cut. And for those who are cranks on getting a diffraction edge, it is possible, with still greater pressure, actually to turn up the edge by means of the procedure, using a lap hard enough to stand it, in preparation for polishing out to a sharp cut-off in the final figuring operations. With ordinary glass, this would most certainly result in dog-biscuit, with the probability of damaging the edge again while getting rid of the dog-biscuit. But more of this later.

A.5.1.8 Tool Deformation

This was mentioned under "Tool Effect," where long strokes or overhang caused excess action at the edge of the tool. During grinding, this need not be considered, as the one-third or shorter strokes used in the final stages will bring the mirror spherical within the limits of the material still to be removed. But in polishing with a normal pitch lap, the effect is also in evidence with the one-third stroke. The mirror passes over the center of the lap quickly, and comes to rest with the greatest weight at the edge of the lap for a comparatively long time, so that the edge of the lap is slightly ahead of the center in the process of slowly sinking down. The effect could hardly be *measured* in extreme cases, and with the one-third stroke it could not even be *seen,* judging from any difference in the appearance when examining the contact through the mirror. But we know from the behavior of the lap that the effect is there, at least in the form of reduced pressure from the marginal facets, so that the only time they are having full polishing action is at the end of the stroke. As a result, the center of the mirror polishes faster than the edge, and it takes several times longer to get a complete polish out to the edge than would be the case if the mirror were polishing evenly all over. This excessive center polishing would tend toward a hyperbolic figure if it were not offset by the evaporation, friction and tool effects, all of which work in the opposite direction, as we saw above. So, with the pitch tool, the mirror will generally come through the preliminary polishing stage very nearly spherical, and the amateur making his first small mirror might go right through the whole process without discovering that any of these effects existed, and believing that the slower polishing of the marginal zone was due to this zone being off the lap during part of the stroke. This is wrong, of course, as will be seen at once if a rigid lap is used, such as HCF cemented tight down to the glass. Using exactly the same stroke, the edge will polish as fast as the center, since HCF will not sag from the weight of the mirror, and its wearing action occurs only when the mirror is in motion.

But now, without the pitch lap's tendency to deform, the various oblate spheroid tendencies mentioned above will cause the mirror to emerge from the preliminary polishing with this type of figure.

A.5.1.9 Tool Plowing

This is a minor effect, but let's not skip anything—half the fun of the hobby is just thinking about it and trying to figure the "why" of everything, no matter how insignificant, as we stumble along. Assuming a sinking speed of the facets of 0.010″ per hour, and 60 strokes per minute, the lap will be sinking three millionths of an inch *per stroke*. No sinking occurs at the edge while it is uncovered during half of the stroke, but when the mirror is slid back across this area, its edge will plow into this slightly higher pitch about the same as sliding the foot sidewise through mud. The result is a turned edge of perhaps one millionth of an inch in thickness, extending in about one quarter of an inch. We can see, from Figure A.5.4, that the slope of such a turned edge would be about the same as the slope of the marginal zone of the 10″ doughnut, so that it would *appear* to be a gross affair. The answer to this one is a harder lap *or* reduced pressure.

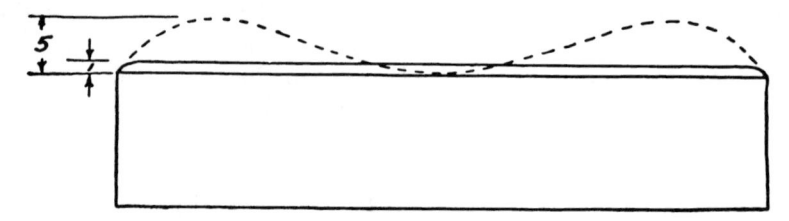

Fig. A.5.4.

A.5.1.10 The Clock Stroke

This is for those who, like the writer, get dizzy walking around the barrel, and for those who suffer the costive effects[3] of long hours in an office chair, the side strokes at three and nine o'clock will be found an efficient form of subdiaphragmatic exercise. Mount the lap at the edge of a firm bench, at the proper height to bring the forearms in a horizontal position. Stand with the feet about 18″ apart. Imagine the lap to be the face of a clock. Make the first stroke to 12 o'clock and back to 6, next to 1 and back to 7, then to 2 and over to 8, and so on around the dial. As the strokes are made, rotate the mirror very slowly in a counterclockwise

---

[3] The author's more accurate language had to be slightly modified, as that puritanical old killjoy, Aunt Sophrony, suffered a spasm.—*Ed.*

A.5.1. From One TN to Another

direction, making from 1/8 to not over 1/4 revolution of the mirror for each cycle of strokes. Rotating the mirror in the same direction as that in which the strokes progress might result in the same side of the mirror getting most of the polishing for a long succession of strokes, making the surface lopsided and no longer a figure of revolution. While getting accustomed to the "feel" of this stroke, try placing a small piece of paper, wet to make it stick, near the edge of the mirror, and see how slowly but uniformly it can be made to go around. Never allow the mirror to be turning at the end of a stroke. Just a slight turning should take place *during* the stroke. In normal grinding and polishing the action should come from the stroke, and not from the rotation of the mirror. This will give the most zone-free and blended results of the various tool and temperature effects.

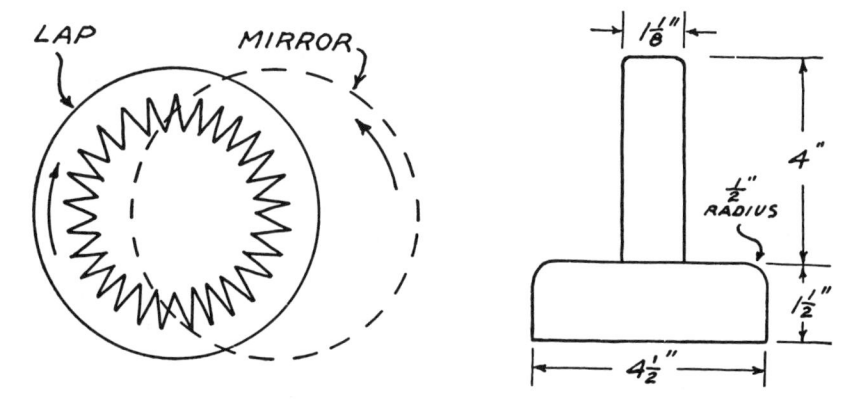

Fig. A.5.5: *The overhanging stroke (left) and the dimensions for a mirror handle (right).*

A.5.1.11 The Blending Overhand Stroke

Except to remove a narrow raised zone, the overhanging stroke should always be performed as shown in Figure A.5.5, at left, where the zigzag line shows the path of the center of the mirror around the lap. This blends the action from the edge of the lap over a rather wide zone of the mirror, instead of producing a narrow depressed zone. The direction in which the strokes progress around the lap, and the rotation of the mirror, should be as shown, for the same reasons stated in the preceding paragraph. The effects of this stroke are particularly noticeable when working on a central bump. With the proper length of stroke, depending on the diameter of the bump, it may be removed with practically no hangover; but strokes parallel to the edge of the lap will result in a hole in the middle of the bump, leaving a crater which is still worse to remove.

A.5.1.12 The Semistroke

Although the reasoning may appear a little farfetched, this stroke may work to advantage in some of the final touch-up, where it is desirable absolutely to prevent tool deformation and keep the surface of the lap somewhat complementary to the surface of the mirror. It is performed by sliding the mirror up to 12 o'clock and back to center, then to 1 and back, 2 and back, and so on around. This brings the mirror to rest at the center of the lap as much as at the edge, so that the central area sinks as fast as the rim, maintaining uniform pressure of all facets throughout the stroke. An alternative that is not so tedious, perhaps, is tapered pressing at the completion of each cycle of full strokes, starting with about 20 pounds additional pressure from the hands of the operator, and tapering down to zero pressure at the end of about 15 seconds, when the strokes are resumed.

Except for rough grinding and zonal correction with small polishers, the above three strokes, of various lengths as the occasion demands, will meet the requirements of amateur mirrors.

A.5.2 Backwoods Technique

A.5.2.1 The Handle

Use a hardwood handle, turned with vertical grain and a flange just large enough to permit a comfortable grip with the finger tips and the balls of the thumbs. The dimensions in Figure A.5.5, at right, are good. Cement the handle with about 30 second pitch, which is hard enough to prevent it from sliding around while in use, yet soft enough to prevent transferring to the mirror any warpage of the wood which may occur. Make certain that the handle is exactly central on the back of the mirror. Use a scale while the pitch is still warm, sliding the handle as necessary. Check this measurement at the beginning of each "spell." If found slightly out of position, loop a stout cord around the flange, as low as possible, tie the cord over to a post at the same height as the mirror and hang on it a weight to exert a steady pull on the handle in the direction wanted. A few minutes of this will do the trick, unless the mirror has been left overnight on the testing stand, in which case it may take half-an-hour. If recementing is necessary during the last fine grinding or the polishing stages, do not work until the heat from the pitch has left the mirror.

A.5.2.2 The Grip

Do not grip the central shaft. This is only for holding the mirror when it is off the tool, and for "sort of" guiding the hands to a uniform grip on

A.5.2. Backwoods Technique

the flange. The grip should be low down, with the finger tips not quite touching the mirror. A high grip will aggravate tool plowing and turn the edge. During grinding, additional pressure may be applied by pushing the palms down on the glass, since thermal effects need no consideration here. But, during polishing, any pressure required should be applied to the top of the flange, and at the first signs of heat in the handle, this should be stopped. Whenever the mirror is being handled while off the tool, hold the free hand under it as if expecting it to drop off the handle. Just a little free insurance.

Fig. A.5.6.

A.5.2.3 Rough Grinding

Hollow out as shown by the solid curve in Figure A.5.6, at right, not as shown by the dotted line. The latter means useless work, as well as wasting valuable thickness of the glass. For the first spell, use an overhang as far out as possible without tipping the mirror off the edge of the tool. The strokes are shown in Figure A.5.6, center. Continue this until the spit test shows a central depression having $1/4$ to $1/3$ the diameter of the mirror and the desired radius of curvature. If properly done, there will be just a few scattered pits outside this area. As soon as the spit test shows that the desired curvature has been reached, make the strokes over chords of the tool slightly farther in for a spell and spit test again. If the curvature is still right, keep going a little farther in with the stroke, the aim being to extend the concavity, as shown in Figure A.5.6, left. At the time the full concave is reached, the strokes should have just come in to the diameter of the tool, and the grinding should have just reached the center of the tool, about $1/3$ strokes having been maintained throughout.

Use all the pressure desired in the rough grinding. The more pressure, the quicker the results. Add fresh abrasive as often as necessary to maintain a loud grinding sound. This will be rather frequent, at the start, as the mirror will push the grains off the edge of the tool nearly as fast as applied. But as soon as the concavity is well started, the abrasive will come to reason; and, of course, whatever is pushed over the side may be reclaimed. Just scrape it up, drop it in a glass of water, stir thoroughly and immediately pour off the gunk. The useful grains will be left in the bottom. If at any time the curve gets too deep, as indicated by the spit test, use strokes over

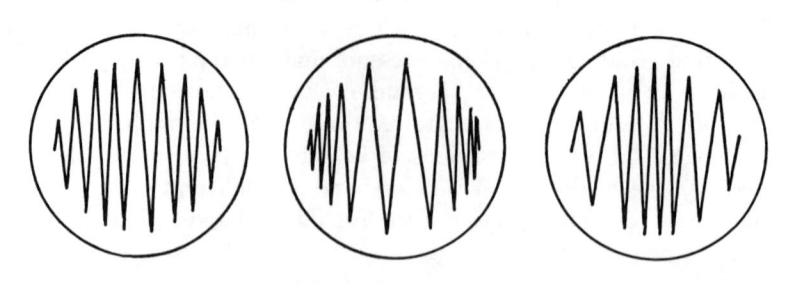

Fig. A.5.7.

chords a little nearer the center of the tool; if too shallow, stay a little farther out. Frequent spit testing while wiping the sweat off the brow will indicate what to do.

If a tool is available which already has the desired radius of curvature, so much the better. Use it. In this case, forget what is happening to the mirror while working, and try to wear the tool uniformly all over, starting with the strokes shown in Figure A.5.7, left. As the concavity in the mirror spreads, shorten the strokes a little, and as it approaches the edge of the mirror, gradually blend into the regular $1/3$ straight diametrical stroke. Spit test often, as before. If the curve becomes a little shallow, work a little more to the outside of the tool, as shown in Figure A.5.7, center; if too deep as in Figure A.5.7, right.

Of course, none of these corrective measures apply to a curve which is found too deep *after it has reached* the edge. With any care, this will not happen. But if it does, it means reversing the tool and mirror. Both of the above methods of rough grinding are easiest performed on the barrel, and either may be depended upon to bring the mirror through very nearly spherical. But test, in order to make sure. If not experienced enough to see the figure as explained under Section A.5.1.3, the bubble test will tell close enough. Grind down the last wet a little, add some water, and then watch the bubbles as a long stroke is used to bring them out to the edge of the mirror. If they remain unchanged in size, the mirror is near enough to spherical for this stage. Watch at the edge in particular. If the bubbles become smaller as they pass out under the edge, the edge is turned. If the mirror is not spherical, continue with the coarse abrasive until it is, using $1/3$ or slightly shorter diametrical strokes. The worst form of turned edge and slow edge polishing is a hang-over from this stage.

A.5.2.4 Fine Grinding

For the fastest action, the fine grinding stages should be used only for the purpose for which they are suited—removing the pits of the previous stage. If the fine grinding stages are deliberately used to deepen the curve, this not only is a very slow process, but it takes several times longer than necessary to get sufficient grinding at the edge of the mirror, since most of it is occurring at the center. Here the professional, with his channeled cast iron tools which hold their shape and spread the abrasive in a uniform manner, has a decided advantage over the amateur with his glass tool which changes shape as easily as the mirror, and which cannot safely be channeled due to the liability of chipping and scratching the mirror. Two methods are open to the amateur for getting as fast grinding at the edge as at the center—shortening down the stroke to an inch or less, or using the regular $1/3$ stroke and reversing tool and mirror in the middle of each stage. For example, with ten wets per stage, grind for five wets with the mirror on top and then reverse for the other five. Leave in this position for the first five wets of the next stage, and then reverse again. And so on with each successive grade of abrasive. This method is a little mussy and requires some special equipment, and so, although it is a sure-fire method of finishing up with a spherical surface, we shall probably choose to stay on top. Even the recommended short strokes will shorten the radius of curvature a slight amount, and an inch or so should be left for this purpose after the rough grinding is completed. From a practical point of view, however, the edge may be considered to be grinding as fast as the center, and the grinding time per stage may be cut in half if the speed of the strokes is increased to offset shortening them down.

As soon as the noise of grinding dies down, add fresh material, first washing out the "mud" if any appreciable amount of this is present. Mud forms a support for the coarser grains of abrasive, slowing down their action to a marked extent. To wash out the mud, slide off the mirror, throw on a tablespoonful of water, replace the mirror and make the regular strokes to squeeze it out. One wet of this is sufficient.

For the last two or three grades of abrasive, use kerosene for the mixture, for the reason explained under "Evaporation effect." Also, during these later stages, be careful when adding fresh abrasive. Add the required amount to the center of the tool, plus three drops about 120° apart near the rim, for "balancers," lower the mirror carefully and parallel to the tool, so that when it touches the abrasive mixture this will be spread out evenly in all directions. Work the mirror very slowly, and actually hold up on the handle a bit so that it will take 10 or 15 seconds before the abrasive takes hold. Listen for the first sound of the coarser grains and let these take their time crushing down until the sound indicates a uniform grinding all over. Then take a few seconds more in gradually applying whatever pressure is to be used. A little care like this will greatly reduce the liability of scratches, which generally occur when the mirror is first placed on the tool. Even an extra large grain will crush up without more than a pit or two, if the pressure is gradually applied as above. Of course, if the sound indicates a chunk of concrete between the surfaces, the mirror should be *lifted* off at once, and both surfaces flushed. And it adds a feeling of security, even with the sealed commercial grades, to settle them first in the water or kerosene, and use only the upper two-thirds.

With regard to pressure, it is safest to taper off to practically zero for the last grade. If the mirror has a weak diameter, any appreciable pressure here will result in an astigmatic surface.

A.5.2.5 The Lap

If the addition of turpentine is necessary, stir thoroughly, and then stir some more. An egg beater is good. Variation in the temper among the tool facets, resulting from insufficient mixing of the pitch and turpentine, will cause some of them to resist the pressure of the mirror more than others, and the action of such a tool cannot be depended upon. The stirring should be done just below the boiling point, so that the bubbles caused by stirring will rise to the surface after the stirring is completed. Bubbles do no harm after the lap is pressed, or while it is in use. But during use, when the pitch is under pressure and seeking an escape, the air in the bubbles is compressed; and during an overnight period of rest, the bubbles will swell out again, warping the surface of the facets. This makes longer cold-pressing necessary before polishing may be resumed. If well-stirred and bubble-free pitch is used, the lap will hold its shape for many days, and require only a few moments' cold pressing at the end of that period.

In cutting the channels, a carpenter's rip saw with plenty of soap suds is a fast worker, after which they may be easily widened into the usual "V" section with a sharp knife. A rubber grid, made as shown in Figure A.5.8, will save some time and eliminate most of those chips which fly around the place. Heat the tool to about 30°F above working temperature, pour and form the pitch in the regular manner to about the thickness of the mat, lay on the mat with the central facet properly located, and push it down in to the pitch with the mirror and plenty of warm soapsuds until all but the marginal facets come up into contact. If the lap chills before the mat gets down to sufficient depth, remove the mirror and reheat the lap in a pan of warm water. After the proper depth has been reached, chill the lap under the cold water tap, remove the mat by pulling up at a corner, rinse

A.5.2. Backwoods Technique

Fig. A.5.8: *Forming a lap with a rubber grid.*

off all traces of soap, dry the surface with an old piece of linen, paint hot pitch on the marginal facets to bring them up to the level of the rest, and then repress for complete contact all over.

The sinking speed of the lap deserves a little thought. This refers to the speed with which the facets reduce in thickness from the pressure of polishing. The lowest possible sinking speed will insure freedom from edge troubles. This involves pitch temper, heat of polishing and the resulting softening of the pitch, pressure used, diameter and thickness of the facets. Pitch hardness is limited to the point where sleeks are likely to result. With fine optical rouge, 20 sec. pitch for plate glass and 40 sec. for Pyrex seem to be about the maximum safe limits. With care in washing the rouge and applying it to the lap, these limits may be exceeded by experienced workers. But, for safety with super-hard laps, it is good insurance to paint on the thinnest possible lamination of about 10 sec. pitch, not over a few thousandths of an inch thickness. Or scrubbing the surface with turpentine, and letting it air-dry, will generally soften the surface enough to remedy a lap which has a tendency to sleek.

The heat of polishing was mentioned a few pages back. A lap of 5 to 10 sec. pitch will behave fairly well for the first 10 or 15 minutes. But after the heat gets well down into the pitch, it will soften enough to cause tool plowing and a rapid closing in of the channels. Use of the maximum safe hardness of the pitch, as suggested above, will generally prevent the trouble. If it doesn't, the pressure should be relieved, and the proper polishing drag secured by correct adjustment of the rouge mixture. To determine this, start with a mixture having the consistency of thick cream.[4] With this the rouge granules will act like so many ball bearings, allowing the mirror to slide easily over the lap and making the facets invisible through the mirror.

---

[4] rouge-water mixture

This, of course is not a true polishing action. Gradually add water, a few drops at a time to exposed areas of the lap, noting how the drag gradually increases as the outlines of the facets begin to appear. Continue until the facets are just plainly visible, with an even red cast, and the lap has a heavy but *smooth* drag. With further additions of water, the facets will rapidly become dark and the mirror start to grab from glass-to-pitch friction. The best point is where the facets have the clear red cast, or just before, and the rouge mixture should be adjusted to produce this effect.

No set ratio of rouge to water can be given, as this varies with the grade of rouge and temper of the pitch. The size of the facets is limited from about 1″ for 6″ and 8″ mirrors, to about $1\,1/2''$ for 10″ or 12″ mirrors. Larger facets have a tendency to produce zones. Regardless of where the center facet is placed, there is never a uniform distribution of facets in all directions from the center of the lap. With too large facets there will be insufficient overlapping of the facet action to prevent zones. The danger is increased by the fact that the pressure of a facet is greatest at the center, tapering to zero at the edge where there is free escape for the pitch. This may be demonstrated by polishing for a few moments without rotating the mirror, keeping the strokes parallel to one set of channels. The surface, when tested, will appear as in Figure A.5.9 at the left, not as shown at the right, same figure. For sufficient overlapping of the facets, experience dictates at least six or seven rows of facets across the lap. This will permit one-third or longer, straight, diametrical strokes without zonal difficulties. For shorter strokes, the necessary blending of the facet action may be obtained by mixing in side, circular and elliptical strokes.

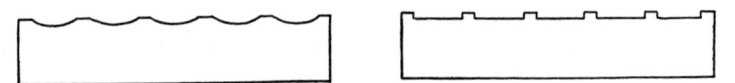

Fig. A.5.9.

In the matter of facet thickness, $3/16''$ is a good starting point, trimming the channels as necessary until the pitch has settled to $3/32''$. At this point the sinking speed will be only one-half what it was at the start, but the diminishing depth of the channels will make it increasingly difficult to squeeze out the surplus rouge mixture after each application, and get the mirror in proper contact with the lap.

The edge facets also need consideration. With the regular method of channeling, these are all undersized, and since sinking speed is a function of facet diameter, these marginal facets will have less resistance to pressure than the complete facets farther in, resulting in turned-up edge. Trimming

A.5.2. Backwoods Technique

the lap or rounding the edges of the marginal facets will only make matters worse and these dodges should never be used in the preliminary polishing, when the aim should be in the other direction. A soft metal strap or plaster of paris dam around the edge of the lap will limit the flow, but these are hard to keep in adjustment just below the level of the pitch as it sinks. Another dodge is to dry up the lap and paint airplane dope around the edges of the marginal facets, allowing it to set hard before using the lap. Perhaps the simplest is to fill in the channels between the marginal facets with pitch, since this will reduce their sinking speed by limiting the flow to two directions, and also help retain the rouge mixture.

A.5.2.6 Polishing

Fine optical practice demands slow intermittent work, with temperature and humidity under strict control, using 50 hours or more to polish a 10" mirror, and an equal length of time to figure. But who has the patience, or the necessary control of temperature and humidity to reap the benefits of such slow work? For the amateur's purpose it is just as well to polish as quickly as possible by any method that will preserve a figure of revolution, giving no serious attention to the figure until the polishing is completed. The mirror will, of course, emerge from this stage looking under test like almost anything but an optical surface. But with a little experience in zonal correction this can be rapidly changed to a spherical surface with practically no evidence of zonal hang-overs, and these disappear in the final figuring. This preliminary correcting not only adds to the fun, but it tunes up mind and muscle so that in the last touching up of the paraboloid, the worker, and not the mirror, is the boss of the situation.

The main thing to watch for is astigmatism, testing for this at frequent intervals early in the polishing. If signs of this are found, work without pressure; if not, give 'er the works. The reasons for the hard lap have been given, and the strokes should be short enough to keep the marginal zone polishing somewhere nearly as fast as the center. The HCF lap, described in Chapter A.3, provides almost a foolproof method of bringing the mirror through this stage with a brilliant, scratchless and sleek-free *visual* polish. But, since the pitch lap will be needed for the final work, many will prefer to go it on pitch from the beginning.

And here one must be ever on the alert to prevent scratches, taking a lesson from the busy spectacle maker who works with grit all over the place but seldom scratches a lens. He is just habitually grit conscious and keeps out of it. So roll up the sleeves, scrub up everything that is to be touched, including the testing equipment and under the fingernails, and never touch anything else. Carbo germs are dead ones and won't hop up on the lap of their own accord. When picking up the rouge jar, bring it around the lap, not over it, etc. Frequent exchange of newspapers on the bench is good insurance, placing anything to be handled on clean sheets of white paper. Keep an eye on the visitors. These are always leaning on first one thing and then another, and just can't be made to understand.

Most scratches occur immediately after placing the mirror on the lap. So be careful in applying rouge. Stir thoroughly each time it is used, wait about five seconds and then draw from the top with a large medicine dropper. Run a narrow line of rouge along the center of each row of facets, and set the mirror down carefully to spread it out. Use ten seconds' tapered pressure, make a couple of small circular strokes to distribute the rouge better, and then press ten seconds again before starting to polish. Not only will this push any coarse grains down level with the rest, but all rouge granules will have a chance of getting a toehold in the pitch rather than being pushed off into the channels.

For uniform action, the direction of the stroke must be changed often. This can easily be seen the first time a pitch tool is used, before the surface gets too discolored. Apply thin rouge so that the facets can be plainly seen, and keep the strokes in one direction. The facets will take the rouge charge in streaks in the same direction as the strokes. These streaks will throw minute thermal bumps on the mirror, resulting in a "lemon peel" surface. Change the direction of the strokes 90° and the first streaks will gradually be replaced by others in the new direction. Now go into the clock stroke and the facets will soon take on an even hue all over. Lemon peel is too insignificant to worry about during polishing, but this was a good time to mention it, with a fresh lap to demonstrate the cause.

A.5.2.7 Correcting

The best way of keeping track of things from now on will be to think at all times in terms of what is seen on the testing stand; imagining any irregularities in surface contour to exist actually as they appear, even while the mirror is on the lap. Instead of millionths of an inch, this will mean seeing and removing material *eighths* of an inch in thickness; even getting a thrill, perhaps, from rubbing off this much with a magic touch of the fingertips around the rim of the mirror. Early in the work it will be essential to form some rather definite notions regarding how fast a certain polishing drag and speed of stroke will remove one of these eighths, or perhaps only a sixty-fourth of an inch, especially where local polishers are used, with the action concentrated in one zone. To complete the illusion, the magnification of the test may also be applied to the sinking speed of the pitch, softening it down so that instead of, say, 0.002" per hour, the facets are pushing down

A.5.2. Backwoods Technique

at the rate of 0.05″ per second. Consideration of what happens when our imaginary mirror is moved over such a lap with various degrees of pressure, and different speeds and lengths of strokes, will show at once the cause of certain peculiarities of lap behavior which otherwise would be hard to understand. And in case there are any misgivings as to when we shall hit bottom, there is a quarter mile to go!

Correction involves the proper setting of the knife-edge, comparing the surface seen with an imaginary sphere tangent to its lowest zone, and polishing off the glass located between the two.

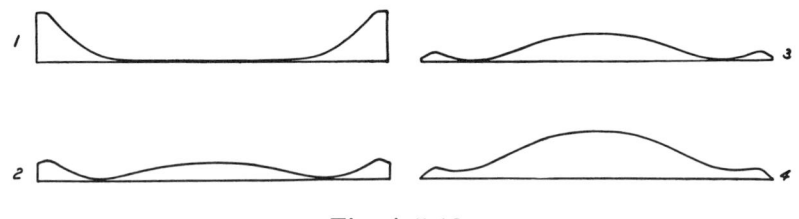

Fig. A.5.10.

If an accurate, predetermined focal length is required, the knife-edge must be set at the center of curvature desired for the marginal zone, the mirror leveled off from this position regardless of the work involved, and then deepened into the paraboloid. Thank goodness, that we amateurs have no need for this. We merely choose a knife-edge setting that will represent the least amount of work. This does not necessarily mean the removal of the least amount of material, since it is always easier to work on the central and intermediate zones than near the edge. We can well afford to push the knife-edge slightly toward the mirror, causing the center to bulge out to quite a considerable extent, if this will reduce the amount of work near the edge. Figure A.5.10 shows, in somewhat exaggerated form, apparent cross sections of the surface usually found after fast, short-stroke polishing on a hard lap, with no regard to figure—oblate spheroid with turned edge. Turned edge goes with this figure and the cause is easily visualized as we picture the edge being dragged over a lap which has gone spherical after a few seconds of polishing. A wet tape around the mirror would have prevented the evaporation edge hook, while judicious scraping of the HCF, or pressing down of pitch facets, would have counteracted the oblate spheroid tendencies. But we were in the usual hurry and didn't do this. The second curve looks like the least amount of material above the sphere, but the third one is the easiest to correct. Here the depth at the edge is practically the same as that of the low zone farther in, giving the least amount of edge correction for any of the four curves. Regardless of the general figure, it is almost an invariable rule, when turned edge is present, to select a knife-edge setting where the extreme edge appears of the same depth as, or very slightly higher than, the lowest zone within. (See Figure A.3.1) In estimating the thickness and radial location of glass to be polished off, the eye alone is sufficient until the surface is reduced to the extent that sensitive adjustment of the shadows is necessary to show any zones. For that third curve, for example, there is no use whatever of attempting to form an accurate estimate of the exact shape of the central bump, and then trying to juggle lap and stroke to remove that exact amount of material. Too many things would happen before the job was completed. And so, for the first trial, it is just as well to take rough cuts to get down to an approximate sphere, the only precaution being to see that we don't go too deep. But after the surface is planed down to the extent that only slight zonal hang-overs remain, it is well to put on the measuring stick for accurate radial location of the high zones, estimate their thickness as near as possible, and then attempt to make the lap or local polishers behave accordingly.

The methods available to bring the preponderance of polishing where it is wanted include local polishers, deformed lap, and special strokes. As a rule, local polishers for edge work, deformed lap for intermediate zones, and overhang for central protuberances will give the fastest and safest results without producing undesirable effects in other parts of the surface. But this will depend also upon whether the surplus to be removed is concentrated in one narrow zone, or is spread out and well blended into adjacent ones. Also, for the first work it is OK to adapt the fastest method possible to get down somewhere near a sphere, although this would be intolerable later on due to the thermal effects produced.

As might be expected, the writer recommends the HCF method where local polishers are indicated, especially for the preliminary rough work. Certainly no other method is faster, or more effective in concentrating the action where wanted. But it must also be said that no other could cause so much trouble if carelessly used. A rigid plaster of Paris and HCF tool should be provided, to hold the strips, as described in Chapter A.3. They would soon wreck a pitch lap beyond any possibility of bringing it back to contact. Strips of various widths, straight ones, arc shapes of various radii, and irregular pieces should be kept floating in a pan of soapy water, ready for use. The soap will make the rouge stick. For the rough work the rouge may be mixed as thick as possible yet still permit the points of the HCF to be seen through the mirror. But, for light touching up, later on, the mixture should be thinned down and the pressure relieved to produce a smoother action.

For small pitch polishers, the glass feet used under furniture legs make

A.5.2. Backwoods Technique

excellent tools, giving a good grip for the fingers. They should be ground to approximate curvature against the mirror while it is still in the rough grinding stage, and of course, no further grinding is necessary. These polishers must be used with the mirror face up, with a fine grade of rouge to prevent sleeks, and they require frequent renewal of the rouge mixture.

Deforming the lap to concentrate its action in certain zones, or remove its action in others, is standard practice with many amateurs. This includes pressing down the marginal facets with the mirror, and pressing or raising facets farther in. Only a slight amount of pressing is required—just enough to cause the facets to take on a slightly cloudy appearance from the greater thickness of rouge mixture between them and the mirror, and to eliminate their action for perhaps five minutes while the rest of the facets are sinking to the same level. In pressing the marginal facets, plenty of pressure should be used while the mirror is slowly pushed around the margin of the lap, the idea being to accomplish the pressing without much polishing. For pressing down facets farther in, lay on squares of paraffined paper and press with the mirror. To raise them up, lay on squares of HCF and press with the mirror; this time, the pitch, seeking the easiest means of escape, will flow up into the HCF depressions. Go easy with this one.

Whenever edge correction is necessary, it is advisable to do this first, so that the slight irregularities left by the local work here will be removed by later work on the full sized lap. The HCF method described in Chapter A.3 is our choice for this, leveling out as close to the edge as possible, and then removing the last quarter-inch raised zone with the finger tips. To do this, place the mirror face up, with the handle down in a milk bottle, paint rouge around the marginal zone, and make about 1" strokes parallel with the edge of the mirror. Use the tips of the first three fingers just inside of, and extending to the edge; using the thumb as a guide against the side of the mirror. Use plenty of pressure and revolve the mirror slowly by grasping the flange of the handle with the other hand, so that each stroke advances about one-quarter inch past the previous one. Three revolutions of the mirror is enough of this without testing to be sure the action is where it is wanted and not producing a ditch.

In wiping up the mirror use two pieces of cheesecloth kept hanging, when not in use, on clean hooks overhead. Use one to sop up the surplus moisture, fan dry, and wipe lightly with the other to clean off the dried rouge.

When removing a central protuberance, the first consideration is to protect the edge. Until the bump is polished down to the proper level, the edge will quickly turn if allowed to drag over the lap. Irregular shaped pieces of HCF of such a size that the stroke will just bring them to the boundary of the bump will quickly take the preliminary rough cut and, of course, cannot possibly harm the edge. To make the mirror properly balance, three small pieces of *clean* HCF may be placed around the margin of the tool, with soapy water applied to make the mirror slide easily over them. If the first attempt starts a hole in the middle of the bump, leaving a crater, never mind. Go ahead until the hole appears to be down nearly level with the marginal zone and then remove the crater with an HCF ring, still using the balancers. In all this work the strokes should be extremely short. When the leveling-off has proceeded to the point of slight zonal hangovers in the form of raised rings, these should be located with the measuring stick, and the stick used to locate the strips on the tool. At this point several zones may be worked on at the same time by placing strips as indicated for each, giving a larger area of support for the mirror and softening down the action.

In using the blending overhang stroke, to remove a central bump, the marginal facets should be well pressed down, using an overhang well out to the edge of the lap. As soon as this is done the overhang should be pulled in a bit to prevent forming a hole. If a hole should start, remove the crater around it at once with a *spinning* overhang, rotating the mirror rapidly with the palms of the two hands against the side of the handle flange, keeping the crest of the crater just inside the rim of the lap and gradually working the mirror to new positions to equalize the facet action. What happens here is easily visualized. The action tapers off to zero at the center of the mirror, in spite of the fact that the center is in complete contact.

In all overhang polishing, after the mirror has been brought spherical, press often. About half-a-minute polishing, and 15 seconds tapered pressing with the mirror centralized, is a good rule to follow here. If the lap is allowed to become of shorter radius of curvature than the mirror, it will produce a hole in the center of the mirror, regardless of the stroke used. Keep the edge of the lap hitting, and learn just what happens with the various strokes under this condition.

In bringing to spherical shape by the overhang method on the full sized lap, zonal hangovers will generally be broad ones blending well into each other, which may be rapidly reduced by pressing in a few facets in some zones, and raising those in others, as the case requires.

After a few optical surfaces have been made, the amateur learns from past experience how to apply the necessary preventive measures to bring the mirror through spherical and to eliminate most of the correction mentioned above. But it does no harm to get into all this mess at least once, so that the tricks will be learned and held up the sleeve to deal with one of those mirrors that just won't behave.

Assuming that we have gone as far as we dare with the local stuff, there

A.5.2. Backwoods Technique

will remain some blending and refining of surface texture to produce an *optical* surface. For the proper humidity to reduce evaporation effects, it is well to sprinkle the cellar floor thoroughly several hours before this is started, and keep it wet thereafter. In the absence of wet and dry bulb thermometers, the presence of large drops of water on cold water pipes will indicate a good condition. It will also be advisable to keep the wet tape around the edge of the mirror, extending down as close to the lap as possible without actually hitting it. This must be removed when testing, of course, and be kept scrupulously clean. Glycerine mixtures will further hold back evaporation, but plenty of soap in the water—enough to cause a layer of fine bubbles on the exposed areas of mirror and lap—works just as well. It is probably the dead air spaces in the bubbles that do the trick, insulating the surfaces for the fraction of a second they are exposed. Next we must watch for oblate spheroid tendencies, since short strokes will be needed to keep full action out to the rim of the mirror. Keeping the facets pressed down wherever the low zone has a tendency to form will accomplish this, starting with two or three and pressing more later if the test so indicates. But don't press them so that it will require more than a few minutes to bring them back to contact, just in case the test shows the wrong ones to have been pressed to get the correct action with whatever stroke is being used. Paraffined paper is cheap and it is better to press often than too little.

The object of this spell is to clean up those hangovers and polish out to a sharp cut-off at the edge. The latter will be shown by the diffraction line that develops during this stage—just a faint spider web of light around the left edge of the mirror at first, but becoming brighter and brighter until both sides have about the same illumination. The main thought here is to gradually taper off the pressure and speed of strokes as the surface cleans up, with frequent change in the variety of strokes, small circles, ellipses, etc., in order to prevent any possibility of throwing up facet zones. And use plenty of pressing, of course—the main reason this time being to equalize any heat present. For those who have been brought up to fear elliptical strokes, we must add that, after the mirror has been once leveled off and is prevented from developing an oblate spheroidal figure, no type of stroke will turn the edge if the recommended hard lap is used. Of course, were we to start a central depression, and then spread this out to within, say, a half-inch of the edge, that half-inch might now be called turned edge. But we can hardly say we have *turned the edge* when we haven't even touched it.

A.5.2.8 Figuring

Here again there is considerable latitude in the methods which may be used. Perhaps the most foolproof is face up with a half-sized star lap, using the strokes shown in Figure A.5.7, left, so that the points of the star just pass over the rim of the mirror. This may be done on the milk bottle, gradually revolving the mirror with the free hand. Every attempt should be made to preserve the paraboloidal *shape* throughout, gradually increasing its intensity until the desired depth has been reached. This should be checked as soon as there are any signs of a figure, measuring the difference between the inside and outside centers of curvature, setting the knife-edge halfway between the two, and then checking the crest for the 70 percent position. If the crest is too far in, shift to the strokes shown in Figure A.5.7, center; if too far out use those shown in Figure A.5.7, right. Try to determine the proper strokes early in the figuring and hold to them.

The distribution of material to be removed can be seen by laying a straightedge across the top of the last curve in Figure B.7.5. It is also important to preserve the flatness of the central area when the knife-edge is at its center of curvature, and not develop a hole. The first curve in the family shows the proper cross section. As the difference between inside and outside radii of curvature approaches r^2/R, work should be more and more leisurely, with plenty of time on the testing stand for the mirror to come to equilibrium. For the final test, the mirror should be left several hours; for Pyrex, one hour will do. For deep curves in the larger sizes, this subdiameter lap method is recommended, but be careful of scratches. For the deepest ones, the facets may be backed with some yielding material such as felt or rubber between glass and pitch, to allow good contact at all parts of the curve. This will make some readers gasp, but it works, and has been used often without throwing the surface out of revolution.

For the smallest mirrors, and up to 12″ with aperture ratios of $f/8$ or so, fine figuring may be done face down on the full sized lap. Long strokes may be used, starting with about $7/8$ length and working back to $3/4$ or $5/8$. Watch as above for the position of the crest, as soon as it can be seen. If too far in, shorten the strokes, if too far out, lengthen them. Absolute contact must be preserved with long stroke parabolizing. Otherwise the deformed lap will put a hole in the middle of the mirror. Fifteen seconds of work, and fifteen seconds of pressing is the rule. The worker must feel that he is pushing material away from the center of the mirror with the edge of the lap. If unexpected high zones develop, they may be treated with the blending overhang. In fact, the final stages of figuring will generally develop into a mixture of long strokes and then overhang, alternated in order to keep the curve smooth. Practiced hands will also include long

A.5.3. The Second Mirror

elliptical strokes, but these require considerable experience, as the pressure must be tapered *during the stroke* to prevent the formation of undesirable zones.

Take plenty of time in figuring. Let it be mostly thinking. Both are good for mirrors. With plenty of testing and thinking, and preventing backtracking by taking the proper corrective measures the moment their necessity arises, the actual labor involved will become almost insignificant.

A.5.3 The Second Mirror

Well, as we read over what we have written, we cannot help but chuckle. While we hope all this will help the beginner to sense his problem, we can hear the question, "Does all this go with making a mirror?" To which the answer is NO! For those who have been through the experience once, and understand what is going on during the various stages of the work, our recommendations would be as follows.

Use Pyrex and forget thermal effects.

Rough grind as recommended, being sure to bring spherical before going to the finer grit.

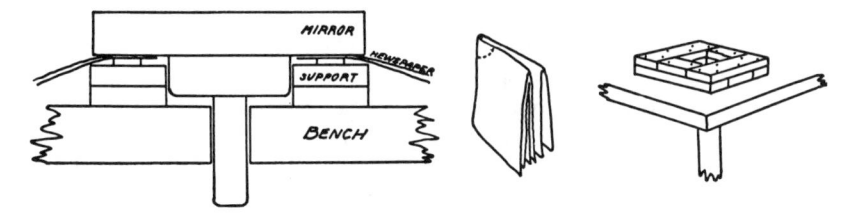

Fig. A.5.11.

Fine grind by the reversal method (our gizmo shown in Figure A.5.11) bringing through strongly overcorrected—about $1/2''$ greater than r^2/R. This will just show in the spit test, and will give just enough material to battle the oblate spheroid tendencies of fast preliminary polishing on a hard lap. To produce this curve, start with about one-quarter strokes for the first stage of fining. Grind down a wet once in a while and, if there is any tendency to grab when the mirror is at the middle of a stroke, lengthen the stroke slightly for the next wet and try again. As soon as the minimum stroke possible is found where no grabbing will occur, use it for the remainder of the wets of that grade. With successive grades it will be necessary to lengthen the strokes very slightly. For the last three stages lengthen the strokes just enough to show a faint hyperbolic figure in the spit test. This will be very nearly one-half strokes.

The kerosene will be of no advantage from a thermal standpoint when Pyrex is used, but it will do away with those central air bubbles which are bump producers.

From now on it is clear sailing. Don't bother to bring spherical but aim for the paraboloid from the beginning of polishing. Get on a 40-second lap and, as soon as it is possible to test, locate the crest with the knife-edge in the half-way position. Start with a stroke which will bring the edge of the lap not quite to this crest and give 'er a spell. On wiping up, the marginal zone will be found to be polishing fast, with the center scarcely touched. Fine. Under test, the diffraction edge will have already put in its appearance and will be a guide from now on.

As the hyperboloid approaches a paraboloid, the point will be found where the left diffraction edge begins to lose its brilliance, with a softening of the illumination just inside—warning that turned edge is approaching. This is the signal to lengthen the strokes just enough to bring the diffraction line back again. The cause of edge turning here is obvious. With short strokes, the edge of the *lap* turns up, and as the mirror approaches a sphere its edge will plow into this high pitch.

This lengthening of the stroke will slow down the speed with which the correction reduces. But remember that, with the hard lap, the strokes must be much longer to produce a sphere, and to hold a paraboloid they must be lengthened to about three quarters the diameter of the mirror. As a result, the correction will keep reducing with one-half or even five-eighths strokes, whatever is necessary to hold the diffraction line at the edge or to bring up the center polish.

Well, that's all there is to the writer's method of making a mirror. As soon as the figure reduces to the paraboloid, the stroke is lengthened in order to hold it there until the polishing is completed. We are so familiar with the method that we once got a little cocky and made a wager that we could produce a mirror by "feel" alone during the grinding, and "appearance" of the polish as it came up; getting a correction within 25 percent over or under without testing at all. After the job was done, we pulled in our horns and bought the cigars without argument, as the mirror was only half-corrected. But, anyhow, it took only ten minutes to finish.

A.5.4 In Retrospect

And now, as we sit back in relaxation after the arduous task of writing all we know about mirrors, and perhaps a little bit more, we must pay tribute to the numerous amateur mirror makers whose correspondence has been a constant source of education, and whose ideas we have freely swiped. And

A.5.4. In Retrospect

Fig. A.5.12: 3 AM and Still at it. Here Porter, the artist depicts the enthusiast as utterly absorbed in the most exacting and interesting part of the work—parabolizing the mirror. The cellar is the best place to work because its temperature is fairly uniform.

if asked to explain the cause of an incurable case of mirroritis, we should blame it about 50-50 upon the continual lashing of Uncle Ephram's whip and the inspiration of early contacts with R.W.P.—*15 Allengate Avenue, April 1, 1936.*

Chapter A.6

Subdiameter Tools—A Composite Chapter of Experiences

A.6.1 Large Mirrors and Subdiameter Tools[†]

Of much practical importance in its bearing upon the making of very large optical surfaces is the fact that, in the case of the 60" glass, grinding and polishing tools of only about one-fourth of the area of the glass were used with entire success in excavating the large concave and in fine grinding and polishing; a flat, full-size grinding tool was used only in the preliminary work of securing a perfect surface of revolution. A circular grinding tool of cast iron 31 1/2" (80 cm) in diameter, was used in all of the fine grinding of the large concave surface. In polishing this concave, and in bringing it to an optically perfect spherical surface preparatory to parabolizing, a 90° sector-shaped polishing tool of exactly one-fourth the area of the large glass was used with the best results. In parabolizing, a circular polishing tool 20" (50.8 cm) in diameter was exclusively used in securing the necessary change of curvature from center to edge of the glass; in addition to this the 90° sector tool, used with long diametrical and chordal strokes, was found to be of great value in smoothing out the paraboloidal surface. With these two figuring tools alone, used with the machine, a very close approximation to a true paraboloid was secured. The figuring was completed with much smaller tools used by hand to soften down several slight high zones.

In figuring the large paraboloid, one modification only was found desirable in the polishing machine described in my Smithsonian paper. The two cranks which give the motion to the polishing tools were remade in such a way that their throw or stroke can now be altered at will while the machine is running. The optician is thus enabled to change the position and stroke

[†]By Prof. G.W. Ritchey, in *Astrophysical Journal,* Apr. 1909.

of the tool with a perfectly smooth progression while parabolizing; these changes are actually made at the end of each revolution of the glass, and a very great improvement in the smoothness of curvature of the paraboloid is at once apparent.

A.6.2 Use of Subdiameter Tools on a 12-inch Mirror

Harold A. Lower, San Diego: Ellison mentions that it is easy to grind and polish with small tools, but does not say how to do it. A 12″ Pyrex was rough ground face down for 9 hours over a 9″ tool. At the end of that time the curve had reached full depth at the center, as determined by measuring the sagitta, but lacked about an inch and a half of reaching the edge of the disk. (This first grinding face down leaves the edge of the mirror untouched—not even scratched.) When the curve had reached full depth in the center, the mirror was turned face up and the same tool used on top. One simply makes large epicycles all around the mirror, working mainly on the edge of the hollow, until the curve reaches the edge, at which time the curve should have become spherical. This grinding with the mirror face up also required 9 hours, but did not deepen the center the slightest bit, so it is important to go to full depth in the first grinding while the mirror is face down.

The fine grinding is all done with the mirror face up. The strokes used are large epicycles around the mirror, alternated with a zigzag stroke across the mirror. Do not permit the edge of the tool to overhang the edge of the mirror more than an inch or so, or a turned edge may result. One can tell when the surface is spherical, as the tool will slide freely in all directions. If it binds at any point, the surface is not spherical, and must be made so by working on the zone that binds until the tool will slide easily.

Polishing was done with a 9″ tool, with the mirror face up. The strokes used were large epicycles, alternated with the zigzag stroke. No difficulties with turned edge or zones were encountered, but it should be understood that good contact is just as important when using small tools as with full size. The handle of the small tool *must* be low and well centered.

A small tool, used on top of the mirror, tends to polish the center faster, and one must work closer to the edge than the zone that seems to need the most polishing. No pressure should be applied to the edge of the tool during normal polishing. It is, however, a very useful trick for removing a raised zone. Like HCF strips, applying pressure to the edge of a subdiameter tool is "powerful medicine" and should be used with discretion. Pressure may be applied to the center of the tool, and will merely hasten polishing.

Figuring was done with a 6″ tool, working with a variety of strokes, mainly over the center. The figure is easily controlled, as one simply applies more abrasion at the points that seem to need it. I would not recommend this small tool method for any except short focus mirrors. For $f/6$ or shorter, it works fine.

Paul Linde, Crossville, Tennessee: The first time I tried a smaller-than-mirror tool was on a 12" $f/4.5$. I used a 7" tool merely because it was one I happened to have. The facets were graduated to fine points at the outer edge. The strokes should be of about the same length and of an elliptical or circular nature while making one round around the pedestal, and should be changed with every round to prevent formation of zones. As a general rule the center of the tool should travel more often over that zone or diameter of the mirror which needs deepening most. Starting with short, circular strokes close to the edge of the mirror and keeping away from the center, the strokes should be lengthened with every other round until they are quite long, after which they can be gradually shortened.

If there is a hill in the center it can easily be reduced by going with the tool across the mirror with slightly elliptical strokes, beginning with short ones and gradually increasing them with every second round. Turned-up edge can be got rid of easily by pulling the tool farther over the edge of the mirror.

Of course, there can be no fixed rules for using the small polisher and one simply has to experiment and test often to see the results. Any mistakes made by the small tool can be corrected in comparatively short time by the use of the full-sized tool to bring the figure back to flat. I find the small polisher method by far the easiest way to get all the zones right, especially with mirrors of short focal length. One word of warning: The mirror, when face up, is much more likely to be scratched.

A.6.3 Use of Subdiameter Tools on a 10-inch Mirror

G.E. Warner, Chicago, Illinois: There is nothing particularly new in the use of the small tool in working optical surfaces, nor is the method of working a mirror "face up" a novel one. In fact, practically all of the larger mirrors are made in this manner. Nearly all machine grinding and polishing is done in this way.

In working mirrors by this method it is the mirror blank, and not the tool, that is fastened to the pedestal. The manner of stroking is much the same as in the conventional method of hand working, but a certain precaution must be observed if dire happenings are not to result. It is imperative that the worker's walking around the pedestal or barrel be very regular. If it is not, astigmatism may be ground in, and it may or may not be ground out in the subsequent working.

One naturally wonders, in first contemplating this process, why the mirror will become concave, when with all of our previous experiences it was the upper member that became concave, while the lower one became convex. Let us suppose that in our grinding we use a straight, center-over-center stroke. Let us commence with a 4" stroke—that is, when we are ready to start a forward stroke the nearer edge of the tool will be over the nearer edge of the mirror. The mirror, being 10" in diameter and the tool 6", a 4" stroke will bring the farther edges of tool and mirror together. In making this stroke we cut a swath 6" wide across the mirror, but leave the 2" margins on either side of the mirror untouched. Also, the swath we have cut is deeper at the center. Now let us take another cut across the center of the mirror, at right angles to the first. Again we see the center of the mirror get the full abrasion, while the margins are untouched. With every change of direction of our stroke the same is repeated, and it becomes easy to see why the mirror will hollow rapidly at the center. In fact, the grinding is so rapid at the center that, except in the earliest stages of roughing out, the straight center-over-center stroke should not be used, as it tends to grind an irregular shape, usually one with much shorter radius at the center.

In practice the stroke used for all "excavating" may be a 4", straight stroke in which the center of the tool passes over a point about 1" to the side of the center of the mirror. This leaves a margin of 1" of the mirror which receives very little work. If we should happen to overshoot the desired depth we can flatten the curve by stroking the mirror with the center of the tool, passing over a point about $1\frac{1}{2}$" or 2" inside of its edge.

When employing this method of work, simply inverting tool and mirror will not cause the curve to reverse, and because we are able to control our curvature so nicely we may rough our curve out to its full depth.

In fine grinding, if we have roughed to the full depth, we shall have to choose a stroke that will not tend either to deepen or flatten our curve. A stroke in which the center of the tool traverses a point about $2\frac{1}{2}$" from the center of the mirror was found to obtain this condition.

As the fine grinding proceeds it becomes increasingly important that our movement about the mirror, and our stroking, become as regular as possible, without employing mechanical artifices.

In work of this kind the abrasive mixture will not have much tendency to run off the edge of the mirror, because the latter is concave upward and liquids tend to flow to the center, hence it is possible to work in a much more clean manner than by the conventional method. The greater cleanliness in working leads to a greater freedom from serious scratches.

Great care must be taken in the fine grinding operations to see that all previous grade pits are removed at the margins of the mirror. The surface of the tool must never be used as a criterion for judging the completeness of any stage of grinding. Many hours of fruitless polishing can be saved at

A.6.3. Use of Subdiameter Tools on a 10-inch Mirror

the expense of a bit of patience and thoroughness in fine grinding.

One of the first things that I am asked, when discussing the use of the small polisher, is "How on earth can you avoid a turned edge?" Let us look into the probable cause of turned edge in the conventional working of a mirror. Suppose our mirror to be at the end of its stroke. The pressure on the handle of the mirror in a horizontal direction tends to press the leading edge into the lap, because of the frictional resistance to the motion. This results in increased polishing at the very edge, and our notorious turn down results. This condition is augmented by poor contact between mirror and lap. Let us invert our full-sized mirror and tool and note the result of our reversal of stroke. If there is any tendency to "toe in," the tool merely tends to put the work on the mirror's face at a point far from the edge where, under ordinary conditions, it can do no harm.

The kind of turned edge we wish to avoid at all cost, because of the virtual impossibility of eliminating it once it is present, is that which is very narrow, yet so severe that it can be seen by casual inspection without applying any means of testing.

The strokes used in polishing are the same as those used during the fine grinding stages. A straight, elliptical, or zigzag stroke of about 4″ length, with a center about $2\,^1\!/_2''$ from the center of our 10″ mirror, may be used, and will result in an even polishing action and a regular figure. The tendency of the beginner is to try to avoid the edge entirely, in his efforts to avoid turning it. It is evident, however, that our tool must pass over the edge if the edge is to be polished. There is practically no danger of a bad turned edge if the tool is in good contact with the mirror, even though the tool may extend over the mirror's edge by nearly 3″.

A gradual turned edge of about 1″ in extent will naturally result from the polishing operation. It is possible to eliminate this by figuring.

In bringing the figure to a sphere from a more or less hyperboloidal curve, it was discovered that, by placing the edge of the tool over the crest of the curve and then applying considerable pressure with the fingers over that edge of the tool, the effect of a very small polisher was obtained. However, because the whole tool was still in contact with the mirror, the use of the tool in this way did not produce zones, as might be expected.

The broad turned edge previously mentioned, was treated in much the same way. The TDE was about 1″ in extent. The tool was used so that its edge was about $^1\!/_4''$ from the edge of the mirror, with the pressure applied to the edge of the tool nearest the mirror's edge. The stroke was about 2″ long. In using the tool in this way it is rotated in the hands—just as in any other polishing operation. The pressure does not, however, move with the tool, but is maintained over the same zone of the mirror.

The local corrections were applied until, when spherical, the diffraction line could be seen all the way around the mirror while under Foucault test, even when the knife-edge had completely darkened the face of the mirror.

The parabolizing was done by means of the same localized pressure on the polisher. In this case the tool was placed on the mirror so that the edge to which the pressure was applied overlapped the center of the mirror by about 2″. The stroke used was about 4″ in length and zigzagged back and forth across the mirror's face. Whenever it was noticed that any zone was not conforming to the rest of the curve as it developed, it was quickly whipped into line by proper application of local pressure to the edge of the tool. This manner of using the tool is not greatly different in effect from the use of honeycomb strips, except that it does not leave the sharp margins characteristic of the latter.

The use of this method of parabolizing permitted deepening the center of the mirror without disturbing the perfection of the edge.

The speed of grinding and polishing with a small tool should not be much different from that of full tool operation. While the area of the tool is less, the pressure which can be effectively applied is much greater. If the practical limit of size of a mirror which it is practicable to produce by hand work with the full tool is 15″ or 18″, it should be possible, by means of the small tool method, similarly to work a 30″ mirror.

For those who are afflicted with "dropsy," this method should have its attraction, for the valuable member—the mirror—is securely fastened down and the hazard of fatal termination due to gravitational attraction is mitigated.

All in all, I would suggest that telescope makers add this mode of operation to their repertoire. If it does not replace the "biblical" manner of operation, it certainly should prove a most valuable adjunct.

A.6.4 Construction of Subdiameter Tools

Horace H. Selby, California: Although subdiameter tools have been used for the past few years by this writer, in the construction of a triplet refractor objective, a paraboloid, two Hindle spheres, 20 flats and two photographic anastigmats, he will not attempt to write on the use or manipulation of such items. One TN will deepen a surface with a stroke which, in another's hands, will make a similar surface less deep. Of course, if all the many variables of any one polishing or grinding operation could be rigidly controlled, evaluated and understood, this operation could be performed and repeated *ad nauseam* and the end results would always be the same. Then art and craftsmanship would change to science, and results could be attained in a straightforward, mathematical and very satisfying manner.

A.6.4. Construction of Subdiameter Tools

Since the *use* of small tools will not be discussed, the only thing remaining as an excuse for these comments is the subject of construction.

Glass tools are more popular with amateurs than are tools of any other material. Professionals, however, use tools of wood, artificial resins, cast iron, sheet steel plus glass, brass, lead and boiler plate. Of these last, wood is cheapest and most easily formed; so consideration of it is most heartily advocated.

Of the physical properties of wood, the most important, from the standpoint of the TN, is probably the elasticity modulus, which is a measure of the rigidity. The approximate moduli for the most suitable woods are:

1. Yellow birch (*Betula lutea*) 2,000,000 lbs. per square inch.
2. Shagbark hickory (*Hicoria laciniosa*) 2,200,000 lbs. per square inch.
3. Black locust (*Robinia pseudoacacia*) 2,100,000 lbs. per square inch.
4. Sugar maple (*Acer saccharum*) 1,800,000 lbs. per square inch.

Of the above, the writer uses sugar maple, well dried, because of its ease of working, small tendency to warp, and availability.

The rough lumber is smoothed, planed to the proper thickness, sawed into circles, mounted on a lathe and cut to curve with a wooden template covered with abrasive paper. These operations are quickly performed by the local planing mill, and entail but small expense. If used for hand polishing, the tool is now ready for the pitch, which is poured, formed and faceted as usual. Since the tools are to be used on top of the work, no paraffin or other treatment is needed. For machine polishing, a metal plate may be screwed to the back, with the usual provision for driving attached. As with metal tools, the force may be applied much closer to the worked surface than when glass is used.

When tools are wanted for grinding, the same construction is used, except that the pitch coating is very thin and squares of glass or of metal are used to cover the tool.

Some workers dip tools of glass into warm water before polishing. This is not feasible with untreated wooden disks; so resort may be had to cheap radiant electric heaters of the reflector type. By holding the heater close to the tool, and revolving the latter, the pitch can be softened quite rapidly. This operation, when used in first forming and cutting the tool or when applying metal or glass facets, is a great saver of time.

Satisfactory dimensions for polishing tools are:

Table A.6.1

	Diameter
Thickness of wood	$1/6$
Width of facets	$1/6$
Width of channels	$1/18$
Thickness of pitch	$1/36$

Sizes will vary, of course. As an example, five tools were used in making an $f/2.3$ paraboloid, $12\frac{1}{2}"$ in diameter: 2", 4", 6", 10" and 12".

The use of wood polishers is by no means new—Ritchey used them years ago at Yerkes, and many professionals use them for some types of work now. To date (1936) no other users of wooden grinding tools are known to the writer, who prefers them for Carborundum No. 400, No. 600 and emery.

Chapter A.7

The Prism or Diagonal[†]

The essential part of the diagonal is an optically flat surface of glass placed at an angle of 45° with the axis of the mirror, in order to throw the reflected cone of light from the mirror out at one side where the image formed at the focus can be viewed with an eyepiece.

This reflection may be produced in one of two ways—either by the use of a right-angled prism, Figure A.7.2, I, which will totally reflect the rays as shown, or by a piece of plate glass silvered on the side facing the mirror. The advantage of the prism over the silvered diagonal is that it requires no silvering, but a 1-inch prism such as will be required for a 5-inch mirror costs several dollars, and perhaps the amateur would prefer to silver his own diagonal and put these dollars into an eyepiece.

The glass for a silvered diagonal can be obtained from a piece of broken windshield or thick plate looking glass. We want to select as flat a portion as possible, so we get a glass cutter and cut the sheet up into several pieces each about $1\,^1/_4$ by 2 inches.

Clean the glass thoroughly, select two pieces, free them from lint or dust and press them together. To bring them into close contact, slide or wring one piece on to the other, using considerable pressure. If they are now held so as to reflect light from a bright area—say the sky—colored bands, or fringes, will be seen, and these will appear to be located on the two surfaces in contact. By squeezing the pieces together near the edges these colored bands may be made to move about or change their form.

Not over half a dozen bands can be seen by sunlight or artificial light, but if a little salt is thrown on the wick of an alcohol lamp, as in Figure A.7.1, the very yellow resulting flame will show many more bands, alternating black and yellow. This must be done in a darkened room.

[†]By Russell W. Porter, M.S. Formerly Optical Associate, Jones and Lamson Machine Co., Springfield, Vermont. Associate in Optics, The California Institute of Technology.

Fig. A.7.1: *Simple setup for viewing interference fringes.*

The shape of these bands, or fringes, tells us the kind of surfaces that are in contact—whether they are convex, concave, warped or flat. The bull's-eye in the pattern at A in Figure A.7.2, II can be moved off the glass and given a pattern like B by pressing the glass together at a. By pressing still harder at a, fringes will begin crowding in from the opposite side of the glass, and the bands themselves will grow narrower (C). With the salt flame hundreds of these fringes may be seen until they become so fine and close together as to pass beyond the resolving power of the naked eye. This crowding in of the fringes on the opposite side from a, means that the surfaces are opening out on the left side, and if the pressure is moved over toward b, the bands will move off to the left until there are only a few left. By varying the pressure, the wedge of air existing between the two plates may be made to take any desired direction, and if the plates are convex or concave to each other, the center (or bull's-eye) of the fringe system may be made to come into view by appropriate squeezing.

What interests us is the fact that these bands may be regarded exactly like contour lines on a map. If we laid plate A on what was known to be a perfectly flat glass, then anywhere along fringe 1 is half a wavelength, or $1/_{100,000}$ inch, above or below any part of the glass along fringe 2; and $2/_{100,000}$ inch above or below fringe 3, depending on whether the surface of plate A is concave or convex. If the rings spread out on lowering the eye the plate is convex; if they close together, it is concave.

However, we have no standard flat, so we must go at it another way. A departure from flatness of one wavelength of light may be tolerated, and since this means two rings, if three different pieces of glass laid one upon the other show no more than two rings, then any one of the three will answer.

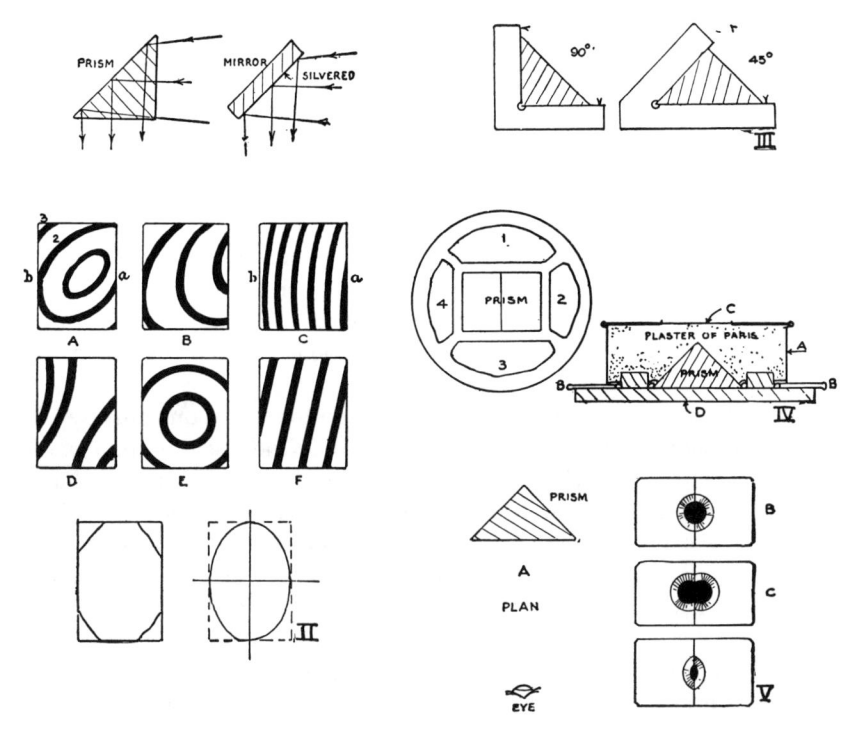

Fig. A.7.2.

They must be tried No. 1 on No. 2, 2 on 3, and 3 on 1. Pick out one of a pair of plates that shows the straightest fringes. The diagonal is so near the eyepiece that any deviation from absolute flatness amounting to less than $1/50{,}000$ inch will not harm the image produced by the mirror.

The corners of the plate may finally be cut off as in II, at bottom, and the edge ground down with No. 200 Carborundum on a slab of iron or glass into the ellipse shown. Silvering should be done on the face of the diagonal as described in ATM3, Chapter C.1.

While a piece of commercial plate glass sufficiently flat to serve our purpose can usually be found, one can purchase a prism blank from the Spencer Lens Company of Buffalo or the Bausch & Lomb Optical Company of Rochester, New York, for less than a dollar and grind and polish it one's self. The prism blank is first rough ground to the approximate shape, then carefully fine ground until it fits the proper templates (III). The prism is then laid, hypotenuse face down, on a slab of plate glass D (broken windshield) as in IV, and four pieces of windshield glass are laid around it as shown, cut so as to form with the prism a roughly circular surface. The end of a tin coffee or baking powder can (*A*) is then laid over the glass, the edge of the can resting on matches (*B*). The hole *C* has previously been cut from the end of the can. Plaster of Paris is then poured into the hole and when set, the whole affair may be slid off the slab *D* and turned over. Cut away the plaster so that the glass will be raised above its surface 1/8 inch, and when the plaster has thoroughly dried, coat it with hot beeswax. The exposed prism and four glass pieces are now to be fine ground on a slab of plate glass and the units polished as though it were a whole disk, just as the mirror was polished. After polishing, the prism is broken away from the plaster, turned over on one of its square faces and the operation is repeated. The other square face is likewise polished, and then the prism is finished.

The right angle of the prism ought to be quite 90° for our purpose. To find out whether it is 90°, the worker should hold the prism in front of one eye, about a foot away, with the hypotenuse face facing him and its longest sides horizontal, as in *V*, *A*. In the prism will be seen the reflection of the pupil of the eye as in *B*, perfectly round if the angle is 90°, but elongated if the angle is less than 90° as in *C*, and drawn together if the angle is more. The displacement is increased as the prism is moved farther from the eye.

Now for a useful application of algebra. Pick out any three pieces of glass and label them *A*, *B*, and *C*. Let us assume that *A* on *B* gives the fringes shown in Figure A.7.3, *I*, and shows three fringes convex:

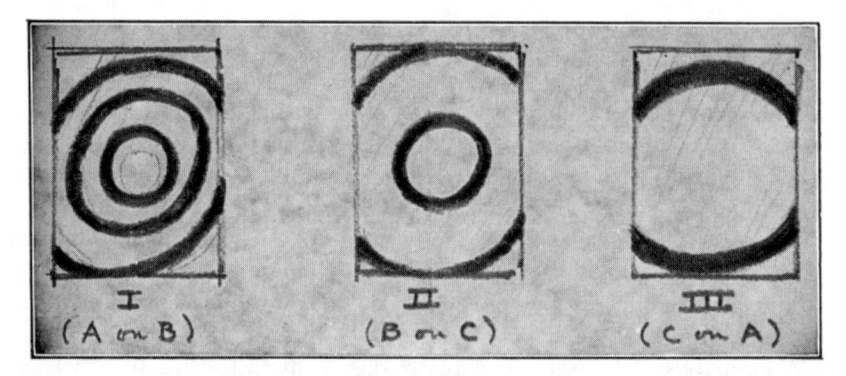

Fig. A.7.3.

Also that *B* on *C* gives two fringes concave, and that *C* on *A* gives one fringe convex. This gives us three simultaneous equations. Then, letting the plus sign denote convexity and the minus sign concavity—

$$A + B = +3$$
$$B + C = -2$$

A.7.1. Editor's Notes

$$C + A = +1$$

Removing B from the first two equations, by subtraction,

$$A + B = +3$$
$$\underline{B + C = -2}$$
$$A - C = +5$$

Combining this with the third and solving for A,

$$A - C = +5$$
$$\underline{C + A = +1}$$
$$2A = +6$$
$$A = +3$$

Substituting this value of A in the first equation,

$$3 + B = +3$$
$$B = 0$$

And substituting this value of B in the second equation,

$$0 + C = -2$$
$$C = -2 \ .$$

Since one fringe of yellow light is $1/100{,}000$ inch, this tells us directly that piece A is three fringes ($3/100{,}000$) inch convex, B is flat, and C two fringes concave.

We can in this manner, with only an alcohol lamp, and without recourse to any expensive master flat or involved mathematics, know how flat a piece of glass is almost to a millionth of an inch.

If one of these pieces of glass were to be increased in size until it was a mile long, a millionth of an inch error in its surface would show up as a bulge or depression only $1/32$ inch high—or deep.

A.7.1 Editor's Notes

A.7.2 Sizing a Newtonian Diagonal

Finding what size prism or diagonal will be needed has puzzled many workers. Unless one is on his guard there may be danger of drawing up a concept like the one in the first of the sketches in Figure A.7.4, being

Chapter A.7. The Prism or Diagonal

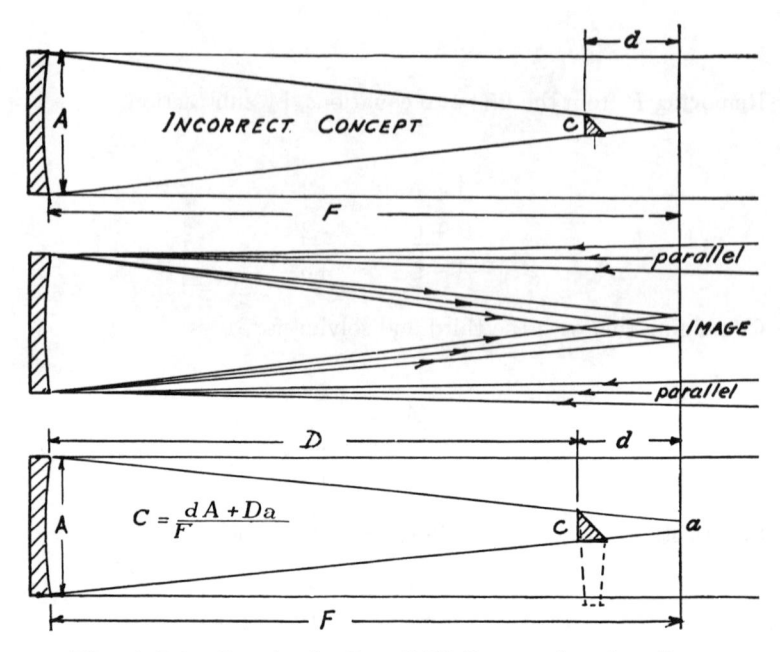

Fig. A.7.4. Drawing by Russell W. Porter, after the editor.

misled perhaps by Figure B.1.1 in ATM2, which are supposed to be only diagrammatic. Here the method would be one of simple proportion between similar triangles, that is, c is to A as d is to F; solve for c. This would be the correct method if all the rays which reach the mirror from the whole field of view were parallel, but they are rarely so. They would be parallel only if they came form a single point; for example, if there were only one in a small angle with the axis of the mirror (up to about 15′ in an $f/8$; more in a $f/6$) and these "out-of axis" rays are reflected somewhat as in the second sketch in Figure A.7.4, combining to form an image. This is the reason why the opening at the top of the tube should be a little larger—say $1/4$ inch or, still better, $1/2$ inch—than the diameter of the mirror, though if it is not, the loss will be small at worst. This image has real diameter, as shown in the second sketch—it is not just a point—and this diameter is *assumed* to be the diameter of the field lens of the eyepiece of longest *e.f.l.* used. In practice this is $7/8''$ for a $1''$ eyepiece. The image in a small $f/8$ mirror will be just about that diameter; in an $f/10$ or $f/12$, or a large $f/8$, it will be larger but the excess will not be caught by the field lens, anyway, unless a rather unusual eyepiece is used, so $7/8''$ is the outside diameter to figure on in all but exceptional circumstances—eyepieces of long *e.f.l.* which in turn have a wide field lens and other undesirable characteristics. This alters the

A.7.2. Sizing a Newtonian Diagonal

triangle of our original sketch to a trapezoid, as in the third sketch, and the formula for finding the diameter of the prism face, as given by John M. Pierce, is shown on the same sketch.

The above is the geometrical method but some may prefer the less highbrow but safe, sure, practical method, which is as follows: The mirror is set up in strong sunlight and adjusted until its own optical axis points pretty closely at the sun, that is, when the image of the sun is caught on a card and at the same time the shadow of the card falls on the middle of the mirror. It will not be simple to explore with the card the whole length of the come of reflected rays and actually see what's what by means of the illumination. It is easy to measure the diameter of the cone at the desired distance from the mirror or from the focus. One way is to lay off on a long stick the offsets and distances required by the mounting and hold the stick beside the cone. Rings of several diameters drawn on a card will enable one to get a close enough measurement of the size of prism face required, as the image will approximately fit one of these rings. If a simple flat of elliptical shape is to be used the diameter of the circle obtained would be multiplied by 1.42 in order to derive its longer dimension.

There has been some confusion regarding whether the sun is a fair criterion for a field of stars. On a mirror having a focal ration of approximately $f/8$, the field of vision will have an angular diameter just about equal to the sun's angular diameter, namely half a degree, and it makes no difference whether it is the sun or a filed of separate stars which subtends the same angel.

The discussions presented above are based on *practical* facts, but in theory there is a certain variation—which, however, will lead us to the same method of working, hence the beginner may safely ignore the following: Strictly speaking, the sharply bounded image depicted in the second drawing does not exists in fact. As Porter points out, this is only that part of a much wider image which is *useful*. We may look at it in this manner: The rays which enter the tube at the edge, and reach the mirror at the center, are reflected out of the tube past its opposite edge. Similarly, others which enter the tube at a small distance from the edge are reflected back and out, unless captured and put to work; though diagonal rays which reach the edge of the mirror will, of course, be reflected against the inner walls of the tube and come to naught. Together, all of these rays, as can be worked out in a simple sketch, will form an image of a fairly wide areas of the heavens—several degrees in width—which extends clear across the tube, and if there is no tube, or if there is a skeleton tube, this image will take the shape of a whole dome, extending right down to the "horizon" of the mirror. We cannot very practically pick up this whole image with our diagonal, because of its size, and in practice we do not even wish to;

first because the best of its rays are already concentrated near the center where we can conveniently capture them, the edge illumination thinning out to minimum; secondly, because the spherical aberration that far away from the axis would be rank. The field lens of a 1-inch eyepiece is about $7/8$-inch in diameter, and this in the average case and, in fact, in nearly all cases, will take in the *useful* part of the image. Hence we get back to our starting point; namely, that the size of the prism may be based on the considerations depicted on the second sketch—the $7/8$-inch diameter of the field lens of the eyepiece is the governing factor. Of course, eyepieces have been made with a field lens several inches in diameter, but only for some very special purpose or as a freak, hence this point has little practical bearing on the issues under discussion. The $7/8$-inch is not merely arbitrary. It is the accepted result of long accumulated experience.

A.7.2.1 Addendum, 1948

Since 1927 when the preceding chapter was written, standards among amateurs have moved steadily upward, leaving the use of plate glass for diagonals at the borderline even when these are carefully selected as specified, and much below borderline when this is shirked as has sometimes been done. On a first telescope—before the maker's observing eye has learned exacting requirements and before his skill of hand and testing eye have become educated to match a first-class diagonal with mirrors of equal refinement, as judged by an old hand and not by the beginner himself—carefully selected plate glass diagonals may serve as a member of the mirror-diagonal-eyepiece-observer team. Thereafter they had best be abandoned.

Even a selected area of plate glass with relatively flat surface across its whole width still will have the "lemon peel" finish of the manufacturer's polish on felt instead of pitch—examine a lemon. These minute local irregularities, which escape the described test with interference fringes, diffuse light.

The effect of plate diagonals is relative, not absolute, and the tyro will be pleased with his first telescope—happier, perhaps, for some months until his tastes have become more sophisticated.

Nearness of diagonal to eyepiece does not in itself improve definition, though it may work indirectly toward that end, thus: the nearer the eyepiece the smaller the cone of rays from the mirror and the better the chances of finding on a large piece of plate glass a relatively flat area wide enough to equal its diameter. Hargreaves has best stated this point: "A small part of a large bad mirror may be good but it does not follow that a small bad mirror is a good one."

Advice to slide and wring pieces of plate glass together when testing

A.7.2. Sizing a Newtonian Diagonal

appears to have been regarded as general recommendations to test all flats by these rough tactics. Small resulting scratches will matter relatively little on cheap expendable plate glass, but elsewhere they will matter more. If pressure is necessary to bring two flats together it is likely that the worker is trying to flatten a piece of lint and should clean the glass. When clean, it should settle without force in the nearly parallel position which will display the fringes. Light momentary pressure should do no harm on small pieces. Holding the finger in position may bring up the fringes but will temporarily deform the glass, giving false appearances.

Chapter A.8

Prism Diagonals Axial Aberration Effects[†]

Let's assume that a perfect telescope which uses a perfect optical flat for diverting the axis 90° to the side of the tube gives perfect definition. If a perfect 90° deviation prism is substituted for the flat, the axial definition may suffer, due to the introduction of spherical and chromatic aberrations. The object of this appendix is to show the magnitude of these aberrations so that the telescope user may decide whether or not he wishes to employ a prism diagonal in conjunction with his mirror or objective.

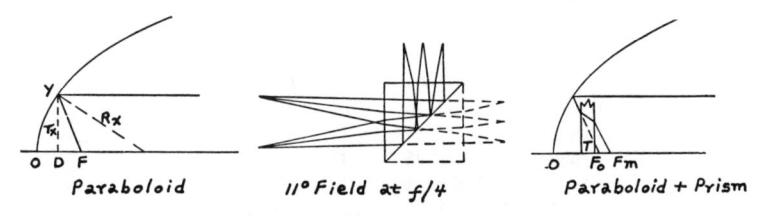

Fig. A.8.1.

The left-hand drawing shows a ray from ∞ being reflected from a paraboloid in such a way that it intersects the axis at F. If the mirror is of perfect figure, all rays from a given point at ∞ will cut the axis at the same point, F, regardless of the value of the zonal radius, r_x, within the limits of diffraction theory. If, now, a prism is introduced anywhere along YF, it will act optically as if it were a plate of glass of thickness T equal to the length of the cathetus or entering face of the prism. Therefore, it is treated here as a plane-parallel plate, as in the central and right-hand

---

[†]By Horace H. Selby.

drawings.

If we know the distance OF_m for every value of r_x, we can easily construct a curve which will represent the spherical aberration introduced by a right angle prism. Also, the chromatic effects of the prism can be determined. In order to find OF_m, we need know the following data:

R_o = the radius of curvature of the central zone or twice the focal length.

r_x = the radius of any zone of the mirror or objective.

T = the length of the prism face, assuming sharp edges.

N = the refractive index of the prism material for the particular color of light to be considered.

In a previous communication (*Scientific American*, April 1941) prism aberration was described directly in terms of these variables plus R_r, the radius of curvature for a particular zone. The result was rather cumbersome. Therefore, the equations have been simplified for presentation here by the use of an angular function derived from the above variables. This function is the sine of angle OFY, which will be called angle α. Since the magnitude of α depends only on the f-numbers or aperture ratio of the telescope, $\sin\alpha$ and $\sin^2\alpha$ have been computed for various f-numbers and are given in Table A.8.1 as a computing aid.

Table A.8.1

$$\sin\alpha = (2r_x R_0)/(R_0^2 + r_x^2)$$

$f/$	$\sin\alpha$	$\sin^2\alpha$	$f/$	$\sin\alpha$	$\sin^2\alpha$
1.0	0.470588	0.221453	5	0.099750	0.009950
1.5	0.324324	0.105186	6	0.083188	0.006920
2	0.246153	0.060616	7	0.071337	0.005089
2.5	0.198019	0.039211	8	0.062439	0.003898
3	0.165517	0.027395	10	0.049968	0.002496
3.5	0.142132	0.020201	12	0.041648	0.001734
4	0.124513	0.015503	15	0.033324	0.001110

A.8.1 Axial Spherical Aberration

Marginal spherical aberration is a commonly-considered aberration in telescopes. It is expressed as the difference between the OF_m for the marginal ray and that of the paraxial ray. If the marginal OF_m is greater than the paraxial, the aberration is considered minus or overcorrected. Mathematically, this aberration equals

$$L_{\text{sph}} = T\left(\frac{1}{N} - \sqrt{\frac{1-\sin^2\alpha}{N^2-\sin^2\alpha}}\right)$$

A.8.2. Axial Chromatic Aberration   159

Zonal spherical aberration may remain after the marginal error is corrected. It is frequently a maximum at the 0.7 zone and can be determined with the above formula by substituting the proper $\sin^2 \alpha$ for the marginal f-number divided by 0.7. In both cases, yellow light is considered when speaking of spherical aberration. Often it is 5555Å.

A.8.2 Axial Chromatic Aberration

The change in OF_m which occurs when two colors of light are considered is called chromatic aberration. The colors are usually near the ends of the visible spectrum. Hydrogen C and F or C and mercury g are popular pairs. The zone chosen for computing chromatism is usually the 0.7 zone although some prefer the margin. If OF_m red is greater than OF_m violet, the aberration is said to be plus or undercorrected. Its magnitude is:

$$L_{\text{chr}} = T \left[\sqrt{\frac{1 - \sin^2 \alpha}{N_{\text{red}}^2 - \sin^2 \alpha}} - \sqrt{\frac{1 - \sin^2 \alpha}{N_{\text{violet}}^2 - \sin^2 \alpha}} \right]$$

For easily calculated, approximate results, chromatism is

$$L_{\text{chr}} \approx T \left(\frac{1}{N_{\text{red}}} - \frac{1}{N_{\text{violet}}} \right)$$

Chromatic Variation of Spherical Aberration: Spherical aberration at a given aperture is not constant for all colors. Therefore, it is sometimes useful to know what effect to expect, for example, when photographing through filters. In order to check this, it is merely necessary to compare spherical aberration results as found for one color with the other colors of interest, substituting the proper N values for the glass used. Unless extreme aperture ratios are used, however, the effect is insignificant, as shown in Table A.8.2.

Table A.8.2
$N_c = 1.514$ $N_D = 1.517$ $N_g = 1.526$

f-number	2	3	4	8	15
T	4.0 inches	3.8 inches	2.5 inches	1.5 inch	1.25 inch
$L_{\text{sph}}C$	−0.04679	−0.01970	−0.00729	−0.00109	−0.00026
D	−0.04684	−0.01969	−0.00730	−0.00109	−0.00026
g	−0.04694	−0.01967	−0.00730	−0.00109	−0.00026
1/3 Rayleigh limit	0.00048	0.0011	0.0019	0.0075	0.026
L_{chr}	0.02092	0.01979	0.01300	0.00778	0.00649
Rayleigh limit	0.00036	0.00080	0.0014	0.0057	0.0195
$\odot_{\text{chr}}$	0.00522	0.00330	0.00162	0.00049	0.00022
Airy disk	0.0001	0.00015	0.00019	0.00039	0.00072

The above arbitrarily-chosen examples may give the erroneous impression that aperture ratio alone determines the aberration magnitude and that focal length is immaterial. The fact is that the second factor, prism size, is greatly affected by e.f.l. and diameter of image field.

To illustrate, compare two $f/4$ telescopes, the first, of 12-inch aperture covering a 2-inch diameter field with the image 8 inches from the axis and the second, of 4-inch aperture, 1-inch field, 4 inches off axis. Applying John M. Pierce's formula (see Chapter C.1, subsection C.1.1.8) the first instrument will require a 3.67-inch prism while the second will need one of 1.75-inch face.

A.8.3 Effect on Definition

Optical designers have found through the years that the Rayleigh limit of $1/4\lambda$ of path difference due to spherical aberration is a rather large allowance, if excellent definition is required in telescopes. Therefore it is customary to consider one third, approximately, of the Rayleigh limit as standard. Considerations of contrast as well as resolving power make this desirable. On this path difference basis, the allowable longitudinal aberration is, for yellow light:

$$\text{marginal spherical} = \frac{0.000029 \text{ inch}}{\sin^2 \alpha}$$

For chromatic aberration, however, the full Rayleigh limit of

$$\frac{0.000022 \text{ inch}}{\sin^2 \alpha}$$

is frequently satisfactory. The reason that the marginal spherical tolerance appears greater than the chromatic when it is in fact smaller is that optical path difference is expressed in terms of longitudinal aberration in each case. A good explanation is by Conrady in Glazebrook's *Dictionary of Applied Physics*, Vol. IV, 1923, pp. 216–227.

The seriousness of the aberration effects can be judged in other ways which do not consider path differences. Knowing the longitudinal chromatic aberration, L_{chr}, the blur diameter will be found to be

$$\odot_{\text{chr}} = \frac{L_{\text{chr}}}{2A} \text{ where } A \text{ is the aperture ratio.}$$

If this approximates the diameter of the Airy disk, which, for yellow light, is 0.000049A, definition will not be greatly affected although slight degradation may occur.

Another way of judging sharpness is to remember that the average eye considers an image sharp if its diameter is less than 0.005 inch at 10 inches

A.8.4. Extra-Axial Aberrations 161

and that telescopic images are usually seen as virtual images some 10 inches in front of the eye. Therefore, if the blur diameter as determined above is of the order of 0.005 inch divided by the magnification of the eyepiece used, the blur will not be noticeable.

Unfortunately, the blur diameter of an image of a point object is not a simple function of the longitudinal spherical aberration. The diffraction nature of images makes simple geometrical treatment of spherical effects meaningless. Under some circumstances, in fact, small spherical residuals can decrease the diameter of the central disk.

A rule of thumb tolerance, which experience has shown to be valid, is to consider 0.0002 inch times the f-number squared as the maximum allowable longitudinal spherical aberration for excellent definition. According to this empirical rule, an $f/1$ system needs marginal spherical correction within 0.0002 inch of perfection whereas $f/8$ telescopes can tolerate 0.013 inch.

Again, if a mirror has been figured to the tolerance given by Wright in Chapter B.3 and the longitudinal spherical aberration is small compared with this tolerance, expressed in terms of $r^2/2R$, it is evident that image blurring will not be serious.

A.8.4 Extra-Axial Aberrations

Astigmatism, coma, lateral chromatism, distortion and field curvature are affected to some extent by the introduction of a prism into a telescope. They are not considered here, however, because the factors of extent of field and type of instrument have significant effects on the results, as does the prism location under certain circumstances, making computation complex. Moreover, when prisms are close to the eyepiece and fields are not wide, the effects are not serious in the average case.

As has been indicated, chromatic aberration tolerances are reached in the more common cases before violation of the spherical tolerance occurs as the f-number decreases; so little overall benefit would be obtained by undercorrecting a mirror to compensate for the spherical aberration caused by a prism. If such compensation is desired, however, it can be obtained by figuring a mirror so that knife-edge movement less than $r^2/2R$ is obtained. The proper amount will be

$$-2\left[L_{\text{sph}} + \frac{R_x\sqrt{R_x^2 - r^2} - R_x^2}{2\sqrt{R_x^2 - r^2}}\right]$$

R_x is the radius of curvature of the edge zone, or $R_0 + r^2/2R$.

A.8.5 Prism Glasses

The usual glass for reflecting prisms which are to cover small angular fields at $f/8$, $f/15$, etc. is a borosilicate crown, $N_D = 1.517$, $N_C = 1.514$, $N_g = 1.526$. This is highly transparent and stable. The critical angle, δ, for yellow light with this glass is 41°. Therefore, if rays strike at smaller angles, total reflection will not occur; so the hypotenuse must be silvered—causing some light loss—or a denser glass must be employed. Flint glasses are undesirable because of their high dispersion which increases the chromatic aberrations and because flints are usually less hard and stable. It is customary, therefore, to use a medium barium crown, $N_D = 1.572$, $N_C = 1.569$, $N_g = 1.584$, in wide field, high speed instruments. This glass has a δ value of 39°.

Referring back to the illustration, it can be seen that some rays, in extreme cases, hit the hypotenuse at incidence angles of even less than 39°.

Chapter A.9

How to Make a Diagonal for a Newtonian[†]

Whilst the use of a totally reflecting prism on a Newtonian is permissible, we must use an optical plane of elliptical contour if we desire the very best results. Various methods, mostly unsound, of obtaining such a plane have been suggested; for example, cutting the ellipse from a piece of stout mirror glass, or by making a circular plane by well-known methods, and cutting the ellipse from that, etc.

A.9.1 Making the Blank

It is almost impossible to pick up a piece of glass which is optically plane, and it is equally difficult to cut an ellipse from a good optical plane without serious deterioration of the optical surface. These two objections practically determine the methods by which a perfect optical plane of elliptical contour must be produced.

Such a plane (Figure A.9.1a) may be considered as a slice cut from the end of a glass cylinder (Figure A.9.1b) which has already been shaped at the end to an angle of 45°. The dotted line shows where the slice of glass would be cut off to give the required shape, which in one particular aspect presents a circle in elevation (Figure A.9.1c).

Several oval pieces of glass should be obtained, sufficiently large to allow for grinding down to the correct size. Note that more allowance is required on the major axis to include the thickness of the glass. This can be easily ascertained from a full size drawing.

[†]By John H. Hindle.

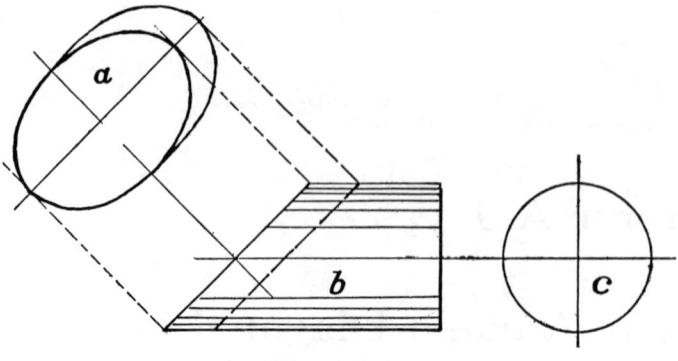

Fig. A.9.1.

We next require a short length of steel shaft (S, Figure A.9.2, at left) about half to three quarters the finished minor axis of the ellipse in diameter. The end of this shaft is cut to an angle of 45°, and the piece of glass firmly attached with melted pitch. We place the shaft in the chuck (C of Figure A.9.2, left). A flat-faced pulley P, say 6" in diameter, is keyed on another shaft, also a vee pulley V, to take an endless rope or cord. This shaft is mounted on the slide rest SR and the drive arranged from above, so that a forward movement of the rest does not affect the tension of the driving cord.

Neither a high speed for the lathe nor the grinding wheel is desirable; 20 r.p.m. for the lathe and 60 r.p.m. for the grinding wheel will be ample. Fine Carborundum is painted onto the wheel, using turpentine as a lubricant. A very slow feed, by occasionally tapping the handle of the screw with a wrench, is advisable. After each feed, allow time for the glass to be ground away by the abrasive. Too much haste will result in chipping of the glass, and possibly forcing it from its seat. When finished with a smooth edge, the beveled shaft is heated, and the perfectly shaped ellipse is detached and cleaned.

A.9.2 Making a "Surround"

It now remains to put an optical face on one side. This can be done perfectly, only by providing a "surround" of similar glass. Assuming our ellipse has a $1\,1/2$" minor axis, then we require a glass disk 6" in diameter or thereabouts. This disk need not be edged, and it must be cut diametrically through the center. We now require to cut a piece out of each half, to accommodate the shaped flat between. The best method is to mount a $1\,9/16$" or $1\,5/8$" shaft in the lathe (Figure A.9.2, at right), and make a frame with

A.9.3. Grinding

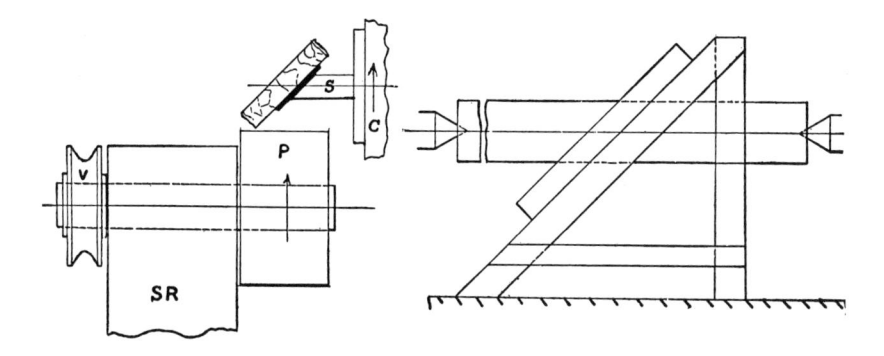

Fig. A.9.2.

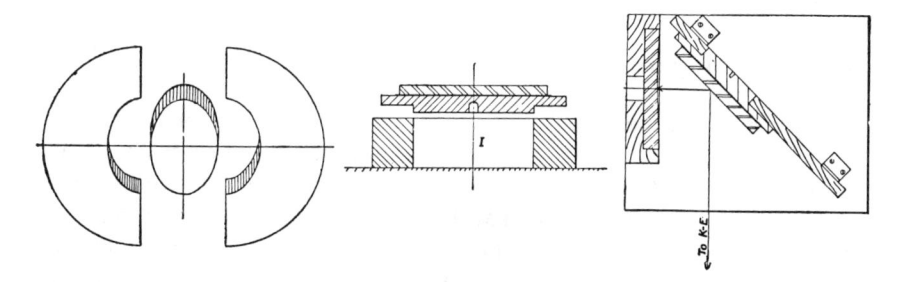

Fig. A.9.3.

a surface at an angle of 45°, permitting the two halves to be forced against the revolving shaft, to which Carbo and water is applied. We finally have three shaped pieces of glass which fit into each other perfectly (Figure A.9.3, at left) when laid on a level surface. These three pieces are actually pitched on, in this position, to an iron plate, preferably of circular contour. When cold, melted beeswax is poured into the interstices, the surplus scraped off, and the surface rough and fine ground to one level. The thickness of glass should be from one-sixth to one-fourth that of the minor axis. Sometimes there may be chipped edges which require grinding out, and a little extra thickness is a wise precaution.

A.9.3 Grinding

Accuracy in fine grinding is advisable, to avoid loss of time in figuring. The well-known method of using three disks, grinding each against the others in turn, may be adopted for getting the surface approximately flat.

Another method is to fine grind on the surface of a much larger piece of glass, which is known to be approximately flat. This leaves the surface of the composite disk slightly convex, which is always preferable, because it is easier to polish away a high center.

A.9.4 Polishing

Before commencing to polish, scrape away the beeswax in the grooves just below the surface with the fingernail, and a smoother action of the polisher results. The polishing may be done either by hand, or on the drill press described in ATM3, Section E.7.1. If the iron plate is made of the section shown in Figure A.9.3, center, and a similar section plate used for the polisher, a much heavier ring of iron, I, may be used to receive either, so that the polisher may be either above or below, alternately.

A.9.5 Testing for Flatness

To test for flatness, we require a small spherical mirror about 6″ in diameter. The radius of curvature of this mirror requires consideration. Theoretically, the longer the radius of curvature, the greater the accuracy obtainable. Practically, for a small flat, too great a distance prevents one seeing the surface of the flat intimately, and diffraction effects around the edge of the flat are somewhat confusing. 60″ radius of curvature is suitable for flats of $1\,1/2''$ to 2″ minor axis.

A suitable arrangement for testing is shown in Figure A.9.3, at right. The spherical mirror is mounted in a circular recess in a block of wood, which is screwed on to a flat board, edge-on to the observer. The composite mirror being tested is at an angle of 45° thereto. It is arranged in a vertical board having a hole of the required size, so that the complete arrangement can be rotated to the correct position, namely, when the major axis of the ellipse is horizontal, in which case the end view will be circular. It is then convenient to provide a close-fitting mask which hides the "surround." Alternatively, the surround may be painted with rouge and water, which quickly dries. The ellipse is approximately flat when the focus, in a vertical plane, coincides with that in a horizontal plane. To be more precise, find the focus moving the knife-edge horizontally, then withdraw the knife-edge just clear, and slide the blade of the pen-knife *up* the knife-edge. If the shadow comes *down* from the top the flat is convex, and vice-versa.

The test described should be used to *approximate* to a level surface. For the greatest possible accuracy, we need a perfectly circular pinhole, and an ordinary eyepiece. We proceed to examine the image of the pinhole with the eyepiece, and expand the image on either side of the best focus. When the

A.9.6. Mounting the Diagonal

images expand identically on both sides of focus, the mirror is an accurate optical plane, and the clearest definition of the pinhole is obtained.

This method of testing is *extremely* sensitive, and in order to memorize it, the word HIVO has been concocted, that is to say, when the expanded image is elongated "horizontally inside focus," and "vertically outside focus," the mirror is *convex*.

At this stage, it may be interesting to inquire whether this test can be used in making a *circular* optical plane. There is only one difference, the expanded images spread into an *elliptical* form, consequent upon the angle at which the plane is viewed. The figures *a*, *b*, *c*, Figure A.9.4, show what might be obtained in the testing of a circular plane, when perfect, and when closely approaching perfection.

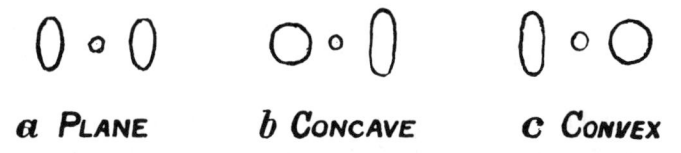

a PLANE   *b* CONCAVE   *c* CONVEX

Fig. A.9.4. *Testing of a circular flat, rather than an elliptical one.*

The pinhole image is shown in the center, the outside expanded image on the left, and the inside expanded image on the right.

Reverting to our original problem. Having now obtained our accurate optical plane, it can be removed from the iron plate and cleaned. The "surround" can be used again and again, with the aid of thin metal packings to raise them to their original thickness, if found necessary.

A.9.6 Mounting the Diagonal

To mount the elliptical plane we require a short length of brass tubing, the interior diameter of which is a few thousandths greater than the minor axis of the mirror, a dimension which can be accurately obtained. This tube is cut at an angle of $45°$, in the manner indicated in Figure A.9.5, at left, leaving four prongs, *a, b, c, d,* projecting, which are bent over to keep the plane from falling through. It is advisable to prepare a solid mandrel, M, with the end cut off at $45°$, on which to bend over these prongs accurately; the brass tube should first be annealed by heating to a red heat, and plunging in cold water. After bending, the prongs can be reduced by filing to a minimum, safe, and symmetrical contour.

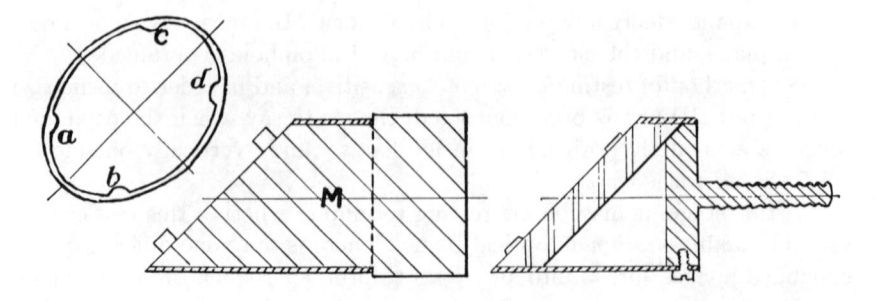

Fig. A.9.5.

Finally, a brass disk to fit the tube, with a projecting screwed stud (Figure A.9.5, at right) is fitted. A 45° bevel is filed on the inner edge in one spot, so that the plane mirror rests thereon. Three screws 120° apart are used for fastening the disk in the tube, and the 45° bevel is carefully eased away with the file, until, with the three screws tightly home, there is just a perceptible "shake," ensuring that the plane is not nipped.

A cylindrical slip-on brass cover may be used to protect the silvered surface of the diagonal plane.

Figure A.9.6 shows how the plane may be mounted in a square wooden tube. A three-armed support has a solid brass disk in the center, with a hole in which the flat is mounted. The lower arm a is fixed in a slot in an optimum position. The other two arms, b and c, are arranged so that they can be individually screwed either forward or backward. This adjustment, along with the possibility of revolving the plane itself on the optical axis of the cell, will give all the adjustment required.

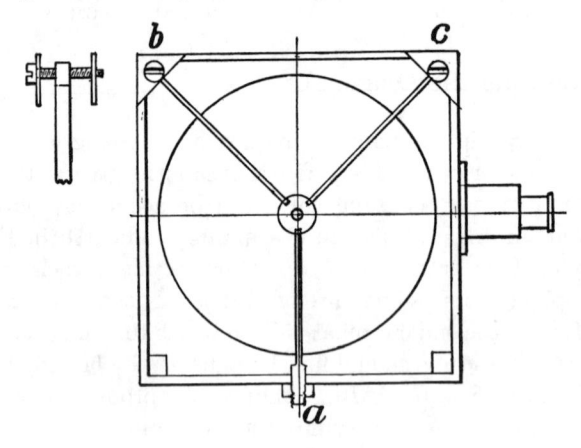

Fig. A.9.6.

Chapter A.10

The Building of a 19-Inch Reflecting Telescope[†]

The department of Astronomy in the University of Toronto, although it has existed as a separate department for over 20 years, has no observatory for research and instruction. Courses are offered to students in general astronomy, spherical astronomy, theoretical astronomy and astrophysics. The University has had the cooperation of the small observatory of the Metrological Service of Canada for the purpose of instruction in the use of the equatorial and transit instruments. It has always been the endeavor of Dr. Chant, the head of the department, to foster a wide interest in the subject, both within and outside the University, and to emphasize the need for an observatory. It is to be hoped that the project will not be delayed much longer.

I suggested to Dr. Chant in 1926, that it would be possible to construct a telescope privately, capable of doing useful research in many lines of astronomical work, and although it would be rather diffident to contribute so much time for purely mechanical work, that I would be willing to undertake the task if we could find means to house the instrument. We decided to build as large a telescope as we felt we could complete in a reasonable time and we were confirmed in this decision because we had in mind the design and construction of a much larger instrument. The firsthand information obtained in building the smaller telescope would be extremely valuable in crystallizing our conceptions of what would be required in a bigger one.

[†]By R.K. Young. Reprinted by permission, from the *Journal of the Royal Astronomical Society of Canada*, January 1930. More and more amateurs are building or planning to build fairly large telescopes. The one described in the present chapter makes appeal because of its ruggedness and clean lines. The author is now Director of the David Dunlap Observatory of the University of Toronto, not built when the article was written.—Ed.

The present article is a description of the design and construction of a 19″ telescope. It is published here because we think it would not be beyond the powers of the ambitious amateur to construct a similar or better one. It has been completed with the aid of a very modest workshop and occasional outside help for such work as could not be done on a lathe. It has taken about two years to construct, using only spare time in holidays and such time in the evenings and weekends as available outside of university duties. The cost of the material and incidentals were roughly as follows:

Optical parts purchased, including plane mirror, finder lenses, eyepieces and disk for large mirror	346.00
Outside work done, gear cutting, patterns, etc.	625.00
Castings, motors, bearings, incidentals	529.00
Total	$1,500.00

There are some very excellent works published on the grinding and polishing of mirrors, among which may be mentioned, *The Reflecting Telescope,* by George W. Ritchey, in the Smithsonian Contributions to Knowledge, being part of Volume XXXIV; *The Amateur's Telescope,* by Ellison; *Amateur Telescope Making,* by the Scientific American Publishing Co. We particularly recommend the last reference for the guidance of the beginner. The descriptions of the mounting in these books are not so satisfactory as for the mirror, and the ordinary amateur is inclined to mount his telescope in a very simple and inexpensive style. There is no doubt that a great deal of pleasure and profit may be obtained with even a crude mounting, but labor spent on mounting will be amply repaid in the pleasure of using the telescope, more especially if the telescope is to be used for research work. As additional references, the beginner might read the very excellent description of the 72″ reflecting telescope of Victoria, B.C., published as the first number of the Publications of the Dominion Astrophysical Observatory, and a book entitled *The Telescope,* by Louis Bell, published by the McGraw-Hill Book Company.

A.10.1 The Mirror

The main mirror consists of a piece of Pyrex glass 4″ thick and $19\tfrac{1}{2}$″ in diameter, which we received from the Corning Glass Works in March, 1926. The simplest material for the amateur to select for small mirrors is ordinary plate glass, the thickness being from $\tfrac{1}{5}$ to $\tfrac{1}{10}$ the diameter. Beginners might try a plate of chromium steel to advantage if they are of an experimental turn. This material will take a high polish and retain it for many years and requires no silvering. It has not, however, been used and would be an experiment. For mirrors more than 12″ in diameter, steel

A.10.1. The Mirror

Fig. A.10.1: *The glass disk after it had been ground to shape and a hole drilled through the center. The glass tool is shown to one side.*

is rather heavy, and ordinary plate glass is not of good enough quality even if it could be obtained of the required thickness. What is desired is a substance with small expansion with heat and of great rigidity. Quartz disks possess these qualities to an admirable degree, but at the time of beginning the 19″ they were in the experimental stage and could not be obtained more than 12″ in diameter. They are available in much larger sizes now. Pyrex glass, as is well known, was developed on account of its small coefficient of expansion with heat and it appeared to me to be the best material. It has the drawback that it is very difficult to melt and consequently difficult to free from bubbles in the pouring. Large disks are usually much marred by these defects. The disk we obtained was rather bad but, as no bubbles of any size would come near the finished surface, I decided to use it.

The disk as it came to me from the makers was about $1/2$-inch thicker at one edge than at the other and the sides were much seamed. The first task was to shape the disk. I constructed a fairly large grinding table with a rotating turntable in the center. In order to true up the sides of the disk, I placed a brass band about 4″ wide and $1/16$″ thick around the glass and fed

Carborundum and water to it as the disk rotated. It proved a very noisy affair but reasonably effective. The turntable is shown in Figure A.10.1, where the disk is also shown after it had been trued and a hole drilled through the center. The turntable was operated by a system of belts and pulleys beneath the table which are not shown in the photograph. The power was furnished from the workshop motor. An iron tool about 12" in diameter and 3/4" thick was used to grind the top flat and parallel with the bottom. By bearing down when the tool was resting on the high side, the disk was gradually trued up. The whole process took quite a little muscular effort, but not a great deal of skill or time. About two weeks after the reception of the glass it was brought to shape. This was a comparatively short time compared to the tedious but interesting months of labor it took to figure the surface.

The final surface of a reflecting telescope mirror is in the form of a parabola, to which the equation in Cartesian geometry is, $y^2 = 2Rx$. The focal length of the mirror is $R/2$. In Figure A.10.2, y is the distance of any point, such as P, from the Axis AD; and x, the depth of the mirror below the chord joining two points on the surface, each at a distance y from the axis. In the present case, the focal length was to be 125" so that the equation to the final parabola would be $y^2 = 500x$. The depth of the curve at various distances from the center is readily computed to be that shown in Table A.10.1.

Table A.10.1

Distance from axis (inches)	Depth (inches)
1	0.002
2	0.008
3	0.018
4	0.032
5	0.050
6	0.072
7	0.098
8	0.128
9	0.162
9.5	0.180

The beginner may not realize how nearly the surface approximates to a sphere or, what amounts to the same thing, how closely the curve $BPHA$ in Figure A.10.2, which is the parabola, can be made to fit a circular arc. In order to show this, imagine a circle described with center at C and radius AC, equal to 250". The equation to this circle is

$$(x - 250)^2 + y^2 = 250^2,$$

A.10.1. The Mirror

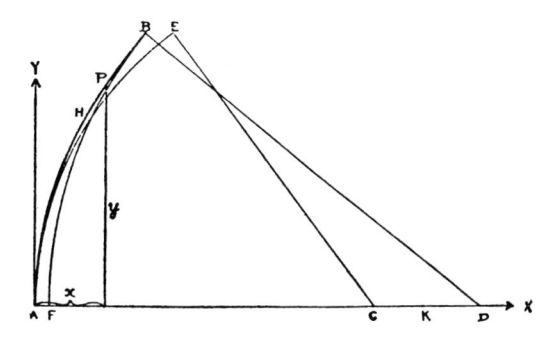

Fig. A.10.2.

or

$$x = 250 - \sqrt{250^2 - y^2};$$

and it may be readily shown that for any given y, the difference between the x for the sphere and the x for the parabola is, to a very close degree of approximation, $8 \times 10^{-9} y^4$ inches. For the outside of the mirror, where $y = 9\,1/2''$, the difference equals BE and can be computed to be $0.000065''$. The small difference between the sphere and parabola is quite below any ordinary method of measurement and the first step in grinding the mirror is to obtain the sphere. The small amount of glass that has to be removed afterwards is polished away by the so-called process of figuring.

The 12″ cast iron tool was used to hollow out the disk until micrometer measures indicated that the depths shown in Table A.10.1 were obtained. All this work was done with coarse Carborundum, grade No. 40, and the grinding stopped a little short of the computed depths, leaving the last few thousandths of an inch to be removed by fine grinding. A glass tool was then substituted for the iron one. This tool is shown in Figure A.10.1, which also shows the manner of operating the tool on the glass. A long arm was pivoted at one edge of the grinding table and the glass tool fastened near its center. The exact position of the tool could be adjusted by means of various holes in this arm, and at the same time it was left free to rotate. The operator, holding the end of the lever arm, could move the tool laterally over the mirror with any desired stroke. The details of the process of fine grinding and obtaining a surface ready to polish are fully described in the references quoted, and I shall not describe them here, further than to say that this part of the work was very satisfactory and the desired result easily obtained and that in a comparatively short time the disk was ready for polishing.

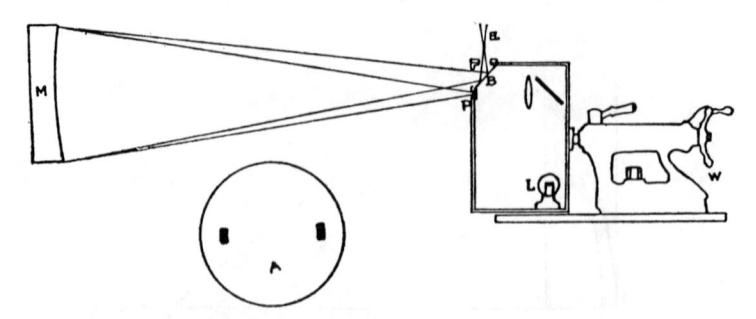

Fig. A.10.3: *The apparatus for measuring the radius of curvature of the various zones. Light from L illuminates the pinhole P, and the image of P is formed at E. A knife-edge slides across the top of the box and determines the exact position of the focus.*

I followed the procedure described by Ritchey in polishing, using resin and turpentine for forming the tool, and rouge as the polishing agent. At first I used a built-up tool from pieces of wood reinforced on the back by strips of angle iron and had considerable difficulty with the tool warping. Eventually, I abandoned this tool and made an aluminum one that was much more satisfactory; the surface was brought to a satisfactory polish.

As soon as the mirror was polished, I tested it to find out how nearly I had succeeded in obtaining a spherical surface.

Figure A.10.3 shows the apparatus for measuring the radius of curvature of various parts of the mirror. The method of grinding ensures the mirror is a true figure of revolution, and so it is necessary to test zones at various distances from the center only. Light from the 24 c.p., 12 volt light L is concentrated on a small piece of opal glass in front of a small pinhole at P. Light from this source, which is placed near the center of curvature of the mirror, is returned to the totally reflecting prism at B and bent upward so that the eye placed at E sees the whole mirror flooded with light. By moving the apparatus forward or backward by means of the screw on the tail stock of the lathe on which the light source was mounted, and sliding a safety razor blade back and forth over the top of the box between the eye and the beam, the position of the focus could be determined. One revolution of the wheel W moved the source of light $1/10''$ and the circumference being divided into ten equal parts, estimations could be made to the thousandth part of an inch. In practice the mirror was covered by means of masks such as that shown at A, and the radius of curvature of the various zones on the mirror determined.

As indicated before, the sphere is changed into a paraboloid of revolution by removing a very small amount of glass, an amount which could be polished away in a very short time, provided it could be done correctly.

A.10.1. The Mirror

There are three ways to effect the figuring, viz, by removing glass from the outside, by removing glass from the center, or some from the outside and some from the center. To see this behavior, examine Figure A.10.2. The parabola is the curve $BPHA$. Let the coordinates of the point B be x_0, y_0. The equation to the normal at the point B is

$$(y - y_0)R + (x - x_0)y_0 = 0,$$

and this cuts the axis Ax in the point $x = x_0 + R$, or D. With D as center and BD as radius, describe the arc of the circle BF, cutting the axis Ax in F. The distance AF is given by

$$AF = x_0 + R - (R^2 + y_0^2)^{1/2}.$$

The sphere may be changed into the parabola by removing the cap of glass BAF and the focal length obtained will be $R/2$. In order to obtain this value we must start with a sphere of radius

$$x_0 + R \text{ or } R + y_0^2/2R.$$

The amount that must be removed is shown by the full curve in Figure A.10.4. It may be shown that the total mass of glass which must be removed is given by

$$M = \frac{\pi y_0^6 \rho}{24 R^3},$$

where ρ is the density. For a glass of specific gravity 2.5, this amounts to the surprisingly small quantity of 0.25 grams or 3.9 grains. In this case, the polishing is done at the center.

With C as center and CA as radius, scribe the arc of the circle AE with a radius equal to R. We could change the sphere into the parabola by removing the cap of glass ABE, and in this case we would polish away the outer edge. The amount of glass which would have to be removed at various distances from the center is shown in curve II, Figure A.10.4, and the total mass of glass which would have to be polished away is just the same as before.

Take a point P on the parabola and draw the normal. The normal will cut the axis in some point K lying between C and D. If we describe a circle with K as center and KP as radius, it will touch the parabola at P and lie between A and F and also between B and E. We could remove some glass from the outer edge and some from the center to change the sphere to a parabola. In this case the total amount to polish away is less than before. If the point P is taken so that its x coordinate equals $x_0/2$, the

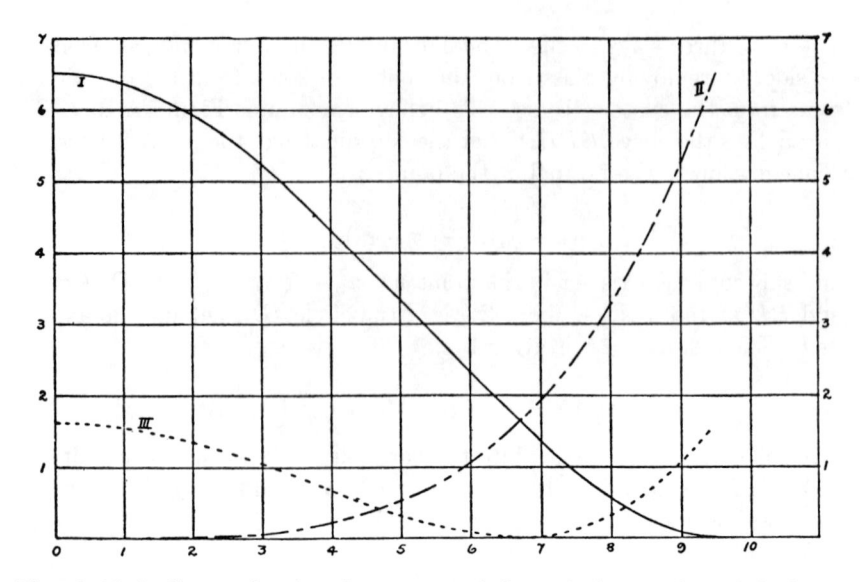

Fig. A.10.4: Curves showing the amount of glass which must be polished away to parabolize a sphere. Ordinates are expressed in hundred-thousandths of an inch. Abscissae are distances from the center of the disk. I, polishing from the center. II, polishing from the edge. III, polishing some from the center and some from the edge.

amount of glass which must be removed is a minimum. The amounts at various distances from the center are shown in curve III, Figure A.10.4. The total amount of glass to be removed in this case is less than one grain.

The usual method of procedure is to polish away the center. As the work progresses we test the radius of curvature frequently for the various zones. It is necessary to be able to interpret these measures in terms of the shape of the mirror. One can readily show that if we test the mirror in zones of width Δs and that if ΔF is the error in radius of curvature of any zone whose center is at an ordinate y from the axis, the distance D by which the inner edge of this zone is high or low, beyond that of the contiguous zone farther out is given by

$$D = \frac{\Delta s \, \Delta F y}{R^2}.$$

From this formula we may construct the actual shape of the surface in relation to the paraboloid required.

In order to illustrate the use of this formula I give a set of readings made in the earlier experimental states, together with the curve representing the shape of the mirror.

A.10.1. The Mirror

Table A.10.2

Zone	Measured Focus	Parabolic Focus	ΔF	$\Delta F \, \Delta sy/R^2 (\times 10^{-7})$
$9\,1/2 - 8\,1/2$	0.000	0.000	0.000	0.000
$8\,1/2 - 7\,1/2$	−0.048	−0.034	−0.014	−17.9
$7\,1/2 - 6\,1/2$	−0.048	−0.064	+0.016	+17.9
$6\,1/2 - 5\,1/2$	−0.082	−0.090	+0.008	+ 7.8
$5\,1/2 - 4\,1/2$	−0.090	−0.112	+0.022	+17.6
$4\,1/2 - 3\,1/2$	−0.120	−0.130	+0.010	+ 5.0
$3\,1/2 - 2\,1/2$	−0.128	−0.144	+0.016	+ 7.8
$2\,1/2 - 1\,1/2$	−0.140	−0.154	+0.014	+ 4.4

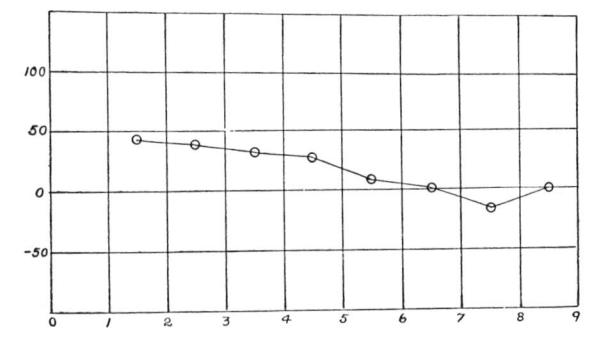

Fig. A.10.5: *The condition of the surface shortly after starting parabolizing. The ordinates are expressed in ten-millionths of an inch above or below the required surface, and abscissae are distances in inches from the center of the mirror.*

The columns in Table A.10.2 are consecutively,

1. The limits of the zone measured, each zone being 1″ wide.
2. The measured position of the focus in inches, the outer zone being the starting point from which the others are measured.
3. The calculated foci for a parabolic mirror, the numbers in this column being obtained from Table A.10.1, because the aberration at the center of curvature for any zone equals the depth of the mirror below that point when tested by moving both the knife-edge and source of light.
4. The amount the measured aberration differs from the computed value.
5. The error in the surface expressed in units of 10^{-7}-inch. The first and last columns are plotted as a graph in Figure A.10.5. At that time there was a low spot about $7\,1/2''$ from the center.

I used local polishers almost entirely to produce the desired figure, smoothing up the surface from time to time with a full-size or two-thirds-size polisher. I shall not describe any of the methods used in making the polishing tools as these are treated very fully in the references given, but it may be of service to someone if I emphasize a few points. Before starting to polish be sure that you know the shape of the mirror. This may involve repeated measures, with precautions being taken that the room is at a uniform temperature and the air steady. See that the tool fits the glass. This is especially necessary with the large polishers. I usually allowed the large polishers to *cold press* for several hours before using. Try to get a room for testing where the air is steady. I had some difficulty in this regard. The workshop was in a basement and during the summer months, when there was no furnace going, the conditions were quite good. As soon as the furnace was started in the fall there was too much difference between the temperature in the basement and outside and I found that air currents coming from the windows made it impossible to get satisfactory readings.

When the figure seemed to be right, as nearly as could be judged by the visual knife-edge tests, I made a photographic test by the method described in the Dominion Astrophysical Observatory Publications, Volume I. The results are given in Table A.10.3, and express the final shape of the mirror.[1]

Table A.10.3

Zone	Measured aberration	Computed aberration	O-C	Equivalent O-C at Focus
9	8.26	8.22	+0.04	+0.01
8½	7.50	7.34	+0.16	+0.06
7½	5.35	5.71	−0.36	−0.09
6½	4.05	4.29	−0.24	−0.06
5½	3.15	3.07	+0.08	+0.02
4½	1.59	2.05	−0.46	−0.12
3½	1.15	1.24	−0.09	−0.02
2½	1.26	0.63	+0.63	+0.16

The aberrations are expressed in mm. The center zone is covered up by the secondary mirror which is 5″ in diameter. Outside of this zone the greatest departure is in the 7½″ and 8½″ zones, but all are fairly small. One can compute from these numbers the angular diameter of the mean circle of confusion of the image of a star, and it comes out 0″.16. The theoretical diameter of the diffusion disk for a 19″ telescope due to diffraction is 0″.24 so that the telescope should theoretically give almost perfect definition.

A.10.2 Mounting

A general view of the mounting is shown in Figure A.10.6. I have chosen the conventional form of equatorial mounting with the tube mounted on one

---

[1] These publications are now difficult to obtain, but may be consulted in some astronomical libraries. However, the test referred to above is the Hartmann test, and is described in Chapter B.11 of the present volume.–Ed.

A.10.2. Mounting

Fig. A.10.6.

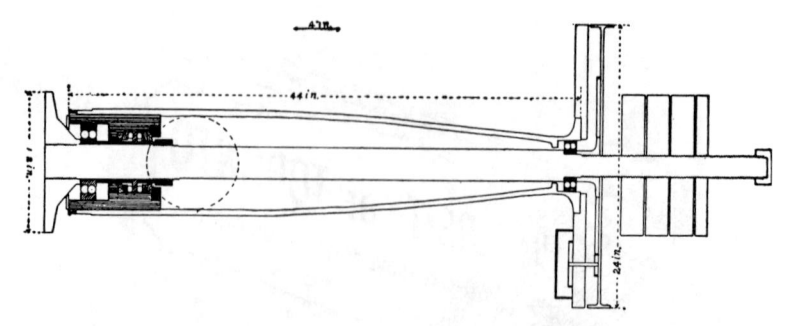

Fig. A.10.7: *A longitudinal section through the declination axis and housing. Note scale.*

end of the declination axis. The telescope is meant to rest on a cement pier rising two or three feet from the floor, so as to give room to walk underneath the mirror.

The tube is of the open type. It consists of the central casting which supports the open tubular construction, and on one side of the central casting is a boss which holds it to the declination axis. The framework of the tube is built up from light steel tubing in sections which are fastened together by steel rings. The advantages of this construction are that the tubes may be made progressively lighter toward the upper end of the tube, and the lower end, when made heavy enough to carry the mirror, naturally balances the longer end. Also the system of braces, which consist of steel rods threaded right and left-hand, can be adjusted to bring the geometrical axis of the tube exactly at right angles to the declination axis. The short tubes that connect the various sections are also threaded right and left-hand and by adjusting these in connection with the brace rods the tube may be made to bend in any direction or to rotate screw fashion, either right or left-hand. If I were making the tube again, I would substitute aluminum tubing for the steel.

The mirror cell at the bottom of the tube consists of a single casting with a flange to bolt it to the bottom ring of the tube. The mirror rests on three disks supported from the bottom of the casting on ball-and-socket joints which can be raised or lowered to effect collimation. There is about $1/4''$ clearance around the edge of the mirror for packing. Although very simple, this method of mounting seems to me to be sufficient. The complicated system of levers and counterweights is quite unnecessary unless the mirror is very thin. The mirror is ground, polished, and figured on a simple turntable, placed on its edge for testing, without any special support. Its position and use in the telescope tube makes no greater demand on its rigidity than the conditions of testing, and if the mirror shows a good figure in the laboratory

A.10.2. Mounting

Fig. A.10.8. *The motor and gear train for slow motion in declination.*

tests it should not be necessary to introduce elaborate methods of support in the telescope.

The secondary mirror at the top of the tube is a 5″ flat made by J.W. Fecker of Pittsburgh. It is supported in a light aluminum casting carried by four strips of saw-steel about 3″ wide that reach to the sides of the tube. The supports for these strips on the tube can be adjusted for focus and the mirror cell can be rotated to bring the image out to the edge of the tube in any position. The plate carrying the eyepieces can be fastened at either side of the tube, toward the pier or away from it as most convenient. A second mirror cell can replace the Newtonian mirror when the telescope is to be used in the Cassegrain form.

The declination axis and housing is shown in Figure A.10.7. The axis proper consists of a cold rolled shafting 3″ in diameter and the housing is a molded casting carrying at its inner end the aligning and double-thrust bearings shown in the figure, and at its outer end a simple aligning bearing. These bearings are SKF and the tube moves very freely

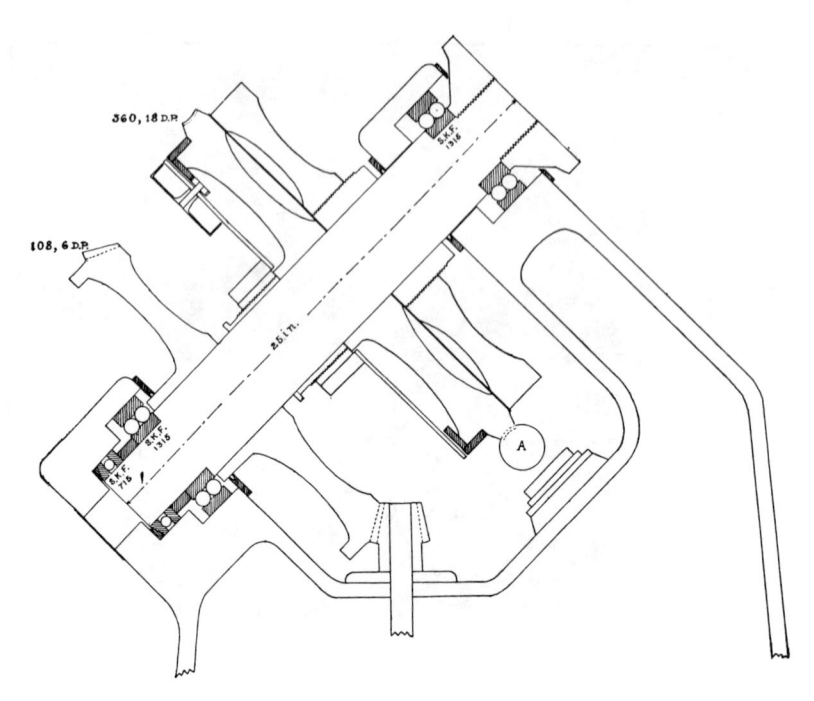

Fig. A.10.9: A section through the polar axis and the top part of the pier, showing the gearing and the various driving wheels and clamp in right ascension.

but without perceptible play. When the tube is balanced, about two ounces applied to the upper end of the tube will start it moving. The main part of the counterweight for the tube, consisting of a large cast iron flange, is fastened to the declination housing, and four cast iron weights threaded on the end of the declination axis proper serve for adjustment. The declination circle consists of a bronze wheel, 2′ in diameter and 2½″ face, rigidly fastened to the declination axis. It is divided into single degrees. In order to be able to read the declination more accurately than this when setting on a star, I introduced the same kind of gear system as used in the 72″ telescope at Victoria, B.C. The declination circle carries a gear wheel of 270 teeth 18 D.P. and this engages by a train of gears to a small drum carried by the counterweight flange. This small drum is a very light aluminum wheel about 7″ in diameter and makes one revolution for a motion of the telescope in declination of ten degrees. Estimation of declination can be made to single minutes of arc when viewing the drum from the position the observer occupies when moving the telescope. The clamp in declination consists of a band of iron on the declination housing which can be clamped

A.10.2. Mounting

Fig. A.10.10: *The right ascension worm gear and the index arms with the drum for accurate readings.*

or freed from the housing by a hand wheel near the eye end of the finder on the tube. Fastened to this band is an arm about 18″ long which has a small arc of a gear wheel cut in its outer end. This engages with the slow motion gear in declination shown in Figure A.10.8. A hand switch carried by the observer can be made to operate the declination motor in either direction for the purpose of fine adjustment. When running at full speed, this motor will move the telescope about one degree in four minutes of time. By giving the switch a quick tap the tube can be moved as small an amount as one-tenth of a second of arc.

The polar axis is made from a piece of 4″ cold rolled shaft and carries at its upper end a flange which bolts it to the declination housing. A cross section through the polar axis is shown in Figure A.10.9. This section shows in order the various wheels carried on the polar axis. Starting from the bottom there is first the fast-motion hour circle wheel, 108 teeth, 6 D.P., which engages a pinion and communicates with the hand wheel shown on the side of the pier in Figure A.10.6. This wheel is rigidly fastened to the axis and a flange on its edge is graduated to indicate the hour angle of the telescope at any time. Above the hour circle wheel is the wheel for the clock drive. This wheel consists of a cast iron center with a bronze ring shrunk on its outer edge, the bronze ring forming the worm gear of the drive, 360 teeth, 18 D.P. It turns freely on the axis but can be clamped to the axis by the single-disk clutch which is shown immediately above it.

Fig. A.10.11: *The electric drive.*

On its lower edge it carries a sidereal circle which can be adjusted at the beginning of the night to read right ascensions directly. The sidereal circle is graduated to four minutes and in order to set it more accurately there is a set of gears which engage with the small drum shown on the index arm in Figure A.10.10. The small drum makes one revolution in ten minutes. The operator standing beside the pier turns the hand wheel until the R.A reading is correct to the nearest 5 seconds and then clamps in the clock drive by the single-disk clutch. The worm screw is indicated at A in Figure A.10.9. It has three degrees of freedom, not shown on the cut, which make it possible to obtain the proper mesh with the worm gear. The worm gear and sidereal circle are also shown in Figure A.10.10. In this plate a temporary axis has been inserted in order to carry the index arms and the small drum for fine reading in right ascension.

A.10.3 The Clock

The astronomical department possessed a gravity driven clock made by Cooke but I was doubtful whether it was large enough to serve. Moreover there is always more or less trouble with a gravity drive in this climate, due

A.10.3. The Clock

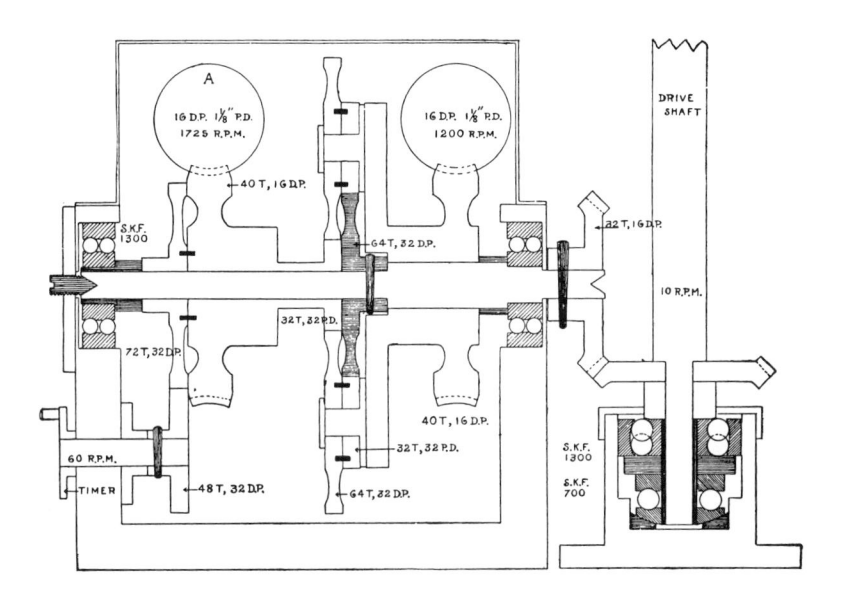

Fig. A.10.12: *A section through the electrical clock gearing.*

to thickening of the oil in cold weather. I decided to build an electrical drive. I wanted to do this as an experiment and the results have been very successful. I chose the system invented by Mr. Gerrish, of the Harvard College Observatory, and described in Bell's book, *The Telescope*. A small motor furnishes the power and the speed of the motor is controlled from a sidereal clock that allows power to be fed to the motor intermittently to keep it running in unison with the clock. Figure A.10.11 gives a general view of the clock and a section through the gearing is shown in Figure A.10.12. The entire cost of the parts was less than $75, the main item being the motors.

The driving motor turns the worm gear A (Figure A.10.12) at 1725 revolutions per minute when running at its normal rate. This gear engages with the timer wheel, as shown in the diagram, which interrupts the current fed to the motor and keeps the timer wheel running 60 revolutions per minute, thus reducing the motor speed to 1600. The current is flowing for about 0.60 to 0.75 of a second and the balance wheel carries on for the rest of the second. A little difficulty was experienced in getting the balance wheel to run smoothly, but by inserting small weights around its edge, as in the balance wheel of a watch, it was finally adjusted so that no vibration could be felt. The electrical circuit and the manner in which it operates are fully described in Dr. Bell's book.

Part B
Optical Testing

Chapter B.1

Curves Found During Figuring[†]

The discussions which follow treat the curves which are found on a mirror at various stages of figuring, and which are frequently mentioned in the special literature of mirror making, in a naturalistic, nonmathematical manner, because there has been a call for such a treatment. They are *not* intended for the worker who has had the good fortune to study analytic geometry. Such workers will find a far superior treatment in their own familiar textbooks. There are, however, to judge by the Editor's mail over a period of years, many workers who either have had no opportunity to pursue higher mathematics or who are not naturally mathematically minded, yet who would welcome some simple attempt at treatment of these curves. It is felt that these same workers will not demand or even desire that such a treatment conform to all of the pedantic peccadillos of the blackboard type of theorist, and they doubtless do not care whether the lingo is that of the corresponding type or not, for they are not preparing to become mathematicians but mirror makers.

As we saw with Figure A.1.15, the curves to be found at different times on a mirror may also be found on a cone cut by a plane. A cone of wood similar to the one shown on that figure may easily be turned-up on the lathe and then dissected to reveal the nature of these curves, and is a nice plaything to possess if the parts are detachably pinned together with dowels.

At some convenient height (near the top, thus leaving room below for other things) a saw-cut is made horizontally. The intersection is a circle. A second cut is taken, beginning anywhere along the side, and parallel to the opposite edge. The intersection is a parabola. A third cut, steeper than the last one, is taken for practical reasons—to favor the wood—some distance below it. This gives a hyperbola. Any cut (less steep than the parabola) between the circle and the parabola will give an ellipse. Any cut (steeper

---

[†]By Albert G. Ingalls.

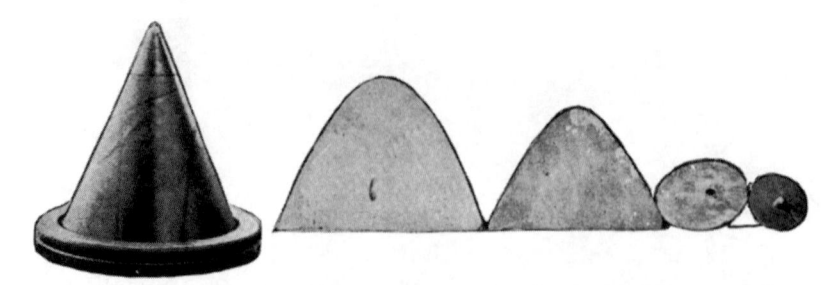

Fig. B.1.1: *A wooden cone at Stellafane, for demonstrating the conic sections. At left is the assembled cone on its base, showing the intersection lines. (The vertical line at the left is an opened joint in the wood and may be ignored.) At right are four parts of the cone—respectively with a hyperbola, the parabola, an ellipse and the circle at intersections.*

than the parabola) on the other side of it, will give a hyperbola.

Thus there are an infinite number of shapes and sizes of ellipses. There are likewise an infinite number of shapes and sizes of hyperbolas similarly derived. Of parabolas we can obtain an infinite number of sizes by cutting an infinite number of cones of increasing sizes, but there is no way to obtain more than one *shape* of parabola for there is only one shape. The same is true of the circle.

It is usually difficult at first to believe this statement regarding the unique shape of the parabola, so let us try to break it. We shift the cutting plane bodily to the left in Figure A.1.15 until it almost reaches the opposite edge of the cone. Manifestly the parabola we now obtain will be long and relatively slender, quite unlike our original parabola in its superficial appearance. Yet it is the same curve and the larger parabola would have a shape analogous to it if its arms were suitably extended. True, the one parabola would be larger in size than the other, but their shapes would be the same. Another attempted dodge would be to make the cut on a cone a mile, or a million miles, high. But, no matter what we do, if the arms of the one curve are suitably extended, or if those of the other are suitably cut off (which is legitimate so long as the shape is preserved), the two will look alike, and if one of the two is then magnified, or if the other is reduced in size without changing its shape, the two will coincide exactly when superimposed. The actual curves on a typical telescope mirror may be regarded as if taken from low squat cones, say 60 feet in diameter and only an inch high, or else as taken from exceedingly limited parts of higher cones near the vertex and magnified—which in either case amounts to the same thing.

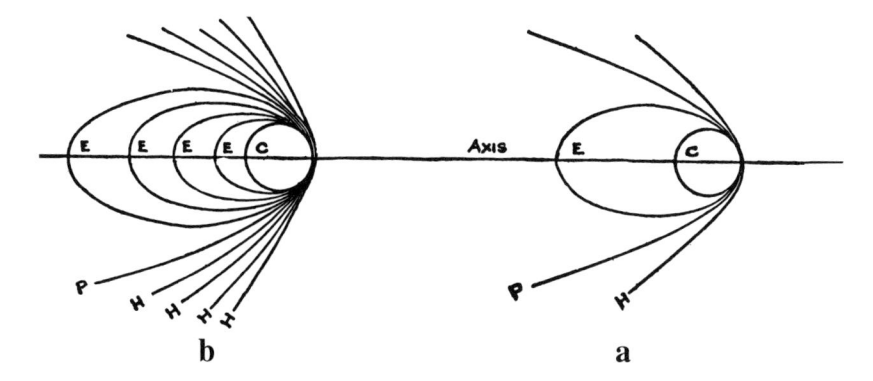

Fig. B.1.2a and b: *Drawings by Russell W. Porter, after the author.*

The parabola is the infinitesimally thin boundary between all the infinite number of ellipses on the one side and the infinite number of hyperbolas on the other.

A set of these curves, if suitably chosen, can be made to nest neatly together like Figure B.1.2a. The point where they touch is the vertex. The same set or "family" of curves may be increased in population, as it were, by slipping in a graduated series of ellipses and hyperbolas, as in Figure B.1.2b.

Starting with the circle, and at the common vertex of the curves, its "arms" may be said to return to one another most quickly of the lot. As we proceed through the ellipses these arms open out wider and wider. This means that the parts near the vertex, the part of the curve dealt with in practice on a mirror, will do the same. On the last of the ellipses, before coming to the parabola, these arms are not far from parallel. On the next curve, the parabola, the closed arms break open and the arms, as they are extended farther and farther, approach more and more closely to a pair of parallel lines, as at Figure B.1.3c. On the first of the hyperbolas the arms are slightly more divergent; they approach, though they never quite reach, two straight lines drawn at a very acute angle, as in Figure B.1.3b and in subsequent hyperbolas they approach straight lines in a constantly widening "V" (see Figure B.1.3a) until the sides of the "V" are themselves finally in a straight line. The radius of all ellipses, that of the parabola, and that of all hyperbolas, is shortest at the vertex and grows longer and longer out on the arms. The arms of a parabola are spread just enough to reflect parallel axial rays to one point (focus) from all points on the paraboloidal mirror. Those of the ellipse are turned in more than just enough to do this and an ellipsoidal mirror of the same family of curves therefore reflects its marginal rays to a point nearer the mirror than the rays from its center

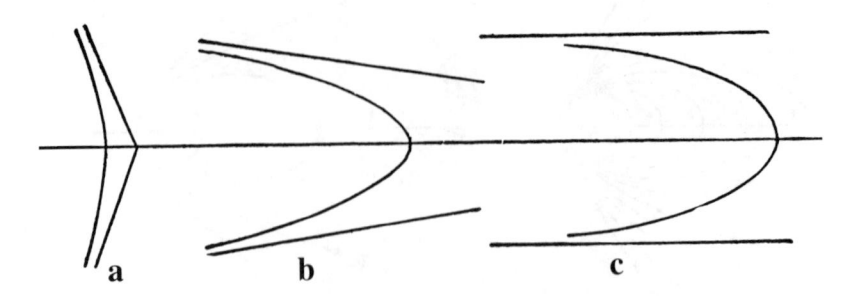

Fig. B.1.3a, b, and c.

(undercorrection; ellipsoid). The arms of the hyperbola are turned in less than just enough to reflect parallel axial rays from all points on the mirror to one point, and here the marginal rays are brought to a point farther away from the mirror than the rays from its center (overcorrection; hyperboloid). If this statement and the one previous seem to be contradicted by the fact the we deepen a mirror's center in parabolizing it, see later discussion.

While these four classes of curves—circle, ellipses, parabola, hyperbolas—have different names and behave in a sensibly or demonstrably different manner if extended widely, they are not extended widely on a telescope mirror, for we use only the flat parts near the vertex, and there the curves all lie very close together. *These curves (except the circle) give the same pattern of shadows in the knife-edge test* because they are all of a kind, varying only in degree, a fact which we have seen in the gradual transition from one to another, as explained above. This variation is especially small near the vertex, as in telescope mirrors. They all give a "parabolic" type of shadow. Why such a shadow, merely because a parabolic mirror will give it, should be termed *the* parabolic shadow is a mystery, for an ellipse will also give the same appearance, and so will a hyperbola. Except in a general way we cannot distinguish with the eye whether the shadow we see on the mirror denotes an ellipse, the parabola or a hyperbola. If the ellipse were an extreme one near the circle (sphere), and if the hyperbola were also extreme in the opposite direction, these might perhaps be distinguishable by their different depths of shade, but not quantitatively. Actually measuring the radii of the zones is therefore the only way to know which curve we have on a mirror. Even if we should learn for one mirror how to judge the approximate depth of shadow for the parabola, another mirror having a longer or shorter focal ratio would throw us off the track, for the shadow of an ellipse on a mirror of short focal ratio may be actually darker than the shadow of a hyperbola on a mirror of long focal ratio. Once more,

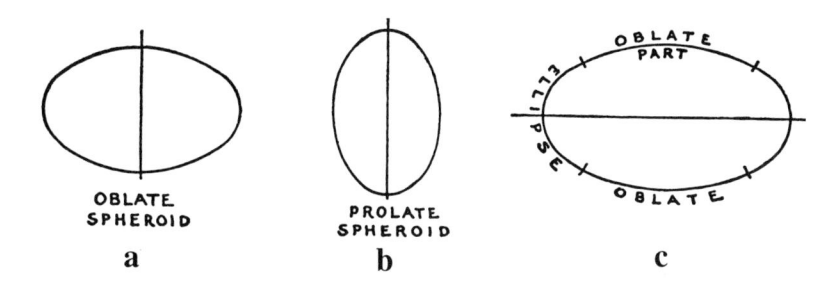

Fig. B.1.4a, b and c.

since there seems to have been some question regarding the matter, it is repeated that the shadows of the ellipses, the parabola and the hyperbolas show no observable difference in *shape, outline and location* on the mirror. They differ only in *degree of illumination,* and this can seldom be judged by the eye.

There is another familiar curve, the oblate spheroid, but this was not included in our family of curves. Mathematicians assert that the oblate curve is not a conic section; this curve is a bastard. However, since our aim is not the study of the conic sections as such, but the study of curves as we find them at times on a mirror, and since the oblate spheroid anyway can be found on a cone suitably cut, and often is found on a mirror, we shall not allow this little quibble to worry us. If an ellipse can be taken off a cone as described earlier and rotated around its minor axis, as at Figure B.1.4a we shall have a distorted sphere, like the Earth or a round sofa cushion. This is an oblate spheroid. Its radius is longer at the center, and grows shorter toward the ends. Thus, it is the opposite of the prolate spheroid, and it gives the opposite kind of shadow. If, instead, the ellipse is rotated around its major axis, as at Figure B.1.4b we shall have a sphere distorted in the opposite manner, like a football. This is a prolate spheroid. The curve usually designated in mirror making merely as an "ellipse" or "ellipsoid" is the same as the prolate spheroid. The oblate spheroid is also derivable from an ellipse, and the two mirror curves, indicated on one ellipse, are shown in Figure B.1.4c. There is no equivalent term to designate the simple two-dimensional curves corresponding to the oblate spheroid, and in general, the inconsistent, unsystematic, slovenly nomenclature for the whole mess of curves, both two-dimensional and three-dimensional, is perhaps as serious an obstacle to acquiring an understanding of them as their own intrinsic differences. To be consistent we should call the sphere a circloid. It would also help if the suffix "oid" were to mean either the solid figure obtained by rotating a two-dimensional figure, or a deformed solid figure (*e.g.*, spheroid), but not both, as it actually does. However, after a time the various significances of the loose nomenclature "soak in," despite the initial difficulty, but they must be learned arbitrarily if learned at all.

After reading all of these attempted elucidations and pondering them well, the worker may turn up with the following question: "If there is only one shape of parabola, and if a mirror is parabolized to a 75 or 80 per cent or some other fractional correction, what shape is it then?" Such a mirror is not a paraboloid but sensibly an ellipsoid!

A point in connection with the curves of mirrors that has evidently kept some well-meaning mathematicians lying awake nights worrying about the blunders which the poor benighted mirror maker may commit is contained in the following typical conversation:

"How," asks the mathematician, "dare you mirror makers tells us geometers, that your method of parabolizing is to *deepen* the central parts of a sphere? As your own sketches show, and as any geometer knows, the parabola is *back* of the sphere, hence you must not deepen your sphere but make it shallower, in order to alter it to a paraboloid." "Well," replies the practical mirror maker, "I am no geometer, but I can make good parabolas, and that is exactly how I do it." In order if possible to alleviate the headaches which this scandal appears to have caused some of the pure or arm-chair mathematicians (few of whom, for some reason, ever actually make a mirror) it may be revealed that the bare conventional statement that we convert a sphere into a paraboloid by deepening the center of that sphere conceals another truth, and is thus in a sense a loose manner of speaking. As John H. Hindle patly explains it, we "do not deepen *that* sphere into *that* paraboloid." Refer to Figure B.1.5a, where S_1 is the sphere before we parabolize and R_1 is its radius. R_2 is the radius of a new and imaginary sphere of reference, S_2. (Of course, we never actually make this sphere on the mirror.) The parabola fits between the two, touching the new sphere at the center. *It is the parabola which belongs to a smaller family.* In this process the radius of the mirror, and hence its focal length, is very slightly shortened.

"Ah, I see it now," says the mathematician, "but why in the name of common sense don't you simply wear away the outsides of your sphere, instead of the central parts, and thus obtain the paraboloid which belongs with its own family, avoiding the complication of new and imaginary reference spheres?" Also, the geometer asserts, this method would be more in accord with the beautiful aesthetics of geometry. This would be the method in Figure B.1.5b, where R_1 is the radius of the initial spherical surface. With a lap trimmed just the opposite of the conventional parabolizing lap the outside zones would be worn down more than the center. When the

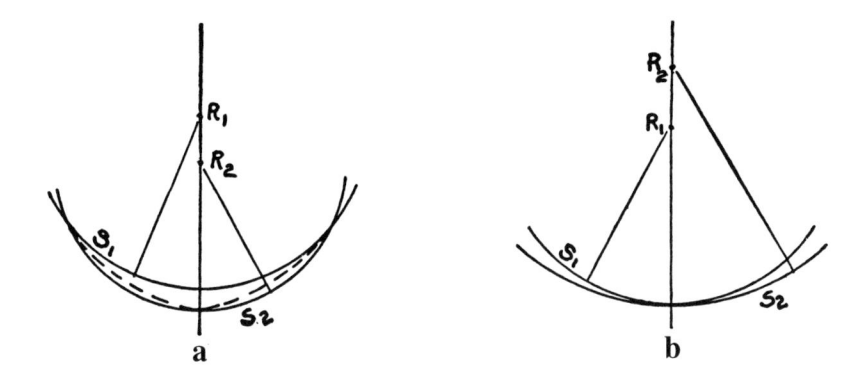

Fig. B.1.5a and b.

knife-edge test shows that the curve has reached the parabola a new sphere of reference, S_2, may in imagination be drawn in, touching that parabola at the center and at both edges. The radius of the mirror will now have lengthened slightly, as at R_2, Figure B.1.5b, instead of shortening as in Figure B.1.5a. (In both sketches the radius, R_1, has the same length.) This method, while more direct and logical, is seldom used, because of the practical difficulty of producing the desired curve right up to the edge without turning the edge.

A third method involves parts of the two just described. The parabola is conceived as lying partly within and partly without the sphere of reference, and is obtained by local abrasion, some from the outer edge and some from the center. Commenting on this last statement, John H. Hindle says, "If one is parabolizing a large, heavy mirror which must be supported face upward, it is practically advantageous to use a combination method, deepening the center by using a part-sized trimmed polisher with moderately long stroke, and reducing the outer zones by means of a full-sized polisher trimmed on the outer edge, again with a moderately long stroke. This produces a perfectly blended surface free from rings, which are usually due to short stroke."

The last-named method involves the removal of much less glass than the other two methods (which remove equal amounts) but is even more difficult than the method shown in Figure B.1.5b; in fact, this is a fine method for all except advanced workers to leave alone. In any case, in parabolizing a mirror our chief concern is not the small labor expended on removing the trifling amounts of glass involved, or even the aesthetics of geometry, but the most practical method of doing the work and doing it well. Ritchey says, "Parabolizing is done by shortening the radii of curvature of all the *inner* zones of a mirror, leaving the outermost zone unchanged. This is a far better and easier method in practice than to leave the central parts unchanged, and to lengthen the radii of curvature of all the outer zones."

The discussion of the same methods of parabolizing, developed in less elementary manner, may be seen in the *Journal of the Royal Astronomical Society of Canada,* January, 1930, pages 20–24.

Once a mirror is finished we may, of course, cease to worry about theoretical reference spheres and, if we wish, we may regard the mirror, under changing temperature, as changing in the same family of curves. Doubtless, however, in different mirrors these changes occur in part near the edge and in part on the central points; perhaps both in the same direction but at different rates; perhaps in opposite directions in the case of some mirrors, pivoting, as it were, on one zone. A familiar case is when a disk is taken off the lap and the radii of its zones will have lengthened in radius but its edge will have lengthened in radius still more. Not all disks behave in this manner.

Chapter B.2

Where Is The Crest Of The Doughnut?[†]

B.2.1 A Study in Shadows

The answer to this question is, "Anywhere, depending upon the position of the knife-edge between the inside and outside centers of curvature." Therefore we shall revise the question and ask, "Where is the crest when the knife-edge is *exactly halfway between* the center of curvature of the inside zone and the center of curvature of the outside zone?" We shall call this the "halfway zone" and the answer is, "70% of the mirror's radius from the center of the mirror." (70.71%, to be exact, this being easily calculated by means of the familiar formula r^2/R; or $r^2/2R$, as you prefer.)

Fix this "70%" firmly in your mind, for the same percentage (the proof this time involves complicated mathematics) is used to locate the points of greatest apparent slope in the central depression. These points are important because they, in turn, locate the boundaries of the three paraboloidal shadows.

Let us inspect these three shadows for the halfway zone. Referring to Figure B.2.1:

- Crests A and E are located 70% of the mirror's radius from the center of the mirror.

- Shadow boundaries B and D are located at less than 70% of the distance AC or EC from the center of the mirror.

- Shadow 1 (on the drawing) advances toward the center from the left-hand edge of the mirror, over crest A to the boundary B.

[†]By A.W. Everest.

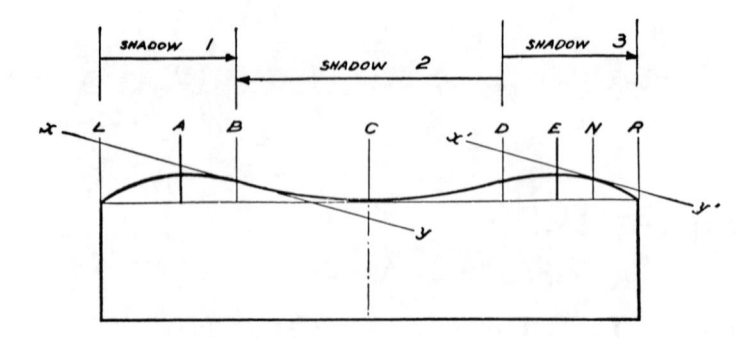

Fig. B.2.1: *(Drawings by the author.)*

- Shadow 2 starts at boundary D and advances toward the left, across the center of the mirror to B, reaching that point at the same time as shadow 1.

- Shadow 3 starts from D at the same time as shadow 2, advances to the right over crest E, and thence "down" off the right-hand edge of the mirror.

- Shadows 2 and 3 start at the time shadow 1 has reached a point between L and A having the same apparent slope as that at D.

- Shadows 1 and 2 meet and disappear at point B when shadow 3 has reached a point N between E and R having the same apparent slope $x'y'$ as the slope xy at B.

Now let us inspect the apparent cross section and the three shadows for the "outside zone" (when the knife-edge is at the outside center of curvature).

Referring to Figure B.2.2:

- Crests A and E are now at the edges of the mirror.

- Shadow boundaries B and D are, as before, less than 70% of AC and EC from the center of the mirror.

- Each shadow travels in the same direction it traveled in the halfway zone, but the timing is different.

- Shadows 2 and 3 start first, in opposite directions from D.

B.2.1. A Study in Shadows

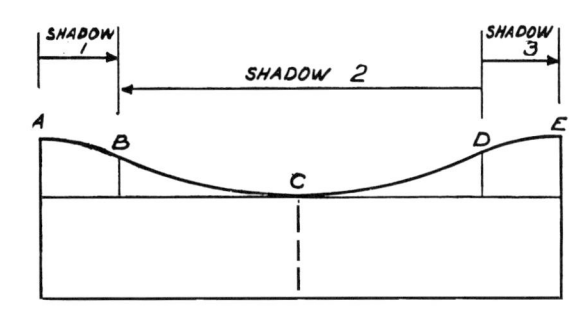

Fig. B.2.2.

- Shadow 3 is just disappearing off the right-hand edge of the mirror when shadow 2 has reached the center. At this instant also, shadow 1 is just coming in from the left-hand edge of the mirror.

- Shadows 1 and 2 then continue until they meet and disappear at B.

In making practical application of the phenomena just described to the testing of mirrors we are concerned only with shadows 1 and 3. To test the halfway zone, or any other intermediate zone, we place the knife-edge in the proper position and hang a piece of wood about $3/16$-inch square across the horizontal diameter of the mirror, with two pieces of pin driven in at the points where the crest should be. If the last wisps of light at the edges of shadows 1 and 3 reach their respective points at exactly the same instant, the zone tests true. In testing the outside zone, we first watch shadow 3 and note the point at which the last wisps of light are just passing off the right-hand edge of the mirror. Then, if the completely darkened edge of shadow 1 is just entering the mirror at the left, this zone also tests true. It is interesting to note that at the time the readings are taken on both of the above tests, the edge of shadow 2 will have just reached the center of the mirror.

Chapter B.3

Accuracy in Parabolizing a Mirror[†]

The question of how accurate the figure should be before a job of parabolizing is considered finished may be approached in this manner: Imagine a sphere and a paraboloid with their central parts and optical axes coinciding, both with the same equivalent focal length F and diameter D. The two surfaces are spaced very nearly at a distance $e = D^4/1024F^3$ apart at the edge. This distance varies widely with the dimensions of the mirror. For an $f/8$ mirror with $D = 6$ inches and $F = 48$ inches,

$$e = \frac{6 \times 6 \times 6 \times 6}{1024 \times 48 \times 48 \times 48} = 11.4 \text{ millionths of an inch.}$$

Theory and experience agree that a mirror will give practically perfect definition if it is figured so as to reduce this distance to within a quarter of the wavelength of light, that is to about 5 millionths of an inch. Taking this as a working limit, our 6-inch mirror must be figured to within $5/11.4$ or 44% of a parabolic figure. That is, it must be figured somewhere between 56% and 144% (preferably below 100% to allow for thermal effects) of the theoretical amount given by the formula r^2/R or $r^2/2R$ depending on whether the pinhole is fixed, or moves along with the knife-edge in testing at the center of curvature. Table B.3.1 gives the corresponding tolerance for various sizes of mirrors, including the example just cited.

As this table indicates, the knife-edge position must be read with great accuracy for very short-focus mirrors, while more rough-and-ready measurements will do with those of long focus. Those having a tolerance of over 100% may be finished spherical, since the sphere and paraboloid are practically coincident in these dimensions.

Nothing in these remarks should be interpreted to mean that long focus mirrors may be measured carelessly. What is meant is that precision mea-

---
[†]By Franklin B. Wright, M.S.

Table B.3.1
Greatest Allowable Deviation from a Perfect Figure

Ratio F/D	$D = 4"$	6"	8"	10"	16"	24"	36"
4	8.2%	5.5%	4.1%	3.3%	2.1%	1.4%	0.91%
5	16	11	8.0	6.4	4.0	2.7	1.8
6	28	18	14	11	6.9	4.6	3.1
7	44	29	22	18	11	7.3	4.9
8	66	44	33	26	16	11	7.3
10	128	85	64	51	32	21	14
12	222	148	111	89	55	37	25

suring devices such as micrometer screws or vernier scales may be dispensed with except with short-focus mirrors. All knife-edge measurement needs to be done with the greatest care and freedom from personal bias. Otherwise no accurate analysis of results is possible.

Long focus mirrors with a perfectly regular appearing figure require measurement of knife-edge position in two zones only, to determine the figure. But as a general rule mirrors with any evidence of zonal irregularities, and all short-focus mirrors, should be measured in many zones in order to make sure of the figure. A convenient zonal testing arrangement for the average mirror is to have the knife-edge stand slide in guides with a permanent scale fastened along one side. All readings are then taken with reference to this scale.

After obtaining a complete set of observations in all zones selected for testing, it is necessary to adjust them for the arbitrary position of the scale. This may be done by subtracting the average of the readings obtained for the zone nearest the center of the mirror; but if this zone appears irregular, by subtracting enough to make the average for a zone a little farther out, check with the knife-edge formula. This is illustrated by Table B.3.2, which shows the final readings on the writer's $6\frac{1}{2}$-inch reflector, compared with the theoretical readings from the formula r^2/R. Upper and lower limits of allowable deviation according to Table B.3.1 are also calculated. (The readings were actually made to tenths of a millimeter, which accounts for the frequent repetition of the same figure in the decimal place when expressed in hundredths of inches.)

Although this table contains all the necessary information, one can get a much better idea of the figure by plotting all the adjusted readings on a chart like Figure B.3.1, with curves for the theoretical readings drawn as a background. The scattering of the spots gives a good picture of how reliable the actual readings are, and a curve representing average results is easily sketched in.

In case the readings for any of the zones lie outside of the limiting curves, it is best to refigure the mirror unless adjacent zones are well within limits

Table B.3.2
Readings in 100ths inches

Center of zone	$r = 1''$	$1\,5/8''$	$2\,3/8''$	$2\,7/8''$	$3\,1/8''$
	8.6	11.8	13.3	14.4	23.1
Original	7.6	12.2	13.3	14.4	22.1
Readings	10.1	12.2	12.2	14.7	21.8
	8.3	11.5	12.6	13.6	21.8
	10.8	12.9	12.6	13.6	22.1
Total	45.4	60.6	64.0	70.7	110.9
Average	9.1	12.1	12.8	14.1	22.2
	−0.5	+2.7	+4.2	+5.3	+14.0
Adjusted	−1.5	+3.1	+4.2	+5.3	+13.0
Readings	+1.0	+3.1	+3.1	+5.6	+12.7
(= above −9.1)	−0.8	+2.4	+3.5	+4.5	+12.7
	+1.7	+3.8	+3.5	+4.5	+13.0
Total	−0.1	+15.1	+18.5	+25.2	+65.4
Average	0	+3.0	+3.7	+5.0	+13.1
Paraboloid	0.8	2.1	4.5	6.5	7.8
30% Correction	0.2	0.6	1.4	2.0	2.3
170% Correction	1.4	3.6	7.7	11.0	13.2

and the excess is not too great. Errors of curvature near the edge affect the surface more seriously than errors of similar amount near the center.

Turning again to the $6\,1/2$-inch mirror, the extreme readings just about fall within the required limits, indicating that the mirror is well-figured. Although it is irregular, and the slope of the curve shows that the parts between $r=1\,1/2$ and $2\,7/8$ inches are only about 30% corrected, the center and edge are overcorrected so that the curve as a whole approaches that of a 100% paraboloid.

Probably nine times out of ten the graph of the knife-edge readings will give all the information desired. However, it is not difficult to go a step further and reconstruct the actual surface compared with a paraboloid. Divide the graph of knife-edge readings evenly into narrow zones. Let a represent the width of a zone, r the distance from center of mirror to center of the zone, and R the radius of curvature. Let b represent the distance between the centers of curvature of paraboloid and actual surface, respectively. This is half the distance between the two curves on the knife-edge reading chart at the center of the zone (if lamp and knife-edge move together b would be the whole distance between the two curves). The actual surface is higher (in relation to the paraboloid) at the outside edge than at the inside edge of a given zone by $a \times r \times b/R^2$. These results are then summed up zone by zone, starting at the center to find how far apart the surfaces are at any point on the mirror. All this is easier to do than to explain, and the calculations are given in Table B.3.3 for the $6\,1/2$-inch mirror for which $R = 125$ inches.

Chapter B.3. Accuracy in Parabolizing a Mirror

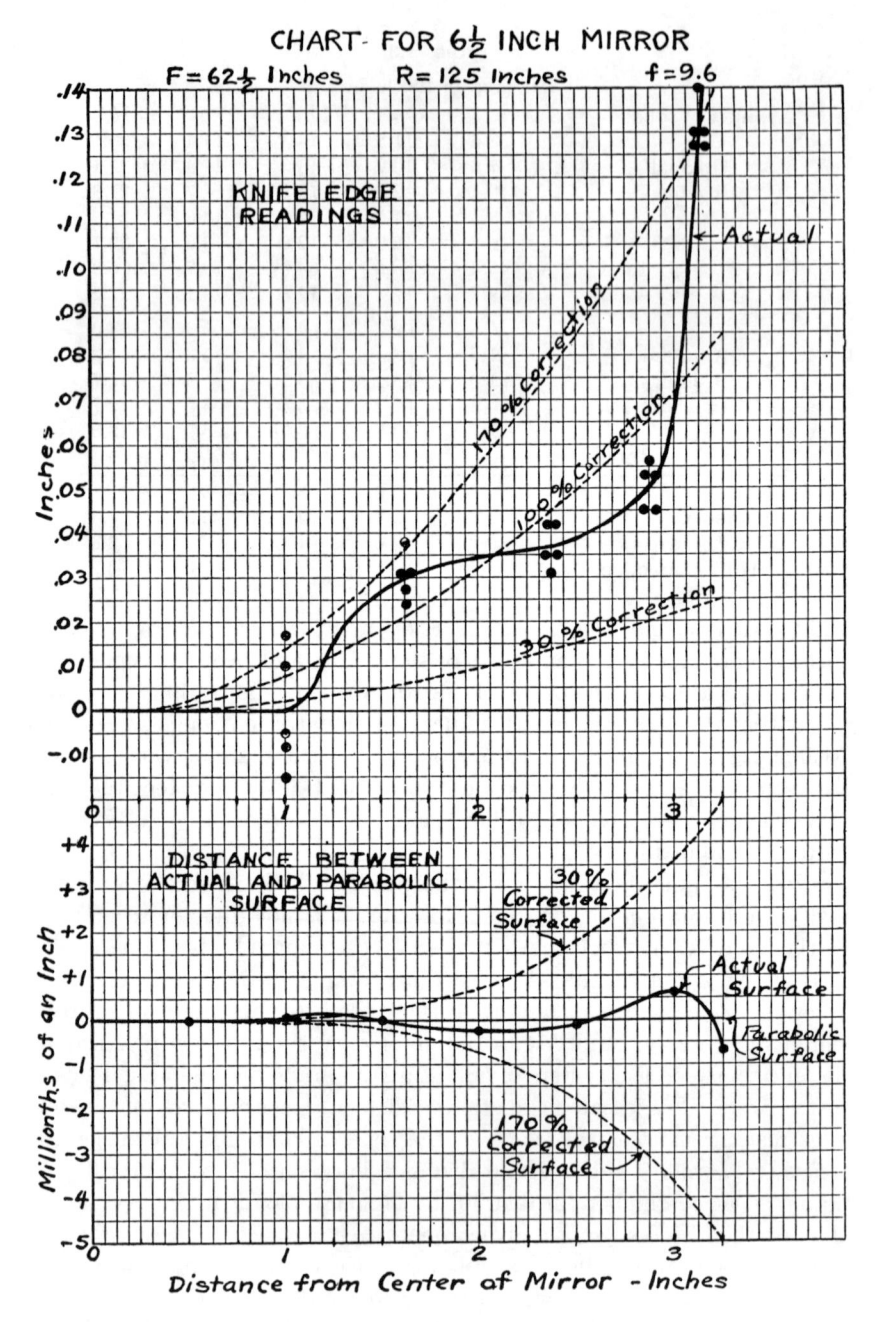

Fig. B.3.1: Chart for 6½-inch mirror. (Diagram by the author.)

Table B.3.3

Width of zone, a (in)	0.50	0.50	0.50	0.50	0.50	0.25
Center of zone, r (in)	0.75	1.25	1.75	2.25	2.75	3.13
Fixed-source knife-edge curves at zone center ...						
... Teoretical paraboloid (in)	0.005	0.013	0.025	0.041	0.060	0.078
... Measured (in)	0	0.015	0.032	0.036	0.045	0.131
$b = 1/2$ difference* (in)	+0.002	−0.001	−0.004	+0.002	+0.008	−0.026
**$a \times r \times b \div (125 \times 125)$	+0.05	−0.04	−0.22	+0.14	+0.70	−1.30
**Surface (at zone outer edge)	+0.05	+0.01	−0.21	−0.07	+0.63	−0.67

* A conversion to true center of curvature, which was doubled by use of a fixed source
** These two lines are in millionths of an inch, other lines are in inches.

The last line is plotted on the chart, and represents to scale exactly the way the mirror would appear if tested at the principal focus with an optical flat. If any mirror is figured well enough so that this curve varies through a range of less than 2 millionths of an inch from the lowest to the highest point, it should meet the most rigid specifications for fine optical surfaces. Even with 5 millionths of an inch the mirror will still be a good one, *provided* the curve moves gradually to this error at the extreme edge, as in the case of the 30% and 170% surfaces shown by the dotted lines, because regular surfaces like these permit most of the error to be "focused out" with the eyepiece.

All of this may seem like a lot of work, but it seems that after one has spent many months of patient work on a mirror it is worthwhile to spend two or three evenings analyzing the results of one's labors.

It has been stated correctly by many writers that the greatest difference in optical path between rays striking various parts of an objective should not exceed $1/4$ of the wavelength of light. In the case of a mirror this corresponds to a surface error of $1/8$ of a wavelength, since the light travels through the space in front of the mirror both before and after reflection.

If it were a fact that a telescope would be used with eyepiece or photographic plate in focus for the parabolic surface which has the same curvature at the center as the actual mirror surface, then the $1/4$ wavelength surface error used in computing Table B.3.1 would be too liberal. However, this is not the case, because one naturally finds the best average focus for the objective as a whole, which corresponds to a paraboloid of shorter or longer focus than the one used for computing Table B.3.1, depending on whether the mirror is undercorrected or overcorrected. Assuming that it tests within the limits of Table B.3.1, and that its figure is regular (*i.e.*, spherical, ellipsoidal or hyperboloidal) the greatest separation from the nearest paraboloid will not exceed $1/16$ of a wavelength, and the greatest actual difference in light path will not exceed twice this, or $1/8$ of a wavelength.

A useful variation of the method of adjusting actual knife-edge readings for the arbitrary position of the scale (see last sentence in second paragraph under Table B.3.1) is to subtract an amount from each reading which will make the average of the adjusted readings equal the average of the values of r^2/R for all the zones measured. This method is for making a comparison of the actual surface with the paraboloid of closest fit. It is suggested that the reader carry out this method completely for the 170% corrected surface whose knife-edge readings are given in the last line of Table B.3.2, in order to verify approximately the writer's statement about the $1/16$ of a wavelength surface error.

B.3.1 Editor's Note

As discussed in Section A.2.15, based on the Rayleigh limit, Wright's tolerances are too exacting by a factor of 2, and a 6-inch mirror is at the Rayleigh limit if left spherical with a focal length of about 50 inches. Wright says, "In the preceding chapter, written in 1933, I based my limits on smooth curves of surface viewed from the position of average focus, then made the limits twice as strict as this calls for to take care of imperfect measurements or imperfect focusing, and from just plain conservatism."

Fig. B.3.2: Franklin B. Wright

Franklin B. Wright was born in Philadelphia in 1891, educated in Lansdowne, PA., public schools and graduated at the University of Pennsylvania with the degree of B.S. in Electrical Engineering in '13. He spent 3 years with the New York Telephone Co., then did graduate work in physics at the University of Pennsylvania, followed by a short time in the Army in '17. He spent 3 years with the Philadelphia Electric Company in distribution engineering work. He was an instructor in physics at the University of Pennsylvania '20–'23 and was granted the degree of Master of Science in Physics in '23. He was in the General Plant Department of the Pacific Telephone and Telegraph Co., in San Francisco, from '23 to '51 and retired in '51 as a Plant Staff Supervisor who specialized more or less in the statistical measurement of plant results.

He became greatly interested in an article in *Scientific American* about '28, made several mirrors and a few telescopes and, as he states it, "has 'played around' with the optical and mathematical problems of telescope design and methods of testing off and on ever since." He is also an enthusiastic advocate of the proposed adoption of "The World Calendar" and contributed an article on this subject to the *Journal of Calendar Reform* in '38 December.

B.3.2 Note Added in 1996

Wright's method is showing its age. By referencing the curves to the hard-to-measure central zone, Wright is forcing the entire analysis to depend on its least reliable measurement. A much better tolerance envelope was published by A. Millies-Lacroix in the February 1976 issue of *Sky & Telescope*. In this article, called "A Graphical Approach to the Foucault Test," Millies-Lacroix adapts a transverse aberration technique to yield fast and easy reduction. The Millies-Lacroix tolerance curve widens toward the center, rather than toward the outside. The measurements in the Millies-Lacroix method are for fixed-source testers

$$\frac{r^2}{R} \pm \frac{2\rho R}{r},$$

or for moving-source testers,

$$\frac{r^2}{2R} \pm \frac{\rho R}{r}.$$

The radius of the Airy disk is $\rho = 1.22\lambda F$, where F is the focal ratio or f-number. The radius of the zone is r.

In the Millies-Lacroix technique, you take an overlay of measurement points and slide them up and down until they fit the narrow outside tolerances. You are no longer tied to a central zone measurement of questionable accuracy. The Millies-Lacroix measurements can be reduced to a wavefront by using the technique described by Robert C. Follett in *Telescope Making #33*.

Which method is more accurate? *Both can be accurate*, particularly if you use Mr. Wright's $1/8$ wavelength tolerances of Table B.3.1. The mirror of Figure B.3.1 satisfies Wright's tolerance curve, but pokes out of the Millies-Lacroix tolerance envelope, exhibiting a slightly rolled edge. I am inclined to give the nod to a test that enforces discipline on the edge rather than the middle (i.e., to Millies-Lacroix), but in Wright's defense, I will admit that diffraction calculations revealed little wrong with the example mirror.

Comparisons aside, the Millies-Lacroix technique will probably result in fewer headaches, simply because it allows you to add or subtract a constant to the measurements until they fit (a move equivalent to changing focus). With Wright's method, you are often scurrying around trying to remove glass that has no more reality than the last uncertain measurement of the central zone.

There is one more sharp lesson we can take from Fig. B.3.1. Notice how the raw measurements swing up sharply toward the edge in the top half of the figure, but when we calculate the shape of the surface (lower half), they curl down. This behavior points out that the knife-edge measurements are not the same as the surface, but only related to its slope. You wouldn't believe how much trouble such measurements have caused beginning TMs, who try desperately to cure the "turned-up edge" detected in the raw measurements by corrective methods that only make the problem worse. (This lesson was first pointed out to me by Mr. Follett.) *H.R.S.*

Chapter B.4

The Ronchi Test for Mirrors[†]

It seems almost undisputed that the greatest single advance in mirror making came with the announcement in 1859 of Foucault's test, making it possible for the first time really to see the figure of a mirror. Unfortunately, as is almost equally undisputed, a great deal of skill is necessary in order to decide by that test when we have very precisely what we want. Only an artist, who is "born and not made," can say with certainty whether a mirror has a *perfectly* regular curve and no turned edge.

With the more recent popularity of the Cassegrainian and Gregorian telescopes having primary mirrors of extremely short focal ratio, this difficulty has become many times greater. The edge of a short-focus paraboloid is so deep in the parabolic shadows that decision is very difficult, and the matter is further complicated by diffraction effects. A customary method of checking upon the regularity of the curve is to make a number of stops and to test five or six zones, which is often disconcerting to the eyesight, besides requiring much time and some genius. How much better it would be if one could at a glance *see* the contour of the reflector in its entirety. Very well, we have it—the Ronchi test.

The Ronchi test for mirrors is not new, and in its simplest form it has probably been in use for years. Although it has been characterized as a modification of Foucault's test, there really is very little resemblance between the two and, when performed in this way, all of the advantages peculiar to the test are lost. The Ronchi effect may, however, be produced with the Foucault test equipment, in the following manner: Move the knife-edge toward the mirror a distance beyond the center of curvature of the central zone. This will cause its shadow to fall entirely upon one side of the mirror, as in shadowgraph B, Figure A.1.10. Irregularities will now be shown by a distortion at the edge of the shadow, the shadow apparently bend-

---

[†]By Alan R. Kirkham, Amateur Telescope Makers and Astronomers of Tacoma.

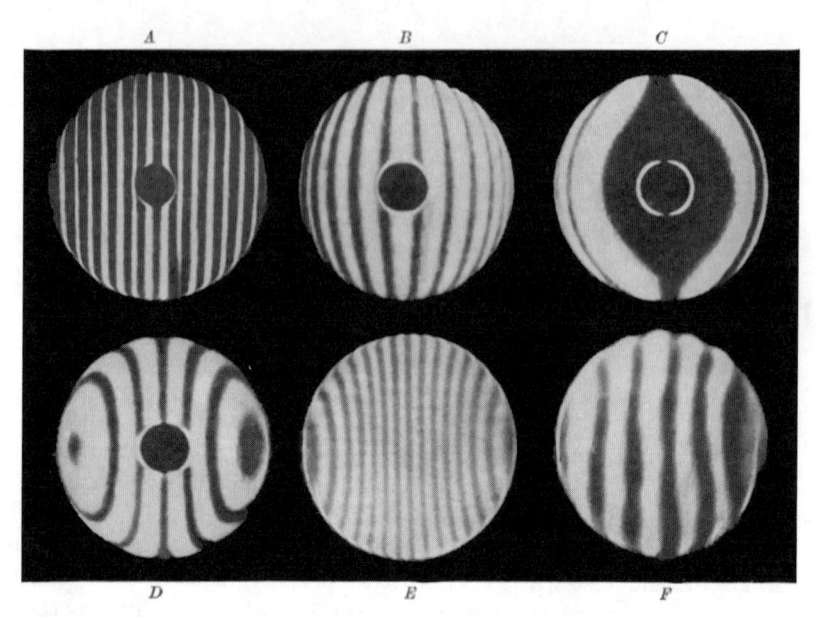

Fig. B.4.1: *Top row. Left: A sphere exhibits straight bands (this particular one also has a turned-down edge). Center: A paraboloid exhibits bands having a parabolical sweep. Right: Same mirror as at center, but with grating closer to focus. Lower row. Left: Outside of focus the bands curve opposite to those inside of focus (compare with upper, right). Center: Oblate spheroid inside of focus. Again note that the bands are the shape of the actual curve on the mirror—in this case, for an oblate spheroid, having a shorter radius at the ends. Right: The unbeautiful result of an attempt to make a Ronchigram by means of a slit and without a lens.*

ing out toward the center into raised areas, and drawing in from low ones. A sphere is the only figure which shows the knife-edge shadow perfectly straight clear across the mirror in these circumstances. If the mirror is an ellipsoid, paraboloid or hyperboloid, the knife-edge shadow will be concave away from the center, that being an area depressed below the sphere. If the mirror is an oblate spheroid, the opposite will be true.

If, in place of the knife-edge, a very narrow obstruction is now substituted, it may not cover the entire mirror at once but, instead, will cause a dark band to appear, with both edges distorted in relation to the figure—somewhat as in *C*, Figure B.4.1. Furthermore, if a great number of very fine obstructions are used, we shall have in effect a number of knife-edges coming both ways at one time, and the mirror will show dark bands similar in appearance (but not, of course, in actual origin) to those produced by interference, in testing a flat against its mate. A still further improvement will come with elongating the illuminating pinhole into a slit, parallel with

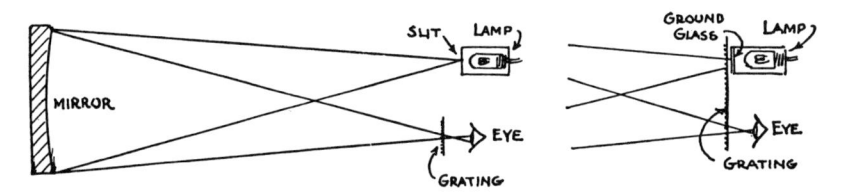

Fig. B.4.2: a (Left) *The original setup as devised by Ronchi.* b (Right)*The Anderson-Porter modification of the original Ronchi setup. The same grating extends over both bundles of rays. Drawings by Russell W. Porter, after the author.*

the obstructing lines. This is the original version of the Ronchi test, Figure B.4.2a.[1]

In practice, a grating is used having 40 to 200 lines to the inch, such as a piece of halftone engraver's grating, or a photographic reproduction of one. Most of the accompanying photographs were made with silk bolting cloth having 135 threads to the inch, and this is very satisfactory. Pieces may be obtained, usually gratis, from flour mills, being fragments of larger cloths rendered useless to the miller because of localized holes, or from dealers in miller's supplies. The threads running horizontally do not show because the slit is longer than the interval between them. The purpose of the gratings of silk cloth is to avoid diffused light which tends to blind the eye and destroy the knife-sharp edges of the shadows. Most photographic reproductions of engraver's gratings with which I have had experience have been objectionable for another reason: they have a strong diffraction effect which may cause prismatic colors and nearly always results in showing two or three mirrors instead of one. The emulsion between the lines on the grating is never perfectly transparent and thus the mirror is always seen as if surrounded by a heavy fog. Those who have access to a lathe may make a very fine grating by threading the edge of a piece of brass with 60 to 100 threads to the inch, winding on it wire (about No. 36, B & S) under tension, soldering at top and bottom and removing the wire from one side, as shown in Figure B.4.3a.

The more lines to the inch in the grating, the more bands will appear on the mirror a given distance inside focus; that, however, being the only effect due to this cause. The farther the grating is moved from the focus, the less will be the sensitivity of the test, hence it is advisable to use a fairly large number of lines per inch in the grating, thus providing from six to twelve bands on the mirror when the grating is still near enough to the

---

[1]Described 1925, in *La Prova dei Sistema Ottica,* by Vasco Ronchi, Bologna. See also *Revue d'Optique* 1926. Ronchi is pronounced Ron′kee.—*Ed.*

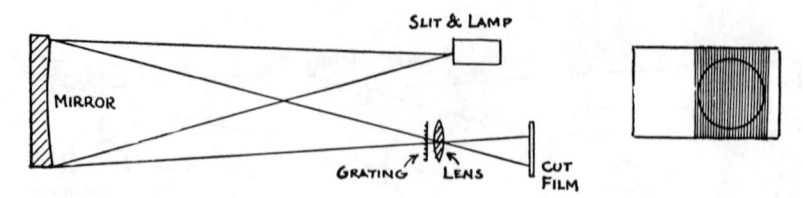

Fig. B.4.3: a (Right) *A grating made of fine wires.* **b** (Left) *The setup for making Ronchigrams.*

focus to get sharp distortion. In order to avoid confusion, it is best to test with the grating always inside of the focus.

The customary and a perfectly satisfactory light source is a slit, parallel with the lines in the grating, Figure B.4.2a. It is difficult to overdo the matter of narrowness of the slit. A good way of producing a suitably narrow one is to press a razor barely through a piece of heavy lead foil or brass shim stock, making sure that the slit is not pried apart in the operation. Another arrangement of the light source, devised by Anderson and Porter[2] allows the grating to extend over the lamp. See Figure B.4.2b. This, in effect, is a multiple slit. It has the advantages that the worker may move his head a bit without causing the mirror to vanish, and that the lines over the lamp must necessarily always be parallel with those before the eye. The eye sees light from only one of the grating slits at a time, and when the eye is moved, it passes from one to another without consciousness of the change. The camera does not make this distinction, and a slit is necessary if the test is to be photographed.

In the accompanying series of photographs, Figure B.4.1, A is a sphere with a turned-down edge. B is a parabola of $f/3$ focal ratio, C being the same mirror with the grating nearer to the focus; note how much more clearly the figure is seen. D is the same mirror with the grating outside focus where the bands curve in an exactly opposite manner. Compare this with E, an oblate spheroid inside focus. A hyperbola looks somewhat like a parabola, the main difference being a greater curving of the bands. An ellipsoid would show less curvature. A mirror having a long focal ratio of $f/7$ or $f/8$ should show scarcely any curvature at all.

While the bands should continue regularly, clear off the edge of the mirror, one must not condemn a mirror for showing as much turned-down edge as A or E, covering less than one-fortieth of an inch of the rim, for this would be altogether invisible by Foucault's test. The remainder of the band should, however, be perfectly regular, and can be made so by perseverance and a strong right arm.

---

[2]See *Astrophysical Journal,* October 1929.—*Ed.*

The Ronchi test may be used to measure very simply the overall correction of a mirror, in the following manner. Adjust the grating so that the two bands at the center of the disk are, say, one inch apart, and mark its position. Now move the grating back so that the bands are one inch apart at the rim. This move should amount to r^2/R, if the correction is right. Intermediate zones may be measured in the same manner, though it is quite easy to judge by sight when the bands are evenly curved, and the overall correction is therefore all it is really necessary to measure.

It is of interest to move the grating slowly and watch the bands roll across. This, in the case of a short-focus paraboloid, gives an appearance like that of a barrel rolling about, while low or high irregularities will make parts of the bands seem to lag or jump ahead. With a pinhole instead of a slit a parabola will resemble a golf ball (only applies to cloth gratings).

The Ronchi test is very useful in figuring secondary mirrors for compound telescopes by Ritchey's, Hindle's or the direct focal methods (see Chapter B.10). The bands should be straight. The same applies to a flat tested against a spherical mirror, an achromatic objective lens by Ellison's autocollimation test, or a paraboloid tested at its focus by means of a flat.

Gratings made from spiderwebs and other very fine material have been proposed. This is neither desirable nor would its use be possible, since the finer the lines of the grating are made, the narrower the slit must be, and the average worker cannot carry this far, even with wider gratings. Gratings having several hundred lines to the inch have also been suggested, but they too would fail for the above reason. It is safe to say that any amateur who succeeds in making a mirror which shows no defects with a grating having 100 lines to the inch is justified in considering himself an expert mirror maker.

In photographing the Ronchi test some precautions are to be observed. While focograms are usually made with the camera lens removed, this cannot be done with Ronchigrams, for the length of the slit would cause them to be gibbous, F, Figure B.4.1, and their sharpness would be lost. Of course, the trouble could be avoided by using a pinhole, but this destroys the brilliant, sharp contrast between dark and light. The ordinary camera lens is satisfactory for photographing tests on mirrors of focal ratio $f/3$ to $f/4$. Above that, the Ronchigrams become very small. Very best results on short-focus mirrors may be had with a camera lens of some 8 or 9-inch focus; while a mirror of $f/7$ or $f/8$ should be photographed with a lens of at least 15-inch focus. In any case, the lens is placed directly back of and almost touching the grating, Figure B4.3b, and the film a little farther back of that than the focus of the lens. It is best to use Eastman cut film and to work in the dark. One can then see the image being photographed. Incidentally, the lens used need not be of high quality, nor even achromatic. Very splendid photographs may be taken with a ten-cent-store toy reading glass, or other low-grade lens. The time of exposure depends on so many different factors that the best way of judging is to make three test photographs of, say, one, five, and fifteen minutes exposure.

B.4.1 Note Added in 1996

Mr. Kirkham's comment about the Anderson-Porter setup, "the eye sees light from only one of the grating slits at a time" is incorrect. Many identical source slits can be in view at a time, all producing identical, superimposed Ronchi patterns. One difficulty with the Anderson-Porter arrangement that Mr. Kirkham did not mention was the relatively wide "slit" produced by use of the other side of the grating. The visible pattern is the width of the bar convolved with the width of the slit, so the contrast of the Ronchi pattern is aided by having a separate, relatively narrow slit. H.R.S.

Chapter B.5

Hindle's Method in the Knife-edge Test

This method consists in becoming accustomed to the appearance of the shadows, not when the knife-edge is halfway between the center of curvature of the respective outside and inside zones, as is the orthodox method of viewing the paraboloid, but at the center of curvature of either the outside or the inside zone. The essential principles are the same as in the more familiar test but, instead of the "Life Saver" curve which changes general direction four times, the curve is a single one. Workers who are already familiar with the regular test may find this variation interesting and instructive, as the interpretation of shadows from different points of view tends to limber up the mind and give added confidence and familiarity with testing in general. The following description is from *English Mechanics*, November 16, 1928:

"Assuming we have a parabolic mirror under test with this apparatus, what do we see visually? The reply is, it all depends where the knife-edge cuts the cone of rays. The longitudinal distance along the axis of the cone to which the knife-edge can be usefully applied is bounded by the radius of curvature of the center of the mirror, on the inside, and the radius of curvature of the outer edge of the mirror, on the outside. Let us first of all examine the mirror at the focal point of its center. See Figure B.5.1A. The knife-edge is presumed to pass from left to right, and in noting the passage of the shadow, the eye follows the same path as in reading these lines. The hatched portion shows the shape of the shadow as it first impinges on the mirror, and the dotted lines the shape its outline assumes as it progresses across the mirror, being vertical, even if somewhat unstable, in the center. The apparent shape of the surface is that of a solid seen in Figure B.5.1B, the center being nearest the observer, and falling away evenly on all sides.

Chapter B.5. Hindle's Method in the Knife-edge Test

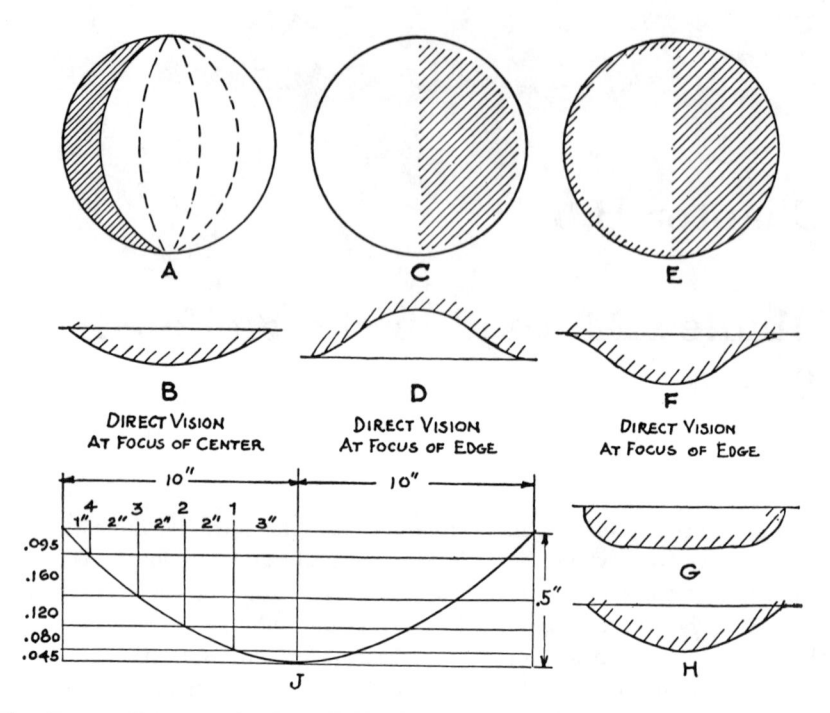

Fig. B.5.1: *Drawings by Russell W. Porter, after John H. Hindle* and English Mechanics.

As we 'fling' the shadow across this projecting convex surface, the slightest irregularities become visible, and any changes of curvature can be instantly seen. The oblique illumination is from the right, and it is impossible to mistake the nature of any irregularity seen.

"When we pass to the focus of the outside edge, we may recognize when the knife-edge is in the precise position in the following manner. A point along the axis is chosen where the shadow forms as in Figure B.5.1C, that is to say, it all but fills the right half of the mirror, being only a little short of the vertical center-line, and not quite touching the right-hand outer edge. The position chosen is a correct one if, with a further minute movement of the knife-edge, the appearance indicated in Figure B.5.1E is obtained; that is to say, the previous shadow expands to entirely fill the right half of the mirror, and simultaneously a narrow and delicate strip of shadow appears around the left-hand rim. This position can be fixed with about the same accuracy as an ordinary zonal test with diaphragms (assuming a normal edge to the mirror free from turn-down). If the knife-edge is too far forward the strip of shadow on the left-hand edge is too wide. If the knife-edge is too far back the strip of shadow does not appear at all before the whole of the right-hand half of the mirror is obscured.

"Bearing in mind that the proper position in which to assume the oblique illumination is, on the side toward which the knife-edge is traveling, the section of the imaginary solid we see is shown at Figure B.5.1D, but it is possible, and indeed, probable, that we may see it inverted, in which case it would appear, not as a depression but as a protuberance, as shown at Figure B.5.1F. I prefer this method of viewing it. The shadows seem to show up the shape of the imaginary solid much better, and irregularities are better estimated. But it is absolutely essential to know that we see the thing inverted.

"Having now ascertained the method of accurately locating the foci of the center and of the edge, we measure the distance between the two points, and this is the well-known r^2/R. It is of course greater than the sum of the zonal differences, but can be made to fit in with the latter, as well as providing an excellent check on their accuracy. If the appearances indicated are obtained with mirrors whose apertures are $f/8$ and upwards the use of diaphragms may be dispensed with. On short-focus mirrors, however, although the total aberration may be right, it may not be correctly distributed from edge to center. Strictly speaking, the shape of the apparent solid at center focus will show this. If the correction is too great in the outer zones we get an appearance like Figure B.5.1G, and if too great in the central zones, more like Figure B.5.1H.

"The apparent solids seen are sketched from a mirror of aperture $f/5$ or rather less, and in consequence the following method is suggested to the beginner of ascertaining what his mirror should look like at central focus (Figure B.5.1B). On a squared paper draw the mirror full size, then calculate r^2/R. The zonal differences are also calculated, enabling us to plot the curve shown in Figure B.5.1J, in which are shown precisely the relations of the various quantities referred. For a 20-inch mirror of 100 inches focal length the figures given below apply.

Table B.5.1

Zone	Radius (inches)	r^2	$r^2/200$	Differences
center	0	0	0.000	
1	3	9	0.045	0.045
2	5	25	0.125	0.080
3	7	49	0.245	0.120
4	9	81	0.405	0.160
edge	10	100	0.500 to center	0.095

"A careful study of these figures and of the diagram will give a clear conception as to what a parabolic mirror really should be. 'Turned edge,' which usually exists between the outer zone and the edge, immediately becomes apparent by measurement. No special attention need be paid to the appearance of the shadows at any other point than the two indicated. There, they have a definite form independent of the relation between diameter and aperture, and can be utilized, not to supplant the zonal test entirely, but to supplement it in a most useful and efficient manner."

Chapter B.6

The Slit Test

This is simply a glorified test with the pinhole. If instead of one pinhole there are two, one above the other, it has been found that the shadows will not be altered. There will, however, be better illumination. This in turn will make feasible a reduction in the size of the individual pinholes and thus it will effect a gain in sensitivity of the test. The above being granted, why not carry the same idea further and use three pinholes in a vertical row? Or use five or ten, or any number, reducing their diameter in proportion and thus gaining that much more in sensitivity? This, too, will work without trouble and the shadows will remain the same.

If this is so, why not interconnect the pinholes of the chain, creating a slit, as this would not affect the principle involved? This is the slit test.

The increased illumination affords a corresponding reduction in width of the slit; or a compromise will permit both some increase in sensitivity and better illumination. One method of making a narrow slit is described in Chapter B.4. The knife-edge must be maintained parallel with the slit. A large pinhole is the equivalent of a slit, as a simple experiment will indicate. Set the knife-edge at the average center of curvature of a paraboloidal mirror, or where a spherical mirror is about 50 percent darkened. Then, with a magnifier, examine the cone of rays at the knife-edge. The cone is not bisected on the vertical median line, as might be expected, but only the edge of it passes beyond the knife-edge. Also, if a larger pinhole is used the width of the passed part will be the same when the conditions mentioned above are arranged but, being from a larger circle, it will be more closely resemble a slit. Despite these facts, a very narrow pinhole will give greater refinement, for the same reason that a very narrow slit would do the same. The slit, of course, would give the better illumination of the two, width for width.

If it is desired to have equipment for special work, such as locating astigmatism with a diagonal cut-off, or when testing a flat in combination with a sphere, some means for rotating the slit and the knife-edge, and at the same time maintaining the two in strictly parallel position, must be provided—either doing this separately with graduated dials on each part, or gearing the two parts together.

The origin of this test is difficult to assign; in a measure it was a growth. Russell W. Porter suggested it in 1918, but did not actually try it until 1931. At about the latter time two amateur telescope makers, James C. Critchett of Julian, California, and Daniel E. McGuire of Shadyside, Ohio, each hit on it independently.

"The slit and knife-edge can be combined by covering the square face of a small right angled prism with a rectangle of tinfoil (see Figure B.6.1)," Russell W. Porter adds. "The slit is formed in the foil at A, and the edge of the foil at B serves as the knife-edge, as shown in the left-hand sketch. Of course B must be cut parallel to the slit. For diagonal cut-offs the whole unit—lamp, prism, slit and knife-edge—must rotate as a unit about A as a center." This is the application of the slit test said to be used by Dr. Samuel Jacobson, in charge of optical work for the Gaertner Scientific Corporation.

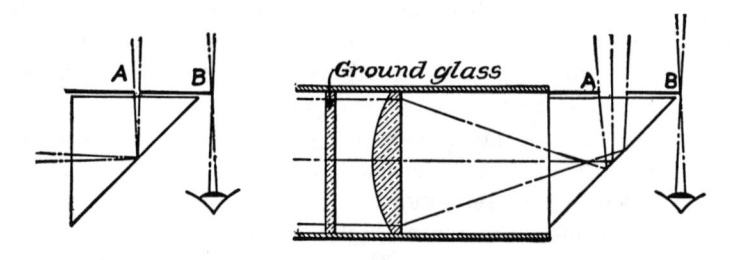

Fig. B.6.1.

To the last statement Professor Arthur Howe Carpenter of the Armour Institute of Technology, Chicago, adds: "Dr. Jacobson has shown that it is not essential to have a narrow slit at A. It should at least be wide enough to give plenty of light, which is a great help in finding the image at the observing stand (right-hand sketch) where a means of getting uniform illumination is indicated. Dr. Jacobson simply ground the side of the quarter-inch right angled prism and allowed the diffused light so obtained to illuminate the whole thing without the use of condensers. That part of the prism to the left of A may be left uncovered and a single piece of tinfoil used, so cut that the two edges are parallel; or the left-hand side may be covered, leaving a wide slit at A. This is better for obtaining uniformity of illumination. When the prism with the illuminated opening and knife-edges moves as a unit toward the image, the latter, of course, moves toward the knife-edge, disappearing behind it until the shadow of the knife-edge (or right-hand side of the slit) at A approaches the knife-edge at B. With a delicate screw control this can be made to give any width of slit desired. Dr. Jacobson states that, in this way, he uses two smooth straight knife-edges in place of the edge of the usual pinhole which is seldom smooth. Fine pinholes are likely to have jagged edges and, even when perfectly smooth, they easily collect dust particles and are troublesome to keep clean, giving erroneous appearances to the surface of the mirror. I have my apparatus so arranged that I can use a slit, a wide opening or a pinhole at A, as I wish, and use a knife-edge at B. My observation is that a clean-cut fine pinhole, properly illuminated, will give everything which these slits will give, up to the very final work; after which a fine slit with very smooth straightedges is an advantage because of the extra light which it affords. The Jacobson scheme, in common with all slit tests, is extremely sensitive in focusing, and is very delicate. Unless the worker is very careful to obtain the exact focus, a slit will give misleading results to one who is accustomed to the pinhole test. The point in the Jacobson arrangement is that no mechanical slit is used or needed. The shadow of one knife-edge meets the real knife-edge and forms a slit. A single safety razor blade may be used, as the edges are straight and smooth." [Use $r^2/2R$ here.—Ed.]

Chapter B.7

Shadow Appearance[†]

A Piece of Glass

He labored late into the night,
At early morn' his task resumed,
To fashion thus a disk of glass
Into a subtle curve, not deep,
But measured only by the shades
    of light
From a simple pinhole made of
    foil,
Revealing to his practiced eye
Imperfections infinitesimal;
Until at last his skill produced
A curve so true the mind of man
Could not discern the wavering of
    a breath.

"Just a piece of glass," 'twas said,
But in that simple disk
The heavenly host
Of suns and stars, yea, universes,
Revealed their glory in the sky
For man to ponder—and adore.

—C.A. Olson
Westwood, N.J.

[†]By A.W. Everest.

Chapter B.7. Shadow Appearance

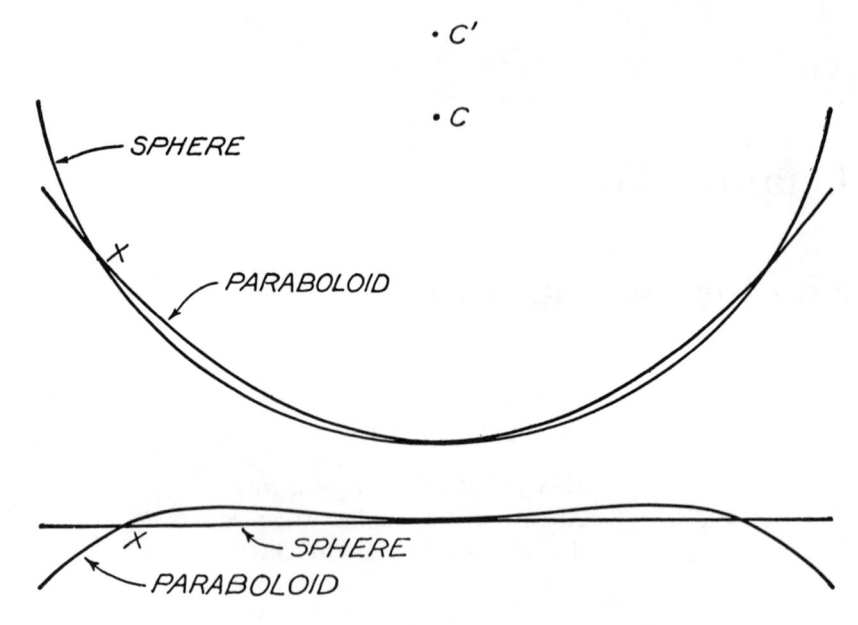

Fig. B.7.1: *(All drawings by the author.)*

The first requirement in figuring a mirror is a clear conception of how it should appear on the testing stand, since it is by comparing what we see with what we ought to see that we determine what to do. This appearance, with the knife-edge in some intermediate position, has been said to resemble that of a doughnut—and the analogy is a good one, since, like amateur's mirrors, there are all kinds of doughnuts—rough ones, smooth ones, doughnuts with holes and doughnuts without 'em.

Well, what is the shape of the true paraboloidal doughnut? With Porter at the other end of the continent, about all we can do in the way of showing this three-dimensional appearance in two-dimensional space, is to draw its cross section only, and leave it to the reader to visualize the surface of revolution this represents. The calculation of this cross section is a simple matter. In fact, the cross section of the paraboloid, as it appears with the knife-edge in any useful position, may be covered in a single equation. Let's figure the thing out under a heading somewhat in keeping with the method of calculation.

B.7.1 Doughnut Mathematics

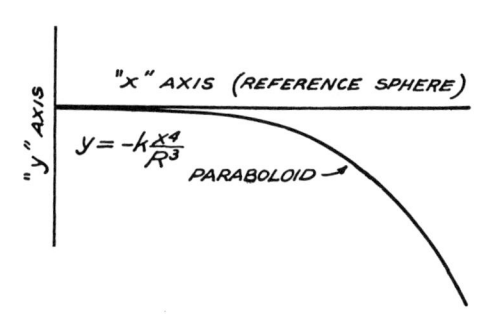

Fig. B.7.2.

B.7.1 Doughnut Mathematics

For reasons which will become clear later on, the logical reference surface for our calculations will be a sphere whose center of curvature is at the point of observation, *i.e.*, where the knife-edge cuts the cone of light. Further to simplify matters, this reference sphere may be made tangent to the paraboloid at its center, as in the upper part of Figure B.7.1, making this point the origin of coordinates and thereby eliminating the constant of integration. In this example we have chosen a sphere whose center of curvature C prime is just outside of C, the center of curvature of the paraboloid's center zone. Starting at the center, such a sphere would lie outside the paraboloid as far as some zone X, beyond which it would lie within. Viewed from its center of curvature, this reference sphere would appear flat, its apparent cross section would be a straight line, and the apparent cross section of the paraboloid would be a curve, somewhat as shown in the lower part of the same figure.

An expression for the deviation of the paraboloid from this or any other reference sphere will involve the following two propositions, which will be stated with a degree of accuracy equivalent to that of r^2/R without the tail end of the formula. This will permit the use of simple terms which will introduce no measurable error in the calculations for our telescope mirror; although, of course, the error would be intolerable in similar calculations for a large searchlight reflector.

Proposition 1: For the condition shown in Figure B.7.2, the deviation of paraboloid from sphere varies as the fourth power of its radius r, and inversely as the cube of its radius of curvature R. In mathematical terms we would write,

$$y = -k\frac{x^4}{R^3}. \tag{1}$$

The minus sign is inserted to indicate that the curve bends away from

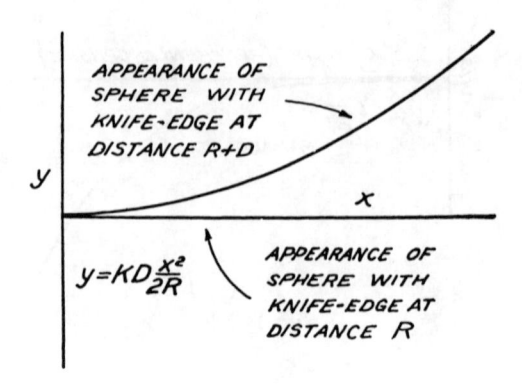

Fig. B.7.3.

the observer. For the slope at any point, the first theorem in elementary calculus (about all we remember) tells us,

$$\frac{dy}{dx} = -4k\frac{x^3}{R^3}. \qquad (2)$$

Proposition 2: When observed from first one and then the other, of two points along the axis near the center of curvature, the slope of any zone changes in proportion to the radius x of the zone, in proportion to the distance D between the two points of observation, and in inverse proportion to R. For the slope in one position compared with that in the other we may write,

$$\frac{dy}{dx} = KD\frac{x}{R} \qquad (3),$$

whence

$$y = KD\frac{x^2}{2R}. \qquad (4)$$

A useful example of this would be a spherical mirror examined first from its center of curvature, when it would appear flat and its apparent cross section would be the straight line shown in Figure B.7.3. If now the knife-edge be moved a short distance D toward the observer, the surface will appear concave, y varying in proportion to x^2 as shown.

We may now investigate what would happen to the curve shown in Figure B.7.2 if the knife-edge is moved toward the observer. The slope of each zone would change in accordance with the law given in the preceding

B.7.1. Doughnut Mathematics

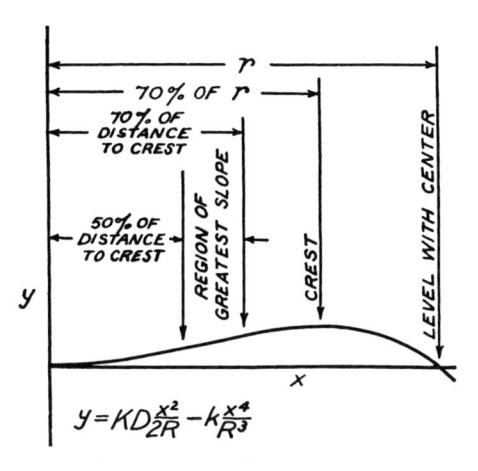

Fig. B.7.4.

paragraph, which means that certain sets of y values determined by Prop. 2 would be added to those determined by Prop. 1. For example, Figure B.7.4 shows the result of adding the actual y values of Figure B.7.3 to those of Figure B.7.2. As a useful proposition, however, the mere addition of the right-hand members of equations (1) and (4), giving

$$y = KD\frac{x^2}{2R} - k\frac{x^4}{R^3} \tag{5}$$

would mean nothing until we express K in terms of k. To do this we may consider that y is to reach its plus maximum at the point where the curve, such as the one shown in Figure B.7.1, at bottom, or the one shown in Figure B.7.4, becomes flat at the crest. This will be when the slopes represented by the $(dy)/(dx)$ values of Propositions 1 and 2 cancel each other, i.e., where

$$KD\frac{x}{R} - 4k\frac{x^3}{R^3} = 0$$

whence,

$$K = 4k\frac{x^2}{DR^2}. \tag{6}$$

For this point in the curve,

$$D = \frac{x^2}{R}.$$

Substituting this value of D in equation (6),

$$K = 4\frac{k}{R}$$

and substituting this value of K in equation (5),

$$y = 2kD\frac{x^2}{R^2} - k\frac{x^4}{R^3}$$

whence,

$$y = k\frac{x^2}{R^2}\left(2D - \frac{x^2}{R}\right) \tag{7}$$

where D is the distance of the knife-edge from the center of curvature of the center zone. For the actual deviation of paraboloid from reference sphere, k would be 0.125, and the whole expression would represent an amount too small to be seen. *But—*

The combination of tin can and razor blade forms a telescope magnifying, roughly, *one hundred thousand times;* the exact amount varying directly with the focal length and depending somewhat upon the sensitivity of the arrangement and individual interpretation. The amount stated is about right for an $f/8$, 10″ mirror, based on the opinions of a number of experienced amateurs, using the testing arrangement to be described. Any error here is immaterial, since it will have no effect whatsoever upon the characteristics of the curve which are to be considered.

Compared with the more familiar telescope, this one has the peculiarity that its tremendous magnification is in *depth only,* that is, the mirror is seen with actual diameter, but with every deviation from reference sphere magnified by the above amount. For example, our 10″ paraboloid, examined with this apparatus from the center of curvature of the center zone, appears to deviate about 2″ from flatness at the marginal zone, although in reality this deviation is only about 20 millionths of an inch. Likewise, the apparent deviations of all intermediate zones bear the same relation to the real.

Assuming a magnification of 100,000, or 625R, to be correct for the 10″ mirror, k becomes about 80R, and equation (7) changes to,

$$y = 80\frac{x^2}{R}\left(2D - \frac{x^2}{R}\right)$$

This is our final "Equation of the Doughnut," from which the values of y may be stated in terms of the markings on a 6″ scale.

Selecting values of D from zero to r^2/R in 10 percent steps, we may calculate coordinates for our 10″ mirror and then put these in curve form, as shown in Figure B.7.5. The top curve shows the apparent cross section of the mirror as it should appear with the knife-edge at the center of curvature of the center zone. The bottom curve shows how it should appear with the

B.7.1. Doughnut Mathematics

THE DOUGHNUT FAMILY

$$y = 80\frac{x^2}{R}\left(2D - \frac{x^2}{R}\right)$$

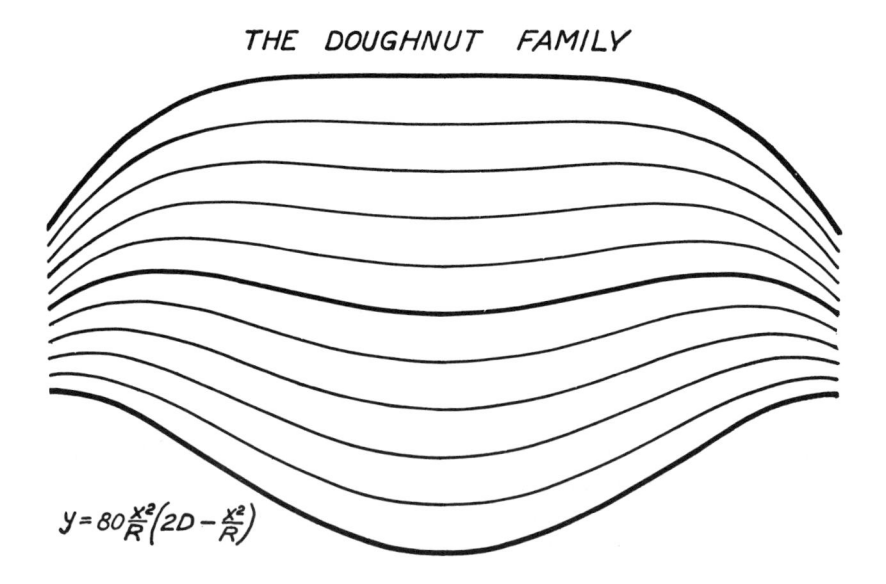

Fig. B.7.5: *The Doughnut Family.*

knife-edge at the center of curvature of the marginal zone. The heavy curve halfway down shows how it should appear with the knife-edge exactly halfway between the above two positions. We call this the "halfway" or 50 percent curve. Starting at the top, the light curves drawn in between the three heavy ones represent the apparent cross sections corresponding to knife-edge positions of 10, 20, 30, 40, and 60, 70, 80 and 90 percent of r^2/R.

The curves have been drawn with sufficient accuracy to bring out their fixed characteristics, that is, those landmarks which remain unchanged regardless of the values of k, r, or R we may select.

First, note that all the curves are *smooth*. There is no sudden change in slope at any point. Next, starting with the top curve and examining the whole series in order, note how the crest spreads from center to end; in large jumps for the first step or two, and then with smaller and smaller hops with each succeeding curve. A simple transposition of the r^2/R formula tells us the location of the crest as shown in Table B.7.1.

A trial with straightedge and scale will show the curves to be in agreement with the above. Note, however, the flatness of crests near the center and the resulting difficulty of determining their exact location compared with those farther out. Note especially the flatness of the crest in the first curve which, of course, is at the center of the mirror. Pay particular at-

Table B.7.1

Knife-edge position 00 in percent of r^2/R	Location of crest
00	Center
10	31.6 percent of r
20	44.7
30	54.8
40	63.3
50	70.7
60	77.5
70	83.7
80	89.5
90	94.9
100	100.0

tention to the distance this flatness extends from the center before there is any perceptible deviation from a straight line.

Now try to locate the regions of greatest slope in the central depressions, *i.e.*, where the curves reverse. This will be more difficult to decide, and impossible with the first two or three curves. But, as nearly as can be judged, these regions of greatest slope will be found to extend from 50 percent to 70 percent of the distance from center to crest for any of the curves. A solution for the maximum plus value of the expression

$$\frac{dy}{dx} = 4k\frac{x}{R^2}\left(D - \frac{x^2}{R}\right)$$

[derived from equations (2) and (3)], would show an intermediate point of slightly greater slope than the rest. But let's not introduce this typesetter's nightmare into the discussion, as no difference could possibly be detected by visual inspection. We are interested only in the practical result, which is shown in Figure B.7.4 and will be discussed in more detail later on.

Other landmarks:

- The depth at the end of the first curve is the same as the depth at the center of the last one.

- The depth of the halfway curve is one-quarter of the above.

- The center and end of the halfway curve are of the same depth, and this is the only curve for which this is true.

- In the 70 percent curve (the eighth one down), the greatest slope in the central depression is the same as that at the end.

There is *only one shape of doughnut*, just as there is only one *shape* of parabola. This is shown in Figure B.7.6, where we have superimposed the

THE DOUGHNUT "SHAPE."

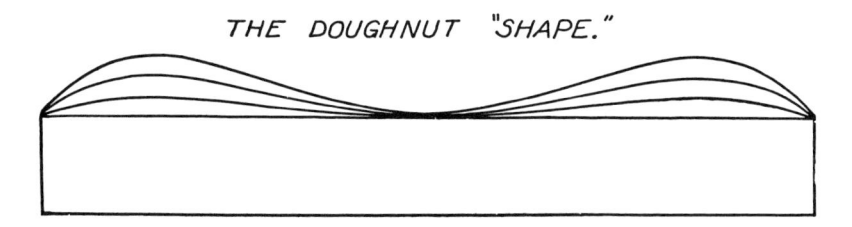

Fig. B.7.6: *The Doughnut "Shape."*

halfway curves for three paraboloids of the same diameter but different focal ratios. All have exactly the same characteristics in regard to the relation between progressive values of y and dy/dx. The only difference is in the intensity of the "bulge."

A glance at the doughnut family will show how the intensity of curvature varies far out of proportion to the diameter of the mirror. For any of the curves after the first one, the portion inside the crests may be used to represent the 100 percent curve for a mirror of that diameter. For the second curve, this diameter would be 31.6 percent of the last one, but the depth is hardly discernible in comparison—actually 1 percent.

And now that we have completed some painful reasoning, let's proceed to forget it and just form a clear mental picture of the curves, especially the three heavy ones, and the surfaces of revolution they represent. The squared background of the graph paper on which the curves were originally drawn has been purposely omitted from the reproductions, since there will be no such thing to guide us in the knife-edge test. Outside of a few zonal measurements, when figuring our shallow mirrors, we must be guided entirely by appearance in deciding when the goal has been reached.

B.7.2 Shadow Behavior

In speaking of a shadow, we refer to the *division between light and shade,* not to a whole *area* of shade. In locating a shadow, we refer to the point where it crosses the horizontal diameter of the mirror, since this is the point where any measurement will be taken. An area of shade which does not extend to the rim of the mirror will, of course, be surrounded by one continuous shadow; but since this shadow will cross the horizontal diameter at two different points, we shall speak of this area as bounded by *two* shadows, one on the left and one on the right.

If we imagine the mirror to be divided in half vertically, zones which have any appreciable slope will be illuminated in one half and dark in the other, when the knife-edge is in the correct position to bring out the grazing incidence effect. In Figure B.7.7, for example, the outer zone will be illuminated in the right half and dark in the left. In the central zone the illumination will be just the reverse. In this illustration we see three shadows as defined above which, for reference purposes, will be numbered 1, 2 and 3, as shown.

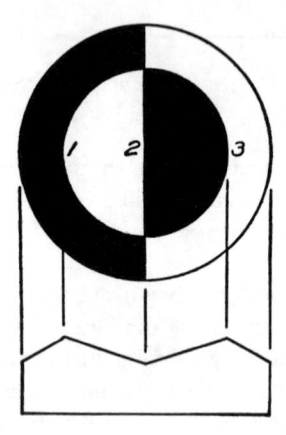

Fig. B.7.7.

It is evident from the law of reflection that shadows will move in the same direction as the knife-edge in zones which appear convex toward the observer, and in the opposite direction in zones which appear concave. In Figure B.7.8, at left, the mirror is convex from center to rim, in spite of the fact that it has a central depression. Therefore, shadows 1 and 3 would move in the same direction as the knife-edge across their respective halves of the mirror. (Shadow 2 in this case would remain stationary at

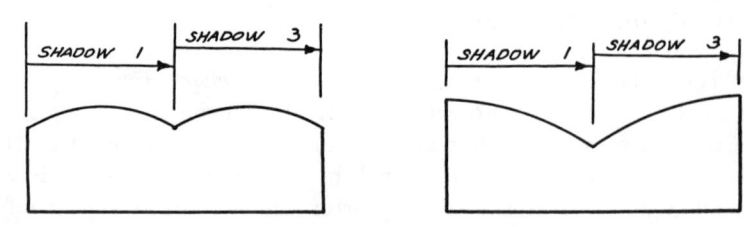

Fig. B.7.8.

the center.) In Figure B.7.8, at right, which represents the same surface as it would appear with the knife-edge just outside the center of curvature of the marginal zone, the shadows would still move in the same direction.

B.7.2. Shadow Behavior

For the zone inside the crest in Figure B.7.8, left, and for the whole surface in Figure B.7.8, right, the direction in which the shadows move contradicts the old rule that "the shadow will move in the opposite direction when the knife-edge is outside the center of curvature." Strictly speaking, this rule applies to spherical mirrors alone, and, for general application, should be restated as at the beginning of this paragraph.

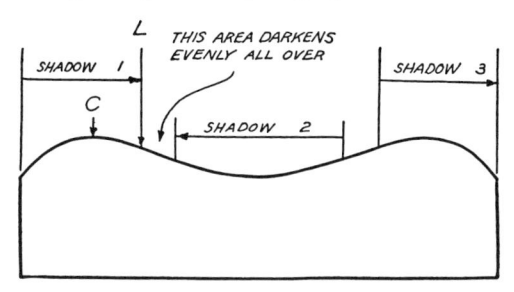

Fig. B.7.9.

Assuming a uniform speed of the knife-edge as it cuts through the cone of light, the speed of a shadow will be an inverse function of the rate of change of slope. In other words, the shadow will move slowly across zones of pronounced curvature, and comparatively fast across zones having little change in slope. In Figure B.7.9, shadow 1 would move slowly to the crest and then move faster and faster as it approaches L, where the curve flattens out at the beginning of the zone of greatest slope in the central depression.

In zones where there is no change in apparent slope with a given knife-edge setting, there will be no shadow motion. Such zones will darken evenly all over. Referring back to Figure B.7.4, we see that, in the true paraboloidal figure, such a zone always exists in the central depression extending from about 50 percent to 70 percent of the distance from center to crest. We are particularly interested in the 70 percent point, L, which we may call "the shadow limit," since this is the point beyond which no motion of shadow 1 can be detected. Picking up this shadow again from where we left off in the preceding paragraph, when it reaches the shadow limit, the zone of greatest slope will have just finished darkening, and the two effects will blend in with each other. During all this procedure, shadow 2 will have been approaching shadow 1 across the concave part of the central depression and will blend in about the same manner, except, of course, from the opposite direction.

In interpreting shadows, we imagine them to be caused by the light source being off somewhere to the right of the mirror rather than by the knife-edge shutting off the light. For a shadow to exist under this condition, it follows that the incident light path must be imagined to be tangent to the

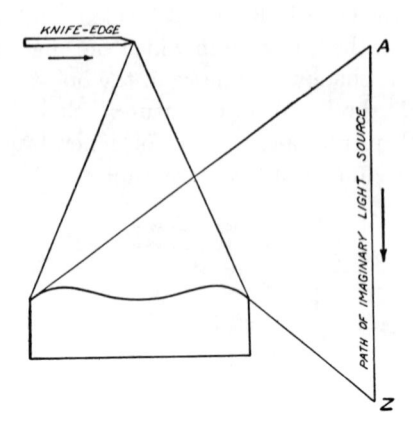

Fig. B.7.10.

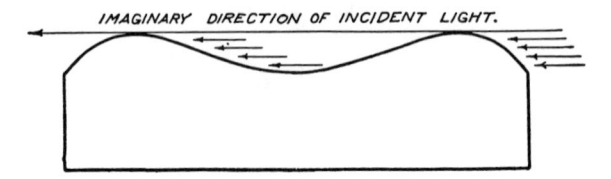

Fig. B.7.11.

surface at the point where the shadow happens to be. And, since shadows move as discussed above, it follows again that the light source must be imagined as moving away from the observer in a direction parallel to the axis of the mirror. In Figure B.7.10, from the time when shadow 1 enters the left side of the mirror until the time shadow 3 disappears off the right, the light source must be imagined to have traveled from A to Z.

Bumps, raised zones, etc., will not cast shadows on adjoining parts of the mirror. This is due, of course, to the fact that the actual direction of the light path is parallel to the axis of the mirror. In Figure B.7.11, when the apparent direction of the light path is at right angles to the mirror's axis, the right crest will not shut off the light from the left side of the central depression, but this area will be illuminated as shown by the arrows. In the case of a bump in the center, as in Figure B.7.12, when the light path appears to be in the direction shown, the left side of the bump will be darkened, but not the area from B to C, as would be the case if the light were actually coming from that direction. This area will have the same illumination as from A to B. This fact must be remembered

B.7.3. The Error of Observation

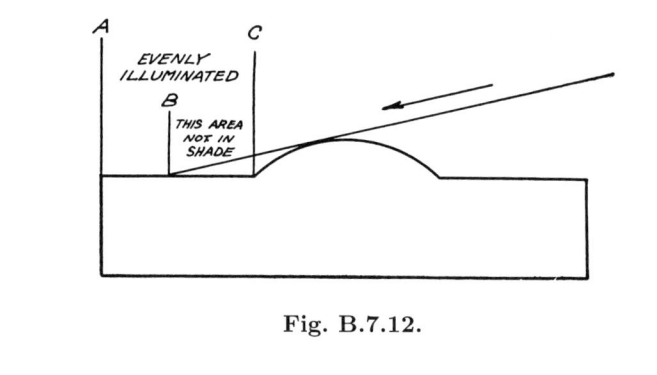

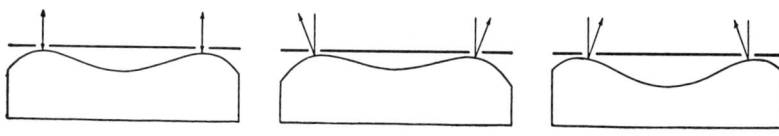

Fig. B.7.12.

Fig. B.7.13.

when locating the left boundary of a bump or raised zone. Otherwise, being accustomed to seeing light behave in the more familiar manner, the mind's eye will unconsciously place the left boundary of the protuberance somewhere within the area of shade, rather than at the left edge of this area where it belongs.

B.7.3 The Error of Observation

If we are using the stop method of testing a zone, both exposed arcs will darken at the same time when the knife-edge is at the center of curvature of this zone. If the stop is removed, the apparent crest, or greatest bulge toward the observer, will be located at this zone. If we are watching shadows 1 and 3 move across the mirror, both will reach their respective crests at the same instant. Assuming, for graphical purposes, that the knife-edge and slit are at the same point, the light striking this zone will be reflected back along the incident path, as shown by the arrows in Figure B.7.13, left.

If the knife-edge is slightly nearer the mirror, we can see from the slopes in Figure B.7.13, center, that, with the stop, the left arc will darken before the right. Without the stop, the crest will appear nearer the center, and shadow 1 will reach the zone before shadow 3. If the knife-edge is slightly farther away, as in Figure B.7.13, right, all these effects will be reversed.

So much for theory. In actual practice, however, we know that in making any fine measurement there is always the possibility of a residual error due to the limitations of the measuring device and, perhaps, to the human element involved. To get some idea of what this might amount to in the knife-edge test, the group of amateurs mentioned above tested the $f/8$, 10″ mirror, first using a stop having $1/4$″ arc-shaped openings at the zone of 4″ radius, to determine just how far the knife-edge could be moved along the axis and still have both arcs appear to darken at the same time; or, to put it another way, the minimum movement of knife-edge before the transition from Figure B.7.13, center, to Figure B.7.13, right, could be detected. The average result was the rather large quantity of 0.02″, with the individual readings surprisingly consistent. All stated that the diffraction around the arcs had a blinding effect on what they were trying to see. Hence, the stop was replaced with a quarter-inch strip of wood across the horizontal diameter of the mirror, with two short pieces of pin driven into the stick to indicate the exact location of the zone. The test this time was to see how far the knife-edge could be moved and still have shadows 1 and 3 appear to reach their respective pins at the same instant. The average immediately came down to slightly less than 0.01″, the main difficulty this time being to decide just where, in the mass of spiderwebs between light and dark, the shadow should be interpreted to be.

The next test was on the center zone, first using a stop with a 2″ hole, and then removing the stop and testing for the appearance represented by the first curve in the doughnut family; that is, with neither a bump nor a dimple in the center of the mirror. This time the results were about the same for either method, the diffraction at the circumference of the opening not bothering due to the larger area of the mirror's surface exposed. The knife-edge could be moved about 0.04″ without detecting any motion of the shadow when using the stop, or seeing any deviation from apparent flatness at the center without the stop.

In these tests, we were sneaking up from both sides, so to speak, so that the maximum displacement of knife-edge from the correct position would be about half the amount stated for either zone. It is also reasonable to assume that, in taking the average of a number of readings for a zone, or taking the mean of the two measurements giving the transition from Figure B.7.13, center to Figure B.7.13, right, we could cut these amounts in half once again. Let's go ahead on this basis, using 0.0025″ as the error of observation for the 4″ zone, and 0.01″ as the error for the 1″ zone.

The factor which determines the interval of time between the darkening of the two sides of a zone, that is, the ease with which this can be detected, is the dy/dx value of the slope. This can be seen in Figure B.7.14, where there are two zones $A - A'$ and $B - B'$, of different radius, but having the

B.7.3. The Error of Observation

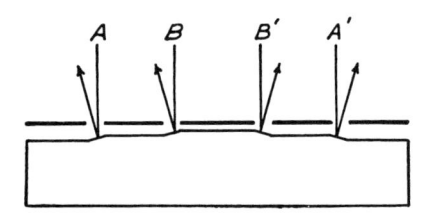

Fig. B.7.14.

same slope. A and B will darken at the same time, as will A' and B', when the knife-edge is farther advanced. Therefore, the interval between the darkening of the two sides will be the same for either zone. Assuming the slope of one zone to be just sufficient to detect this difference in time of darkening of the two sides, the same will hold true for the other. In other words, the error of observation involves a minute fixed value of slope, regardless of the radius of the zone. We saw in Doughnut Mathematics, Prop. 2, that any change in slope of a zone, resulting from a movement of the knife-edge along the axis, was represented by the expression

$$\frac{dy}{dx} = KD\frac{x}{R}.$$

And since, here, dy/dx has a fixed value, as stated above, we may combine the constants and rewrite the expression,

$$K = D\frac{x}{R},$$

whence

$$D = K\frac{R}{x}. \tag{8}$$

Figure B.7.15 shows another way to approach the problem. Here we have two mirrors of different focal length, but the same focal point f, being tested for zones $M - M'$ and $m - m'$ of the same aperture ratio, R/r. If the knife-edge cuts through the cone of light at some distance E from f, but just close enough so that no difference in the time of darkening of M and M' can be detected, the same will hold true for m and m'. In other words, the error of observation is a direct function of the focal ratio, for which we may write,

$$E = k\frac{R}{r}.$$

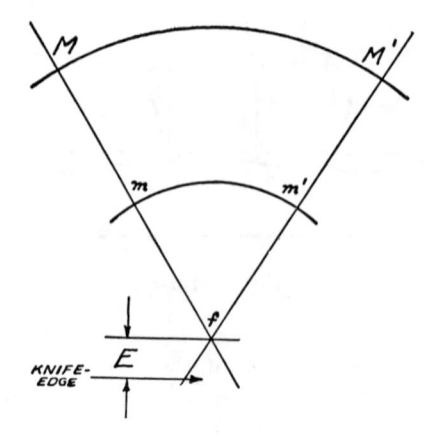

Fig. B.7.15.

This, of course, is the same thing as equation (8) with a change of characters. Solving for k, by using the value of E determined for either the 1" or the 4" zone of the 10" mirror, the equation for the error of observation becomes,

$$E = 0.00006\frac{R}{r} \qquad (9).$$

B.7.4 Accuracy of the Knife-edge Test

The answer to the question, "How accurate is the knife-edge test?" depends upon which class of errors is under consideration—those that can be measured but not seen, or those that can be seen but not measured. Strange as it may sound, the latter can be detected much the closer of the two, in terms of thickness of glass to be removed.

Referring to Figure B.7.6, and assuming that the three curves represent different intensities of figure on one and the same mirror, it would be impossible to tell from visual inspection under the knife-edge test, which one was correct. Zonal measurement for radius of curvature would be required. In Figure B.7.16, however, such measurement would be impossible; it would be necessary to estimate the intensity of these various protuberances by visual inspection alone.

The measurable errors, then, are those in overall correction, and the accuracy obtained depends upon the zones selected for test. Theoretically, the exact center of a mirror cannot be tested for radius of curvature. Also, the extreme marginal zone is difficult to read, owing to the diffraction at

B.7.4. Accuracy of the Knife-edge Test

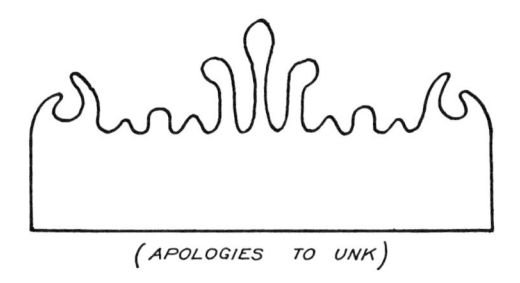

(APOLOGIES TO UNK)

Fig. B.7.16.

the edge; or if a diffraction edge is missing, the glaring illumination of the turned edge at the right is a worse source of trouble. So, in practice, it is better to select knife-edge positions of, say 10 percent and 90 percent of r^2/R, bringing the zones to be tested at 32 percent and 95 percent of r, as shown in Table B.7.1. Although the ease of reading the difference in knife-edge positions will be reduced 20 percent, this will be offset many times by the gain in sensitivity and reduction in the error of observation. There are better ways than zonal measurement to tell when the curve is true right in to the center, or out to the extreme edge.

Table B.7.2

	80 percent of r^2/R (inches)	knife-edge error (inches)	percent error	$r^4/8R^3$	error in millionths of an inch
6-inch f/8	0.075	0.008	10.7	0.0000114	1.2
10-inch f/8	0.125	0.008	6.4	0.0000191	1.2
10-inch f/5	0.200	0.005	2.5	0.0000781	2.0

In Table B.7.2, $0.00006R/r$ has been calculated for the 32 percent and 95 percent zones for three different mirrors, and since it is probable that the error would be made in one direction for one zone and in the other direction for the other, the two values have been *added together* to get the total error in overall correction which might be made.

(Note: Your mentor confesses the responsibility for Figure B.7.16, the result of early experiments with HCF strips.)

The above shows that measurements down to 1 millionth of an inch in thickness cannot be made with any degree of certainty. But it will be seen, by referring to Chapter B.3 "Accuracy in Parabolizing," that the probable error of observation is well within the tolerance given—all of which means that, if a mirror is figured to this degree of accuracy, brilliant performance may be expected, provided the numerous other conditions necessary for this

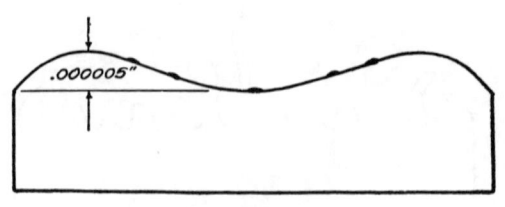

Fig. B.7.17.

performance are correct.

The other class of errors includes close zonal irregularities, dog-biscuit, lemon peel, bumps, etc. One illustration with the 10″ mirror will serve to show how closely the thickness of these protuberances may be estimated by visual inspection, although it is impossible actually to measure it. Figure B.7.17 shows the apparent cross section with the knife-edge at the halfway position, bringing the crest at 70 percent of the distance from center to rim. Referring to Table B.7.1, it will be seen that the deviation of this crest from the reference sphere is one quarter of $r^4/8R^3$, or about 5 millionths of an inch. Several slight protuberances have been placed on the curve, having a depth of $1/20$ of this amount, or one quarter of 1 millionth of an inch. These could most certainly be seen with a sensitive slit.

B.7.5 Testing Equipment

Let this be simple but substantial. Micrometer screws are unnecessary, and if the outfit is mounted on anything less rigid than a solid concrete foundation, they are useless. If the mirror and knife-edge are mounted on separate supports, a slight pressure of the finger tips on the side of the bench holding the knife-edge will give more sensitive control of the shadows than any micrometer arrangement the amateur will be likely to make. The only care necessary is to see that this pressure will result in motion at right angles to the mirror's axis alone. And the more solid the bench, the better this will work.

If there are any perceptible air currents between the knife-edge and mirror, a testing tunnel will be required for close reading. This may be a simple wooden framework covered with heavy cloth, and having a cross section about 50 percent greater than the diameter of the mirror. The cloth may hang down at one side, in the form of a flap which may be raised to insert the mirror.

Provide the mirror support with some means of adjusting to bring the light rays back to the proper point when the mirror is in position. Three wooden wedges will serve the purpose. After the adjustment is once made,

B.7.5. Testing Equipment

screw or clamp the support to the bench. If the front and back of the mirror are not parallel, paint an index mark on the edge, and always have this mark in the same position when the mirror is on the rack. The above will save time getting lined up for those frequent tests during the final stages of figuring.

The light source may be an inside-frosted incandescent lamp bulb inside a small tooth powder can, in one side of which a $3/8''$ square or circular opening has been made.

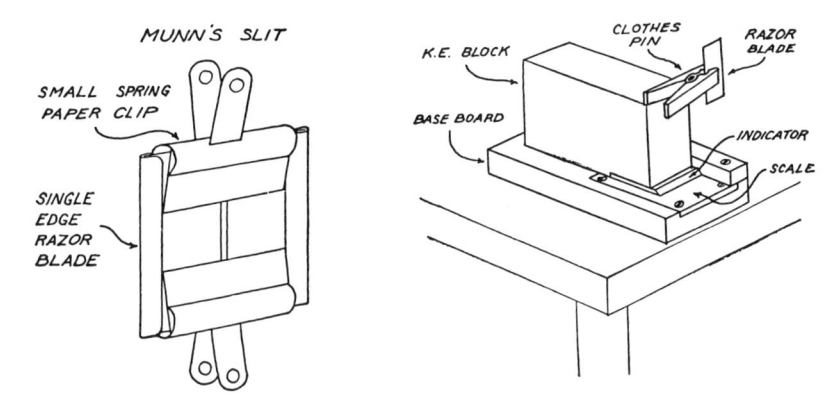

Fig. B.7.18.

The slit may be two safety razor blades clipped between two spring paper clips, as shown in Figure B.7.18, at left, with the opening between the blades set to about the thickness of ten sheets of typewriter paper. An extremely narrow slit is not advisable when used with the knife-edge, due to the diffraction effects which result. Sensitivity depends on parallelism of knife-edge and slit. True, when the mirror is just beginning to darken, there will be a marked reduction in contrast between light and shade with the wide slit, making the overall shape much easier to see or photograph. But the actual shadow, as defined above, will be as sharply defined as if produced by an infinitely narrow slit. In fact it *is* produced by just such a narrow strip of light, bounded on one side by the right edge of the image of the slit and on the other side by the knife-edge itself. For the same reason, the wide slit will be just as effective as the narrow one in detecting close zonal irregularities just before the mirror completely darkens. And for comparative freedom from diffraction effects in the vicinity of the shadow, the wide slit is easily the choice. Also the rectangular aperture of the slit is the only light source which will illuminate the various zones of the mirror in direct proportion to their slopes, giving the correct appearance to the curve. It is essential that knife-edge and slit both be at right angles to the axis, otherwise a marked reduction in sensitivity will result.

Remember that the pupil of the eye is well-dilated when testing and takes in a considerable portion of the knife-edge and image of the slit.

For Ronchi testing, the slit must be closed down to the thickness of one or two sheets of paper. For example, if gasoline screen is used for the grating, the slit must be somewhat narrower than the thickness of the wires in the screen, to get good definition of the lines.

For quick exchange from razor blade to screen, a spring clothespin may be cemented to one end of the knife-edge block, as shown in Figure B.7.18, at right. This will also permit rotation of either one around the mirror's axis, to get parallelism with the slit. To get the knife-edge parallel, push it about 1" inside the center of curvature, bring the shadow in to the center of the mirror, and then rotate the knife-edge one way or the other until the point is found where the shadow doesn't move sidewise when the head is bobbed up and down. With the Ronchi grating, do the same thing until the lines will stand still. The recommended length of slit, or even longer, will be found a decided advantage in getting accurate parallelism by this simple method.

The knife-edge block should slide freely on its base board. It is rather difficult to construct parallel ways for the block to slide between with the necessary freedom, so it is better for the amateur's purpose to provide just one cleat, as shown in the illustration, using a little side pressure when moving the block, in order to keep it against this cleat.

The measuring devices will include a scale on the knife-edge base board, and a measuring stick hung across the horizontal diameter of the mirror. The scale we prefer, because of its simplicity and ease of reading, is shown in Figure B.7.19. This is a flat plate of brass screwed to the base board, with parallel cross lines at tenth-inch intervals, and diagonal and vertical lines as shown; all accurately scribed with a fine, sharp engraver's tool. The indicator is another piece of brass screwed to the underside of the knife-edge block, as shown in Figure B.7.18. With this device, the nearest half of a hundredth inch may be easily seen, which is about as accurately as we may expect to read. The measuring stick is a quarter-inch-square stick of wood with a loop of wire to hang it across the horizontal diameter of the mirror, as shown in Figure B.7.20. Our preference for the spacing of the pins is to locate the zones given in Table B.7.1, those at the center and at the 70 percent points being slightly longer than the rest. Others, depending upon how their mental processes work, may prefer to place them at $1/2''$ intervals.

For the reasons given in "The Error of Observation," stops are taboo for testing parabolic mirrors in at least one outfit we can name. They are used only for finishing spherical mirrors and flats, where shadows are lacking to

B.7.6. Testing Routine

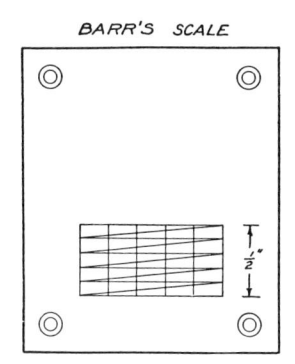

Fig. B.7.19.

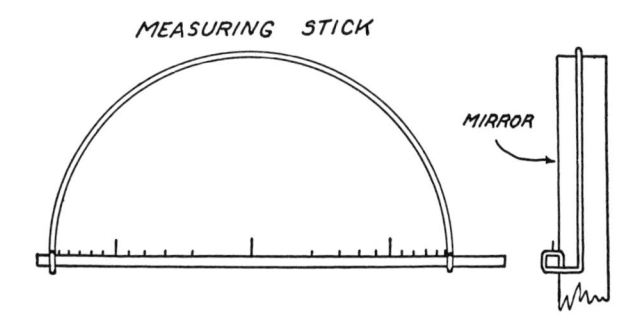

Fig. B.7.20.

test both sides of a zone. And we are not so sure that even here we cannot see all there is to see without them on small surfaces. When used, the openings are made 1" wide.

B.7.6 Testing Routine

The tests for a true parabolic curve will include the following routine.

1. Look for the diffraction effect at the edge, as explained in Chapter C.1, subsection, C.1.13.8. Theoretically, the left edge should be as bright as the right, but in practice this will seldom be obtained. However, a clean hairline of light should follow the left edge of the mirror, and the illumination at the right side should not persist to any noticeable extent longer than the diffraction around the edges of a screw driver or similar object placed vertically before the mirror.

2. Inspect the surface as a whole for zonal irregularities, that bumpiness called dog-biscuit, lemon peel, etc. If these are absent and the surface appears velvety smooth, it has something greatly to be desired, or in more highbrow terms, it is an *optical surface*.

3. Set the knife-edge at the 90 percent position on the scale, and then slide the whole assembly including the base board along the bench until the point is found where shadows 1 and 3 reach the 95 percent zones together. Here is where the value of the recommended spacing of the pins on the measuring stick will be seen—they leave nothing to guesswork. Take a squint along the side of the cleat to see that it is pointing at the center of the mirror, since this is the key setting for the whole test. If accurately lined up, the knife-edge may be pushed toward the mirror, causing the crest to roll in toward the center without the illumination otherwise dying out, or the mirror becoming wholly illuminated. After making any adjustment necessary to produce this effect, check the 90 percent position against the 95 percent zone once more and then leave the base board in this position. Before leaving this zone, check the shadow limit, which should be at 70 percent of the 95 percent. No guiding pin will be found here, but the point may be estimated closely enough from the pins at either side. And don't forget that homely method of pushing lightly on one side or the other of the bench to bring the shadows exactly where wanted. Also decide right here what you are going to *call* the shadow; probably the point where none of the remaining spiderwebs of light cross the pin.

4. Push the knife-edge in to the 50 percent position and test as above on the 70 percent zone, first seeing that shadows 1 and 3 reach their respective pins at the same time, and then that the shadow limit for shadow 1 is halfway from center to rim. Also note, while the mirror is only partially darkened, that the *depth* at the center appears to be the same as that at the rim.

5. Push the knife-edge to the 10 percent position and test the 32 percent zone. This time there will be some difficulty in determining just when the shadows reach their respective pins, due to the flatness of the crest. The general appearance of the whole central area of the mirror must be taken into consideration, based on the 10 percent curve of the doughnut family. There should be just a slight resemblance of a depression inside the zone, coming to a crest at the 32 percent point, as nearly as can be decided. If no slope can be detected one way or the other at this zone, it will never scatter light in a star image.

B.7.6. Testing Routine

6. Push the knife-edge to the zero position. The last trace of a central depression should just disappear without raising any resemblance of a bump at the center. In other words, the whole central area should appear *flat*. See the first curve in the Doughnut Family, Figure B.7.5.

7. Push the knife-edge 1/2″ past the zero position and bring the shadow halfway in from the left edge to center. The shadow should have a slight but smooth curve inward, and run off the edge clean, with no change in curvature. If it bends inward at the edge, the edge is turned. If it straightens out or bends outward, the edge is undercorrected or turned-up. All this, of course, is a simplification of the Ronchi test. But it tells the whole story as far as the edge and marginal zone are concerned.

8. Bring the knife-edge back to the 70 percent position, with the crest at 84 percent of r. As can be seen from the note on this curve (see Figure B.7.5), shadow 1 should just be entering at the left edge of the mirror when shadow 3 is breaking out in the right side of the central depression. Also, shadow 1 should reach the shadow limit when shadow 3 is just passing off the right side of the mirror.

With a little experience, all of the above tests may be made in a few minutes, and, in our opinion, a mirror which passes them may be rated 100 percent. We believe the so-called personal equation associated with stop testing is due mainly to the blinding effects of diffraction around the openings, and to the necessity of seeing something happen at two different places at one time. These troubles are not present with crest reading. The diffraction at the pins is just sufficient to enable them to be readily seen. Shadow 1 is brought to its pin and *stopped there,* so that we may take all the time we wish to see whether shadow 3 has just reached the correct point. If both are where they belong for the various settings of the knife-edge, they just can't be anywhere else—that is, within the limits of the error of observation.

By now we can see the advantages of testing in equal steps of knife-edge position rather than in equal steps of the radius of the mirror. Each zone may be measured with the same degree of certainty, and all are of equal importance, since each involves the same area of the mirror.

Well, we have tested a good mirror. Let's test some of the others. If a mirror has the true paraboloidal appearance but, with the knife-edge settings mentioned above, the crests are not found in the correct location, the inference is obvious. If the crests are too close together, the mirror is overcorrected, and vice versa.

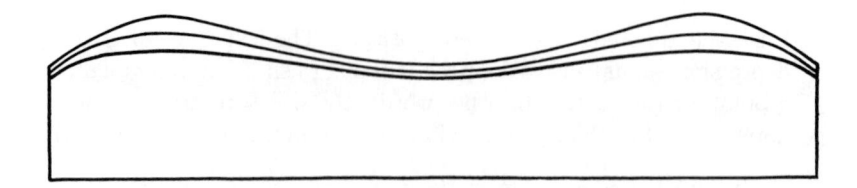

Fig. B.7.21.

For irregular surfaces, such as represented by the upper and lower curves in Figure B.7.21, the three zonal measurements would not tell the story. The center curve is the only correct one, yet it is evident from the slopes of the center, 70 percent, and marginal zones, that any one of them would test the same for all three mirrors, and if these were the only tests made, any one of the three mirrors would be pronounced correct. Yet it is obvious that the upper curve represents about 50 percent overcorrection, and the lower one about the same amount under. While it is true that the light reflected from these three zones could be brought to a focus, the waves arriving from the 70 percent zone would be somewhat out of phase with those from the other two zones, resulting in an actual *loss* of light. The best means of detecting these errors will depend on the size of the mirror and the depth of the curve. For shallow mirrors, which will include most amateur sizes of $f/8$ or thereabouts, the general appearance of the doughnut will tell the story. For the upper curve, the crest will appear too sharp, as in Figure B.7.22, at the left, and for the lower curve, it will appear too flat. The correct appearance will be about as in the same figure, at the right. For mirrors of more pronounced curvature, where the depth of the shading makes it difficult to see the overall correction at a single view, the shadow limit will spot the trouble. For the upper curve, the shadow limit will be too near the crest; for the lower one, too far away. And if we have developed a sensitive touch in pressing on the side of the bench, the shadows for the upper curve will slow down too much at the crest; for the lower curve, they will jump across too fast. For still larger mirrors, which takes us up out of the amateur class, actual measurement for center of curvature of successive zones, in not over ten percent steps, is required to determine the figure with the necessary degree of accuracy.

A frequent error in amateur mirrors having the correct difference between inside and outside centers of curvature, is misplaced crest when the knife-edge is in the 50 percent position. The crest is too far in or too far out. These are shown by the dotted lines in Figure B.7.23.

B.7.6. Testing Routine

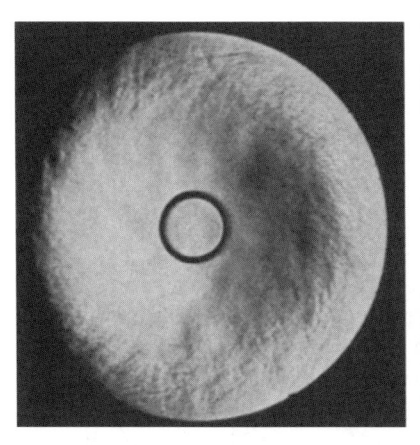

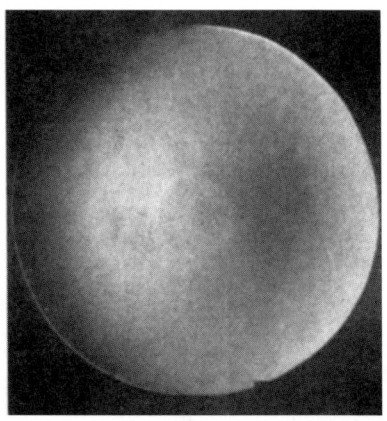

Fig. B.7.22: *Note the slight depressed ring at the center of this mirror which was later made into the perforated primary of a small Cass. The mirror was drilled from the back to within $1/16''$ of the face with a copper cylinder, after the rough grinding had been completed. It was polished and figured in this condition, after which the plug was knocked out by tapping lightly on its face with a hammer. During the polishing this ring, $1/16''$ thick, could not dissipate its heat as rapidly as the remainder of the surface. As a result, the ring was swollen out, polished off and, after coming to equilibrium, became a depressed zone. (The mirror on the left was not made by the technique described in this chapter, or by its author, but the one on the right was made in its entirety by Mary A. Everest, the wife of the author, under his oral instruction but without manual assistance. It speaks for itself regarding the results obtainable from the technique described, as do many fine mirrors made by the author.—Ed.)*

With the exception of the final test to prove the perfection of figure, the purpose of all testing is to determine the comparative thickness and radial location of the surplus material which is to be removed. In most cases this may be estimated at a glance, with the aid of the measuring stick. In other cases, after measuring up, it may be advisable to draw on a piece of paper the cross section of the surface seen, superimposing this curve upon the curve of nearest fit from the doughnut family. Then the area between the two curves will represent the material to be removed.

For the upper curve in Figure B.7.21, it can be seen at a glance that the greatest thickness of surplus material is at the crest, gradually tapering off into the adjoining zones. For the lower curve, however, it might be advisable to draw underneath a true 50 percent curve, such as the one shown in the doughnut family, but with the y values cut down to get a closer fit. Removing the material indicated would bring the surface to a true paraboloidal *shape*, but undercorrected, after which the crest could be brought up by regular parabolizing methods. A similar process is indicated for both cases

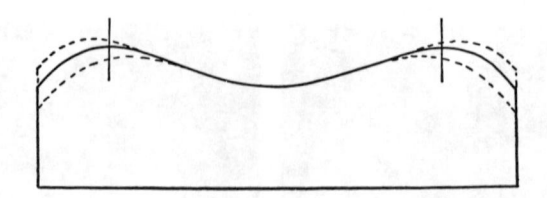

Fig. B.7.23.

in Figure B.7.23, reducing the figures to the correct shape, but undercorrected. With a little experience, this process of superimposing the surface seen on the correct surface, will generally be a mental one.

For the medium deep curves, we can get along in similar fashion. But, for the big fellows, where all but the crest is so brightly illuminated or so deep in shade that the overall figure cannot be seen under any conditions, complete zonal measurement in not over ten percent steps is required, after which some laborious graphical work based on these measurements is necessary in order to determine the exact condition of the figure.

B.7.7 Note Added in 1996

Recall that the "shadow edges" represent the first and last appearances of light as the knife edge is introduced into the beam. They look like "eyespots" cartooned on the face of the mirror. Everest uses the position of the shadow edges as confirmation that the knife is at the proper place longitudinally. He correctly says early in this article that the perceived radius of the shadow edge is between 50 and 70 percent of the radius of the crest. However, he decides to approximate the shadow edge as occurring at one end of this range, i.e., at 70% of the crest radius. By this argument, if the crest were at the 70% zone, the shadow edge would be about at the 50% zone.

Why Everest uses 70% of the crest radius is uncertain, although it may be because he does not wish to contradict his earlier published descriptions. That would confuse the people he was trying to reach in this article. It may also be a "personal equation" type of bias. In any case, for the slow mirrors that were typically made at this time (with their wide and delicate shadows) it would not make any difference. Indeed, the shadow edge is difficult to see at all for inner zones or shallow mirrors.

In *Telescope Making #33* (Kalmbach Publishing Co.), R.C. Follett pointed out that TMs following Everest's instructions to the letter may be confused if they are making a fast mirror. The shadow edges of these sharply-curved mirrors don't seem to appear at the right fraction of the

B.7.7. Note Added in 1996

radii of the crests. What they have encountered is a limit to the 70% approximation. Follett presented the following table (which extends Table A.7.1), with the exact position of the shadow edge calculated at 57.7% the radius of the crest. By conceptually moving the knife beyond the crossing point of the outer zone, we can calculate that the "shadow edge" finally reaches the outer lip of the mirror at $3(r^2/R)$, just as the caustic test says it should (see Section B.8). Thus, we see that Everest was using caustic test concepts very early on. *H.R.S.*

Table A.7.3

Knife Movement (fraction of r^2/R) x	Radius of Peak (crests) $\sqrt{x}$	Radius of Max. Slope (shadow edge) $\sqrt{x/3}$
0	0.0%	0.0%
0.1	31.6%	18.3%
0.2	44.7%	25.8%
0.3	54.8%	31.6%
0.4	63.2%	36.5%
0.5	70.7%	40.8%
0.6	77.5%	44.7%
0.7	83.7%	48.3%
0.8	89.4%	51.6%
0.9	94.9%	54.8%
1.0	100.0%	57.7%
⋮		⋮
1.5		70.7%
2.0		81.6%
2.5		91.3%
3.0		100.0%

Chapter B.8

The Caustic Test[†]

In 1859 Leon Foucault publicly described his test at the center of curvature of a mirror, which has since been used in making thousands of excellent mirrors. Despite its great usefulness, certain disadvantages show up in applying the test to nonspherical surfaces, such as the paraboloid, that make it difficult to use and which set a practical limiting accuracy of zonal measurement at about 2 millionths of an inch. Some improvement results when the zonal mask is discarded, testing then being done by observing shadow edges and the overall shape of the surface (Everest, Gaviola). However, shadow edges are never very sharp and the overall curve, such as the doughnut, can be studied only in the case of moderate focal ratios. Hence a zonal mask is needed for a short-focus paraboloid and for any other aspherical surface.

Let us start on familiar ground with a description of the Foucault test. An illuminated slit is placed a little to one side of the center of curvature of a spherical mirror, so as not to obstruct the returning cone of light reflected from it. To simplify the discussion let the knife-edge be mounted on the same block that carries the slit, so that both are the same distance from the mirror and move as a unit. (In actual use the slit will be kept fixed in the caustic test, as is the usual practice in the Foucault test.)

The knife-edge is cut across the reflected cone of light and the center of curvature of the sphere is located as follows. If the knife-edge and the shadow on the mirror travel in the same direction, the knife-edge is inside the center of curvature; if they travel in opposite directions it is outside. In the case of a perfect sphere the mirror appears to be perfectly flat, and darkens evenly all over without any trace of shadow motion when the knife-edge is at a distance from the mirror equal to its radius of curvature R. Thus, for a sphere, this is a null test. The accuracy to which a spherical

---

[†]By Irvin H. Schroader, Applied Physics Laboratory, The Johns Hopkins University, Silver Spring, Maryland.

surface can be figured then does not depend upon measurements of any kind but depends only upon the sensitivity of the test setup and the worker's skill.

When the above outlined procedure is carried out we soon learn by experience several facts. Very minute zonal errors, perhaps $1/5$ of a millionth of an inch high, can be seen by delicate adjustment of the knife-edge as it cuts across the optical axis. On the other hand, if the problem is to measure the radius of curvature of a sphere (by moving the knife-edge along the optical axis until the mirror looks flat and darkens evenly with no shadow motion either to left or right as the knife-edge cuts across the cone of light) it is difficult to judge just where to "call" it. Even the average of several settings can easily be off as much as 0.01 inch, which corresponds to an uncertainty of several millionths of an inch on the mirror surface (Ritchey). Thus we conclude that the Foucault test is very sensitive when used to *detect* zonal error but is much less sensitive when used to *measure* error.

For any mirror surface other than spherical, a paraboloid for example, simplicity is replaced by complexity, for no setting of the knife-edge can be found anywhere along the optical axis that will cause more than small areas of the surface to darken evenly. In fact the test is now characterized by an apparent deviation of the surface from flatness and by moving shadow edges over most of the surface. Everest has described this behavior in great detail for the case of a paraboloid(Chapter B.7), so we shall here discuss it only in sufficient detail to enable the reader to understand later developments. Briefly, the explanation is that the mirror surface has been deliberately distorted by figuring so that there is no longer a single center of curvature for the whole surface as in the case of a sphere but, rather, now there are many centers of curvature spread out in a small space near the center of curvature of the center zone of the mirror.

The problem is to determine the amount of correction that has been given the mirror. Since for nonspherical surfaces the test is no longer a null test, a scheme of measurement must be resorted to, which in turn raises the question of how accurate our measurements are.

Although some read shadow edges in testing paraboloids, most workers use a zonal mask in which a row of holes is cut out across the horizontal diameter of the mirror. The idea is that each hole exposes a part of the surface which can be considered very nearly spherical if taken small enough. But then their size is too small to make possible the accurate measurement of the radius of curvature of each of these small spheres individually. A more accurate procedure is to take a pair of these openings, a and b, Figure B.8.1, exposing the two sides of the mirror zone ab; the centers of both a and b are at a distance r from the optical axis. The focal length of this zone is measured by trying to locate with the knife-edge the point where

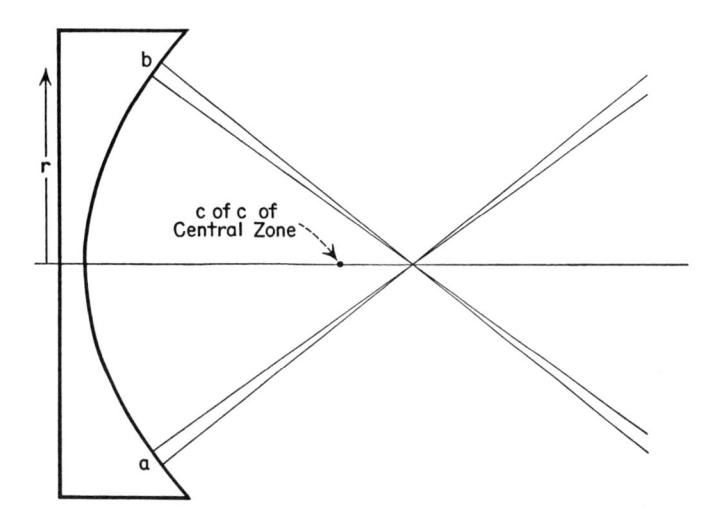

Fig. B.8.1.

the cones of reflected light from a and from b cross over the optical axis (the crossover point).

A great deal of confusion has grown up in the literature describing this quantitative form of the Foucault test with regard to this measurement. The assumptions usually are (see Figure B.8.1): (1) the two sides, a and b, of a zone ab are spherical, for all practical purposes; (2) the center of curvature of sphere a and that of b lie on the optical axis at the crossover point, no matter what curve the mirror surface has—paraboloidal, ellipsoidal, etc. In other words, it is assumed that the little spheres a and b are segments of a single sphere whose center is at the crossover point. Therefore, if a knife-edge intercepts the two cones of light at this crossover point, the two sides a and b should act the part of spheres by appearing to be flat, and by darkening evenly all over simultaneously, with no moving shadow edges. In the case of a perfect paraboloid, the knife-edge should then be at a distance $R + r^2/2R$ from the vertex of the mirror (from here on R is the radius of curvature of the central zone of the mirror, unless otherwise stipulated).

Let us check these two assumptions against experimental results. If one starts with fairly large holes in the screen, no such behavior can be observed, so our first guess is that assumption No. 1 has been overdone—the holes are too large. To avoid this mistake we go to the opposite extreme and cut slots only $1/4$ inch wide in the mask. Now, however, we are on the other horn of the dilemma for it is impossible to decide whether these narrow zones are flat or whether there are any moving shadow edges. In other words, the test is now reduced to judging a photometric balance between two widely separated slots. All will agree that here the proper setting of the knife-edge is very difficult to judge accurately, the more so for lower f ratios. (Note also that it would take many such slots to test the whole surface of a mirror.) Moderate sized holes in the mask, next chosen in an attempt to avoid the dilemma just described, only add confusion to the uncertainty since now confusing details can be seen in the shadow behavior. A narrow slit gives rise to prominent diffraction effects, whereas a wide slit is less troublesome but is also less sensitive. Just what goes on in one of these holes is hard to see experimentally because the eye is blinded by the bright diffraction glare around the edge, and it is even harder to see theoretically because of the mathematics required by diffraction theory. Linfoot has worked it out (*Monthly Notices,* Royal Astronomical Society, No. 6, 1948) for holes $R/600$ inches wide, ignoring the diffraction fringes due to an assumed slit width of 0.001 inch. He concludes that under these conditions the accuracy of the test still depends essentially upon the ability of the worker to judge a difference in brightness between two widely separated holes.

The final clincher by way of experiment would be to cut two holes in a mask, about $3/4$ inch in diameter, each with its center say $2/3$ of the distance from center to edge of the mirror along the horizontal diameter. Set the knife-edge so that (as nearly as can be judged) the two sides of the zone darken evenly and simultaneously. Now remove the mask and it will be easy to see that the shadows on the mirror move in the same direction as the knife-edge across both sides of the zone previously exposed by the two holes; that is, the knife-edge is inside the center of curvature of these two areas of the mirror surface. We can only conclude that experimental facts do not verify our two assumptions. Let's examine them.

In *Popular Astronomy,* Volume 10 (1902) pages 337–348, the late Professor F.L.O. Wadsworth demonstrated that our second assumption, that the centers of curvature of both sides of a zone lie on the optical axis, is incorrect and that Figure B.8.1 should be redrawn as in Figure B.8.2. If the latter is correct, it would explain why the knife-edge was inside the center of curvature of both a and b when it was set at the crossover point in the experiment just described.

That something like Figure B.8.2 must be correct can be argued from the method used in parabolization. Glass is worn away in amounts increasing from none at the mirror's edge to a maximum at the center; in other words, as the center is approached the radius of curvature of the surface is gradually shortened. In doing this the tilt or slope of each zone is also changed. It therefore would be an extraordinary coincidence if the change in slope and the shortening of the radius were to keep exactly in step for all zones, permitting all their centers of curvature to remain on the optical axis.

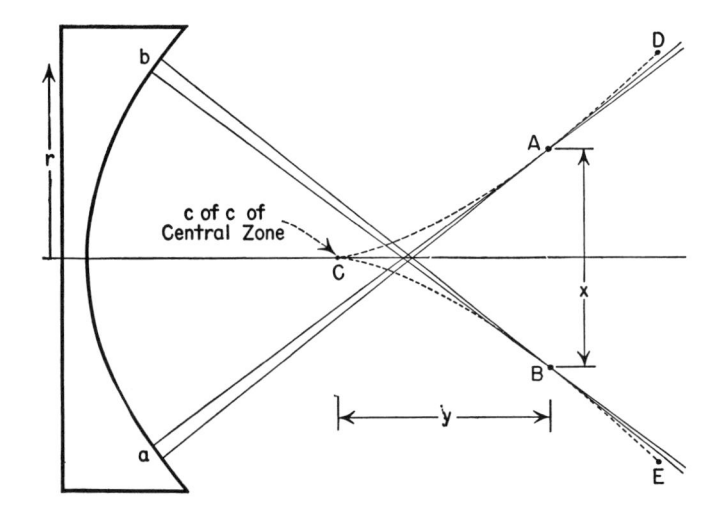

Fig. B.8.2.

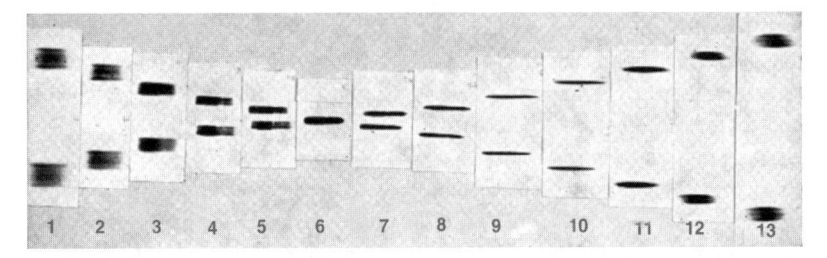

Fig. B.8.3: Images of a slit about $1/635$ inch wide and $1/25$ inch high, taken near the average center of curvature of a 6-inch paraboloidal mirror. The mask openings were $7/8$ inch wide and their centers were 2.16 inches from the center of the mirror, which was at the left of the images. Slit and plate were moved as a unit along the optical axis distances of $1/15$ inch from numbers 1 to 3, $1/30$ inch from numbers 4 to 8, and $1/15$ inch from numbers 9 to 13. The principal fact revealed by this series of photographs is that the sixth image, which was taken at the crossover point on the axis, where the knife-edge is placed in the Foucault test, is not sharp, while the ninth pair, taken about $1/8$ inch farther from the mirror and at the true foci of the areas isolated by the mask, are sharp. These photographs were originally published in an article by Platzeck and Gaviola, of Argentina, in the Journal of the Optical Society of America, Volume 29 (1939), pages 484–500. Prints from the original plates were numbered and kindly furnished by Dr. E. Gaviola.

Again let us approach the situation from the purely experimental point of view. Figure B.8.3 is a series of photographs of the images of a slit, taken at 13 points spaced along the optical axis near the "average center of curvature" of a paraboloidal mirror, masked so as to expose the two sides of a single mirror zone. The ninth photograph from the left shows the images at A, B, Figure B.8.2. The sixth shows the image at the crossover point, that is, where the two cones of light from a and b cross the axis as in Figure B.8.2. (This is the *so-called* center of curvature of zone ab in the Foucault test.) The image on the sixth is not sharp because neither the cone of light from a nor that from b has yet come to a sharp focus at the crossover point, where this photograph was taken. On the ninth, the images are sharp because it was taken at the *actual* centers of curvature of the two spheres a and b (Figure B.8.2 at A, B).

These images speak for themselves about testing along the axis. To the tyro they may even speak too loudly, exaggerating some relations and throwing undue doubt on the Foucault test, which will not be supplanted by the test to be described. Briefly, they were made with a short-focus mirror and then further enlarged to make the effect large enough to show up clearly. It is easier now to see the why of some of the odd shadow behavior of the Foucault test. Although each side a and b of a masked zone is very nearly spherical, the centers of curvature A and B (Figures B.8.2 and 3) of these two spheres are separate from each other and do not lie on the optical axis. In performing the Foucault test, then, we placed the knife-edge at the crossover point (that is, *inside* the actual centers of curvature) and then struggled to force the shadows to behave as though it were at the center of curvature of both a and b simultaneously; a little like trying to force a nut when the thread is crossed.

If now anyone has read these revelations and worked up a temperature let him relax. In practice the Foucault test made with masks as described by Ellison, Porter and others, is adequate for a large majority of surfaces, where it makes possible measurements in good hands accurate to about $1/10$ wavelength, which is better than the standard tolerance of $1/8$ wavelength of good optics. Thousands of good mirrors made by it in the past 90 years or more provide tangible proof of this fact. Yet, for the more experienced worker a test that will not reliably measure more closely than the needed tolerance, thus having no reserve, leaves something to be desired. For short-focus mirrors, where many zones would have to be measured, a finer test is almost a must, as is also the case for other and more strongly aspherical surfaces.

The preceding paragraph may seem contradictory. Briefly, the situation is this: In the case of a 6-inch $f/8$ mirror, one would make knife-edge settings at the focus of the central, (0.707) and edge zones and in addition would check the surface between these zones to make sure that it was smooth. Since the total departure from a sphere in this case amounts to only about $1/2$ wavelength, it is very reasonable to assume that three measurements accurate to $1/10$ wavelength made on a smooth surface would assure adequate mirror performance. But now consider a 12-inch $f/5$ mirror, which will depart so far from a sphere that it is not easy to be sure by inspection that the surface has a smoothly flowing curve. It is easy to overlook zonal errors in the diffraction fringes preceding the main shadow. Here measurement of many zones is necessary and, for reasons previously explained, they must be narrow to attain $1/10$ wavelength accuracy. It is generally assumed that the errors committed in setting a knife-edge are *accidental,* that is, the displacement measured along the optical axis is just as likely to be too large as to be too small and thus errors cancel out in the average of several settings. On the other hand a *systematic* error is committed when the knife-edge is set always too close or always too far from the mirror, due perhaps to a repeated mistake in judging vague shadow behavior. Suppose then that five zones are measured on the 12-inch mirror, each with a systematic error of $1/10$ wavelength. The total error at the mirror's edge would be 5 times $1/10$ wavelength, or $1/2$ wavelength, a very considerable amount.

Conclusion: Either errors must not be systematic or they must be kept much less than $1/10$ wavelength; otherwise claims regarding a mirror's accuracy may prove illusory. Since $1/10$ wavelength accuracy for each zone measured has been shown to be about the best consistently obtainable with the Foucault test, and since judging a photometric balance between two widely separated holes in a mask has been shown to be definitely subject to systematic error, a more accurate test is needed. (The zonal Foucault test itself has no systematic error.—Linfoot, *loc. cit.* It is human fallibility that may introduce systematic error into the test results).

To summarize, then: We have tried to show that

1. the premises on which the zonal Foucault test is based are not exact;

2. nevertheless the Foucault test remains the basic testing technique to be mastered (and perhaps better understood) before attempting advanced methods;

3. and a better test is needed as soon as the worker's figuring skill is capable of taking advantage of its superior accuracy.[1]

---

[1] Editor's Note: Nearly all more advanced mirror makers will be able to recall their own wide grins after going back to test their first mirror after completing their of course practically perfect second; and the same on testing their second after completing the much superior third; and so on and on, up to the point where their curve of improvement flattened off to any asymptote of perfection (or else dipped downward again). Yet did not each mirror at the time it was made seem to give satisfactory performance on the stars at least until the observer's eye and mind had become more sophisticated and exacting? In 1929 Russell Porter was in your editor's shop just after mirror No. 3 had been finished and, after testing it, suggested a test of No. 2. The test was made, glances and chuckles were exchanged.

In the preceding part we saw that Foucault's test was a null test for a sphere and that it therefore was natural to try to divide an aspherical surface into a number of small, practically spherical segments and to test each little sphere separately. Lacking an accurate method of doing this, we then resorted to testing these little spheres in pairs by locating their foci. But with the knife-edge at the crossover point it was still inside the center of curvature of both little spheres, so no use was actually made of the nice properties of a sphere under the Foucault test. Rather, testing was reduced to reading a poorly designed photometer (one with very wide spacing between holes).

In that which follows the technique of testing at the real center of curvature of a zone will be introduced one step at a time.

Consider a mask of height sufficient to cover the mirror and more than twice as wide as the mirror, with a half-inch hole in its center. Start with the mask so placed that the left edge of the hole is just even with the left edge of the mirror as viewed from the testing stand. The average center of curvature of the small portion of the mirror exposed by the hole will be at D, Figure B.8.2. Then, as the hole is moved across the horizontal diameter of the mirror, the center of curvature traces out the dotted curve DCE. Since this is called a caustic curve, the test to be described will be called the caustic test.

Thus there are two modes of expression—at the center of curvature, or on the caustic curve. It will be convenient when dealing with a single zone to speak of its two centers of curvature, say A and B, Figure B.8.2, which of course lie on the caustic curve; whereas when reference is made to the mirror as a whole it will be convenient to speak of all its centers of curvature as a whole, which form the caustic curve. The process of parabolization starts with a spherical mirror whose caustic curve is a single point (since a sphere has but one center of curvature). It proceeds with the center of curvature of each zone slowly being changed by figuring until the mirror produces a caustic curve of the proper shape.

The simplest kind of testing rig, a razor blade on a block of wood, will suffice to introduce caustic testing. However, knife-edge settings instead of being made along the axis as in Foucault's test, are made along the caustic curve. The knife-edge is first set at the center of curvature of the central zone and then is pulled away from the mirror a small distance. It will be noticed that as it cuts across the axis from left to right the very first part to darken is a small patch on the right half of the mirror, and the very last part to darken is at the same distance from the center of the mirror on the left half. In Everest's technique these two areas, the first and the last to darken, are called the regions of greatest slope. The following theorem will hold for any smooth curve, not necessary paraboloidal. *When the first faint shadow appears on the right-hand half of the mirror, the knife-edge is at the center of curvature of the shadowed area; when the last glimmer of light fades from the corresponding area on the left half of the mirror, the knife-edge is at the center of curvature of that area.*

An important application of the theorem is made in a qualitative form of the caustic test (hereafter called the smoothness test) to reveal very small zonal errors that may be lost in deep shadows when short-focus paraboloids are tested by conventional methods. Starting at the center of curvature of the central zone, the knife-edge is moved in small steps away from the mirror, attention being paid at each step only to the modest sized area that first darkens as the knife-edge cuts across the axis. In effect this amounts to dividing the mirror into a string of small areas across its diameter (without using a mask), each area small enough so that it darkens evenly all over. Any small irregularities will be shown up with maximum sensitivity. Here a reasonably narrow slit and a delicate touch on the knife-edge will help.

This qualitative test may be made quantitative if a stick with brads driven in at measured intervals is hung in front of the mirror to measure shadow location and a scale is provided on the knife-edge to measure its displacement. The knife-edge is first set at the center of curvature of the central zone. Then it is moved a distance y along the axis so that, as it is cut across the axis, the first and last areas in shadow are centered on the first pair of brads, one on either side of the center of the mirror. This is repeated for each pair of brads. For a perfect parabola it turns out (if the slit is fixed) that the displacement should be $Y = 3r^2/R$ where r is the distance from the center of the mirror to a given brad, and R is the radius of curvature of the central zone. This displacement is just three times as large as that measured in Foucault's zonal test. Thus the displacements measured in the caustic test may be divided by 3 and then compared with the old familiar formula r^2/R. A suitable testing rig, a little more refined than a block of wood but no more complicated in principle, will be described in a later section.

The advantages of this test over Foucault's test are:

1. masks are not used;

2. attention is fixed on only one side (then on the other) of a zone while making settings, with no comparison between them required;

3. the position of the shadow is essential rather than its comparative brightness;

4. measured knife-edge displacements may be interpreted as in Foucault's test after they have been divided by 3 (for a paraboloid);

5. it serves to introduce the workers to testing along the caustic curve;

6. it is particularly useful for very short-focus mirrors, say $f/1$ to $f/2$.

The principle disadvantage is the difficulty in estimating the center of the first and last areas to darken. After a little practice it has been found to give accurate results and has the advantage of requiring less fancy equipment than the advanced (and more accurate) form of caustic testing now to be described; though the latter will divorce the test procedure from all vague shadow behavior, much to its advantage.[2]

In the *Journal of the Optical Society of America, 29* (November 1939), appeared a paper by Richard Platzeck of the LaPlata Observatory and E. Gaviola of the Cordoba Observatory, Argentina, entitled "On the Errors of Testing, and a New Method of Surveying Optical Surfaces and Systems." This new method is a caustic test with a different measuring technique than that described above, so we will backtrack a bit.

It is shown, by a derivation we will dodge, that if a hole (say a, Figure B.8.2) has a diameter of about $R/100$ inch the small segment of the mirror thus exposed can be fitted to within $1/100$ wavelength by some sphere; and this sphere will have its center of curvature at A, Figure B.8.2. This is a very large hole compared to those used in Foucault's test. For a 12-inch $f/5$ mirror it would be $R/100 = 120/100 = 1.2$ inch! However, in order to take advantage of this fact, it is necessary to test at the actual center of curvature A.

The next step is to set up a test procedure that will measure the amount of correction a mirror has received. In Figure B.8.2, the two cones of light reflected from the mirror at a and b will have directions determined by the simple law of reflection; that is, incident and reflected rays make equal angles with the normal to the surface. We can therefore make use of the cones as long optical pointers that will be very sensitive to small changes in the shape of the surface (which changes the direction of the normal), provided the cones can be accurately located by measurements.

Roughly speaking, one can locate the two cones by making measurements anywhere along their lengths, but in practice only a few positions are useful. Foucault's test uses the crossover point, which looks good until it is tried. The best place of all is at the actual centers of curvature A and B where the cones of light have narrowed down to a sharp focus (Figure B.8.3, No. 9) and the two holes a and b, Figure B.8.2, actually test like

---

[2]The caustic test described above is neither new nor original, though the writer worked it out independently by comparing Everest's material in Chapter B.7 with the paper referred to in the next paragraph. Subsequently a paper by M.G. Yvon in *Revue D'Optique 4*, page 8 and following, has been found in which this quantitative form of caustic test is described, and which has proved invaluable.

spheres. Two measurements will be made—this distance y measured along the optical axis from the center of curvature of the center zone out to where a line joining A and B, Figure B.8.2, crosses the optical axis; the distance x from A to B, measured perpendicular to the optical axis.

For making this measurement a knife-edge is poorly suited since it is a one-sided device, a circumstance that subjects settings on A and B to systematic error. It can, however, be made a two-sided symmetrical device by substituting for the one-sided knife-edge a simple vertical wire. One can see around both sides of a fine wire, and thus tell when it is exactly centered.

There is a good reason to expect this form of testing to give more accurate results than Foucault's test. In the latter test the knife-edge is moved *along* the optical axis to locate by trial and error the point where two out-of-focus cones of light (Figure B.8.2, Figure B.8.3 No. 6) cross over each other at an angle of about 5° or less. Accuracy depends upon the correctness of this rather uncertain setting. In practice, in the caustic test the fine wire is displaced along the optical axis merely to locate it *near enough* to a line joining the centers of curvature A and B, Figure B.8.2, so that a transverse or cross-the-axis motion will cut the wire almost at right angles *across* the two cones of light where they are in sharpest focus. Centering a fine wire on a sharply focused image of a narrow slit can be done very precisely. Figure B.8.3 shows that in this rather extreme case "near enough" means at least $1/10$ inch, the distance between No. 9 and No. 10 which are almost equally in sharp focus. For paraboloids the easiest way to take care of this matter is to calculate the theoretical value Y, set the wire at exactly this distance from the center of curvature of the central zone and measure x. As figuring progresses, the centers of curvature draw nearer and nearer the test wire, and if perchance a perfect mirror is produced, the last test will be run with the wire exactly at the center of curvature.

In summary, testing is done with a fine wire, near or at the true center of curvature. The wire is displaced a calculated distance Y from the center of curvature of the central zone, so that the critical measurement is that of the distance x between the two sharp images of the slit; this distance can be measured very precisely.

Benefits result as follows:

1. The accuracy of measurement can be increased as much as five to ten times.

2. Only one hole in the mask is observed at a time, the wire being set so that the light from that hole is a minimum; and this kind of observation is much easier than matching two widely separated areas.

3. Or, alternatively, two other observing methods entirely eliminate the observation of shadows on the mirror, with an increase in precision.

4. Considerably wider holes can be cut in the mask.

Such a promising array of advantages makes a tryout of the test seem almost imperative.

In the following sections a step by step procedure will be given, followed by an example. Then a simple but suitable test rig will be described. If at first these instructions look formidable, it is because they are given in sufficient detail to help the uninitiated to get started. The actual steps involved are italicized and are seen to be concise. In fact, steps 1 through 3 and the calculation in 9a are done once for all at the beginning of testing. With this simplification and with increasing familiarity by use, the whole test can be run and the results analyzed in less time than it takes to study through this material. The tolerances specified hold the error in the final results to $1/100$ wavelength or less for each measurement in question. Essentially this makes the accuracy of the test depend largely on the accuracy with which the distance x is measured, which can be adjusted to changing needs without starting from scratch again.

NOTE: FROM THIS POINT ON, THE SLIT WILL BE FIXED, WITH ONLY THE TEST WIRE BEING MOVED.

B.8.1 Caustic Testing Procedure

1. *Measure the radius of curvature R of the central zone* of the mirror with an error less than $1/2$ inch for an $f/8$ mirror, less than $1/8$ inch for an $f/5$.

2. *Cut a mask* out of thin cardboard. The holes may have a diameter as large as $R/100$ inches if the mirror surface is smooth. The distance r from the optical axis to the center of each hole must be known within 0.05 inch for an $f/8$ mirror, within 0.02 inch for an $f/5$. There should be an odd number of holes so that one exposes the central zone. If the holes are carefully laid out with a compass and the cutting is neatly done, the value of r for each hole is known. If narrow zones are present, the holes will have to be smaller so that they will show up in the test results.

3. *Calculate the theoretical value Y* for each pair of holes corresponding to a mirror zone. $Y = 3r^2/R$. Note that this is just three times the displacement in Foucault's test.

B.8.1. Caustic Testing Procedure

4. *Line up the test rig* so that the fine wire stays on the optical axis as it is moved away from the mirror.

5. *Set the wire at the center of curvature of the central zone* and read the scale. While watching only the central hole in the mask set the test wire so that as it cuts across the optical axis the shadow moves neither to the right nor to the left—the zone darkens evenly, exactly as in Foucault's test. (For a better procedure see Section B.8.5.)

6. *Move the wire away from the mirror a distance y*, along the axis, as calculated for the first zone to be tested. y must be correct within 1.5/1000 inch for an $f/5$ mirror, within $2/1000$ for an $f/8$. (The slit remains fixed.)

7. *Measure the distance x.* Set the wire so that the light from one hole of the zone is a minimum and read the scale; move the wire until the light from the other hole is a minimum and read the scale. The difference in readings is x. To get a sharp minimum, the illuminated slit should be narrow and the wire about the same size as the sharp image of the slit. After a little experience more accurate results will be obtained by use of the alternate setting techniques described in the section on Accuracy, Section B.8.5. x must be measured to within 0.0002 inch.

8. *Repeat steps 6 and 7 for each zone in turn.*

9. *Interpret the results.* This may be done in at least three ways.

    9a. Compare the measured values x with the theoretical values X for the corresponding zones given by $4r^3/R^2$. If the measured value x is larger than the calculated value X for a given zone, that zone is undercorrected, and vice versa.

    9b. Best of all, the actual shape of the mirror surface can be easily calculated. This is done by calculating the amount and sign, + or −, of the deviation from correct for each zone and adding them algebraically. Thus: deviation of a zone = $K(X-x)$ where K = width of hole in mask/$4R$. K is a constant having the same value for all zones. Since a mirror is figured by leaving the edge alone and wearing away the central zones we will consider the deviation to be zero at the edge. Then at the inner edge of each zone the deviation of the mirror surface from a perfect parabola will be the algebraic sum of all errors from the mirror's edge up to that point.

9c. Since most workers are used to thinking of mirror correction in terms of the longitudinal displacement of the knife-edge in Foucault's test, it may be more convenient at first to put the results of the caustic test in the same form. This can be done by finding an equation which will transform an error in the measured value x for a given zone into an equivalent error in the longitudinal displacement of a knife-edge along the axis in Foucault's test. Without showing the derivation, this can simply be stated as knife-edge error $= R/2r(X - x)$, where the measured values x are taken from step 7 above, and the theoretical values X are given by the equation in step 9a. Thus if x is measured to be too large, the equivalent longitudinal displacement would be less than r^2/R and the zone is undercorrected.

An example will serve to tie all these directions together. Take a 12-inch diameter $f/5$ mirror and run through our procedure.

1. Make $R = 121$ inches, measured to the nearest $1/8$ inch.

2. Maximum diameter of holes in the mask $R/100 = 121/100 = 1.2$ inch (approximately).

A mask is laid out with centers 1.1 inch apart, giving 11 holes. The holes are cut out a little smaller than this to separate adjacent holes a little, or two masks are made, as suggested. Record values of r in column 2, Table B.8.1.

3. For the first zone $Y = 3r^2/R = 3(1.1)^2/121 = 0.030$ inch. Since $3/121$ appears in each calculation of Y, it will save time to calculate at the start. Thus, for zone 2, $Y = 0.0248(2.2)^2 = 0.120$, and so on. Record the values of Y in column 3.

4. Line up the test rig.

5. Set the wire at the center of curvature of the central zone.

6. To measure the first zone, move the wire along the axis a distance $Y = 0.030$ inch from its position found in step 5 (move away from the mirror).

7. Measure x several times and record the average in column 5.

8. Repeat steps 6 and 7 for each zone.

B.8.2. Interpretation

9. Calculate the theoretical value of X for each zone, putting them in column 4. For zone 1, $X = 4(1.1)^3/121^2 = 0.0004$ inch. In calculating X the factor $4/121^2$ appears each time; calculate it separately to save time and mistakes. For zone 2, $X = 0.00027(2.2)^3 = 0.0029$ inch.

Table B.8.1
All dimensions in inches.

zone	ra- dius r	calcu- lated Y	calcu- lated X	meas- ured x	raw error $e = (X-x)$	adjusted error $e = e - cr$	deviation of zone h	deviation of mirror from true parab. sum of h's
0	0	0	0	0	0	0	0	0
1	1.1	0.030	0.0004	0.0004	0.0000	−0.0002	−0.0000004	−0.0000001
2	2.2	0.120	0.0029	0.0021	+0.0008	+0.0005	+0.0000011	+0.0000003
3	3.3	0.270	0.0098	0.0082	+0.0016	+0.0011	+0.0000025	−0.0000008
4	4.4	0.480	0.0232	0.0224	+0.0008	+0.0001	+0.0000003	−0.0000033
5	5.5	0.750	0.0455	0.0463	−0.0008	−0.0016	−0.0000036	−0.0000036
edge								0.00000

B.8.2 Interpretation

A detailed analysis of our example is given under 9a and 9b for illustrative purposes. A short cut for routine testing is given at the end of 9b which will decrease the numerical labor involved.

9a. *The mirror is figured until the measured values x in column 5 match the corresponding theoretical values X in column 4 within the desired tolerance.* As a rule of thumb for any mask having holes about $R/100$ inch in diameter the error $(X − x)$ for each zone (recorded in column 6) can be about 400 times the allowed error per zone of the mirror surface. For example, if the mirror's surface must be tested to the nearest $1/2$ millionth of an inch for each zone, then the allowable error is $(X − x) = 400/2(0.000001) = 0.0002$ inch. For this tolerance, values in column 5 may differ from those in column 4 by as much as $1/5$ of $1/1000$ inch. This would by a very tight tolerance if it were not possible to do some juggling of figures. (Skip the next paragraph if the alternative procedure was followed in Step 5.)

The values of X in column 4 were calculated for a certain radius of curvature R for the center zone. In practice R will not be exactly the actual or true value for the mirror under test, and if it were the wire probably would not be set at exactly this distance from the mirror under Step 5. Furthermore, some other paraboloid having a slightly different radius of curvature might fit the mirror better. Perhaps the most instructive method of adjustment is to plot a graph of the errors $(X-x)$ as in Figure B.8.4a. The horizontal zero error line corresponds to a paraboloid whose radius of curvature R was used in calculating Y and X. When R is changed to a different value this zero error line rotates about the point $r = 0$ as an axis. A clear plastic straightedge is very convenient for drawing in a new line such that the average error is zero; that is, using the new line as a reference the errors are made as small as possible. This line is shown dashed in the figure. Those who prefer numbers will note that this method of adjustment corresponds to subtracting r (multiplied by a constant) from raw error, the constant being 0.00015 in this example. In practice the constant must be found by trial and error, so the graph is simpler. The adjusted errors are recorded in column 7.

9b. It is possible to plan the strategy for the next spell of figuring by a study of the adjusted errors in column 7, just as has been done all along with knife-edge settings in Foucault's test. Strategy is much simplified, however, by a knowledge of the actual shape of the mirror surface in relation to the desired paraboloid. Deviation of zone $= K \times$ adjusted error. $K =$ width of holes$/4R = 1.1/(4 \times 121) = 0.0023$. Since an important point of interpretation is involved, several sample calculations will be given.

B.8.3 Sample Calculations

Deviation of zone $5 = 0.0023(-0.0016) = -0.0000036$ inch. Assume that the desired parabola and the actual surface to be coincident at the very edge of the mirror. This means that at the inner edge of zone 5 the mirror surface is 3.6 millionths of an inch too low (low because of the minus sign). This value is recorded in both columns 8 and 9.

Deviation of zone $4 = 0.0023(+0.00014) = +0.0000003$. Record in column 8. This deviation, added to that already existing where zone 4 joins zone 5, makes the surface 3.3 millionths of an inch too low at the inner edge of zone 5. Record -0.0000033 in column 9. Similarly adding the deviation of zone 3 leaves the mirror 0.8 millionths of an inch too low at the inner edge of this zone. To summarize, values in column 8 are K times those in column 7; column 9 is obtained by writing 0 for the edge of the mirror and summing up algebraically the values in column 8. Column 9 is plotted in Figure B.8.4b, showing the distance of the actual surface above (plus sign) or below (minus sign) the paraboloid chosen when column 7 was worked out, which is represented by the dashed line.

B.8.3. Sample Calculations

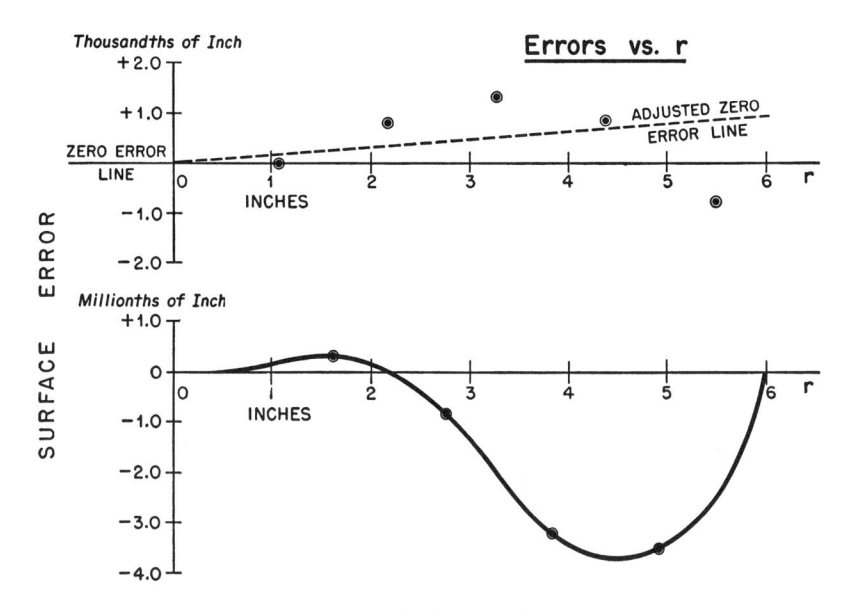

Fig. B.8.4: a *(top)* and b *(bottom)*.

From Figure B.8.4b, it is evident that the adjusted zero reference line of Figure B.8.4a was not drawn in a manner calculated to make the next stage of figuring easier, even though it looked like the thing to do under step 9a of the procedure. It would be much better to adjust so that the error for zone 5 is reduced to zero by drawing the dashed line sloping downward through the outermost point in Figure B.8.4a; in Figure B.8.4b, this would result in the actual surface following the paraboloid over zone 5 and then rising steeply to a central bulge about 11 millionths of an inch high. Thus we see that rotating the dashed line of Figure B.8.4a about the point $r = 0$ corresponds to rotating the curve of Figure B.8.4b, about the point $r = 6$. A little time spent in juggling these two curves will give an insight into the problem and will emphasize the value of drawing the curve of shape.

Having gone through the whole process in great detail, as one might do it during final stages of parabolization, a short cut is in order for the early stages when superior accuracy is not needed. The numbers in the first four columns of Table B.8.1 will not change during the parabolization process unless R changes by a significant amount—they need not be recalculated for each test. Values of x can be measured with a gradually tightening tolerance, starting at 0.0005 inch. The errors $(X - x)$ are found and recorded, without attempting an adjustment for the time being. With sufficient accuracy $K = 0.002$. Thus the approximate deviation h of a zone (in millionths of an inch) is simply twice the error of the zone (in thousandths of an inch). The surface is given by summing the deviations algebraically. After a look at the plot of this surface, a try at adjustment can be made and new deviation calculated from the adjusted errors as in Table B.8.1.

The whole analysis—the table work—can be completed in five to ten minutes. It goes without saying that care must be exercised in keeping plus and minus signs where they belong. Further, there is no need to carry along useless decimal places; record X, x, and the errors to the nearest 0.0001 inch and deviations and surface to 0.1 millionths of an inch. It will be noted that even in the detailed analysis two significant figures sufficed to give the required accuracy in practically every equation.

9c. Although the writer feels strongly that the method outlined above is by far the best, some may wish to throw the results of the caustic test over into the more familiar longitudinal displacements-along-the-axis of the Foucault test. For zone 5, taking $(X - x)$ from column 6,

$$\text{knife-edge error} = \frac{R}{2r}(X - x) = \frac{121}{2(5.5)}(-0.0008) = -0.0088$$

The calculated displacement of the knife-edge for Foucault's test would be one third the value given in column 3, or 0.250 inch. Thus the equivalent Foucault test reading for this zone would be 0.2500 − 0.0088 = 0.241 inch. This calculation enables us to make at least a rough comparison between the caustic and the Foucault test. The difference between measured and calculated displacements is about at the limit to which most workers can make knife-edge settings, and that only with mask holes much narrower than 1.1 inch. From column 6 the deviation of this zone is 1.8 millionths of an inch, about four times the easily-obtained limiting accuracy of the caustic test using the test platform now to be described.

B.8.4 Test Rigs[†]

Three methods of caustic testing have been described. The first, being used for studying surface smoothness, involves no measurements and so requires no elaborate test equipment beyond that which the worker already possesses. The second (Yvon) method is a quantitative version of the first method and requires a simple but carefully made test rig. The third test method developed by Gaviola and Platzeck requires an accurate measuring device. In this section several forms of test rig will be described that can

---
[†]One of which is also suitable for the Foucault test.

B.8.4. Test Rigs

be made with a minimum of tools and yet will perform with the required accuracy. At the beginning it is emphasized that the examples described are offered to illustrate certain design principles rather than to be copied slavishly. The individual worker is then left to his own ingenuity as has been the tradition in telescope making.

B.8.4.1 Test Rig for Second Method

Except for small mirrors of long focal length the old trick of pushing sidewise on the test bench will not work in the caustic test. In the worked out example the distance x between the two centers of curvature for zone 5 was nearly 0.05 inch. Hence some means of moving the knife-edge perpendicular to the optical axis must be provided. A device similar to that shown in Figure B.7.18, would be satisfactory if the transverse motion were provided. This may be provided by a lengthwise arm on top of the knife-edge block, pivoted at the end farthest from the eye and carrying a knife-edge at the end nearest the eye, with a machine screw working against a spring to supply the cross motion at the side. The cleat on the lower piece serves to keep the block sliding parallel to the optical axis and the screw adjustment makes easy the examination of the first and last areas to darken. An accurately made Barr scale may serve to measure displacement along the axis, or a steel rule (such as a Brown and Sharpe or a Lufkin) reading to 0.01 inch may be fastened to the slide and an index provided for reading. With the help of a magnifying glass settings may be read to about 0.002 inch. Another simple arrangement involves fastening a thin sheet of brass to the lower board and making a short scratch mark along the straightedge with a needle for each knife-edge setting. The needle can be stuck into a stick for a handle; a few trials will show how to make the scratch mark so that no variation is introduced into the readings by the way the needle is held. A steel rule graduated to 0.01 inch used with a magnifier can be used to estimate the distances between marks to the nearest 0.002 inch.

There may reasonably be some question whether a test rig can be homemade without special machinery capable of measuring to the tolerances indicated in our example. The answer is affirmative in the same sense that it is possible to make a mirror to a tolerance of $1/4$ wavelength without special equipment. In both cases it is a matter of intelligent application of proper procedure, considerable work, and of pride over a job well-done.

B.8.4.2 Test Rig for Third Method

As outlined in the section on test procedure, a short piece of wire about 0.005 inch in diameter (B & S gauge 30 to 40) must be held parallel to the slit while moving along the optical axis a distance y and moving perpendicular to the axis a distance x. Imagine the test wire to be held at first by hand on the optical axis near the center of curvature of the central zone and parallel to the slit. It can be moved to any other position by some combination of these three motions:

1. along the axis,

2. perpendicular to the axis, left or right,

3. perpendicular to the axis, up or down.

Furthermore, the wire can be made no longer parallel to the slit by some combination of three rotations, one rotation about each of the above-mentioned three mutually perpendicular directions as axes. *Thus the wire is said to have six degrees of freedom.* The plan is to add four *constraints* to the wire in the form of a test rig so that it can move only in two directions—along the axis and perpendicular to the axis left or right. The wire will then have only two degrees of freedom. If any motion other than the two desired be attempted by the wire, some *restoring force* equal to the disturbing force must be brought to bear to inhibit such undesired motion. These principles go by the name of "kinematic" or "geometrical" design, which is well-known among instrument makers.

A few pictures will be worth reams of words in explaining the application of kinematic design. Note the four steel balls in Figure B.8.5 are each seated in a hole in the base plate and a fifth ball is by itself in a hole on the block. The slide has been temporarily removed and inverted to show construction details. In use it is replaced so that the group of four balls forms a guide on which the rod can slide, and the fifth ball supports the farther side of the plate. This slide is thus supported, that is, it is constrained, by five points of support. There is only one manner in which it can slide (the balls do not roll) while maintaining contact with all five balls—that is, along the length of the rod. Gravity holds it in contact with the balls and acts as a restoring force if the rod tries to ride up out of contact with any of them. The working principles are then:

1. use of supports that are points, or at least very small in area, so that the point of support is definitely located;

2. 6 minus the number of support points equals the number of degrees of freedom remaining, provided that no useless or "redundant" points are used;

3. gravity (or sometimes a spring) is used as a restoring force.

B.8.4. Test Rigs

Fig. B.8.5.

The part of the plate sliding on the fifth ball has been ground flat, an operation calling for a special machine. Figure B.8.7 shows how a piece of commercial flat-ground stock can be bolted on to accomplish the same purpose.

The slide just described is mounted on top of another slide with their directions of travel at right angles so that the test wire will have the desired two degrees of freedom. The slide underneath in Figure B.8.5 was built on the same design principles carried out in a different manner, as shown in Figure B.8.6. There the rod slides on four cones turned in a lathe. The cones could as well be filed and polished with fine emery cloth, or, simpler still, a V notch can be cut in each of two blocks with the inside of the V rounded somewhat so that point contact is made with the rod. Together the two slides make a rugged instrument whose main disadvantage is a rather large amount of friction on the lower slide.

Figure B.8.7 shows how sliding friction can be replaced by rolling friction. Shown also is the piece of flat-ground stock on which the fifth ball rolls.

The top slide of Figure B.8.8 shows still another arrangement that has, however, proved rather unsatisfactory because of its small size and weight—the restoring force is too small. Three rods are clamped in position side by side. Two of these support the micrometer head and form a way in which slide two balls which are fastened to the underside of the slide. A fifth

Fig. B.8.6.

Fig. B.8.7.

B.8.4. Test Rigs

Fig. B.8.8.

point of support is provided by the third rod and a small rod mounted at right angles to it on the underside of the slide. In principle this might be called an upside down version of Figure B.8.5 and B.8.6.

Some means of pushing the slides and measuring the distance moved must be provided. A screw pushing each slide gives delicate control. For all purposes except the most exact testing, the distance y along the optical axis can be measured by mounting a steel rule graduated to 0.01 inch on the slide, with a fine index mark attached to some fixed part of the instrument. A magnifying glass will help in estimating fractions of divisions. Another trick is to scribe a special scale on sheet brass, using the micrometer screw next discussed, with marks at the calculated values of Y from the first mark. For all around convenience and accuracy a micrometer screw is better, and for measuring x it is a necessity. Such a micrometer screw is commercially available in the form of a micrometer caliper which sells for about $8.00 and is accurate to a fraction of 0.0001 inch. For those already in possession of such a caliper the slide described can with a little ingenuity be arranged so that they can be pushed by the spindle (unthreaded end of the screw); the distance the slide is pushed can be read off to the nearest 0.001 inch and estimated to the nearest 0.0001. A neater arrangement for those not already outfitted uses the so-called micrometer head (Figure B.8.7) which is built just like a micrometer caliper except that the horseshoe shaped frame is absent. Where the frame usually joins the nub of a micrometer caliper the micrometer head has a cylindrical shank that can be clamped in a $3/8$-inch hole. The writer has purchased two for about $8.00 each. Cheaper still is a model having a screw $1/2$-inch long which is enough for measuring x.

In the first two slides discussed, the micrometer head is clamped in a hole in line with the rod it pushes; a disadvantage shows up in trying to line up the rod and screw. If they are not in line the slide does not move the distance indicated by the micrometer division; this is a systematic error. Figure B.8.8 shows how to take care of the objection automatically. A U-shaped piece of metal is cut out too small and carefully filed on its inner sides until it just slips over the two rods and holds them in place without shake or binding. Similarly the inner side of the bottom of the inverted U is filed very carefully so that the micrometer head, which lies in the groove formed by the two rods, is just clamped in place when the screws are tightened down. This trick again illustrates the idea of *designing errors out* of the system, rather than depending too much on super-accurate workmanship.

Also shown in Figure B.8.8 is a weight attached to the slide by means of a heavy thread running over a small pulley. This serves to keep the rod tight against the end of the micrometer spindle, thus minimizing *backlash*. Actually it is not the best practice to allow the flat end of the spindle to push against the rod since it might not be cut off exactly square and it may not be exactly flat, introducing a small error in the readings which repeats with each full turn of the screw—*a periodic error*. This is of importance only in measuring x and then perhaps only for the finest measurements. In Figure B.8.8 a short sleeve can be seen on the end of the spindle. It was made from a piece of rod by drilling a hole in it to make a tight sliding fit over the end of the spindle; a steel ball of the same diameter is then pushed in the end. Thus contact between spindle and the rod it pushes is limited to a point. A similar device (Starrett) can be purchased with the micrometer head.

The following are a few practical points concerning use. The writer works right-handed; others may prefer to put the x micrometer on the left side. In any case, if a micrometer head is used to measure y, it will almost have to be on the side toward the mirror to be out of the way of one's nose when the eye is brought close to the test wire. Considerable thought should be given to the arrangement of parts for convenience in use, and in relation to one's facial protuberances, for one prerequisite for accurate measurements of any kind is a reasonably comfortable position for the observer. On the other hand the test wire should be as close as possible to a line drawn through the length of each micrometer screw for the sake of accuracy. Under usual basement humidity conditions the breath will condense on the metal. It is therefore best to make all the flat plates of brass, which is also much easier to work than steel. Coat all steel parts including the micrometer with a thin film of Vaseline.

B.8.4. Test Rigs

The rods used are drill rod, which is very straight and accurately round in sizes over $1/2$ inch. It is probably best selected for straightness and freedom from nicks, and it can be obtained from machine shop or machinery supply houses. Handle it carefully while cutting, to avoid marring it. The steel balls were two bits a dozen at a hardware store. Brown and Sharpe, Providence, R.I., and L.S. Starrett, Athol, Mass., are among the reliable makers of steel scales, micrometer calipers, micrometer heads, flat ground stock and other small tools. Purchases can be made through machinery supply houses or directly. Investigate discounts, given by some, by others not. Both companies named put out small tools catalogs which are more interesting than a mail order catalog to the tool-minded and a source of helpful ideas.

The writer has the following equipment, listed only as suggestions since similar equipment by other makers or the same maker are equally useful.

- Six-inch steel scale calibrated in $1/16$, $1/32$, $1/64$, $1/100$ inches. L.S. Starrett No. 607 No. 7 graduation.

- Micrometer head, one-inch thread, with vernier reading to 0.0001 inch. Brown and Sharpe No. 295 RS.

- Micrometer head, one-inch thread, without vernier, estimate to 0.0001 inch. Lufkin.

Brown and Sharpe has a micrometer head with $1/2$-inch thread, No. 290 which is $1.00 less than the one-inch thread model. Some may wish to purchase a regular micrometer caliper and modify the slide for use with it, in order to have a regular shop tool; No. 8 and No. 11 are typical among a wide choice. Another suggestion is to use a micrometer depth gauge. The long rod running through the screw could be clamped so it can rotate but not move longitudinally. Then, as the screw is turned, the base will travel like a slide if prevented from rotating by letting it slide along a rod placed parallel to the screw on one side.

Tools used include hacksaw, file, several small drills, hand drill (or drill press if available), hammer, center punch, 6-32 tap and holder. The tap and holder can be dispensed with if the work is bolted together with screws and nuts. Those having more tools than these can perhaps do a fancier job, those with less may get around on ingenuity or ability to borrow. It is emphasized again that careful workmanship plus attention to design features will produce a good job, more than fancy tools used without imagination.

Perhaps some new justification might be needed for attempting the test rig described. Besides its use for caustic testing it is also useful for Foucault testing if a razor blade is mounted near the test wire. It can also be used for shadow analysis testing as described by Gaviola in the *Journal of the Optical Society of America*, Volume 29, page 484–500, November 1939, and for the Zernike test. If two slides are made somewhat larger than required for testing purposes alone and a low-power microscope with cross hairs is set up over it, the worker will have a combination toolmakers microscope; starplate coordinate measurer, spectrum plate comparator, measuring microscope; or the microscope might be mounted on a slide to make a traversing microscope. The eyepiece for a microscope can be borrowed from a telescope, the objective of the microscope can be made, bought second hand, improvised from available or surplus lenses or purchased at a modest price from Wm. Gaertner Scientific Corp., 1201 Wrightwood Ave., Chicago, Ill. A Hastings triplet will serve quite well as an objective, especially if stopped down. The micrometer heads can be removed from the stage and used in a spherometer or perhaps in a double-star micrometer such as described in ATM2, subsection A.3.3.2. In fact they can be used anywhere an accurate screw is called for.

Only a few possibilities have been discussed above. Much help will be gained by studying the section on kinematic or geometrical design in such books as

1. *Fundamentals of Optical Engineering,* Jacobs, McGraw-Hill Book Co., New York, 1943, pp. 286–295.

2. *Procedures in Experimental Physics,* Strong, Prentice-Hall, Inc., New York, 1943, pp. 585–590. (Reprinted by Lindsay Publications Inc. Bradley, Il.)

3. *Kinematical Design of Couplings in Instrument Mechanisms,* Pollard.

4. *Design and Use of Instruments and Accurate Mechanisms,* Whitehead.

5. *Optical Measuring Instruments,* Martin.

6. *Dictionary of Applied Physics,* Glazebrook, National Bibliophile Service, Gloucester, Mass.

The first two are especially good and at the same time concise. Strong's book has thought-provoking diagrams. The others are more detailed and show diagrams and photographs of helpful examples. Also included are discussions on error of such devices as we have described, how measured and corrected. To the latter subject we now turn for a short discussion. While the Pollard and the Whitehead books are out of print they may be consulted in some libraries.

B.8.5 Accuracy

After making the test rig, most makers will have to lift themselves by their own bootstraps in checking and perhaps improving its accuracy. This

B.8.5. Accuracy

obviously is not impossible for it merely puts one in the same position as the originator of such devices. Several of the books listed above discuss at length the procedures involved. The author had the good fortune to have access to a Gaertner toolmaker's microscope; the lower slide of Figure B.8.8 (which is used to measure x) checked correct to within ± 0.00005 inch over a distance of $3/4$ inch and the upper slide within ± 0.0001 inch. The upper slide of Figure B.8.5 checked as well against one of Rowland's ruling engine screws. These tolerances give an idea of the accuracy of the commercial micrometer heads used and of results obtainable by kinematic design.

The drill rod should be reasonably straight. A length of it can be placed in two V notches and rotated to detect any wobble. It should not be bent when clamped in place on the slides; hence the short hold-down blocks. If the assembly has been done properly, the test wire will remain parallel to the slit during its displacement along the optical axis.

While it is likely that the test rig is good enough as it is, it is fun to check up on it. Those having faith in their own handiwork may skip this paragraph. There will be no doubt about the micrometer, but the slides may not move the distance indicated by the screw. Either systematic or periodic error may be present. Systematic errors are most likely to arise from the screw and the rods not being parallel and they can be checked for measuring a known distance. For example, the width of some object about $1/2$ to $3/4$ inch wide, having sharp edges, can be measured with a micrometer caliper and the measurement checked by placing the object on the test dingbat and measuring it with the help of an improvised microscope. Gaertner makes a glass scale especially for this purpose, a luxury if one has the price. Periodic error can be checked for by making two fine scratches about $1/10$ inch apart at the edge of a slide and arranging a movable index mark. With the help of a magnifier, set the scratch mark opposite the index, read the micrometer, set the second scratch mark opposite the index and read the micrometer again. With a little practice the settings can be accurately made. Now move the index over about 0.005 inch and repeat; this makes the readings fall at a different place on the micrometer scale. Repeat until the readings have traveled around the scale several times. Now the distance measured is always the same; the measurements would be expected to differ somewhat, due to errors in setting against the index mark, but the error should not repeat with each complete turn of the screw. A repeating error is a periodic error. If present, the trouble may be in the coupling between the end of the screw and the slide it pushes.

Some idea of whether the test results are up to the standard of accuracy desired can be obtained from the measured values of x. For a given zone make 3 sets of 5 measurements of x, taking the average of each set. All the readings in a set will differ from their average and the size of this difference has significance. Under step 9a in the example, it was shown how to find the allowed error in measuring x for a given error on the mirror. As a rough rule of thumb, half the readings in a set of 5 should differ from their average by no more than this allowable error in x. If the three averages are compared with the average of all 15 readings, the differences will be much smaller; the size of the differences will give some indication of the ultimate accuracy of the test in the individual worker's hands. From this another rule of thumb can be derived. During rough figuring take the average of two readings per zone, the second serving as a check against gross error. As more accuracy is required, increase the number of readings per set on each zone up to 5, beyond which there is little advantage. For the final test, use the average of 3 sets of 5 readings each. Of course no amount of averaging will remove systematic error, which we have tried to design out of the test.

Accuracy can be increased by changing the manner in which the test wire is judged to be centered in the reflected cone of light. A method already mentioned is to set the wire so that the light from a given hole in the work is a minimum. A slit whose width is adjustable is a help in easily getting the maximum sensitivity, but several different wire sizes can be tried with a fixed slit. A movement of a few ten-thousandths of an inch of the wire should make a noticeable change in brightness of the hole. Two other methods entirely eliminate shadows on the mirror from consideration, a very considerable help when wide holes are used on short-focus mirrors. Testing is done on the images of the slit.

Method 1. A positive eyepiece of about $1/2$ to 1 inch focal length is mounted on the slide that measures x so that the test wire is in sharp focus. The test wire then serves as a cross hair to be centered as precisely as possible on the sharply focused images of the slit at A and at B, Figure B.8.2.

Method 2. This is a little more complicated at first, but becomes very precise with practice. The eyepiece is fixed about 6 inches back from the test wire so that the two beams of light from a and b, Figure B.8.2, are out of focus. Then, paying attention to only one of these two blobs of light, set the test wire so that the diffraction fringes observed are symmetrical. What happens is that as the test wire is centered in the beam at its focus it does not simply block off the light with a clean shadow but, rather, diffraction fringes are set up on either side in the shadow. If the wire is precisely centered, the pattern of fringes will be the same on both sides of the shadow of the wire; a slight error in centering shows up as a noticeable lack of symmetry in the pattern. Better than many words is to try it, moving the wire

B.8.5. Accuracy

very slowly across the beam of light a number of times. Again the relative size of slit and wire is of importance in producing good fringes.

A trick that will eliminate much of the adjusting called for under step 9a is this: Instead of starting the test as described in step 5, by setting the wire at the center of curvature of the central zone, set it so that the measured value of x for the edge zone is as close as possible to the theoretical value X, a cut and dried method. Move the test wire toward the mirror a distance equal to the value of Y for the edge zone and proceed with step 6. This assures that the error $(X - x)$ for the edge zone will be small, and the setting easier to judge accurately.

If a mirror is to be tested with a mask, the error allowed for each zone must be less than the desired $1/8$ wavelength tolerance usually allowed for the surface as a whole. In addition, some allowance must be made for possible errors on the Newtonian flat or the secondary mirror in the completed telescope. As a rule of thumb, if a mirror is tested in N zones, the allowed error per zone is wavelength/$11\sqrt{N}$ if errors are accidental and is wavelength/$11N$ if systematic error is present. Since wide zones are allowable in the caustic test, probably most amateur mirrors can be tested with not more than 9 zones, making the allowed errors per zone wavelength/33 and wavelength/100 respectively. The advantage of minimizing systematic error is obvious. If wavelength/40 per zone be adopted as a fair goal (about $1/2$ millionth of an inch), and the holes are $R/100$ inches wide, then, from step 9a, x must be measured to within 0.0002 inch, which is easily within the tolerance of the equipment described. With care, the error can be cut in half.

In Section B.8.1, a set of approximate tolerances was given on the allowable error in measuring the quantities involved. For the most precise results, more accurate tolerances can be derived by calculus or by simply varying the size of each term in equations for X and Y by small amounts. In either case results close to the following should be obtained: (all quantities in inches)

$$d_{\max} = 5.5 \left(\frac{\Delta h}{r}\right)^{1/3} R \text{ inch} \qquad (1)$$

This is the diameter of a spherical zone element whose center of curvature is at the average center of curvature of the actual zonal element under test—the separation of the two surfaces being Δh around the rim of the element.

$$\Delta x = \frac{2R}{d}\Delta h \text{ inch} \qquad (2)$$

$$\Delta R = 16m^3 \Delta x \quad \left(m = \frac{\text{focal length}}{\text{mirror diameter}} \right) \qquad (3)$$

$$\Delta r = 16m^2 \Delta x \qquad (4)$$

$$\Delta y = 4m\Delta x \qquad (5)$$

As a check on what limit can be reasonably reached, it is suggested that the rig be first used to make the best possible sphere, for in this case the errors can be seen with great sensitivity and compared with the test results, right down to the point where the caustic test measurements no longer show up the errors.

B.8.6 Nonparaboloidal Surfaces

The only change required to test other surfaces than the paraboloid is to calculate the proper values of Y and the corresponding values of X. The width of the mask holes may have to be decreased for more strongly aspherical surfaces.

For example, consider the sphere. It has only one center of curvature; for all zones $Y = 0$ and $X = 0$. Therefore the test wire is set at the desired radius of curvature R from the mirror and the values of x measured, all of which will be zero if the mirror is perfect. Caution: In this case some of the crossover points will probably fall in front of the test wire as was the case for the paraboloid, but some may fall behind it, in which case a minus sign should be placed before the equation for the deviation of a zone in step 9b.

For any other curve, whose equation is known, the values of Y and X can be calculated by means of the formulas for the center of curvature given in calculus books. There is an out for those not familiar with calculus, and for those cases where an equation is not known or is too complicated to use if known. In many such cases (for example the Wright telescope in ATM2, Chapter E.2) the theoretical knife-edge displacements along the axis have been given. The caustic test data can then be handled in a manner similar to that described under 9c under the procedure. Since the theoretical values of Y are not known in such a case, the test wire cannot be preset. The trick is to find the center of curvature of the center zone, record the test wire position, and then hunt for the center of curvature of each zone to be measured—i.e., the position of the test wire when the shadow in one of the holes exposing a given zone moves neither to right nor left as the wire cuts across the beam, but rather darkens evenly. The difference in these two positions is recorded as y, and x is measured at this value of y. The equivalent longitudinal displacement of the knife-edge in Foucault's test would then be given by knife-edge displacement $= y - (R/2r)x$ inch.

The writer was interested to learn that a part of the testing of the 200-inch telescope was carried out by a photographic version of the caustic test. A set of photographs, each like No. 9, Figure B.8.3, was made, one at the center of curvature of each of 13 zones, and the distance x was measured directly from the plates. Some idea of the size of the mirror is given by the largest value of y which was 21.4 inches and of x which was 2.04 inches, approximately. Some workers may wish to try this method on short-focus mirrors in place of one of the micrometer heads.

Attention is called to the ingenious test devised by E. Gaviola and described in the *Journal of the Optical Society of America,* Volume 29, pages 480–483, November 1939, for testing Cassegrainian secondaries, which can be run by the caustic method. A cheap lens is used to supply a converging beam of light on the convex mirror, the errors of this lens canceling out in the result.

The author has found the caustic test of immense value and hopes it will prove helpful to many others. He wishes to acknowledge the considerable help of John Strong in the project described, as well as the use of the facilities of this laboratory, both of which were freely extended.

B.8.7 Editor's Note

Reproduced here by permission directly from the dusty files of *Popular Astronomy*, 1902 August-September, are four compacted fragments from the classic article by the late Professor F.L.O. Wadsworth, entitled "Some Notes on the Correction and Testing of Parabolic Mirrors" cited earlier in this chapter. Only a small part of that 12-page article dealt with the virtually forgotten fact that the center of curvature of any off-axis area of a paraboloid is not on the axis, as most of us have assumed when Foucault testing. The article was buried in the earlier files of the periodical named, and this particular fact was buried within its more general discussions. It remained for Enrique Gaviola of Argentina to disinter this fragment and to prove its worth experimentally. In 1939, November, he, with Ricardo Platzeck, another Argentinean astronomer, described in the *Journal of the Optical Society of America* their method of surveying optical surfaces, based on Wadsworth's observation. Then, apparently, their own article became similarly buried in the monthly flood of periodicals. A contributing factor may have been the fact that its 17-page treatment was so mathematical that it may have scared away the working optician. Now that it appears in a book it is hoped that the preceding lucid exposition of the caustic test in several applicable forms will result in its use. It is true that it is at its best on short-focus mirrors but short-focus mirrors are important. The test enjoys an advan-

F. L. O. Wadsworth. 341

Under actual conditions of test, however, there will usually be a small difference between the above theoretical value and the actual quantity measured, even when the surface measured is a perfect parabola of revolution. The reason for this is that the point N of intersection of the normals is not the true focal point of the rays from either of the elements A or A' independently. For a radiant point at N the focus of the rays reflected

342 On the Correction and Testing of Parabolic Mirrors.

from the element A is at O, for the rays from A' at O'.

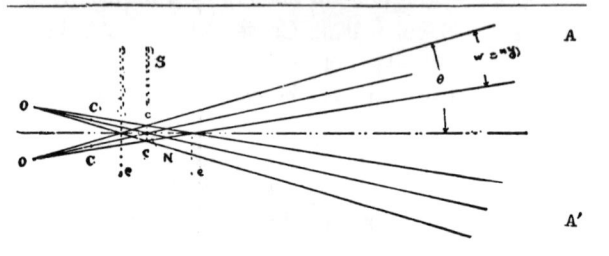

Fig. B.8.9.

tage in the matter of visibility; using the eyepiece technique of measuring, a mirror could be examined half a block away since the testing is done on the image. It is also of use on the high f ratio Cassegrainian, a significant advantage. It is an elegant method whereby the lover of high precision and the perfectionist may go about as far as he wishes, also have fun in building the beautiful kinematic stage and using it.

Irvin Schroader was born in Louisville, Ky., in 1918. At the susceptible age of 12 he was bitten by the TN bug when extensive use of a friend's $20x$ mail order telescope stimulated a desire for one of his own. Through a combination of influences—articles in *Scientific American*, ATM, and the advice of Ward deWitt—he was guided through to completion of a crudely figured but usable 6-inch reflector. A second attempt, a 10-inch, started a search for a test more satisfactory than Foucault's, ending in the adoption of Gaviola's caustic test.

Extracurricular activity in college consisted of assisting in the physics laboratory, figuring two mirrors for the college observatory, photographing and courting his wife to be (she says he did it with mirrors). Graduated B.A. in physics '42.

Following two months in the optic shop of the Gaertner Scientific Corporation, 3 years were spent as instructor in the Army Electronics Testing

B.8.8. Note Added in 1996

Fig. B.8.10: *Irvin Schroader*

Center at Harvard. He is now (1953) at the Applied Physics Laboratory, Johns Hopkins, as a senior physicist, and a member of the American Physical Society.

Driven by the continual lashing of ye ed's blacksnake whip (formerly cracked by the late Mr. Simon Legree), he wrote up the caustic test while playing hookey from graduate school. His present optical equipment consists of a 6-inch reflector mounted with circles and motor drive, and a temporarily mounted 12-inch, with a 16-inch telescope being slowly worked up. He admits piddling at photography, amateur radio (ex W3MFP) and with young sons, Jimmy and Bobby.

B.8.8 Note Added in 1996

A review of the terminology used in this chapter may help keep things straight. Capitalized values of X and Y are the perfect paraboloid quantities, while x and y are the measured values. Y is the value of the longitudinal shift (toward or away from the mirror) of the off-axis center of curvature of a zone from the center of curvature of the central zone. X is the sideways shift between the center of curvature of the left hole and right hole of a zone-pair. The origin of the coordinate system is at the center of curvature of the central zone. There is no column in Table B.8.1 for the measured value y, because the analysis technique does not rely on it. However, recording the y-values would be interesting because it would allow one to graph the horn-shaped region of the caustic. It would also be interesting to compare the "inferred" values of the knife-edge error calculated using $(X - x)R/2r$ with the more directly measured values of $(y - Y)/3$.

One other point should be made. If you put the light source on the moving platform of the kinematic stage, the motions analyzed in sections B.8.1 through B8.3 will be only half as large. The easiest way of then converting measurements to the data-analysis system used by Schroader is to multiply the measured values by two. Measurement errors are also multiplied by two, so the required accuracies of the micrometer and motion stage are emphasized. *H.R.S.*

Chapter B.9

Quantitative Optical Test for Telescope Mirrors

Several methods have been proposed and used in the testing of optical surfaces.[†] Among these the Foucault knife-edge test is the most popular as it is visual and requires only simple apparatus. The Hartmann test, being photographic, is used where a final analysis of the performance of an optical instrument is desired.

About 1923, Vasco Ronchi proposed a method of testing, which has the appearance of an interference method but involves only simple apparatus.[1] J.A. Anderson and R.W. Porter have found the Ronchi test to be fully as sensitive as the Foucault test and, under favorable conditions, capable of showing an error as small as $1/10$ of a wavelength in the wave surface leaving the principal plane of the instrument.[2] However, the Ronchi test is only qualitative as it does not indicate the amount of aberration present.

While considering the theory of the Ronchi test, and wishing that it were also quantitative, the writer discovered a simple scheme using this principle with the addition of a very simple auxiliary apparatus, which resulted in a quantitative test.

B.9.1 Quantitative Failure of Test

The feature, making the test quantitative, consists in superimposing two fine parallel line images, which serve as gauges, on the lens or mirror under test. The Ronchi grating is reduced to a single wire aligned with

---

[†] J.H. King, Member Technical Staff, Bell Telephone Laboratories, Inc. Reprinted by permission, from the *Journal of the Optical Society of America*, Sept. 1934.

[1] Vasco Ronchi, *La Provi dei sistemi ottica,* Bologna, 1925.

[2] Anderson and Porter, *Astrophys. J.* 70, 175 (1929).

the gauge lines and the optical slit. By moving the wire axially its shadow image is symmetrically expanded until it crosses the fixed gauge images at the zonal radii under test, the displacement necessary being a measure of the aberration existing between these zonal radii.

The exact method of forming the line images on the mirror is not important so long as it does not introduce errors larger than those of measurement. The method that first occurred to the writer was to stretch piano wires immediately in front of the speculum and view them by the diffraction around their edges or illuminate them by a light of contrasting color from one side. In the case of testing mirrors where the back (not the optical surface) of the mirror is polished and reasonably plane, a cardboard screen, placed against the back of the mirror, having the gauge slits and markings indicating the zonal radii cut in it and illuminated from behind, would be very simple and perhaps could not be surpassed. The method to be described produces virtual gauge line images against the speculum by the use of a small glass diagonal, and has the advantage that it is more generally applicable especially where the lens or mirror may be mounted in an inaccessible place. The distortion produced by the use of a small parallel glass diagonal in having the cone of rays from the speculum pass through it is probably small, since the diagonal need be nothing larger than a microscope slide cover glass and can be placed very close to the eye. It may even be placed behind the test wire right next to the eye.

B.9.2 Apparatus and Procedure

The setup of apparatus is shown in the sketch of Figure B.9.1. Here we have at B an optical slit from which the rays of light emanate, strike the speculum A (under test) and return to focus at G, the position of the eye for the test. Immediately in front of G is placed a tensioned wire C, arranged symmetrically in the returning cone and parallel with the optical slit. (The writer has used a wire about 0.006″ in diameter.) An unsilvered diagonal D reflects the image of two parallel gauge slits F, allowing them to be viewed at G, and appear symmetrically superimposed on the speculum A along with the shadow image of the test wire. The scale E is used in measuring the displacement of the wire C along the axis of the returning cone.

The procedure consists of adjusting the gauge slits until they appear as separated about $1/4$ or $1/3$ the diameter of the speculum and then leaving them fixed. The test wire is then adjusted along the axis until its shadow image appears to just touch the gauge images as in a in Figure B.9.2. The reading on the scale E, Figure B.9.1, is then noted. The wire is again moved along the axis until its shadow image intersects the gauge slit image

B.9.2. Apparatus and Procedure

Fig. B.9.1.

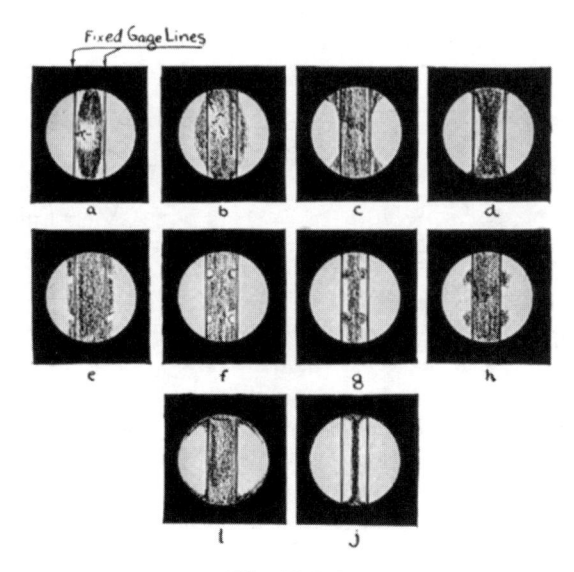

Fig. B.9.2.

symmetrically at zone r_2 as in b of Figure B.9.2. The reading on the scale E is again noted. The difference between the two scale readings measures directly the aberration existing between zone r_1 and r_2. Likewise, the remainder of the mirror outside zone r_1 may be tested with the gauge slits fixed. Should it be desired to test quantitatively inside zone r_1, the gauge slits should be moved closer together and the above procedure repeated.

B.9.3 Geometry of Test

Figure B.9.3, a, shows a horizontal plan view (not the conventional vertical section through an optical system) of the test when adjustment is made for zone r_1 as in a in Figure B.9.2 and shows only the rays proceeding from zone r_1 to focus. Similarly, b of Figure B.9.3 shows the plan view of the adjustment for zone r_2. In a of Figure B.9.3, y_1 is the distance that the test wire C of Figure B.9.1 is situated in front of focus to appear to just touch the gauge slit images. The distance that the wire C, Figure B.9.1, is situated in front of focus for zone r_2 is designated by y_2, the shadow image of the test wire C appearing, in this case, as in b of Figure B.9.2. Δy is the distance test wire C is moved to effect these two conditions. Since in a and b of Figure B.9.3, angles α_1 and α_2 are equal (within limits affecting this test), it follows from substantially similar triangles that $y_1 = y_2$ and Δy measures directly the aberration existing between these two respective zonal radii. In the case of testing a spherical mirror at its center of curvature, Δy is equal to two times the difference in radius of curvature of the zones providing the optical slit remains fixed.

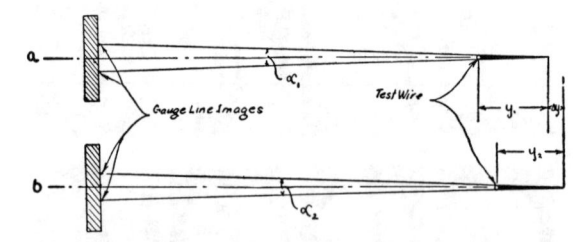

Fig. B.9.3: Horizontal plan views—looking down on test from above.

Before discussing the appearance of some common forms of aberration in this test, it might be well to state that the shadow image is not quite as shown in the sketches. With a wide optical slit, the shadow image of the test wire is not sharp on the edges, while a narrow slit causes secondary diffraction effects to appear. The fine slit adjustment seems preferable, since it gives the sharper shadow image to the test wire. Too fine a test wire (0.001" diameter) does not apparently improve the sensitivity of the test, since then the wire has to be brought very close to focus to expand its shadow appreciably, again making it difficult to preserve a sharp shadow image. However, in spite of the above, the writer has been able to duplicate measurements at least as accurately as with the Foucault test. The test is much faster than the Foucault test when the latter is used with zonal stops and at the same time the qualitative feature of the Ronchi test is always observable.

B.9.4 Appearances of Common Forms of Aberration

In *c-d*, *e-f*, *g-h* and *i-j* of Figure B.9.2 are shown some common forms of aberration as they appear in this test. Aberration of the opposite sign to *a-b* of Figure B.9.2 is shown in *c-d*, the former being the case of outer zones too long in focus, and the latter, outer zones too short. The appearance of a raised zone is shown in *e-f* of Figure B.9.2, and *g-h* shows a depressed zone at the same place on the speculum. Turned edge, one of the worst enemies in the figuring of optical surfaces, is easily detected and can be accurately analyzed by this test. In all other tests the wire C, Figure B.9.1, has been kept just inside of focus but in the case of turned edge the sketches Figure B.9.2 *i-j* show the appearance of placing the wire just outside the focus as this gives more prominence to the effect. In quantitatively testing parabolic mirrors of short focal ratio at the average center of curvature, the Foucault shadows become very deep, necessitating the use of a large number of zonal stops. The present test would seem more suitable in such cases since the use of stops is eliminated and photometry is unnecessary.

B.9.5 Editor's Note

Alan R. Kirkham of Tacoma, Washington, Franklin B. Wright of Berkeley, California, Loren L. Shumaker of Dayton, Ohio, and possibly others, were working on the same problem—making the Ronchi test quantitative—at about the same general time as Mr. King, and each developed his own method. Mr. Wright's was published in a *Supplement* to the *Bulletin of the Eastbay Astronomical Association;* Mr. Kirkham's was offered the *Scientific American* and was accepted, but publication was delayed through the present writer's dilatoriness; Mr. Shumaker's was distributed privately in a multigraphed circular.

B.9.6 Notes on the Ronchi Band Patterns[†]

The Ronchi band patterns shown in Figure B.9.4[3] are the appearances of a 3 3/4-inch mirror of 36-inch radius with a grating of 175 lines per inch interposed between the mirror and the observer. The light from a 2-volt flashlight bulb, first diffused by ground glass, passes through one half of the grating on its way to the mirror, returning to the eye through the other half.

The mirror was not truly spherical, having a central area of 0.016 inch longer radius than that of the marginal zone. The Ronchigram at i shows the knife-edge appearance of the mirror if the knife-edge were placed at the focus of its outer zone.

The Ronchi patterns seen through an engraver's screen are not perfectly sharp around the circumference. Various diffraction orders tend to blur the right and left-hand edges of the disk. In drawing the patterns reproduced here I intentionally omitted the blurred areas, in order to show more clearly the character of the patterns produced by the grating when placed at stated distances from the focal plane of the mirror (near its center of curvature, not at the focus).

The drawings of these patterns start with the grating about 1 1/2 inches inside the focus (a), and show the progressive changes up to the time the grating is at the focus (i); and then on through seven more positions ($j-p$), where the lines become too fine to be seen. It will be noticed that the most sensitive, and therefore most desirable, position of the grating is just inside the center of curvature of the mirror, as at g and h. The patterns here shown are separated by intervals of confusion where the superposition of different line series do not give characteristic figures.

As the center of curvature is approached the bands spread out rapidly and reveal the features which indicate the character of the mirror surface. The positions chosen between f and i are arbitrary, as there are here no intervals of confusion.

At h, 0.07 inch inside focus, the first wide band just covers the apparently raised central part of the mirror. Were this the only criterion for detecting surface errors the interpretation would not be clear. But, by moving the grating slightly nearer the center, the band continues to spread out until it covers the whole mirror. If then the grating is shifted laterally so that the broadened band advances from the left, we see the shadow pattern i, which is exactly the appearance under the knife-edge

---

[†]By Russell W. Porter.
[3]Reproduced by permission, from *Astrophysical Journal*, October 1929. "Ronchi's Method of Optical Testing," by Anderson and Porter.

B.9.6. Notes on the Ronchi Band Patterns

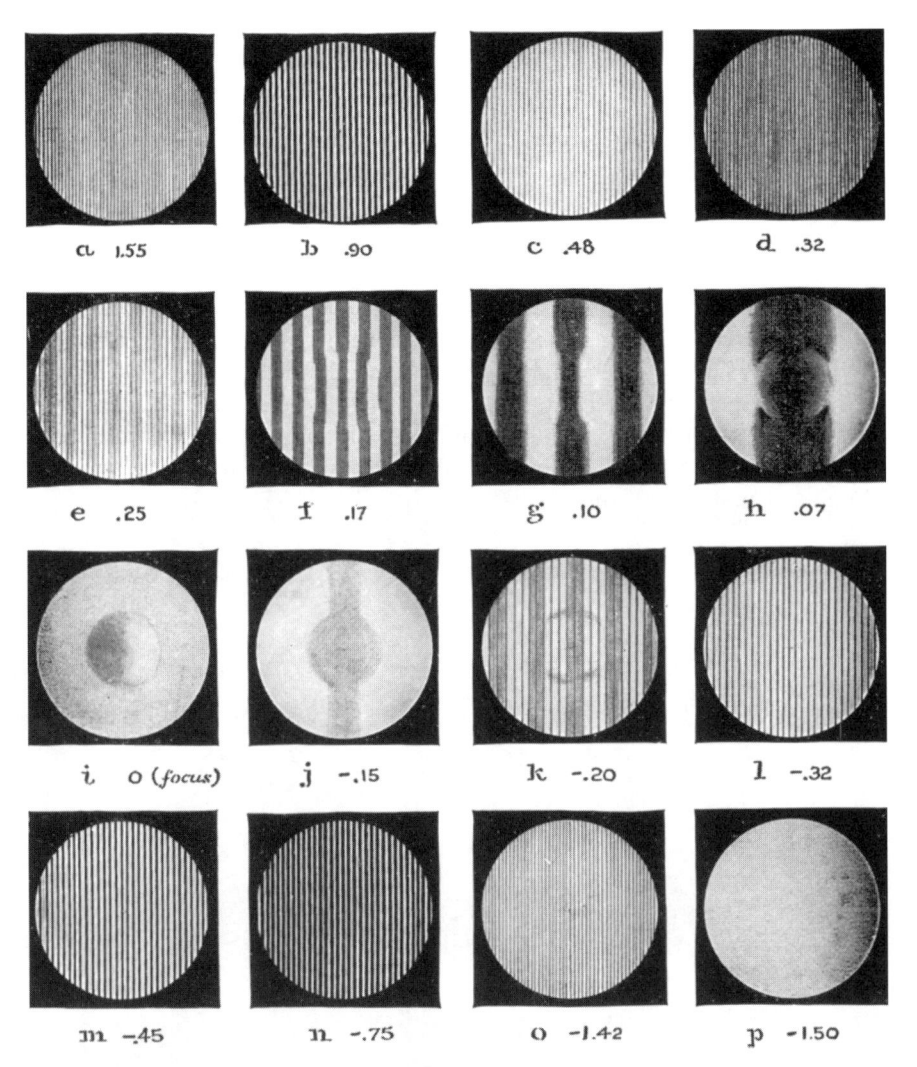

Fig. B.9.4: *Appearance of grating patterns of a concave silvered mirror, as seen at various distances inside and outside the center of curvature. The distances are given in inches.(Drawings by Russell W. Porter).*

when the latter is at the center of curvature of the outer zone. Hence, for this particular position of the grating, the shadows may be interpreted as in the Foucault test. The contour is determined by remembering that, if the band advances from the *left,* the resulting shadow models the surface of the mirror as though illuminated by light coming from the *right* at grazing incidence.

It was not anticipated, when developing this modification of the Ronchi test, that its application to mirror inspection would be found as useful as that for lenses. But amateurs seem to be adopting it as a supplement to the knife-edge, and with the improvement of Kirkham in the use of woven mesh screens or wire gratings, the method may come into quite general use.

B.9.7 Note Added in 1996

The Ronchi test as modified by Mr. King should not be used as the only test method for fast optics (focal ratios below 8) because the tolerances for error are so narrow (see Wright's article in section B.3).

King's diagram in Figure B.9.1 showing the parallel-slit stand within easy reach of the tester should be understood to be only an explanatory device. The slits should be backed off until both they and the mirror are in approximately the same focus. *H.R.S.*

Chapter B.10

The Direct Focal Test for Gregorian Secondaries[†]

Much of the acclaim which rightly belongs to the compound telescope has been withheld as soon as it becomes evident to the amateur that one must make a second mirror at least as large as the primary paraboloid in order scientifically to figure the little secondary. In any case, before deciding to make a Gregorian telescope, it is well to recognize that the field in this type of telescope is small, and that in all circumstances the telescope will be worthless for daytime use. On the other hand, this type of telescope may be built and correctly figured, without the trouble of making a flat or a spherical mirror.

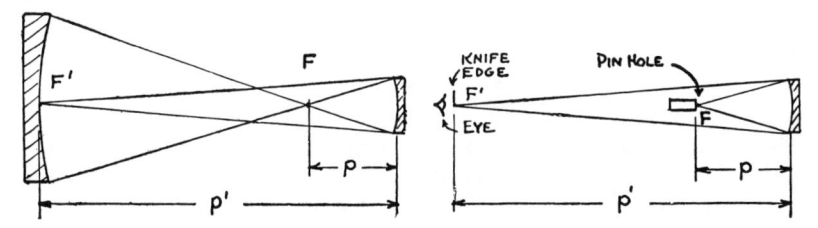

Fig. B.10.1:a (Left) and b (Right). *Drawings by Russell W. Porter, after the author.*

In order that the testing, adjusting, and use of a Gregorian telescope may be more intelligently considered, it is well to investigate at some length the function of the two mirrors, and we shall perhaps obtain the best perspective of the matter by beginning with the paraboloidal primary. The purpose of this mirror is to collect from a distant object as much light as its aperture permits. It brings this light to a focus at F, Figure B.10.1a. Here there exists a *real* image of the object, rather small if the paraboloid

---

[†]By Alan R. Kirkham.

is of short focal ratio.

The primary mirror has now performed its function, and may hereafter be completely disregarded, and the real image regarded during the remainder of the investigation as if it were a little photograph placed at F.

It is now desired to enlarge this little image, exactly as in making a photographic enlargement from a small negative. A concave mirror of correct radius is placed a given distance p from the primary image F, and reflects this image to another focus F', at a distance p'. The amount of enlargement (amplifying ratio) is p'/p.

Now it is an interesting and a useful geometrical property of the ellipse that light originating at one of its foci (such as F or F') will be reflected back to the other. Therefore the secondary mirror must be an ellipsoid. This immediately suggests a simple and delicate test to be used in figuring the ellipsoid, making direct use of its two foci. Simply by placing the pinhole at a distance p from the mirror, and the knife-edge at the distance p' (that is to say, at F and F'), the knife-edge should cause the mirror to darken evenly all over, just as it does in the case of a sphere at the center of curvature; or by the Ronchi test, it should show straight bands.[1]

The arrangement is shown in Figure B.10.1b, all values being the same as in Figure B.10.1a. Since the image at F' of the pinhole at F is enlarged, the test is more delicate than a center-of-curvature test. Incidentally, it is advisable to use a smaller pinhole, or slit if a slit is used, than is customary. In this test there is only one reflection of light, and this is much easier to find than is the case when a greater number of reflections are involved. Also there is no opportunity to make errors due to the figure of the test mirrors.

The light source may be a radio dial lamp operated from a small stepdown transformer, or from batteries. The small size of these lamps makes it possible to place the entire lamp housing and pinhole directly in front of the mirror, provided the latter is 2 inches or more in diameter. Smaller sizes can be tested by using a quarter-inch right angled prism illuminated from the side. In this case the pinhole should be on the square side of the prism, and the one which faces the mirror. The knife-edge and its adjustments are entirely orthodox.

A common tendency when designing compound telescopes, one which seriously reduces the effective field of good definition, is to make the secondary mirror too small in diameter. If this mirror is placed one-fourth of the primary focal length outside of focus, its size should be considerably larger than one-fourth of the diameter of the primary. Only the truly axial cone of light falls concentrically upon the secondary, all others falling

---

[1] This sentence is referred to in the last paragraph of this chapter as easily overlooked, but contains important instructions on where to place knife-edge and pinhole when testing the ellipsoid.

somewhat to one side. The secondary should therefore be large enough to catch all of the light from all parts of the useful field. A good rule to follow, if the telescope is for visual use, is to make the secondary one fourth larger than seems necessary, and still larger if the telescope is to be used for photography—for example, to show the entire Moon. The effect of making it too small is simply to reduce the aperture of the telescope to out-of-axis cones of light.

It is interesting to note that the figuring of the secondary, which can be done so perfectly and simply, is not nearly so important as that of the primary, which must be fully and perfectly corrected. An error giving rise to only 0.01 inch of longitudinal aberration will result in a good fraction of an inch aberration in the final combination. Errors of $1/10$ of an inch are not uncommon, and such would result in a final aberration of 2 or 3 inches along the axis. Of course, the equivalent focus of a Gregorian generally is very long, and the circle of confusion is not increased as much in proportion to the longitudinal aberration as is the case with the shorter focus Newtonian. Naturally, every effort should be made to arrive at the highest possible degree of perfection in figuring the mirrors, especially the primary.

Do not attempt to make a Gregorian with a long focus primary mirror. About $f/5$ is the upper limit, and $f/3$ or $f/4$ is better. Otherwise the equivalent focal length of the finished telescope will be unduly long—over $f/30$, and no eyepieces of the characteristics required for such a telescope are in existence; neither would they be desirable if they were.

The following discussion is for the advanced amateur who desires to know how things happen and it may be skipped by those whose intention is to make but one Gregorian telescope, especially if made to more or less usual specifications.

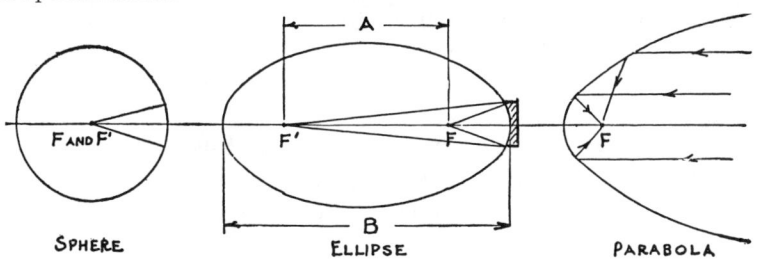

Fig. B.10.2: F' for the parabola lies at infinite distance to the right.

A sphere is a figure having no eccentricity. It has but one focus, that being at the center of curvature; or it might also be regarded as having two foci, but superimposed at the center of curvature. (Figure B.10.2.) A parabola has two foci, one where the eyepiece is placed in the telescope, the other infinitely beyond. Between these two figures, there exist an infinite number of shapes under the general classification of ellipses, having two foci which are separated more or less, this separation determining their eccentricity. In Figure B.10.2, center, the distance A, divided by the distance B, represents the eccentricity. When the foci F and F' are separated by an infinite distance (as in a parabola), the eccentricity is said to be 1. Thus it is seen that the eccentricity of a Gregorian secondary is

$$\frac{p' - p}{p' + p}$$

Where it is desired to know the distance p or p' with a mirror of given solar focus f, the equation $(p - f)(p' - f) = f^2$ is solved for p, or p'. Thus:

$$p = \frac{f^2}{p' - f} + f \tag{I}$$

$$p' = \frac{f^2}{p - f} + f. \tag{II}$$

These formulae also apply in the case of the Cassegrainian secondary, though here it is necessary to remember that p becomes a negative number. It is this fact which makes the Hindle spherical test mirror necessary. Its function is to supply a negative incident cone, a thing which a pinhole cannot do.

It is easy to calculate the effect of errors in the primary upon the secondary focus, by adding or subtracting one fourth of the longitudinal error of the primary, tested at the center of curvature, to p in the formulae (I), and solving for p'. The final aberration is the difference between this result and the result derived from solving the same equation for p' without introducing the error into p. If the error is toward undercorrection, one-fourth of its value is subtracted from p. If it is toward overcorrection, the value is added to p. By working out a few examples one will realize how necessary it is to have the primary fully and perfectly corrected. Any very serious deviation will render the telescope completely useless.

Fig. B.10.3: *The author's setup for making the direct focal test for Gregorian secondaries.*

B.10.1 Editor's Note, 1948

Perplexed readers have inquired how to find p and p' in order to place the pinhole. G.P. Arnold, State College, Pa., suggests the following:

Equations I and II are merely our old friends

$$R = \frac{2pp'}{p+p'},$$

or

$$\frac{1}{p} + \frac{1}{p'} = \frac{2}{R'} = \frac{1}{f}$$

(see C.1.22.2 and ATM3, E.7.6). They are one equation with two unknowns, p and p', and therefore tell us nothing if we are trying to *find* p and p'. Their purpose is to illustrate the two paragraphs that follow them. By the time one has a Gregorian secondary ready to test, he presumably knows p, p' and R', otherwise he would have no system. Use Hindle's chapter to design the system; then, concerning where to place the knife-edge and pinhole, Kirkham is quite clear, though the statement may easily be read over unnoticed (its importance is now highlighted by a footnote in this edition). A possible source of error in all this may be confusion between the elliptical secondary of the old Gregorian and the elliptical primary of the modified Cassegrainian, which came in with Dall. That is, the Dall primary is tested as an ellipse and used like a paraboloid, while the Gregorian secondary, if zones were measured (but why?) would be tested like a paraboloid and used as an ellipse.

Chapter B.11

The Hartmann Test[†]

The standard observatory method of grading the performance of large mirrors and objectives is the Hartmann test. In general procedure and simplicity, the Hartmann and Foucault tests bear a close resemblance. Both seek to find to what extent the light of a star can be brought to a point; to do this, both locate the points of intersection of various pairs of light beams coming from definite points on the surface of the mirror. It should be emphasized that the Hartmann test is primarily a critique of the finished mirror rather than a guide during the process of figuring. Being more laborious than the Foucault, it is frankly less adapted to the needs of the amateur. Nevertheless, as a slightly different approach to the central problem of mirror making, it is of more than general interest.

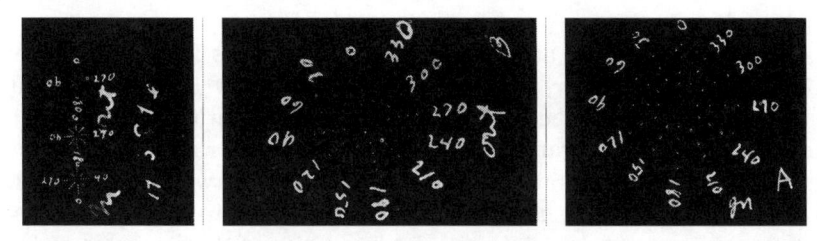

Fig. B.11.1: Left: Plate made photographically from 24″ mirror. Center: Plate for 61″ mirror—extra-focal position: pattern compressed toward center. Right: Plate for 61″ mirror, inside of focus: pattern expanded. In each case, note the extra hole at the top; this is cut in the screen which is placed against the mirror, to identify the zero position on the photographic plate. For a fairly large mirror the holes are made about $3/4''$ or so in diameter.

Let us consider first the simplest case, namely, the test of a parabolic mirror under actual working conditions, the light source being a star on the

---

[†]By William A. Calder.

optical axis of the telescope. With a perfect mirror (and perfect "seeing") the light from all parts of the mirror comes to one focus, the individual rays all intersecting at one point on the axis. Now suppose that we cover the mirror with an opaque screen in which a series of holes is punched along various diameters of the mirror. The holes will be in pairs, so that each pair is accurately centered with respect to the mirror. If we place a photographic plate considerably inside the focus, upon exposure to the star we shall obtain a pattern similar to the array of holes in the screen, since the individual bundles will have not come to the intersection. Similarly, we shall get another pattern with the plate placed out beyond the focus.

Figure B.11.1, at left, shows the patterns obtained with the Harvard 24" reflector under this kind of experiment.

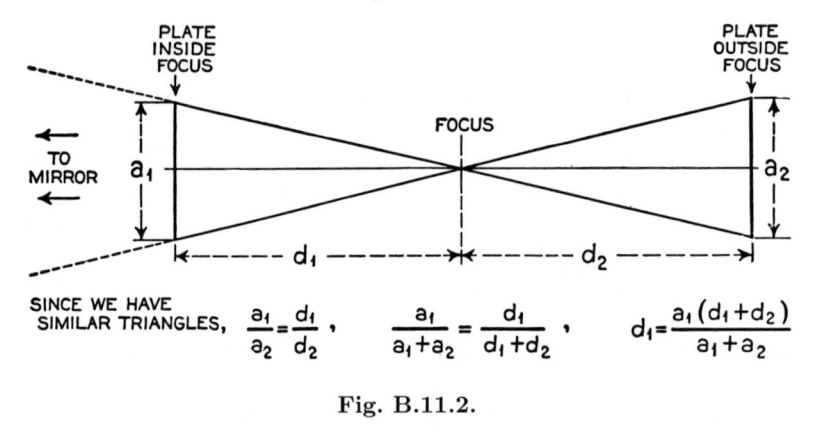

SINCE WE HAVE SIMILAR TRIANGLES, $\dfrac{a_1}{a_2} = \dfrac{d_1}{d_2}$, $\dfrac{a_1}{a_1+a_2} = \dfrac{d_1}{d_1+d_2}$, $d_1 = \dfrac{a_1(d_1+d_2)}{a_1+a_2}$

Fig. B.11.2.

We now confine our attention to the light bundles from a single pair of holes on one diameter, say the holes for a radius of 12", where a_1 and a_2, Figure B.11.2, are the separation of the images on the two plates, as measured with a comparator (microscope micrometer or measuring engine) and $d_1 + d_2$ is the separation of the emulsion in the two positions of the plate. d_1 is thus the focus of the two areas of the mirror with which we have been concerned, referred to the inner position of the sensitized surface of the plate. We next compute the position of focus for all other pairs of apertures along the same diameter, so that we may find the mean (or average) focus. But since the area and hence the effectiveness, of each zone is proportional to its radius, we "weight" the mean[1] accordingly. Thus if

---

[1] In taking a representative average, it is always necessary to give emphasis to various elements in accordance with their effect on the result. The simple average focus of all zones of a mirror is found by adding the individual foci and dividing by the number of zones. But the inner zones are of relatively little importance. Hence, in finding significant

there are n zones the "weighted mean" focus[2]

$$F_0 = \frac{R_1 f_1 + R_2 f_2 + R_3 f_3 + \cdots R_n f_n}{R_1 + R_2 + R_3 + \cdots R_n} = \frac{\Sigma R f}{\Sigma R}$$

If there are imperfections in the mirror, it will be impossible to obtain a point image in any position of the plate. The optimum position will be at the mean focus derived above. By similar triangles, the radius of the image disk at any focus is

$$r_d = \frac{R|f - F_0|}{F_0}$$

R being the radius of the mirror zone in question. The weighted mean of the radius of these scatter disks is

$$r = \frac{\Sigma R r}{\Sigma R} = \frac{1}{F_0} \frac{\Sigma R^2 |f - F_0|}{\Sigma R} \quad \text{(See footnote 2)}$$

As seen from the center of the mirror this average disk radius subtends an angle

$$\frac{r_d}{F_0} \quad \text{or} \quad \frac{1}{F_0^2} \frac{\Sigma R^2 |f - F_0|}{\Sigma R}$$

in radians[3] or

$$\frac{206,265}{F_o^2} \frac{\Sigma R^2 |f - F_0|}{\Sigma R}$$

in seconds of arc (the factor 206,265 is the number of seconds of arc in one radian).

"Hartmann's criterion" is defined as

$$\frac{200,000}{F_0^2} \frac{\Sigma R^2 |f - F_0|}{\Sigma R}$$

---

value of the average focus, we "weight" the value of the individual zones in this case by multiplying each by the radius of the zone.

[2] The symbol Σ (Sigma) is the short notation for "sum of." There is nothing highbrow about it, and its meaning is readily inferred from the first equation in which it occurs. Likewise, the vertical bars mean "absolute value" or "always positive,"

$$\frac{\Sigma R^2 |f - F_0|}{\Sigma R} \quad \text{means} \quad \frac{R_1^2 |f_1 - F_0| + R_2^2 |f_2 - F_0| + \cdots + R_n^2 |f_n - F_0|}{R_1 + R_2 + \cdots + R_n}$$

[3] For convenience in mathematical treatment, it is frequently better to measure angles in radians rather than in degrees. A radian is the angle subtended by an arc equal to the radius; one radian is approximately $57\,1/3°$. One radian is likewise equivalent to 206,265 seconds of arc. An object observed to subtend one second of arc must have a size which is $1/206,265$ its distance, and proportionately.

from which it is seen that the criterion is approximately the angular size of the radius of the circle of least confusion. We can at once compare the size of the best *possible* (with the given mirror) image with the *theoretical* resolving power or evaluate it in terms of the scale of the plate. (By scale of the plate is meant, for example, that 1 mm on the film equals 1 minute of arc in the sky.) It is generally considered that, for a good mirror, Hartmann's criterion must be something less than 0.5 arcseconds.

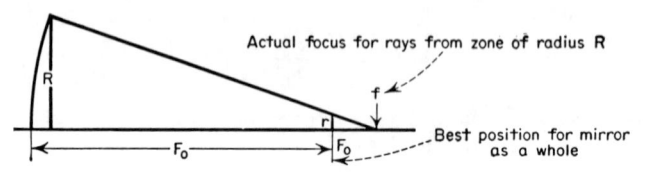

Fig. B.11.3.

By comparing the mean foci along various diameters of the mirror the astigmatism is determined. The Hartmann test may also be made at the center of curvature with an artificial star, just as the Foucault test. In this case, the patterns on the plate are not exactly similar to the array of holes in the screen, being compressed toward the center in the extra-focal position, and vice versa (Figure B.11.1, middle and right, respectively). This is due to the distribution of crossing points along the axis, as given by the

$$2F_0 + \frac{R^2}{2F} + \frac{R^4}{16F^3}$$

equation. Comparison is made of the observed and theoretical distributions, care being taken to make the fit at a weighted mean position. Deviations in this case are four times as great as those made at the primary focus with parallel light, which must be taken into account in computing the Hartmann criterion.

The Hartmann test is applied to object glasses, but color filters must be used, since chromatic aberration would render the images too diffuse for measurement. The performance of the lens in various spectral regions is effectively studied in this manner.

To summarize: In testing a mirror by the Hartmann method, a zone plate (mask) with radially symmetrical holes is placed over it. A photograph is taken inside and outside of focus, with carefully measured separation of the two positions of the film. By measurement of the images on the plate, the intersections of the various pairs of rays are found by simple geometry. The angular size (in seconds of arc) of the best possible image is derived, and this is called the "Hartmann criterion."

That this is not as complicated as might seem will be shown by an example where only three pairs of rays along one diameter are considered. We will suppose that we are testing a 15" mirror of 90.0" focus. A cardboard mask is laid on the mirror and on one diameter are cut three pairs of 3/4" holes whose centers come 3", 5" and 7" from the center of the mirror. A photographic plate is placed about 1/2" inside the focus and a one-minute exposure is made on a bright star (with the telescope following the star during exposure). The plate is then shifted back 1.00" and a similar exposure is made. It is important to know the amount of the displacement to within 0.01" (inch), whereas the exact distance from the film to mirror need not be known better than to 0.1".

Measurement of images on the developed plates gives these results:

	Separation of holes with plate inside focus		Separation of holes with plate outside focus	
7" radius	0.0934"	} a_1 in Fig. B.11.2	0.0622"	} a_2 in Fig. B.11.2
5" radius	0.0678"		0.0443"	
3" radius	0.0410"		0.0254"	

The actual intersections of the rays from the 7", 5" and 3" zones referred to the position of the emulsion inside focus (d_1 in Figure B.11.2):

$$d_7 = \frac{.0934 \times 1.00}{0.0934 + 0.0622} = 0.601"$$

$$d_5 = \frac{.0678 \times 1.00}{0.0678 + 0.0443} = 0.605"$$

$$d_3 = \frac{.0410 \times 1.00}{0.0410 + 0.025} = 0.618"$$

The weighted mean

$$d = \frac{(7 \times 0.601) + (5 \times 0.605) + (3 \times 0.618)}{7 + 5 + 3} = 0.606"$$

Hartmann's criterion is

$$= \frac{200,000}{(90)^2} \left[\frac{(49 \times 0.005) + (25 \times 0.001) + (9 \times 0.012)}{7 + 5 + 3} \right]$$

$$= \frac{200,000}{8100} \left(\frac{0.378}{15} \right) = 0.62$$

Chapter B.12

Notes on the Optical Testing of Aspheric Surfaces[†]

The amateur telescope maker has at his command methods of testing spheres, paraboloids, hyperboloids, ellipsoids and planes which are amply precise and thoroughly satisfactory. Many such methods, however, depend on visual acuity, ocular contrast sensibility and personal judgment of geometric forms. These apparently new test methods are different, in that they give results in terms of measurement and are, therefore, in some cases preferable to other methods. No crying need for these methods exist—they are not necessary for most work. They are published here as possible aids to the advanced amateur in solving specialized problems, and as "different" methods by which other methods may be checked by the hobbyist whose earthly joys are mathematics and optical experimentation.

B.12.1 The Ellipsoid

The ellipsoid is, by definition, the conic of revolution which will bring all rays emanating from a point to a focus at another point. The secondary mirror of a Gregorian should have an ellipsoidal figure. The common methods of testing ellipsoidal surfaces (Kirkham's, Hindle's and Foucault's) yield results of high precision and are satisfactory. The criterion of perfection is the degree to which the surface approaches apparent flatness under the knife-edge, or the degree to which the bands approach straightness and parallelism if the tests of Jentsch or Ronchi are used.

It seems, however, that many experimenters find the judgment of flatness and straightness difficult and prefer a method of measurement. Also,

---

[†]By Horace H. Selby.

instances may be found in which a definite amount of under- or overcorrection is desired. Such a method is outlined below, with the usual mathematical development. The only variables considered are the easily-measured distances, P, P', r_n and D, the distance of the source.

In Figure B.12.1, $H_b S_o H_b'$ is an ellipse symmetrically described about o, the intersection of the x axis with the y axis. F_1 and F_2 are the foci. $V_n T_n$ is the tangent at H_n. $H_n S_n$ is the radius of the H_n zone and is identical with r_n. As is usual in the description of compound telescopes, SoF_1 is called P, SoF_2 is termed P'.

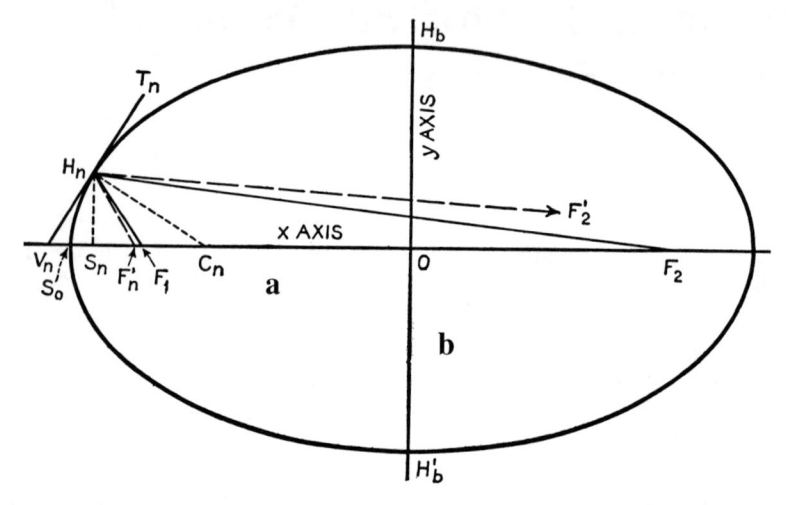

Fig. B.12.1.

With the source at F_2 and the knife-edge at F_1, or vice versa, the surface appears flat if perfect. If, however, the source is moved to F_2', the surface appears to be changed and the knife must be moved to a different position, F_n', in order to "focus" each zone, H_n, of the mirror. Each zone requires a setting different from every other and the object of this paper is to develop simple formulae in P, P' and r_n by which this focal shift can be determined. This shift, $\Delta F'$, will usually be the focal difference between two zones, one near the center, the other near the edge. With special equipment, the center and the extreme edge can be used, but there is no advantage in this procedure.

In the following, the distance of the light source is D, which equals SoF_2', the distance, axially, from the surface tested.

B.12.1. The Ellipsoid

Equation of curve:

$$\frac{x^2}{a^2} + \frac{y^2}{b^2} = 1 \quad y = \frac{b\sqrt{a^2 - x^2}}{a} \quad x = \frac{a\sqrt{b^2 - y^2}}{b}$$

Where

$$a = OS_o \quad b = OH_b \quad x = OS_n \quad y = H_n S_n \quad \frac{dy}{dx} = \frac{xb}{a\sqrt{a^2 - x^2}}$$

In terms of P, P' and x,

$$a = \frac{P' + P}{2} \quad b = \sqrt{P'P}$$

$$x = \frac{(P' + P)\sqrt{P'P - r^2}}{2\sqrt{P'P}} \quad \frac{dy}{dx} = \frac{2\sqrt{P'P}\sqrt{P'P - r^2}}{r(P' + P)}$$

In the figure, $H_n C_n$ is $\perp V_n T_n$ therefore $\angle C_n H_n F_2 = \angle C_n H_n F_1$. Also $\angle S_n H_n C_n = \angle H_n V_n S_n$. Therefore $\angle S_n H_n F'_n = 2H_n V_n S_n - S_n H_n F'_2$ then

$$\tan S_n H_n F'_n = \frac{2\tan H_n V_n S_n - \tan S_n H_n F'_2 + \tan^2 H_n V_n S_n \tan S_n H_n F'_2}{1 - \tan^2 H_n V_n S_n + 2\tan H_n V_n S_n \tan S_n H_n F'_2}$$

but

$$\tan H_n V_n S_n = \frac{dy}{dx} = \frac{2\sqrt{P'P}\sqrt{P'P - r^2}}{r(P' + P)}$$

and

$$\tan S_n H_n F'_2 = \frac{D - S_o S_n}{r},$$

or

$$\tan S_n H_n F'_2 = \frac{2D\sqrt{P'P} - (P' + P)(\sqrt{P'P} - \sqrt{P'P - r^2})}{2r\sqrt{P'P}}$$

and, since

$$\tan S_n H_n F'_n = \frac{S_n F'_n}{r},$$

$$S_n F'_n = \frac{r^2(P' + P)\left[8\left(\sqrt{P'P}\right)^2 \sqrt{P'P - r^2} - 2D\sqrt{P'P}(P' + P) + \sqrt{P'P}(P' - P)^2 - \sqrt{P'P - r^2}(P' + P)^2\right] + 4\left(\sqrt{P'P}\right)^2}{2\sqrt{P'P}\left\{(P' + P)^2 \left[r^2 - 2\sqrt{P'P}\sqrt{P'P - r^2} + 2\left(\sqrt{P'P - r^2}\right)^2\right] + 4\sqrt{P'P}\sqrt{P'P - r^2}\right.}$$

308   Chapter B.12. Notes on the Optical Testing of Aspheric Surfaces

$$\frac{\left(\sqrt{P'P-r^2}\right)^2 \left[2D\sqrt{P'P} - \sqrt{P'P}(P'+P) + \sqrt{P'P-r^2}(P'+P)\right]}{\left[D(P'+P) - \sqrt{P'P}\sqrt{P'P-r^2}\right]}\Bigg\}$$

$$S_o S_n = OS_o - OS_n = \frac{P'+P}{2} - \frac{(P'+P)\sqrt{P'P-r^2}}{2\sqrt{P'P}}$$

$$= \frac{(P'+P)(\sqrt{P'P}-\sqrt{P'P-r^2})}{2\sqrt{P'P}}.$$

Then $\Delta F'$ (from center to edge) $= P - S_n F'_n - S_o S_n$
and $\Delta F'$ (from zone A to zone B)$= S_B F'_B + S_o S_B - S_A F'_A - S_O S_A$
where A is the inner, and B the outer zone.

Although the formulae are lengthy and appear to be complex, application requires nothing more than arithmetic and the average example can be applied in an hour or two, once the distances P, P', r and D are known.

The ellipsoid can be tested at "center of curvature," also, in which case the lamp and knife-edge will move together from C_n toward S_o to C_o through the distance ΔC, which equals $OC_o - OC_n$.

In all cases,

$$\Delta C = \frac{(P'-P)^2(\sqrt{P'P}-\sqrt{P'P-r^2})}{2\sqrt{P'P}(P'+P)}$$

B.12.2 Testing The Paraboloid On Near Objects

At times it may be desirable to test a telescope mirror or a complete telescope on a light source not situated at either center of curvature or at infinity. The formulae developed below make such testing simple.

In Figure B.12.2, $V_n T_n$ is tangent to the curve at H_n. $H_n C_n$ is normal to the tangent. $\infty H_n F$ is the normal ray path. $F_2 H_n F'_n$ is the new ray path.

Let $S_n H_n = r_n$, the mean radius of zone n; also $= y$.

$S_o F = P$, the focal length.

$H_n C_n = R_n$, the axial intercept distance of the normal at zone n.

$S_o F_2 = D$, the axial distance of the source from the mirror.

x, y the coordinates of the point H_n.

B.12.2. Testing The Paraboloid On Near Objects

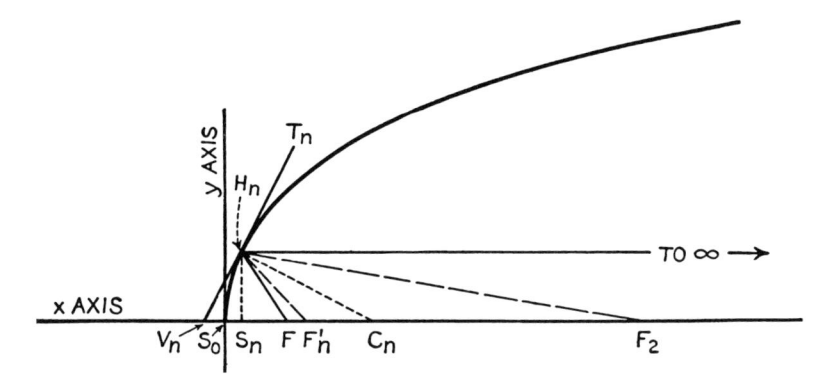

Fig. B.12.2: *Paraboloid.*

Equation of curve:

$$y^2 = 4Px; \quad \frac{dy}{dx} = \sqrt{P/x} = \tan H_n V_n S_n = \frac{r}{V_n S_n}$$

then

$$\tan H_n C_n S_n = \sqrt{\frac{x}{P}} = \frac{r}{S_n C_n}$$

and

$$S_n C_n = \frac{r\sqrt{P}}{\sqrt{x}}$$

but $x = r^2/(4P)$. Therefore, $S_n C_n = 2P$ and $S_o C_o = 2P$.
Also, since

$$S_o S_n = x = \frac{r^2}{4P}, \quad C_o C_n = \Delta C = \frac{r^2}{4P},$$

which is the usual formula giving the movement of the knife-edge and source together. Here it is interesting to note that the mirror depth is equal to the knife-edge travel.

If the source is at infinity, the true paraboloid will appear to be flat and evenly illuminated if observed from F. When the source is moved toward the mirror, the apparent figure changes and the blade must be moved to a new position for each zone. The difference of position for two zones is denoted ΔF and can be determined by use of the formula derived below:

$$\tan S_n H_n F = \frac{S_n F}{r} = \frac{4P^2 - r^2}{4Pr}$$

$$\tan FH_nF'_n = \frac{r}{D - S_oS_n} = \frac{4Pr}{4DP - r^2}$$

Therefore

$$\tan S_nH_nF'_n = \tan(S_nH_nF + FH_nF'_n) = \frac{S_nF'_n}{r}$$

$$= \frac{4P(4DP^2 - Dr^2 + 3Pr^2) + r^4}{16P^2r(D - P)}$$

then

$$S_nF'_n = \frac{4P(4DP^2 - Dr^2 + 3Pr^2) + r^4}{16P^2(D - P)}$$

and

$$S_oF'_n = \frac{16DP^3 + 8P^2r^2 + r^4}{16P^2(D - P)}$$

when $r = 0$, $S_oF'_n = S_oF'_o = (DP)/(D - P)$. Therefore, $\Delta F'$ (between center and edge)

$$= S_oF'_n - S_oF'_o = \frac{r^2(8P^2 + r^2)}{16P^2(D - P)}$$

Unless special equipment is used, however, the center is not available, being blocked by a prism or by a diagonal. It then becomes necessary to determine the focal difference of two extra-axial zones. Where A is the outer, and B, the inner zone, the equation becomes (between zones A and B):

$$\Delta F' = \frac{r_A^2(8P^2 + r_A^2) - r_B^2(8P^2 + r_B^2)}{16P^2(D - P)}$$

B.12.3 Quantitative Test of Hyperboloidal Mirrors

Several experimenters have recommended testing hyperboloids from the back, among them, Ellison back and the Lowers. The test has been but a visual examination for smoothness of figure, however, and measurement of zonal foci has not been attempted.

In 1935, Robert Russell, in unpublished work, attempted to make such testing practical. It is to Russell that the writer is indebted for the idea of developing a quantitative test. Such an idea had never occurred to him before he was allowed to criticize Russell's work.

In Figure B.12.3, V'_nV_n is the tangent to the hyperbola S_oH_nA at H_n. H_nX_n is the normal. S_oI_o is the axial thickness of the denser medium. H_nS_n is the perpendicular to the axis from H_n. As is usual with Cassegrain telescopes, $F_3S_o = P'$ and $S_oF_1 = P$.

B.12.3. Quantitative Test of Hyperboloidal Mirrors

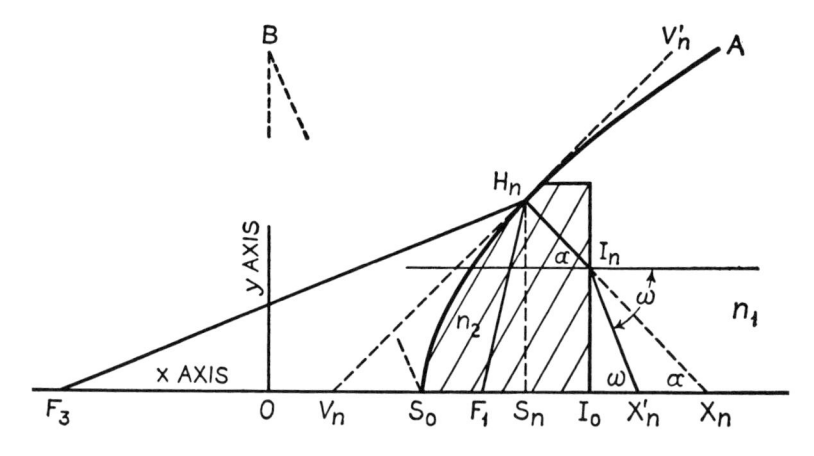

Fig.B.12.3.

Let $H_n S_n = r_n$, the radius of zone n, $= y$ and $S_o I_o = T$, the thickness, axially of glass (or of glass and liquid, as in the King Test.)

$$OS_o = a = \frac{P' - P}{2}$$

$$OB = b = \sqrt{P'P}$$

$x, y =$ the coordinates of H_n

Equation of curve:

$$\frac{x^2}{a^2} - \frac{y^2}{b^2} = 1$$

$$y = \frac{b\sqrt{x^2 - a^2}}{a}$$

$$\frac{dy}{dx} = \frac{xb}{a\sqrt{x^2 - a^2}}$$

$$x = OS_n = \frac{(P' - P)\sqrt{P'P + r^2}}{2\sqrt{P'P}}$$

$$\frac{dy}{dx} = \tan H_n V_n S_n = \frac{r}{V_n S_n} = \frac{2\sqrt{P'P}\sqrt{P'P + r^2}}{r(P' - P)}$$

therefore,

$$\tan H_n X_n S_n = \frac{r(P' - P)}{2\sqrt{P'P}\sqrt{P'P + r^2}}.$$

Since
$$OS_n = \frac{(P'-P)\sqrt{P'P+r^2}}{2\sqrt{P'P}},$$
$$OS_o = \frac{P'-P}{2}$$

and
$$S_n X_n = \frac{2\sqrt{P'P}\sqrt{P'P+r^2}}{(P'-P)}$$

therefore
$$S_o S_n = \frac{(P'-P)(\sqrt{P'P+r^2}-\sqrt{P'P})}{2\sqrt{P'P}}$$

$$S_o X_n = S_o S_n - S_n X_n = \frac{(P'-P)^2(\sqrt{P'P+r^2}-\sqrt{P'P})+4P'P\sqrt{P'P+r^2}}{2\sqrt{P'P}(P'-P)}.$$

When
$$r=0, \quad S_o X_n = S_o X_o = \frac{2P'P}{P'-P}$$

therefore ΔR for zero thickness
$$= S_o X_n - S_o X_o = \frac{(P'+P)^2(\sqrt{P'P+r^2}-\sqrt{P'P})}{2\sqrt{P'P}(P'-P)}$$

Using the above as a basis, ΔR can be calculated for hyperboloids of known thickness and refractive index (n_2) as follows:

$$\sin\omega = \frac{n_2 r}{\sqrt{r^2+\overline{S_n X_n}^2}},$$

$$I_n I_o = \frac{r(S_o X_n - T)}{S_n X_n},$$

$$I_o X'_n = \frac{(S_o X_n - T)\sqrt{r^2+\overline{S_n X_n}^2-n_2^2 r^2}}{n_2 S_n X_n}$$

and
$$I_o X'_o = \frac{S_o X_o - T}{n_2}.$$

Then
$$\Delta R = I_o X'_n - I_o X'_o = \frac{\left[(S_o X_n - T)\sqrt{r^2+\overline{S_n X_n}^2-n_2^2 r^2}\right] - S_n X_n(S_o X_o - T)}{n_2 S_n X_n}$$

B.12.3. Quantitative Test of Hyperboloidal Mirrors

Substituting and rearranging,

$$\Delta R = \left\{ \frac{\left[(P'+P)^2\sqrt{P'P+r^2} - (P'-P)\sqrt{P'P}(2T+P'-P)\right]}{\dfrac{\sqrt{r^2\left[(P'+P)^2 - n_2^2(P'-P)^2\right] + 4\overline{P'P}^2}}{\dfrac{-4P'P\sqrt{P'P+r^2}(2P'P - TP' + TP)}{4n_2 P'P(P'-P)\sqrt{P'P+r^2}}}} \right\}$$

Fig. B.12.4 Horace H. Selby

Horace H. Selby was born in Redkey, Indiana, in 1906, grew up in Sheridan, Wyoming, and later went to California where he has lived since '26. As a youth he was an eager beaver with the books. Reading of Charles Steinmetz he determined to become an electrical engineer, but accounts of Pasteur's researches and those of Koch made him want to be a chemist-bacteriologist-physician. However, he thought math and optics had much in their favor as careers when he pored over the biographies of Descartes and Newton.

When 16 he began learning the rudiments of clinical laboratory work, instructed by his surgeon father who found in him such an apt pupil that after two years he was given much of the actual routine work to do. This work made necessary the frequent use of the microscope, so its innards had to be understood, even to the curves and types of glass employed—naturally, since he's Selby.

He has studied at the University of Wyoming, the University of California, Caltech, and San Diego State College, and has taken courses from several schools including the Army's Chemical Warfare Service School.

He's a regular member of the Optical Society of America and has maintained memberships in the American Chemical Society and the Society of American Bacteriologists for many years. As a member of our prism gang he made roof prisms for the Army during World War II. He's designer for an optical company, has been consultant in optical design for another firm and, during the war, for the Navy.

Chemistry has been his official profession since 1928. He is chief chemist and an officer of a chemical firm that makes agar-agar, a substance widely used in bacteriology.

What does Mrs. Selby do? She's also a chemist—in charge of the control laboratory of the same company. That's her, in one of the pictures.

B.12.4 Note Added in 1996

Alert readers of both section B.11 and Mr. Selby's article probably noticed something very strange. The formula for the crossing points of the various zones of a paraboloidal mirror when tested near the center of curvature is given as

$$2f + r^2/2f + r^4/16f^3$$

(This expression is buried in the second formula from the end of section B.12.2, with $D = 2P$ and $P = f$. In section B.11 it appears just after Figure B.11.3.)

The first term is merely the distance from the center of the mirror to the center of curvature of the central zone. The second term is the well known Foucault formula. But what is the source of the third term? Is there some sort of correction to the Foucault expression that advanced TMs have been keeping as a secret?

Well, yes and no. When the tester source is at the center of curvature of the central zone of a paraboloid, the outside zone is farther away than the value given by the simple Foucault relation. The fact that the crossing points of all of the zones is beyond the position of the source results in a slight magnification of motion. That magnification is just the very small factor $r^4/16f^3$. If you move the source a little farther away (by $r^2/8f$) you will find that the knife will shift the normal amount. A moving-source tester also yields correct answers without any extra terms.

Since few people pay any attention to the position of the slit, exaggerated care in including the small correction term is usually a waste of time. H.R.S.

Chapter B.13

Null Test for Paraboloids[†]

The Foucault or Ronchi test applied at the center of curvature is probably still the most commonly used method of testing paraboloids for astronomical mirrors. That this is so is more due to the convenience and simplicity of the setup than to the ease of interpretation of the results. While no particular difficulty of interpretation of zonal measurement of the parabolic shadows is experienced for mirrors of aperture ratio $f/8$ upward, the test becomes increasingly difficult for short-focus mirrors from $f/7$ to $f/3$. Zonal errors of appreciable magnitude may remain undetected, and in particular the outer zones are frequently found to be faulty even though the knife-edge shift r^2/R between the center and outer zone is correct. It is just these outer zones which contribute so greatly to the formation of the final image, and if a short-focus mirror is to equal in performance its longer focus counterpart, it is essential that the grading of curvature in these zones be correct to a close tolerance. Hence the need, felt even by the most experienced mirror maker, for changing to a null method of testing.

Where equipment is available this need is fulfilled by the well-known autocollimation null test utilizing a large flat mirror and a smaller flat or prism for deflecting the beam to a convenient viewpoint. A large silvered flat of high quality is not always available and, even if it is, the test rig is sensitive to careful squaring on, thus that test cannot be compared in simplicity with a test made at the center of curvature.

The method shown diagrammatically in Figure B.13.1 is carried out near the center of curvature with a simple rig consisting of a plano-convex lens spaced apart from an illuminated pinhole, the light from which is made approximately monochromatic with a red filter. The spacing of these is best done with a piece of tube, which in use is directed to the mirror. The knife-

---

[†]By H.E. Dall, From *The Journal of the British Astronomical Association*, Vol. 57, No. 5, 1947 November, by permission. Revised 1952 December by the author.

edge is applied in the ordinary way to the pinhole image, which should be located as close as possible to the center of curvature of the mirror. The aberration of the lens is opposite in sign to, and of the same character (within limits) as the aberration (r^2/R) of a paraboloidal mirror with the pinhole near the center of curvature. Hence they can be made to nullify as in this test.

The principal requirement in applying the test is to space the lens correctly from the pinhole at a distance appropriate to the lens in use, and the data to enable this to be done are given with sufficient simplicity for the less mathematically minded to follow. A perfectly regular paraboloid will give a null test, i.e., will darken uniformly over the entire disk as the knife-edge cuts across the pinhole image at focus. Defective zones show up clearly and are identified with the same certainty and precision as a spherical mirror that is tested at the center of curvature in the normal manner.

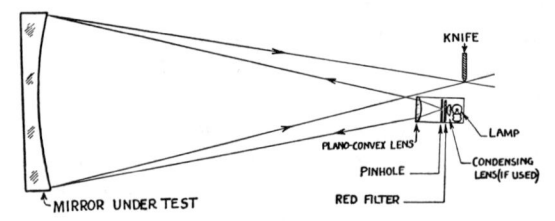

Fig. B.13.1.

The only addition to the normal Foucault testing equipment that is required by the new test is a red filter and a plano-convex lens and holder. The filter is required owing to the dispersion of the simple plano-convex lens. The latter, being of crown glass, has a very low dispersion at the red end of the spectrum, hence a simple red gelatin filter of the type "Tricolor Red" is highly economic of visual light while being sufficiently monochromatic for the purpose, and is sold in most photographic stores. Ruby glass will also serve but will pass barely 10 percent of the visual light compared with the 30 percent of the Tricolor Red. Colored glass, gelatin or cellophane film of the bright red type will generally function quite well. A small piece of the filter material is placed in the lamp, preferably between lamp and pinhole. The writer has used a monochromator giving one percent spectral purity, but finds no appreciable gain in sensitivity compared with the simple red filter when used with lenses up to 12-inch focal length. Lenses of longer focus than this would be required for only the largest sizes of mirrors outside the usual amateur's range, and these would perhaps need and warrant a higher degree of monochromatism. The plano-convex lens is used with its plane side toward the mirror under test. This is the direction of maximum aberration for which the data given are calculated.

It is necessary for the optical axis of the lens to be aligned within a degree or two of the center of the mirror. Sighting along the tube of the holder will generally insure this. Serious errors of alignment or centering will result in astigmatic effects in which the knife-edge shadow fails to advance parallel to the knife-edge movement. A similar type of error also results from too great a lateral separation between pinhole and knife-edge. This should be arranged if possible not to exceed two percent of the focal length of the mirror, and the use of small diameter housings for spotlight torch (flashlight) type bulbs is strongly recommended for all mirror testing. Alternatively, use can be made of prisms or beam splitters to reduce the separation.

All types of optical testing are facilitated by the use of brilliantly illuminated pinholes, perhaps more so with the null test, owing to the losses in the red filter. The use of a short-focus condensing lens inside the pinhole is recommended, although not essential; moreover a short vertical slit may be used instead of a pinhole, although the writer has preference for the latter. If a slit is employed its length should not be greater than 0.03 inch. The pinhole diameter or slit width recommended is from 0.001 inch to 0.002 inch, the smaller size for preference.

The plano-convex compensator lens should have a focal length between $1/5$ and $1/20$ of that of the mirror being tested. This is a long range to choose from, and many amateurs will find a suitable one in their stock of lenses. If the mirror is of wide angle from $f/4$ to $f/6$, preference should be given to lenses at the longer focus end of the range, say with foci half the mirror diameter. The field lens of a low-power Huygenian microscope eyepiece will often supply the need for mirrors up to 10 inches. To take an example, if a 12-inch diameter 60-inch focus ($f/5$) paraboloid is to be tested, a 6-inch focus lens would be quite suitable. For this lens F/f would be $60/6 = 10$. (The precise focus f should be measured as closely as possible, remembering that this is the distance from the vertex of the lens curved surface to the screen when the image of a distant object is formed on that side. Allowance should be made for the considerable aberration by stopping the lens down to, say $f/10$ or, if used without a stop, by measuring the maximum distance at which a sharp image can just be recognized inside the aberrational halo.)

Next refer to the curve, Figure B.13.2, to find the ratio of the pinhole-lens separation B to the focal length f of the lens. This is shown plotted against the ratio of the foci of mirror and lens. In the example quoted the latter ratio is 10. Hence from the curve the B/f ratio is found to be 0.535. The distance B is thus $0.535f = 3.21$ inches. The utilized aperture of the lens is a little more than B divided by the f number of the mirror; in this

Chapter B.13. Null Test for Paraboloids

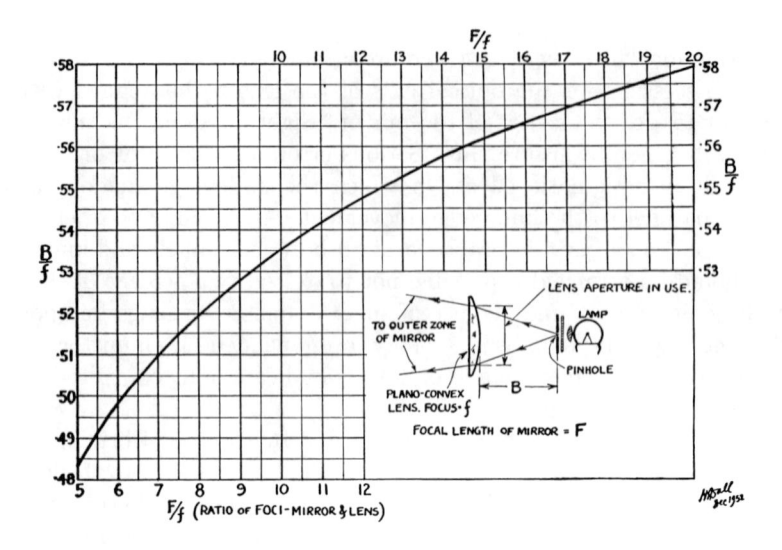

Fig. B.13.2.

example approximately 0.72 inch, and the lens should thus not be less than $3/4$ inch diameter. It can be larger without affecting the test. The curve is calculated for the range of focal ratios and lens foci mentioned, which is a region in which the higher order aberrations are small or negligible. The curves were obtained by rigorous ray tracing for a glass of refractive index 1.52 and a reasonable lens thickness. Departures of normal crown glass from this assumed value will not seriously affect the result. The lens chosen is fitted in a holder, which is preferably a tubular extension from the test lamp, and carefully measured to give the required separation between vertex of curved surface of the lens and the pinhole. The test rig is then ready for use and, upon adjusting its axial position so that the pinhole image comes as close as possible to the center of curvature of the mirror (distance R from the mirror), a null test should be obtained with the knife-edge if the mirror is perfect.

A few further examples are given in Table B.13.1, some of which may fit individual requirements and obviate the necessity for further calculation for those less practiced with slide rule or logs.

If the mirror is regular but overcorrected, it will appear to have the usual paraboloidal shadings though less deep than if no compensating lens is used. A regular undercorrected mirror will appear with the shadings characteristic of an oblate figure tested by standard methods. Zones which depart from the paraboloid will show up clearly as shaded rings, and can be interpreted for further action in the usual way.

Table B.13.1

Mirror aperture D	f-number of mirror	Suitable focus of lens $= f$ (inches)	Distance B (inches)	Minimum diameter of lens (inches)
6	f/5	3.0	1.605	0.4
6	f/6	3.0	1.64	0.4
8	f/4	4.0	2.075	0.65
8	f/5	4.0	2.14	0.5
8	f/6	4.0	2.19	0.5
10	f/6	5.0	2.74	0.6
12	f/4	6.0	3.115	1.0
12	f/6	6.0	3.28	0.7
18	f/5	10.0	5.28	1.2

B.13.1 Editor's Note

The following is from a later personal communication from Dall:

> George Hole, a British Astronomical Association member whose profession is telescope making, has just finished a 24-inch $f/5$ mirror for his own use. Being well-equipped he was able to test the mirror both by autocollimation and by null test. Results of the two tests were indistinguishable.

In 1954, before the second printing of ATM3, the author learned that the French optician André Couder described another form of null test (*Revue D'Optique,* Vol. 6, p. 49, 1927). This uses a compound lens in a similar setup, but gives no data for presetting the required correction.

Chapter B.14

Testing Convex Spherical Surfaces[†]

Concerning the polishing of convex spherical surfaces, Ellison states in ATM2, Section A.2.3, "No question of their figure can arise at this stage of the proceedings, as it is impossible to test it." Professional makers sometimes test convex spherical surfaces by interference methods, referring the convex to a standard concave. This method is hardly practical unless a number of similar convex lenses are to be made.

Judging from the number of times the desire has been expressed for a test for convex lenses which would be as simple as that for concave mirrors, it would seem that such a test, if available, would be very useful. Therefore, the writer proposes a test for convex spherical surfaces requiring no auxiliary optical surface, and one which is simple and as rigorous as the mirror test at center of curvature.

If, for the sake of illustration, we imagine a spherical surface consisting of only a skin of silver of practically no thickness, which would at the same time remain optically true without support on either side, one side would be a convex mirror and the other a concave of the same radius. Then, in order to test the convex mirror, one would merely have to go around to the other side and test the concave at the center of curvature.

However, practical optical surfaces are generally formed on glass but if, as in the case of convex spherical surfaces on lenses, we could eliminate the lens optically and leave the surface to be tested, we could again go around back of the convex and test it as a concave mirror at the center of curvature and that test would be the equivalent of a test of the convex surface.

To do this, we make use of a simple principle employed for many years in inspecting optical glass, but to the writer's knowledge never applied in this manner. In examining optical glass for striae and general uniformity of index, when it is in crude broken chunks, it is placed in a large container

---

[†]By J.H. King and first published in *Scientific American*, Feb. 1935, page 100.

having glass windows at either end. Liquid is introduced having the same refractive index as the glass and then, if the glass is homogeneous, one is able to look clear through the liquid and the chunk of optical glass, and the rays will suffer no deviation. In other words, we have optically eliminated the glass. Bell's *The Telescope*, page 61, gives an account of this method of inspecting optical glass.

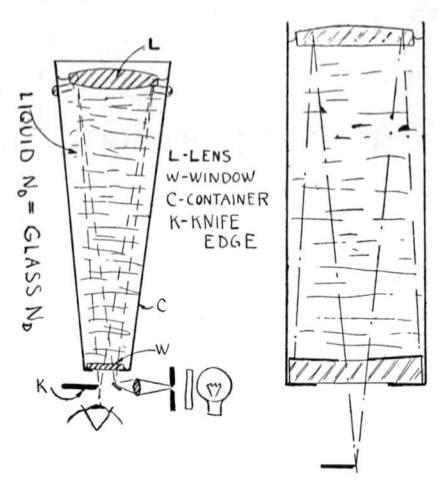

Fig. B.14.1: *The King test.*

Figure B.14.1, at the left, shows a sectional view of the setup which allows us to test a convex as a concave by introducing a fluid equal in refractive index to that of the glass. The fluid optically eliminates the lens. Since the upper convex surface faces air, the light proceeding from the pinhole suffers a partial reflection and some of it returns to focus again adjacent to the pinhole and the test becomes merely that for a spherical mirror at center of curvature. The test is rigorous because it is conducted to all practical purposes entirely within the liquid medium, and the small amount of air between the eye and the window is too close to focus to be detrimental. Several solutions have come to the writer's attention as having about the refractive index of crown glass when near room temperature. Toluene [obtainable from dealers in chemicals; for example, Eimer and Amend, Third Avenue and 18 Street, New York City.—*Ed.*] seems to be the best commercially obtainable liquid, since it is homogeneous and, though inflammable, does not have a low vaporization temperature. A very strong word of caution should be urged against the use of benzene or any other inflammable liquid which vaporizes at room temperature. The worst explosion due to chemical silvering would be very mild indeed compared with that due to a gallon of benzene properly vaporized and ignited in a closed cellar.

Below is a given list of various liquids and their refractive indices for the sodium line at given temperatures. However, the refractive index usually does not vary widely with a slight change in temperature.

	Oils (20°C)		
Castor oil	1.480	Turpentine	1.47
Olive oil	1.470	Kerosene	1.45
Nujol	1.475	Acto oil	1.520

Two refracting telescopes have been made with kerosene as the testing liquid and, although the ref. index is only 1.45, these instruments were both successful.

1. Aniline 1.586 at 20°C (Very poisonous if breathed)
2. Glycerine (Glycerol) 1.474 at 25°C
3. Toluene 1.495 at 20°C (Very poisonous if breathed; inflammable)
4. Carbon tetrachloride 40 percent, Ethylene bromide 60 percent, 1.4989 at 25°C

Special Oils, Crystal Clear (20° C)	
Barjol-F	1.4564
Wirol-F	1.4725
Primol-D	1.4817

The three last-named "special" oils are distributed by Stanco Distributors, Inc., 1 Park Ave., New York, N.Y. However, Nujol, an "internal lubricant," being water white and easily obtainable locally, is a very good substitute for Primol-D, although not quite so high in refractive index.

Some may raise the objection that the dispersion of the liquids may not be equal to that of the crown glass. Of course, testing in sodium light would remove this objection completely, but the writer believes that testing with white light is about all that is necessary and the refractive index does not have to be exactly that of the crown. This has also been borne out by experiment.

As a matter of convenience, a small prism may be used in place of the window, and the testing funnel mounted on a wall. This allows the observer to assume a comfortable posture looking horizontally instead of lying on his back as in the left-hand drawing, Figure B.14.1.

This principle is also applicable to testing a convex hyperboloidal surface by using a small spherical mirror of scarcely larger dimensions than the convex hyperboloid. The right-hand drawing shows the setup. The spherical mirror should be silvered and lacquered and the silver removed in the center, leaving a small transparent hole. Using this method, it is not necessary to construct a large optical flat or a large spherical mirror when building a compound telescope of the Cassegrain type. However, it would be well to construct the small secondary hyperboloid of optical crown in order to insure freedom from striae.[1]

---

[1] Editor's Note: Shall the test described above be called the King test?

Chapter B.15

A Bilateral Slit Mechanism[†]

In setting up practically any kind of optical instrument, the question of securing a good slit is one of the first encountered. Usually this slit must be constructed so that it is bilateral; so that its exact width may be read off from a convenient scale; and so that its jaws close perfectly and move accurately parallel to each other.

It is the purpose of this paper to describe in detail a slit mechanism that may be constructed easily and that satisfies these requirements. Because of the extreme simplicity of this design, the authors thought that it had probably been used before; however, an examination of the literature failed to reveal any previous use of this principle in a slit mechanism. The details of the mechanism are shown in Figure B.15.1. A thrust, caused by the rotation of the screw H, forces the cone C downward (without rotating it) between the two steel pins E_1 and E_2, thus forcing the slit jaws A and B apart. The strong spring G holds the steel pins E_1 and E_2 firmly against the cone C, and the swinging plate B_2 can be set so that C makes contact with both E_1 and E_2 simultaneously. The height of the slit may be varied by means of the fishtail slide F.

Curves on the performance of a slit of this design are given in Figure B.15.2. The readings, from which these curves were plotted, were taken on a comparator of such accuracy that the error in any of the values is not more than ±0.001 mm. Each figure given is an average of at least five readings. Observations of the same accuracy failed to detect any deviation of the slit jaws from perfectly parallel motion. Curve (a) in Figure B.15.2 is the calibration curve of the slit. From curve (b) it is seen that the slit is accurately bilateral for slit

---

[†] By R. Bowling Barnes and R. Robert Brattain, Palmer Physical Laboratory, Princeton University. Reprinted by permission, from the *Review of Scientific Instruments*, March 1935.

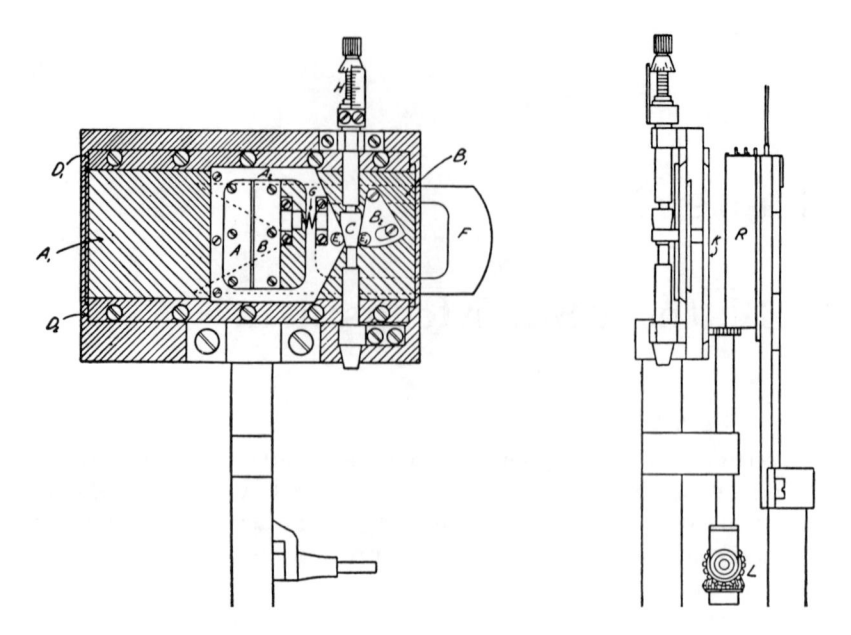

Fig. B.15.1.

widths between 0.10 and 2.6 mm but that it fails to be bilateral for smaller slit widths. This failure is due to the frame A_2 springing slightly downward when a thrust is first applied to the cone C, thus causing an unsymmetrical motion of the slit jaws. As soon as the frame has sprung as far as it can, the tension of the spring holds it in this position, and from then on the motion is truly bilateral, and upon closing the slit it comes to its original zero. Although this defect was not anticipated in the original construction, it could be corrected easily by making A_2 more rigid. If this were done, judging from the performance of the slit between 0.10 and 2.6 mm, the slit would also be strictly bilateral for widths in the range 0 to 0.1 mm and the calibration curve (a) would go linearly through zero. Other precautions which should be observed in the construction of such a slit are:

1. That the axis of the screw ensemble H be accurately perpendicular to the slit ways D_1 and D_2;

2. that the axis of revolution of the cone C be identical with the axis of H; and

3. that the length of the slides A_1 and B_1, to which the slit jaws are fastened, should be at least $1\,1/2$ times the height of the slit opening.

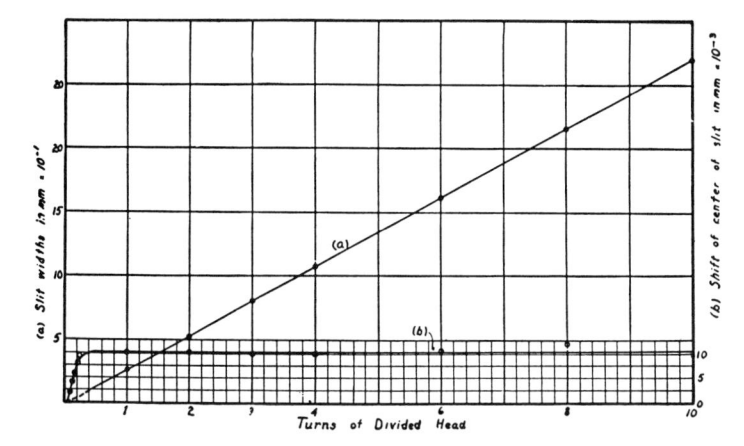

Fig. B.15.2: Curves showing the performance of the slit. (a) turns of the screw H as abscissae, plotted against actual slit widths as ordinates. (b) turns of the screw H as abscissae plotted against the shift of the middle of the slit as ordinates.

The slit as shown in Figure B.15.1 was built for use in an infrared spectrometer of the type recently described in *Review of Scientific Instruments*, by one of the authors, hence it was made with a maximum width of 3.0 mm and a height of 50 mm. The pitch of the screw H and the slope of the cone C were matched so that one revolution of H would open the slit 0.25 mm. By a proper matching of these two units practically any desired sensitivity of adjustment may be obtained. This design of slit may be built in any form or size and as rugged or as delicate as needed. Slits of this design have just been completed having a maximum width of 1.5 mm and a maximum height of 10 mm. In these slits the frame A_2 is relatively much stronger than in the original slit; and measurements analogous to the ones from which Figure B.15.2 was plotted show a shift of the center of the slit of less than 0.001 mm. This result verifies the prediction made above that the shift in the center of the original slit would be eliminated by making A_2 heavier. As an added convenience for infrared work the slotted filter slide K, controlled by rotations of the gears L, and the filter and shutter rack R were added.

This mechanism was designed and constructed with the help of Mr. W.C. Duryea and Mr. C.R. Stryker. For their generous and skillful assistance, the authors wish to express their gratitude.

Chapter B.16

Small Pinholes[†]

B.16.1 Determining the Optimum Size

What size star ought we to use for the knife-edge test? Contradictory opinions have been published. Ellison, in Section A.4.5, "has known grotesque errors result from using too fine a hole." The user who did this "saw a series of diffraction bands inside the margin of his mirror, and took them for a turned-down edge." G.W. Ritchey, on the other hand, writes: "When the knife-edge test is used with an extremely small pinhole between $1/250$ and $1/500$ of an inch in diameter, illuminated by acetylene, or what is much better, oxy-acetylene or electric arc light, minute zonal irregularities are strongly and brilliantly shown, which are entirely invisible with large pinhole or insufficient illumination." (See subsection C.1.14.4.) What in fact determines the delicacy of the knife-edge test?

If light traveled in straight lines, a given zone would show quite black on a fully bright mirror (or conversely) if its tilt with respect to true figure was just enough to divert the rays reaching it from every part of the pinhole, so that they fell completely clear of the image of the pinhole formed by the rest of the mirror. The delicacy of the test would then be inversely proportional to the diameter of the pinhole (provided enough light could be gotten through the pinhole). If, on the other hand we ask, not that a given zone should show fully bright, but merely that it should show *appreciably* bright on a fully black mirror, the delicacy of the test will be limited only by the intrinsic brilliance of the source feeding the pinhole. On the "ray" theory of light, then, we should make the source as bright as possible, and the pinhole only just large enough to let through a reasonable amount of light.

[†]By C.R. Burch, F.R.S., Warren Research Fellow, H.H. Wills Physics Laboratory, Bristol University, Bristol, England.

But, as we hope to test correctly to small fractions of a wavelength, we must take into account the wave properties of light. We can then argue that since the mirror cannot produce a "point" image of a "point" pinhole, but produces an "Airy disk" of finite width, we shall not gain much in delicacy by reducing the diameter of the pinhole below that of the Airy disk produced by the mirror—which depends on the focal ratio at which the mirror receives the light from the pinhole.

Thus, for testing a spherical mirror of diameter D, radius of curvature $R = 2F$, we should use a pinhole of the order of smallness of $0.001F/D$ mm diameter. So, for example, a spherical mirror of $D = 6$ inches, $R = 8$ feet, should be tested with a pinhole not bigger than 0.008 mm = 0.00032 inch diameter—a hole 6 times smaller than the smallest hole advised even by Ritchey.

This raises immediately the questions of whether one will see a series of diffraction rings around the edge of the mirror—for if such rings are seen, much of the advantage of the small pinhole would presumably be lost. You cannot do good optics by guessing what things would look like if they were not surrounded by diffraction rings.

When I first used the knife-edge test, I used a hole 0.013 mm diameter, and saw about 6 diffraction rings inside the edge of the mirror (no knife-edge being present). It was difficult to know what to do about these rings, and in desperation I decided to work out theoretically how they should behave. My result was that they should be invisible, because their spacing on the retina should only just equal the resolving power of the eye lens, and the retina would then be incapable of resolving them. I looked at the mirror again, and was able to count 6 distinct rings. I then remembered that Lord Rayleigh had made a similar calculation many years ago. On referring to it, I found that he also concluded (in effect) that no rings should be visible. I looked at the mirror for the third time, and could not see any diffraction rings ...!

It was all very mysterious: I could understand the mirror showing its contempt for my theoretical treatment, but it must have known that Lord Rayleigh made the same calculation before. A year later, I found the explanation: a visitor, looking at the mirror, complained of diffraction rings, which I could not see, and it occurred to me to look at it through his spectacles, which he had taken off. I then saw the rings, and when he wore his spectacles he saw no rings. Theory predicts that, if you focus your eye not on the mirror but in front of it or behind it, you will see diffraction rings around its edge. (The extreme example of this is the series of rings seen in the extra-focal image of the star, which is after all, also the extra-focal image of the mirror!)

The reason why anyone with normal sight tends to focus his eye any-

B.16.2. Illuminating the Pinhole   331

where but on the surface of the mirror is simply that, if for a brief moment he malfocuses the mirror, he sees diffraction rings, and his eye automatically tries a further change of focus in the direction that enabled it to see something—the rings—which it had not seen before. Therefore the proper advice to give to one who complains of diffraction rings, may be cast in the epigrammatic (if exasperating) form, "If you don't look at them, you won't see them!" If he still sees them, there is nothing for it but the right spectacles, and—in the last resort—a friendly oculist to paralyze the focusing muscles of his eye with the appropriate "dope," so that he cannot unconsciously vary his focus. But it is worthwhile, before adopting so drastic a measure, simply to use the test for some hours. Refraining from malfocusing one's eye is like riding a bicycle—most of us have to learn it by practice but, when learned, it is quite automatic.

B.16.2 Illuminating the Pinhole

There remains the question of how to get enough light through a hole as small as 0.008 mm diameter. It is necessary to use a source of very high intrinsic brilliance, such as a gas-filled lamp, or a Pointolite, or an arc, and since these sources are too small to provide the required solid angle of light even when placed as close to the hole as is practicable, it is necessary to use a condensing lens, to form an image of the source on the hole. One need not, however, correct the aberrations of the condensing lens completely, or even at all. These aberrations will cause the image of the *hole* which the lens forms on the source to be larger than it would otherwise be, but provided this image is not larger than the source, the aberrations will not reduce the light issuing from the hole, nor will they falsify the test by appearing on the mirror. I use a 12-volt, 16-watt gas-filled lamp, the filament of which is coiled in a very close helix, and a simple biconvex lens of 1-inch focal length to image it (with unit magnification) via a prism on to the pinhole. To adjust this star, one puts one's eye very close to the pinhole, and moves the lamp (the base of which is fixed on three "leveling" screws) until the blurred patch of light seen through the pinhole is as wide and as bright as possible. The resulting cone of light is wide enough for testing a system imaging at $f/8$: if a wider cone is needed, either the lens must be corrected, or a wider source must be used. As the lateral aberrations of a lens are proportional to its focal length (for given aperture), the focal length of the lens should be as short as the size of the lamp-bulb permits.

B.16.3 Making the Pinhole

One can sometimes find natural holes of the order of 0.01 mm diameter in "tinfoil": the following account of the technique of *making* even smaller holes is quoted from the *Monthly Notices* of the R.A.S., March 1936 (p. 452).

> The principal difficulty I found in making pinholes 0.0025 mm diameter, by piercing tinfoil with a needle, lay in making the needle really sharp. But Dr. J.M. Dodds, of the Metropolitan-Vickers Co., solved this difficulty by the following honing technique. The needle, mounted on a rod, is first honed on a fine stone until the diameter of its 'point' does not exceed 0.01 mm. This is not difficult, provided one inspects the point frequently during honing with a microscope. The final honing is done on glass. Lay the needle on a glass plate, and press heavily with one finger behind its point. Lift the rod in which it is mounted slightly so that the needle is bent through several degrees. Then, still pressing with the finger, simultaneously twist and withdraw the slightly lifted rod, so drawing out the needle under the pressing finger. This process is repeated until on inspection under 500 diameters magnification the needle looks quite 'sharp'—that is, its 'point' is of the order of 0.001 mm diameter or even less. Fifteen minutes honing on glass usually suffices.
>
> The 'tinfoil' in which cigarettes are wrapped forms an excellent material for the pinhole. The foil is tacked (at its edges only) to a glass plate; the needle point is then placed very gently on it, and the needle rotated without pressure through at least one revolution and then lifted off. Some practice is necessary to avoid dragging the point sidewise in lifting it off. If the needle is not rotated the hole will not be reasonably circular. The needle requires rehoning after a few piercings. Not every pierce is successful, but one can in this way make reasonably circular pinholes 0.002 mm diameter, and occasionally even smaller.

B.16.4 Differences in Usage Between Small and Large Pinholes

There are two main points of difference in the behavior of the knife-edge test when it is carried out not with a large pinhole, but with a pinhole—an "artificial star" as some term it—smaller than the resolving power of the mirror (say, 0.001 f/D mm dia. for a sphere tested at its center: 0.0005

B.16.4. Differences in Usage Between Small and Large Pinholes

f/D mm for a paraboloid tested at its focus, with a flat).

1. The test becomes noticeably more delicate, especially for slow errors of curvature—simple spherical aberration, and more particularly coma and astigmatism. For example, one obtains appreciable "paraboloid" shadows on a 12-inch paraboloid of $f/18$. The test becomes more sensitive without diaphragms than when diaphragms are used, but zonal focal measurements taken without diaphragms can give hopelessly wrong results—as was predicted by Lord Rayleigh, many years ago. A simple (though admittedly loose) way of explaining this is to say that when the knife-edge is not at focus, its shadow is preceded by diffraction fringes (not to be confused with eye-malfocus diffraction rings seen in the absence of the knife-edge). Thus a large error, on one part of the mirror, by putting the knife-edge out of focus for that part, can produce diffraction fringes on an error-free part of the surface. Accordingly, the only part of the error which one can be certain of interpreting correctly is the *largest* error present. This diffraction fringe difficulty automatically decreases as the errors are reduced and vanishes when the mirror is error-free. The mirror then shadows symmetrically as the knife-edge is advanced, but not uniformly, the center darkening before the edges. If the test were interpreted on a "ray" basis one would then say that, since the edges shadow last, one edge must be turned-up, and the other turned-down—in fact, that the mirror is comatic. One can, however, check whether coma is really present, for if it is, on bringing in the knife-edge in the opposite direction, the edges will shadow before the center.

2. The delicacy of the test—expressed in wavelengths—becomes independent of the focal ratio of the mirror, provided that the test is made null-fashion, without diaphragms. When diaphragms must be used—as in testing a wide aperture paraboloid at its center of curvature—the limit of observational accuracy always corresponds to an uncertainty of the tilt of each zone amounting to a given fraction of a wavelength at the edge of the zone. It is extremely difficult to reduce this error to one hundredth of a wavelength per zone. Now, the effect of such an error is cumulative: if one zone is measured wrongly by $\lambda/100$ the resulting height calculated for the 10th zone from it, will be in error by $\lambda/10$—a by no means negligible amount. That is why it is preferable to test wide aperture paraboloids, for which many zones would be necessary, at the focus, with the aid of a flat.

Diffraction effects can complicate the "paraboloidal shadows" seen at the center of curvature so much that even at the modest aperture of $f/8$

there may sometimes be difficulty in locating irregular error from the "general run" of the shadows. It may then be preferable to gain ease of interpretation at the expense of sensitivity by increasing the diameter of the pinhole "till it is as big as a porthole" and obvious diffraction troubles disappear. We have to fall back on "ray" theory to determine a reasonable size for the "porthole." On this basis, if we wish to see the "paraboloidal shadows" ranging in contrast from fully black to fully bright, we should make the "porthole" diameter equal to that of the geometrical circle of least confusion produced by the mirror. That is, for a paraboloid of diameter D, focal length f, we should use a "porthole" of diameter $(D/64)(D/f)^2$. So, for example, a 12-inch paraboloid of $f/8$ would require a "porthole" 0.003 inch in diameter.

But to detect a given error we should then have to look for differences of contrast considerably smaller than those with which the same error would show in a spherical mirror tested at its center with a really small star: this is especially the case with the three errors of longest period: coma; astigmatism; and "error" of absolute focal position.

B.16.5 Note Added in 1996

C.R. Burch may be calculating the size of the pinholes based on the radius rather than the diameter of the Airy disks of interest. The choice appears to be deliberate conservatism rather than an error. Realizing that amateurs would tend to make overly fat holes, Burch probably decided to err in favor of a small size and calculate the radius rather than the diameter.

For example, early in the article, the author comes up with a width of $0.001F/D$ [mm] for the diameter of the pinhole used in testing a sphere at the center of curvature. Actually, the diameter of the Airy disk produced at the center of curvature is $2.44\lambda R/D = 4.88\lambda F/D$, with λ representing the wavelength. The quantity 4.88λ is $(4.88)(0.00055) = 0.0026$ mm for yellow-green light. Thus, the limit for the hole size, if it were taken right to the edge of the Airy disk, is $0.0026F/D$ [mm]. *H.R.S.*

Chapter B.17

Astigmatism[†]

The term astigmatism is applied to an optical fault which causes images of points not to be points. The word "point" is used in the sense that a dot is a point—not referring to an arrow shape.

Astigmatism is normally inherent in many types of optical equipment. Most commonly it is present in the off-axis images of telescope objectives, especially reflectors. This chapter, however, deals not with the normal astigmatism, but with that class which is due to abnormalities of the instrument brought about by faulty conditions of making.

To begin with, the astigmatism of a speculum is always such that the radius of curvature as measured across various diameters is not the same for all. It is very much as if we took a perfectly figured mirror and bent it slightly across one diameter.

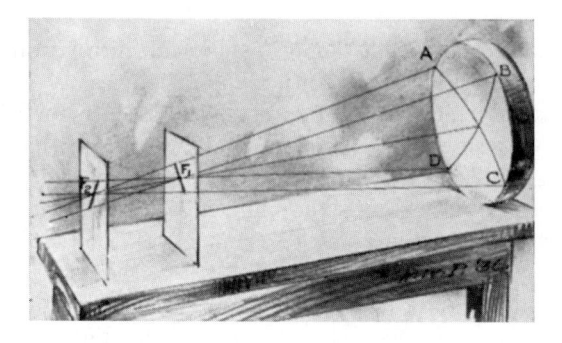

Fig. B.17.1: *Drawings by R.W. Porter, after the author.*

In the illustration, Figure B.17.1, the scale is, of course, much exagger-

---
[†]By G.E. Warner.

ated. The mirror is supposed to be deformed so that across the diameter AC its radius is at F_2, while across BD it is at F_1. With a mirror of this shape it is impossible to get a true point image of a point source. Suppose we examine the shapes of the images which we would see if we used an eyepiece while the mirror was set up à-la-Foucault.

If we are sufficiently far outside of the plane through F_2, our first view of the out-of-focus disk will show it to be quite circular in shape. As we move our ocular toward F_2, however, the disk will become very elliptical as it contracts. When the plane of F_2 is reached the image will be a line of the width of the pinhole of our source, but considerably elongated. The reason for this appearance is evident from the illustration. At F_2 we are in focus for the diameter AC, but we are out of focus for the diameter BD. As we progress from this point, moving our eyepiece toward the mirror, we shall see the image become more or less circular, but increasing in its short dimension from the minimum previously observed. When the image is most nearly circular, at a point about halfway between F_1 and F_2, it is still far from perfect. It is many times the size of a true anastigmatic image. Because of this it can easily be seen how very ruinous astigmatism can be to telescopic definition.

When the ocular is moved toward the mirror, the image again assumes an elliptical form, narrowing in one dimension while it lengthens in the other. This elongation is at 90° from that first seen. The ratio of the length to the breadth of the image again reaches its greatest value when the eyepiece is focused on the plane of F_1.

When the eyepiece is pushed still farther toward the mirror, the image is again seen to expand, first being very elliptical and gradually losing its eccentricity as it grows.

We can analyze how an astigmatic spherical mirror would look under Foucault test by referring again to the illustration.

Suppose we intersect the light coming from the mirror with a vertical knife-edge. If we insert it from the left at F_1, we can see that it is the light coming from A that will first be cut off, hence that part of the mirror will be the first seen to darken. As the knife-edge advances into the cones it will, when it reaches the center of the beam, cut off the rays from B and D. The ray from C will still be uncut. We see, then, that the shadow will first appear at A and will have its edge parallel to BD, as at A, Figure B.17.2. As we cut in, it will advance diagonally across the mirror's face, following the direction of AC of Figure B.17.1. The blade is of course vertical, but the shadow appears like the one in Figure B.17.2 at B.

Choosing the point midway between F_1 and F_2 in which to insert the knife-edge, we can see that it will be the rays from A and B that are cut, and how important it is to provide every kind of assurance that no trace of

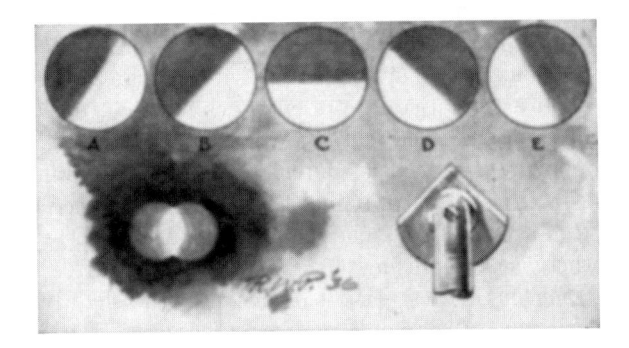

Fig. B.17.2.

it is discoverable in a mirror while those from C and D are unscathed. The appearance then will be like Figure B.17.2 at C.

When inserted in the plane of F_2 the knife-edge will first encounter the ray from B, and then, when the center is reached, it will cut those from A and C, giving us an appearance like that of Figure B.17.2 at D.

If the knife-edge cuts the cone of light much inside of F_1 or outside F_2 the shadow's edge will be more nearly vertical, as in Figure B.17.2 at E. The sequence of shadow shapes that we would see, if our knife-edge was mounted on an accurate carriage, aligned with the mirror's axis and cut into the cone just halfway, would be like those in Figure B.17.2.

This, of course, can cover only the case of a mirror that is spherical except for the astigmatism. Other shapes of surfaces give other appearances. The most common of these is that exhibited by an astigmatic paraboloid, hyperboloid or ellipsoid—the monad.

Again referring to Figure B.17.1, if we should orient either the mirror or the knife-edge so that, when the latter is cut into the cone at F_1 it would be parallel to the focal line at that position, it would then cut all rays coming from the mirror simultaneously and the mirror would be seen to darken evenly all over. With the knife-edge still parallel to F_1 let us cut into the plane of F_2. The ray from B is the first cut; then, as the center is reached, the rays from A and C are intercepted and finally the D ray is stopped. The shadow seen is one whose edge is parallel to AC and progressing from B to D, opposite in direction to the cutting of the knife-edge. This is a normal indication, in a perfectly spherical mirror, that the knife-edge is cutting the cone outside the C of C. Thus, if we should happen to set up even a severely astigmatic mirror and accidentally have either axis of the astigmatism parallel to the knife blade, we would get no indication of this defect from the shadow test.

The infallible test for the presence or absence of astigmatism is one performed with the mirror set up in a perfectly collimated telescope. The stars' images produced by it are examined with as high-powered an ocular as is available. If astigmatism is present, racking the eyepiece slightly in and out of focus will show the telltale elongations.

In the shop a tiny pinhole as a source, not too brightly illuminated, and a short-focused eyepiece for examining the image, will suffice to detect most ordinary cases of astigmatism. The ocular must be correctly aligned with the mirror, otherwise a confusing effect is sometimes seen. The astigmatic indication as seen in this test—the elongation of the slightly out-of-focus image—must follow a rotation of the mirror, otherwise it is a spurious effect due to the test setup.

The matter of determining the axes of the astigmatism is quite important to the worker of specula, inasmuch as any corrective measures to be applied will have to be carefully directed so that the figure be corrected and not further distorted. Also, the discrepancy between the F_1 and F_2 points is of practical importance. There are several possible methods for gaining the desired information.

First, the eyepiece method—noting the two focal points F_1 and F_2 and observing the direction of elongation of the image at either focus. Referring to Figure B.17.1 again, we can see that if we are in focus with the shorter curvature, as at F_1, the elongation of the image will be in line with the long curvature axis. If the separation between F_1 and F_2 is not very great (under $1/16''$) it is difficult to determine the exact location of these points and the measurement is subject to large errors. Mirrors having several zones, or in which the parabolic corrections are present, offer another difficulty, in that the image seen in the eyepiece is subject to aberration.

The second method, a modification of the first, is one in which a mask is placed over the face of the mirror while it is under test. This mask has two equal apertures, diametrically opposite each other, over the marginal zone of the mirror. With the mask in place the eyepiece can be very accurately refocused because, when out of focus, the image will appear double, like Figure B.17.2, lower left. Also, if our pinhole is sufficiently small and our eyepiece high enough in power, interference fringes will be seen when the focus is reached.

The focus of our starting position of the mask can be taken as zero and then, rotating the mask about the mirror's center, the focal length checked at each of several orientations and compared with that of the original position. We may tabulate our resulting figures or make a graph to represent them. The graph of a simple flexured mirror will be sinusoidal. After local corrections have been applied it will usually be found that the curve of our graph becomes somewhat irregular. Our graph will then indicate where

Fig. B.17.3: *The author.*

and the amount of the further correction needed.

A third method, not exceedingly accurate, depends on the fact that the Foucault test fails to show the evidence of astigmatism when the knife-edge parallels one of the axes of the distortion. For this test the writer has used a knife of the form shown in Figure B.17.2, at right. The two upper edges are sharpened, and the angle between them is 90°. The blade can be rotated in a vertical plane about a pivot. In the test the knife-edge is rotated about on the pivot until, when inserted into the average center of curvature, none of the curious distorted figures appear. The knife-edge, without its angle being changed, may then be moved across through the image until the other edge just intersects the beam. If the first edge was inserted at the average center of curvature of the mirror, the second edge will, if moved at right angles to the beam, appear to be inside or outside of the average center of curvature. The amount it is necessary to move the second edge to or from the mirror to obtain the same "depth" appearance as the first edge gave, will be the discrepancy between the foci of the two astigmatic axes.

This method, of course, is not accurate because of the difficulty in determining the exact angle and the center of curvature if the mirror is not spherical.

Of the causes of astigmatism, improper support of the mirror in working "face up"—not in the polishing as much as in the grinding stages—is probably the most prolific source. This is because, in grinding, there is little or no occasion to move the mirror about on its support, so that all the grinding is done while the mirror is flexed in one direction. While being polished, the mirror is frequently removed from its support for the purpose of testing and, as a result, its position on the polishing pedestal or machine is periodically changed, so that the effects of the various flexures are practically sure to average themselves out.

Working in the conventional manner—mirror on top—there is less likelihood of encountering severe cases, but they can and do show up occasionally. Their principal causes are improper motion during the grinding or polishing stages, usually failure to rotate the mirror in the hands, and the use of too large a handle. A wooden handle will usually warp because of the liquids involved in the operations. The pitch fastening will yield, but not fast enough. If work is done while the warping is in progress sufficient stress will be transmitted to the mirror to cause trouble.

Note by John H. Hindle, Witton, Blackburn, Lancs., England: Astigmatism is such an insidious thing that I don't believe one amateur in a hundred knows how to recognize it. Only the greatest care can keep it out. If the mirror is face up during grinding or polishing an improper support will undoubtedly cause astigmatism. I do not bar the use of small polishers, along with that of a full-size, or a nearly full-size polisher to maintain a surface of revolution, and I use small polishers on my drill press polisher for faulty zones, but afterwards use the large polisher to finish with. Much depends upon the definition of a "small" polisher. I should say, one less than half the diameter. But to *entirely* produce a mirror face to the accuracy necessary, and entirely free from astigmatism, by promiscuous *hand* polishing with small polishers, does not seem to me to have any probability about it.

Chapter B.18

Optical Bench Testing[†]

B.18.1 Introduction

An optical bench is an instrument on which optical elements and systems can be mounted and held in proper and variable relationship to one another for the purposes of analysis, demonstration and testing. There are many types and sizes, varying from simple saddle clamps riding on a wooden meter stick to intricate, massive arrangements weighing a ton or two.

By means of a nodal slide optical bench, complete telescopes and binoculars, as well as eyepieces, telescope objectives, camera lenses, enlarger objectives, magnifiers, projection lenses and lens elements can be examined for general performance. Also, many of their characteristics and properties can be determined, their various aberrations can be estimated and the radii of curvature of surfaces can be measured.

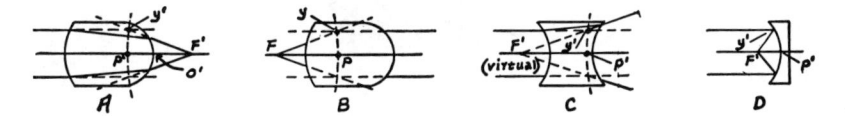

Fig. B.18.1: *All drawings by the author.*

B.18.2 Lens Characteristics

About 1820, Gauss discovered that a lens or a lens system could be accurately described in terms of axial points and that from these the foci could be measured, no matter how complex the system, so long as it produced an image of an infinitely distant object. When a lens receives parallel rays and joins them at a focus, as in Figure B.18.1A, the refracted rays behave

---

[†]By Horace H. Selby.

as if they reached F' from the surface $P'Y'$. Similarly, rays propagated in the opposite direction Figure B.18-1B, appear to have reached F from PY. P and P' are called the first and the second principal points, respectively, of the lens and the surfaces PY and $P'Y'$ are called the principal surfaces. Geometrically, the principal surfaces are the loci of the junctions of the projected rays—dashed lines—and the principal points are the intersections of the principal surfaces with the axis. Mirrors and negative lenses have similar principal points (Figure B.18.1, C and D).

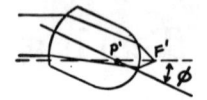

Fig. B.18.2.

When a lens is rotated slightly about its second principal point P', the focus, F', does not leave the original axis. (Figure B.18.2) If, then, a lens is mounted in such a manner that the image does not shift laterally when the lens is oscillated about a line perpendicular to the axis, this line must pass through P' and the focal length can be determined by measuring the distance from the axial focus to the center of rotation. When P' is used for such focometry and F' is in air, P' is referred to as the node of emergence or the exit node. P is called the entrance node. Thus, a sliding lens holder which can be rotated about a vertical axis is called a nodal slide. The use of a nodal slide is by no means limited to the determination of focal lengths.

Fig. B.18.3.

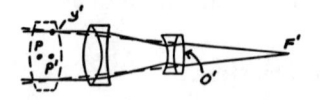

Fig. B.18.4.

A system of lenses, such as the photographic objective of Figure B.18.3 or Figure B.18.4, has the same nodal property as has a simple lens and can be treated accordingly. The single lens corresponding to a system is called

B.18.3. A Nodal Slide Optical Bench

the equivalent lens—dashed in Figures B.18.3 and B.18.4—and the focal length of all systems is therefore stated in terms of the focal length of this imaginary lens. This explains the term equivalent focal length (efl or f').

When a lens receives divergent instead of parallel rays, as in Figure B.18.5, it can also be oscillated about a point without causing transverse image movement. This point is called the null point and its position differs from that of the exit node. The use of the null point enables photographic objectives, erector systems, process lenses and other systems used on near objects to be tested for performance but not for determination of efl. For efl determination, the exit node is used.

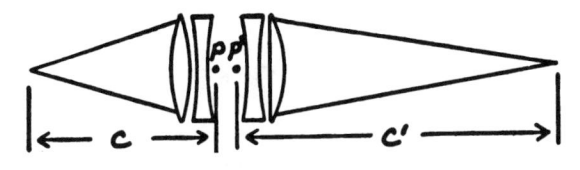

Fig. B.18.5.

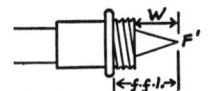

Fig. B.18.6.

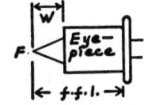

Fig. B.18.7.

B.18.3   A Nodal Slide Optical Bench

A general purpose nodal slide optical bench is shown in Figure B.18.8. The collimator comprises a paraboloidal mirror M and a pinhole illuminator P with its pinhole at the axial focus of M. S is the nodal slide, which carries the lens holder W. S can be rotated about K and the amount of rotation is indicated on J. V is a combination microscope-telescope, having a reticle C at the eyepiece focus. U is a reading microscope with cross hairs used to read the position of the carriage X with respect to the scale N. Z is an illuminator, which may be a penlight bulb. Y represents the bench holding the apparatus. It should be sturdy. T is a transformer or battery supplying

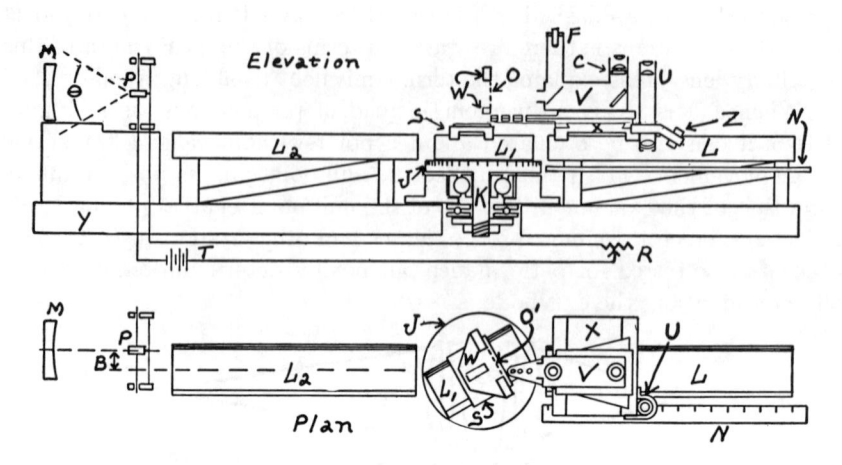

[Fig. B.18.8.]

P, the brightness of which is regulated by means of the rheostat R. L, L_1 and L_2 form the optical bench proper, the right portion of which carries the slide X, to which U and V are rigidly attached. X may be provided with a graduated cross-slide similar to the crossfeed carriage of a lathe.

For the most accurate work, something as elaborate as shown is practically necessary. Fairly good work can, however, be done with a precision of ±0.02 to 0.05 inch on an optical bench which has a pointer in place of U, a snug bushing instead of preloaded ball bearings to hold K and parts L, S and X of wood. As in telescope making, almost anything can be made to serve if the essentials are present and if basic theory is not violated. For testing the alignment of binocular field glasses, the cross slide at X is desirable.

The basic parts and their arrangement are:

B.18.3.1 Illuminator

The illuminator P should be compact. A 2.5-volt penlight bulb (Mazda No. 222), a sleeve socket, a clear, symmetrical 5 mm (0.2-inch) glass bead[1] and a very fine pinhole in a copper disk can be mounted in a short piece of 1/2-inch diameter copper tube. This can be soldered to four pairs of hard copper wire (Figure B.18.8, P) which form a spider support similar to that of the diagonal in a Newtonian telescope. The supporting wires carry the bulb current. M represents any excellent paraboloid, larger than any

---

[1] Kimble No. 13500, obtainable from laboratory supply firms.

B.18.3. A Nodal Slide Optical Bench

specimen aperture. P is placed accurately at the axial focus of M.

Nodal slide S is any arrangement incorporating a sliding member exactly perpendicular to shaft K, which must be accurately vertical and capable of rotation with a minimum of wobble. J should have a circular scale, graduated in degrees, for measuring its angular rotation. The nodal slide should be leveled, securely mounted on the bench and offset from the illuminator axis by an amount, B, equal to $1/2$ the outside diameter of P plus $3/8$ inch.

Fig. B.18.9.

B.18.3.2 Telescope-microscope

The Telescope-microscope, V should have a spirally focusable ocular of 0.5 to 1-inch efl, a reticle for measurement, three eyelens filters of red, yellow-green and blue-violet and two or more interchangeable objectives. A 90° prism diagonal is not essential, but very convenient. The ideal reticle, Figure B.18.9, is a net micrometer disk. Second choice is any type having parallel, equally spaced rulings. Excellent filters are Eastman's Wratten filters No. 29 (F), No. 61 (N), and No. 49 (C4) not mounted in glass. The telescope objective should be of 6 to $7\,1/2$ inches efl and have an aperture of $3/4$ to $1\,1/2$ inch. It should give a sharp image when used with the ocular and focused on infinity. No one microscope objective will serve for all purposes, for some tests require that the objective accommodate the entire cone of rays from a specimen which has an aperture ratio of $f/1.5$, for example, and others require that the objective have a long working distance. These requirements conflict. To illustrate the point, assume that Huygenian, Airy or Mittenzwey oculars, the focal surfaces of which are within the eyepiece proper, are to be tested. In order to reach the foci, an objective with an efl of approximately $3/4$ the efl of the ocular, as a minimum, will be needed. Say that a 2-inch ocular is the lowest power to be tested. The efl of the microscope objective must then be about 1.5 in or 37 mm. 40 and 48 mm objectives are standard and they work in the neighborhood of $f/6.3$ (numerical aperture 0.08). Since Huygenian eyepieces rarely perform well beyond $f/6.3$, either objective is good. However, if a photographic objective of $f/1.5$ is to be examined critically, the nearest standard microscope objective which will suffice has an efl of 8 mm and works at $f/1.2$ (N.A. 0.50). Its working distance is only 1.5 mm (0.06 inch); so it cannot possibly reach the interior of a "negative" eyepiece of efl greater than about 0.1 inch. The reader must choose for himself from the list in Table B.18.1 the objective or objectives which best suit the work which he contemplates.

War surplus achromatic doublets can be found and at times can be pressed into service in lieu of the expensive standard objectives for fairly accurate work. The better short-focus doublets (12–20 mm) work well if stopped down to $f/6$. Two such doublets, crown to crown, give fair results at $f/4$ in exceptional cases. Due to the lack of uniformity among war surplus items, no useful figures can be given here.

The various objectives should be mounted so that they can be interchanged rapidly in such a way that the telescope objective is always focused for infinity and each microscope objective focuses on a constant individual object distance. Standard microscope objectives are corrected for a tube length of 6.3 inches and perform well when from 6 to 7.5 inches from the eyepiece focus.

Table B.18.1
Standard American Achromatic Microscopic Objectives[1][†]

efl	Numerical Aperture	Limiting f-number	Magnification ($C' = 6.3$ inch, Fig. B.18.5)	Working Distance	Approx. Cost
48 mm	0.08	$f/6.3$	2.0×	52 mm	$20
40	0.08	$f/6.3$	2.5×	40	20
32	0.10	$f/5$	4×	21–38	25
16	0.25	$f/2.1$	10×	5–7	35
8	0.50	$f/1.2$	20×	1.5	50

This is why a telescope objective of 6 to 7.5 inch efl is desirable. Parfocal mounting is easier.

Although microscope objectives are designed for use either with or without cover glasses, the difference is slight for the average optical bench determination. However, if new objectives are purchased, it is well to specify that they be corrected for use on uncovered objects.

V should be mounted on X so that V can be rotated in the horizontal plane about a vertical pivot, the projected axis of which coincides with the microscope objective focus. One method of doing this is shown in Figure B.18.8, where holes for three pivots corresponding to low, medium, and high power objectives are shown.

The vertical illuminator F consists of a good quality, thin, semireflecting diagonal which can be turned either parallel to the axis of V or at 45° to it and a pinhole which can be similar to P but of larger aperture. The pinhole-to-objective distance must be optically equal to the reticle-objective path.

[†]Numbers in brackets refer to references at end of chapter.

B.18.3.3 The Optical Bench Proper

The optical bench should be in two sections, L and L_2, with the nodal slide between them. Each section should be as long as the efl of the longest specimen to be tested plus the length of slide X. It may be economically made of 4 or 6-inch channel iron, properly planed, one lip flat, the other a 60°–80° angle. A short length of similar material, L_1, may be mounted on J to support S. A 3-inch channel-iron bench is commercially available with accessories [2] at moderate cost.

B.18.3.4 The Slides X and S

Slides X and S may be of any stable material, shaped to slide on L without side play. A cross slide similar to the crossfeed of a lathe is practically essential on X or S for testing binocular alignment and is convenient if on both.

The scale N should be an *accurate* steel tape under tension or a steel rule, preferably with decimal graduations.

The reading microscope U, if used, should have cross hairs at the eyepiece focus. It need be neither an erecting microscope nor well-corrected. A fine pointer and a magnifier, both rigidly attached to X, are nearly as good, especially if N is tilted outward to give headroom. A simple pointer will give sufficient accuracy for approximate work.

B.18.4 Alignment

As with Cassegrain telescopes, so there are many ways of aligning this test bench. The following is satisfactory.

I. Square P and its ring support with the bench and with a level and clamp it permanently in position, being sure that its height is approximately that of the axis of V. Be sure the pinhole is illuminated when viewed throughout angle θ. Realign bulb and bead, if necessary.

II. Arrange V as a telescope and verify its infinity focus by examining a very distant outdoor object. Permanently correct, if necessary. Mark eyepiece focusing mount and do not disturb until all alignment operations are completed.

III. Verify the perpendicularity of K with a level on S for various angles of rotation. Clamp W on S and stretch a fine wire or thread O vertically across the exact center of the lens-mounting hole in the face of W. Slide S until O is approximately over the axis of K.

IV. Arrange V as a microscope, mount it on L and observe O through V, sliding X to obtain a sharp image. By systematically moving L toward and away from the end of Y and by sliding S on L_1, find a position in which J can be rotated as far as possible on both sides of zero without any movement of O being discernible through V. If this condition cannot be obtained, it indicates that the hole in the face of W is not centered, but is to one side and must be corrected. When this operation is complete, O is accurately over the center of K.

V. Mark the position of X on L, slide X to the right, remove the objective of V and slide X to its former position. Temporarily but firmly mount a magnifier or other lens of $1/2$ to 1-inch efl above the eyepiece of V so that O is slightly out of focus (the magnifier being somewhat too close to the eyepiece) and centered in the field of view. Mount a cross wire perpendicular to O on W with adhesive tape so that it, too, bisects the field of view. Slide X to the right end of L and observe any shift in the apparent position of O or O'. Rotate and shim V on X so that O and O' remain centered for all positions of X on V. Clamp X permanently and recheck. The axis of V is now parallel with the ways of L. Remove magnifier.

VI. Arrange V as a telescope, light P and have an assistant move M until the image of P is visible through V. Have M adjusted to center the image of P. Change V to a microscope, observe O and O' and center them in the field by shifting L. Keep L level at all times and maintain dimension B.

VII. Repeat VI until the images of O, O' and P remain centered in their fields and the image of P is sharp and absolutely symmetrical. Clamp M and L permanently. M and P are now coaxial, the axes of V, P and L are parallel and the axis of V intersects the projected axis of K.

VIII. Remove O'. Using V as a microscope, focus on O. Rotate J through large angles on both sides of zero, sliding S on J, if necessary, to the point where the image of O remains stationary. Slide X on L until O is in sharpest focus. Adjust N parallel with L until the zero of N is seen coincident with the cross hairs in U. Clamp N in place temporarily. N may have to be shifted due to shrinkage or warping or to a change of microscope objectives.

IX. Set J on zero, illuminate the face of W and examine with the microscope, using only the cross-slide of X. The face on both sides of the hole should be sharply focusable without moving X on L, proving it

B.18.5. Testing Methods

perpendicular to the axis horizontally. Check vertical perpendicularity with a square level. Correct if necessary.

X. Light F, arrange V as a telescope and mount a first surface mirror on S or on W, facing V. Turn mirror until the pinhole image of F is seen in V and adjust F to make image needle sharp. Change V to a microscope and focus sharply on some flat, fine-textured surface, such as the ungraduated portion of a steel rule held against W. The image of the pinhole should be centered and sharp. If it is not, adjust F and its reflector and recheck with the mirror and telescope. The image of F must be sharp and centered both with the telescope mirror and with the microscope-flat surface combination.

XI. Light pinhole P and examine its image with V as a telescope, using all three eyepiece filters in turn and adjusting the eyepiece to maximum sharpness. Mark each position so that the focus for each color can be used at will later, or record it in a notebook. Change V to a microscope and stretch a fine white thread across the opening of W, illuminating it with F. Again use all three filters and mark the sharpest focus positions on another portion of the eyepiece focusing mount or record them as before. The reticle should appear reasonably sharp with all colors. It is the writer's practice to use a graduated sleeve on the ocular and to record the positions as follows: $M_{40v} = 0.32, M_{40y} = -0.06, M_{40w} = 0.00$ and $M_{40v} = 0.34$ for the microscope with a 40 mm objective, $M_{8r,y,w,v}$, for the 8 mm objective and $T_{r,y,w,v}$ for the telescope, red, yellow-green, white and blue-violet being the colors indicated by the subscripts.

XII. Sit back and admire your handiwork. The gadget is ready to use.

B.18.5 Testing Methods

The methods to be presented are applicable to all refracting optical elements and systems as well as to some reflecting elements and instruments. They are not necessarily the best methods, however, in all cases, for all mechanical tests are subject to error and optical bench errors are larger for some types of work than those inherent in purely optical methods or in other mechanical procedures.

Bench methods are rapid, direct and of a fundamental nature, however. They are of wide application and of sufficient inherent accuracy for practical purposes.

In order to judge the precision of his results, the beginner should repeat the first determination of any type several times from the very beginning.

The spread of the values obtained will serve to indicate the reproducibility of the particular method employed.

B.18.5.1 Tests on the Axial Image

B.18.5.1.1 Equivalent Focal Length ($efl, f', P'F'$ in Figure B.18.1)

1a. *Well-corrected Positive Systems:* Mount specimen in W with surface which normally faces infinity or the major conjugate (C', Figure B.18.5) toward P. Eye lenses of eyepieces send out parallel rays (face ∞); so they should face P. Using appropriate microscope objective in V, light P and observe its image, having checked alignment operation VIII. Rotate J slightly from zero. If image moves against rotation, slide S away from V and conversely. Continue oscillating J across zero while moving S until image shows no lateral movement whatever for a 5° oscillation. Set J to zero, move X to sharpest image and read efl directly on N.

1b. *Poorly-corrected Positive Systems and Simple Lenses:* Mount specimen in W with shorter convex radius towards P. Mount a small diaphragm with an $f/30$ aperture close to specimen. Use the yellow-green filter on the eyepiece and focus it accordingly. Proceed as in 1a, reading the paraxial efl for the middle of the visible spectrum on N.

1c. *Negative Lenses and Systems* (Barlow lenses, telephoto attachments, microscope amplifiers, etc.): On L_2 mount a positive lens of longer efl than the specimen. An achromatic refractor objective is ideal but not essential. Arrange V as a microscope, set U to the estimated efl of the specimen on N and move the positive auxiliary lens so that a sharp image of P is seen through V, using a diaphragm at the auxiliary if necessary. Mount specimen in W on S, change V to a telescope, use the yellow-green eyelens filter and observe the image of P, sliding S to obtain sharpness. Oscillate J across zero and slide S and auxiliary approximately equal amounts in the same direction to the points which enable J to oscillate a few degrees without lateral image shift. Set J to zero, slide auxiliary only to sharpest focus, closing diaphragm if necessary and remove specimen. Change V to microscope, slide X to position of sharpest focus and read efl directly on N.

1d. *Alternative for Huygenian and Similar Eyepieces:* It is preferable to test all systems in the proper rather than in the reverse position on W because settings are most precisely made on images of wide aperture

B.18.5. Testing Methods

and reversed systems usually have worse definition, necessitating reduced apertures for good definition. This aperture reduction increases depth of focus which reduces setting precision.

However, an occasional eyepiece with an inaccessible focus can be checked for efl (but not for definition and field aberrations) when a microscope objective of suitably long focal length is not at hand by reversing the eyepiece, for the paraxial efl of any system is an invariant, not being affected by the direction in which light traverses the specimen.

Procedure: Mount specimen with field lens toward P and determine efl by method 1b.

B.18.5.1.2 Back Focal Length (bfl, $O'F'$, Figures B.18.1, B.18.3, and B.18.4)

2a. *Telescope Objectives, Camera Lenses, Elements, etc.:* After determining efl and without disturbing any part of the arrangement, place talcum on surface of specimen nearest V. Light F and slide X to obtain sharp focus on talcum particles. Read N. Subtract from efl. Difference is bfl.

2b. *Direct Method:* Set U to zero on N. Set J on zero. Mount specimen in W as for efl. Place talcum on specimen surface nearest V, light F and slide S to sharp microscope focus on powder. Blow powder away with syringe, light P, extinguish F and slide X to sharp focus on image of P. Read bfl directly on N.

2c. *Eyepieces Only:* Mount specimen in W with field lens facing collimator. Proceed as in 2a or 2b.

B.18.5.1.3 Working Distance (W, Figure B.18.6, Figure B.18.7)

3a. *Magnifiers, Eyepieces:* Mount specimen in W with surface which normally faces object or image which is magnified facing V. Set J to zero. With cross slide of X, move microscope in line with the most projecting portion of the specimen. Move S to left, set U to zero on N and slide S to focus of V on specimen projection. Move V back to axis with cross slide and slide X to sharp focus of image of P. Read quantity on N. If no part of specimen projects beyond lens surfaces, cross slide is unnecessary and center of projecting lens surface is used to obtain reference point.

3b. *Microscope Objectives:* Mount entire microscope tube, including a microscope ocular and the specimen in W with ocular facing P. If microscope tube is adjustable, set it to 160 mm. Proceed as in 3a.

3c. *Microscope Objectives:* On the left section of L mount an illuminated pinhole, slit, cross wires or other small target. Mount specimen in W as in method 3a and proceed as there stated until projection is in focus. Now move illuminated target to a position 150 mm (5.9 inches) to the left of the Society screw shoulder of the specimen. Finish method 3a, using target image instead of the image of P.

B.18.5.1.4   Flange Focal Length (ffl, Figures B.18.6 and B.18.7)

4. *Motion Picture Camera Objectives, etc.:* Use method 3a, except that the zero point is adjusted with the use of the mounting flange instead of the projecting portion of the specimen.

Caution: Specimen should be tested at full aperture and the objective of V should have suitable numerical aperature or f-number to receive full cone from specimen. White light should be used.

B.18.5.1.5   Axial Critical Aperture Ratio or Aperture Tolerance

5a. *Positive Systems Used on Distant ($> 10f'$) Objects, also Eyepieces:* If the specimen has no adjustable diaphragm, provide an iris diaphragm or a series of circular stops between P and the specimen and close to the specimen. Mount specimen in W with surface which normally faces distant object or which faces the eye toward P. Using proper microscope objective and with J on zero, examine image of P with V, employing a small specimen aperture. Increase aperture until image at its best focus deteriorates to the point where specimen barely performs satisfactorily as far as definition is concerned. Measure aperture. Divide by efl of specimen. Quotient is f-number of critical aperture for axial use. For extra-axial images, use combination of 5a and 8a or 8b.

5b. *Positive Systems Used on Near ($< 10f'$) Objects:* Use combination of methods 5a and 8c or 8d.

5c. *Negative Systems:* Use method 1c and place iris diaphragm to right of specimen. Calculate results as in 5a. Auxiliary lens must be well-corrected and preferably the one with which specimen is to be used.

B.18.5. Testing Methods

B.18.5.1.6 Axial Chromatic Aberrations[†](Δ_{chr})

6a. *Positive Systems Used on Distant Objects:* Proceed with method 5a, stopping at the critical aperture. Close the diaphragm until definition is sharpest. Use the red eyelens filter, slide X to sharpest focus and read N. Repeat with other filters. Red reading minus violet reading equals primary axial chromatic aberration. Average of red and violet readings minus yellow-green reading equals secondary axial chromatism. Pri. Δ_{chr} is usually slight in good systems. Sec. Δ_{chr} usually runs about 5 to $20 f' \times 10^{-4}$ in achromats (from 5 to 20 ten-thousandths of the focal length).

6b. *Positive Systems Used on Near Objects:* Use methods 6a and 8c or 8d.

6c. *Negative Systems:* Use method 1c with auxiliary lens intended for use with specimen and diaphragm as in 5c. Use filters and calculations of 6a. The Δ_{chr} obtained will be that of the positive-negative combination. Therefore, the auxiliary lens should be well-corrected.

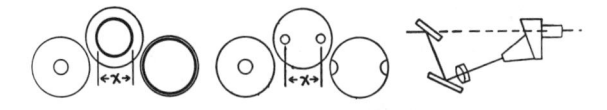

Fig. B.18.10–12.

B.18.5.2 Axial Spherical and Zonal Aberrations (Δ_{sph}, Δ_{zon}, or LA')

7a. *All Systems:* Use appropriate method for Δ_{chr} but, instead of an iris diaphragm, use a series of stops (Figure B.18.10 or B.18.11) between specimen and illuminator. Axial opening should be $f/20$ to $f/50$. Total area of usable opening should approximate area of axial opening. Minimum number of stops is three: axial, 0.7 of full aperture for dimension X and 0.98 of full aperture. Figure B.18.10 masks are theoretically proper and the dimension X and the width W of the annulus to match any axial stop can be calculated from the efl (f) of the specimen as in Table B.18.2.

Table B.18.2

Axial stop size	$f/20$	$f/30$	$f/40$	$f/50$	$f/60$
W	$f^2/1600\times$	$f^2/3600\times$	$f^2/6400\times$	$f^2/10,000\times$	$f^2/14,400\times$

[†]For a description of the aberrations, see ATM3, Chapter B.3 on eyepieces.

They are not practical, however, for small lenses, for when W_{marginal} becomes less than 0.02 inch, diffraction becomes bothersome.

Figure B.18.11 (Hartmann) masks are practical and much easier to use, for extra-focal images are doubled and diffraction is less. The centers of the holes should lie on the zones to be tested. Each zonal hole may be 0.707 of the axial hole diameter and marginal holes may each equal the axial hole in diameter because they are approximately divided in area by the rim of the specimen.

Δ_{sph} is the difference between the readings on N for the axial and the marginal stops. Yellow-green light is usually employed.

Δ_{zon} is the difference on N found between that stop giving the smallest N reading (often the 0.7071 stop) and the axial stop.

Δ_{zon} has little significance unless Δ_{sph} is small.

Δ_{sph} of which Δ_{zon} is only a part, can be plotted as a curve, as in Chapter B.3, Figure B.3.8, ATM3, on "Telescope Eyepieces," by using a series of stops. It varies with color, this being called "Chromatic Variation of Spherical Aberration." By estimating Δ_{sph} for the various colors, the chromatic variation of Δ_{sph} can be found.

B.18.5.3 Tests Using Extra-Axial Images

B.18.5.3.1 Curvature of Image Field

8a. *Systems Used at Infinity: By Computation.* Use method 1a. When axial infinity focus is found and its position noted on N, turn J to any desired angle and move X to sharpest image, rotating V about the appropriate pivot to maintain maximum image brightness. Read N. Continue for other angles. Field is flat if every reading on $N = efl \div \cos\phi$ where ϕ is angle from zero on J. Deviations from flatness are determined by subtraction. Results are accurate for object distances as close as $10f'$. For good results, repeat on opposite side of axis.

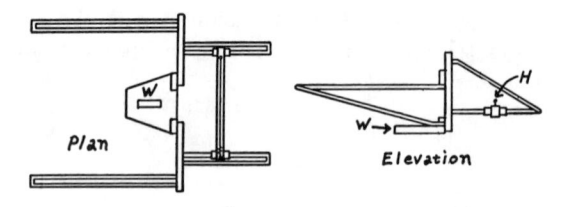

Fig. B.18.13.

B.18.5. Testing Methods

8b. *Systems Used at Infinity: Direct Method.* Arrange a fine wire under spring tension in such a way that it crosses the axis horizontally, moves with W and can be placed at varying distances from it. (Figure B.18.13 illustrates one method of accomplishing this.) Mount specimen as in method 1 and center on exit node, precisely as for determining efl. Wire H must be between V and specimen. Slide H to point where it is sharply focused slightly above or below image of pinhole. H should be perpendicular to axis. Slide X to right and rotate J to right, say exactly 5°. Slide X left to focus sharply on wire where it is nearest to pinhole image. Without disturbing X, turn J past zero to exactly 5° left. Examine wire. If sharply focused as before, F is sufficiently perpendicular to axis. If not, move ends of H in opposite directions enough to correct. Repeat entire process until the axial pinhole image lies on the same vertical plane as the wire and the wire is simultaneously perpendicular to the axis. Measurements can now be made with confidence, using the technique of method 8a. If, at each angular position of J, the best possible pinhole image is found to be contiguous to the wire, particularly if both sides of the axis are checked, the field of best definition of the specimen is proved to be flat. If not, the actual field can be charted by taking a reading on N when the wire is focused and another when the pinhole image is sharpest and noting the difference, which will be the deviation from flatness. This difference is determined at suitable intervals throughout the field of view.

8c. *Systems Used on Near Objects: By Computation.* Mount an illuminated target with vertical and horizontal cross lines on L_2 at the proper distance from the specimen. This distance will be $f' + Mf'$, where M is the magnification at which the specimen is normally used. ($M = 2$ for enlarger objectives, for example, in this case.) Arrange specimen to rotate around its null point as directed in 8d, below. Note the position of the axial focus of the target on N in its relation to the projected axis of K. Call this distance C'. Measure the distance of the target from the projected axis of K. This distance is C. (C and C' are determined with J on zero.) Now, for each angle, ϕ, from zero, the target must be moved away from the specimen to a new position $C/\cos\phi$ from its original location. If the field is flat, V must be moved to the right $C'/\cos\phi$ from its original position to focus the sharpest image. Deviations from flatness can be calculated arithmetically.

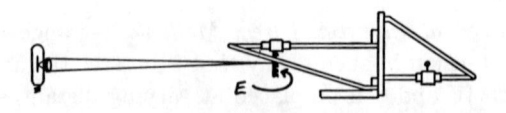

Fig. B.18.14.

8d. *Systems Used on Near Objects: Direct Method.* Process lenses should be tested when the ratio $C/C' = 1.0$ to 1.5 (Figure B.18.5). Enlarger objectives are best examined at $C/C' = 2.0$. Erectors and transfer lenses should be tested at their normal conjugates. In order to facilitate such testing, the standard illuminator is not used. Instead, the part E of Figure B.18.14 is employed. E is a strip of 400-mesh bronze wire cloth under tension, backed with a piece of opal glass. It is held on a sliding carriage similar to that holding H and is illuminated by a spotlight placed on the axis of V on L_2. The illuminated area of E should be no wider than $1/16$ inch. This is conveniently arranged by using a line filament bulb,[2] the filament of which is imaged near E by a high speed projection microscope objective or war surplus achromat, as shown. Examine E or a remnant of the same wire cloth with V and note the size of one screen opening in comparison with the reticle. Record this reticle calibration value.

Mount the specimen on W and place E at such a distance that the image of E, seen in the microscope V, is the proper fraction of the size noted when calibrating the reticle. This fraction is C'/C (Figure B.18.5) for any specific application. Adjust the specimen to rotate about its null point by sliding S to the place which enables J to be oscillated without causing a lateral shift of the image of the narrow illuminated portion of E. Align E so that the image position 10° to the right of zero occupies the same position, measured on N, as it does precisely the same angular distance to the left. Slide H to the place where the wire is sharply in focus with the image of E when J is on zero. Align H at $+10°$ and $-10°$, just as E was aligned. Return J to zero and recheck the image of E to be sure it is coplanar with H. Field curvature can now be measured exactly as in method 8b.

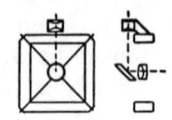

Fig. B.18.15.

---

[2]Central Scientific Co., Chicago 13, Ill., Cat. No. 86615 approx. $3.

B.18.5. Testing Methods

For determining field curvature and estimating the extra-axial aberrations (see later) of mirrors, V must be removed from the axis to avoid obscuration. The arrangement of Figure B.18.15 is useful for this purpose. The microscope objective of V, with a small elliptical flat or a 90° prism, is mounted on a spider support similar to that of P. The eyepiece and reticle are mounted on the spider ring. The wire representing a plane image is stretched across the specimen tube, so that it moves with the mirror. The open tube microscope support ring can be affixed to X of Figure B.18.8 and used on L_2. With this Rube Goldberg contrivance, the principal methods of the optical bench can be applied to mirrors, although computation yields the results with less time and expense if the equation of each surface is accurately known.

B.18.5.3.2 Lateral Chromatic Aberration

9a. *Specimens Used at Infinity and Corrected for Axial Chromatism:* Determine efl by 1a for red and for blue-violet light. If the two efls are equal, lateral chromatism is absent. If not, the longer efl will yield a larger extra-axial image and "edge color" will be evident. Example: $f'_{red} = 7.86$ $f'_{violet} = 8.03$. Lateral chromatism is overcorrected by 2 percent. An image 4 inches from the axis will therefore have a color spread of 0.085 inch, red toward the axis, blue-violet toward the edge.

9b. *All Specimens:* Examine extra-axial images with V while determining field curvature. Use a small aperture to reduce the effects of coma and astigmatism. If lateral chromatism is corrected, image color will be symmetrical or absent. If undercorrected, the axial side of the image will be violet and vice versa. The amount can be determined with the reticle of V.

B.18.5.3.3 Distortion

10. *All Specimens:* Proceed as for field curvature. Use small aperture and yellow-green eyelens filter. If the image leaves the axis laterally by the time the edge of the field is reached, distortion is present and is proportional to the amount of image shift at any given angle.

Fig. B.18.16: *The collimator.*

B.18.5.4 Comatic, Lateral Spherical or Sinical Aberration

11. *All Specimens:* Proceed as for field curvature. The comatic flare caused by this aberration can be recognized instantly, if not masked by other aberrations. Its amount and sign can be estimated fairly well even in the presence of other aberrations, if chromatic effects are minimized with a yellow-green filter at the eyepiece of V and if observations are made both at full aperture and "stopped down." Coma will be greatly reduced at small apertures because it varies as the square of the aperture, while the size of the blur due to astigmatism will be less markedly changed due to its characteristic of varying approximately as the aperture. Coma is zonal in nature. The entire image field should be examined.

B.18.5.5 Astigmatism

12. *All Specimens:* Proceed as for field curvature. This image defect (see ATM3, Chapter B.3) can be measured very accurately by determining the positions of the two focal surfaces which it forms. If method 8a or 8b is used, start at the edges of the field and chart the positions at which the pinhole image is (a) a short, fine, vertical line and (b) a similar horizontal line. The difference between the positions of these two line foci is a measure of the astigmatism at any one angle. Move inward toward the axis. Use a yellow-green filter to avoid chromatic confusion and a small aperture to minimize coma. If method 8c or 8d is the one employed, plot the sharpest foci of the horizontal and vertical elements of the target used.

Fig. B.18.17: *Detail of pinhole illuminator. Maximum diameter 0.48 inch.*

B.18.5.6 Testing Complete Telescopes

Astronomical telescopes, transits, sextant telescopes, individual sides of field glasses and the telescopes of spectrometers, cathetometers, etc., can be checked as units with their eyepieces on the nodal slide or on separate pivots, if they are too large. The following are some of the many possible determinations:

13. *Magnification:* Whiten the face of P, Figure B.18.8, facing M and illuminate it from the side. Using V as a telescope, note the diameter of P in terms of reticle divisions. Place specimen on L_2 or on S, depending upon length, and again determine size of P image on reticle. Size of image divided by image without specimen equals magnification. Refractors and Cassegrainian and Gregorian reflectors can be checked so. Newtonian reflectors can be checked by mounting V in front of the eyepiece at the side. This method is so cumbersome for large telescopes, however, that it is far preferable to employ conventional tests and calculations.

14. *True Field of View:* With specimen on L_2 or, if small, on S, and pivoted over P' (Figure B.18.1C) of the objective or mirror, examine pinhole image with the eye. Swing specimen around P' until image touches edge of field. Read angle on each side of zero on J or measure traverse of tube and length of tube from P' and calculate angles from tangents. Example: A point 40.0 inches from P' was moved 0.265

Fig. B.18.18: *Compensating object and image-plane holder (front).*

inch past center before image was lost. Tangent of the half angle was therefore 0.265/40, or 0.00662. Half angle was therefore 0°22'.8. Total angle = 0°45'.6.

15. *Eyepoint Distance and Exit Pupil Diameter:* These may be determined by illuminating the mirror or objective strongly with a spotlight or floodlight off the axis and receiving the image of the objective or mirror which the eyepiece forms on a ground glass, where its distance from the eyepiece (eye relief) and its diameter can be approximately measured. For greater accuracy, V, used as a microscope, can be employed.

16. *Character of Axial Definition:* The axial image can be examined with the aid of V used as a telescope, employing diaphragms corresponding to the eye pupil diameters which would exist in practice. In this way, errors are magnified directly as the magnification of V, which is normally $5 - 10\times$.

17. *Testing Binocular Instruments:* If M, J and S are sufficiently large and the entire apparatus is sturdy and accurate, binocular field glasses and microscopes can be checked for alignment and the field glasses can be checked for definition. For this work, the pinhole of P is best replaced with a slit. If the ocular tubes are parallel, the instrument can be clamped anywhere on J with the tubes parallel with the axis. If the tubes converge, the intersection of their axes should be on the projected axis of K. The objective of a microscope should be focused

B.18.5. Testing Methods

Fig. B.18.19: *A convenient auxiliary bench with holders for two or four specimens.*

on the infinity focus of a substage condenser capable of filling both ocular tubes with light. (Double-objective microscopes of the Greenough type need large-diameter condensers.) The stage should hold a cross hair slide, or something marking the center of the field. Inclined eyepiece tubes should be horizontal, the objective axes being brought on the same plane by means of two first-surface mirrors, attached to the condenser, as in Figure B.18.12. (Condenser and/or mirrors remain fixed, leaving the instrument free to rotate in the case of convergent or inclined tubes.) The entire instrument in question is mounted in the proper position on J with its objective or condenser facing M. It is leveled and squared with the axis as well as possible, using whatever surfaces seem best. V is changed to a telescope and the image of P is examined through one eyepiece by means of V, using the cross slide of X or of S for parallel tubes and rotation of J for convergent tubes. The image of P is brought to the center of the field by shifting and shimming the specimen slightly and keeping the objective of V centered on the eyepiece of the specimen. X is now moved along L to check the parallelism of the specimen axis with that of the bench. If the image does not move, fine. If it does, shift slightly and repeat until no movement occurs and V is still centered on eyepiece. Parallel tubes **ONLY**. Measure interpupillary distance accurately, using edges of lens cells, if possible. Move cross slide precisely this distance and examine image through other eyepiece. If centered

Fig. B.18.20: *Counterpoised cross slide.*

and if it remains so along L, alignment is perfect. If not, correct first by rotating entire instrument about axis of first tube and then rechecking first tube. If alignment cannot be perfected thus, complete correction by moving prisms, objective or other adjustments. Repeat for at least one other interpupillary distance to check stability of pupilary distance adjustment mechanism (hinges, rack and pinion or pivot). At same time, check focusing of both sides and correct if necessary. Definition can also be judged.

B.18.5.6.1 Convergent Tubes

Change to other side by rotating J until eyepiece is precisely aligned with objective of V. Reflections from F are useful here. Image should be centered and should remain so along L. If not, correct as above, under Parallel Tubes.

B.18.5. Testing Methods

Fig. **B.18.21**: *Compensating object and image plane holder (rear).*

B.18.5.6.2 General Caution

The correction of misalignment is not easy and the novice is advised to proceed with deliberation and the greatest of care. In fact, it is much wiser to entrust such work to the maker of the instrument unless the worker knows himself to be adept and capable.

The methods given above are suitable for the determination of efl, working distance, eyepoint distance (bfl) and the axial aberrations of eyepieces. If a diaphragm is placed at the eyepoint and varied in aperture from 1.5 to 7 mm (0.06 to 0.28 inch), valuable information can be obtained concerning axial definition and the extra-axial aberrations. However, the oblique errors are profoundly influenced by the corresponding aberrations of objectives and mirrors with which eyepieces are used. A considerable amount of experience in the testing of eyepieces, objectives and mirrors is necessary before accurate judgment of the extra-axial aberrations of eyepieces alone on the optical bench is possible. For this reason, the beginner is advised to test his eyepieces for extra-axial performance in conjunction with the system with which it is used.

B.18.5.7 Approximate Refractometry

The refractive indices of simple lenses for various colors of light can be determined on the optical bench when the radii, thickness and efl of the specimen are known.

18. Determine efl by 1b or 1c, using an $f/30$ diaphragm, repeat for other colors, if desired, and enter data in the appropriate formula in Table B.18.3.

Table B.18.3

Second Surface Plane A. Exact	Both Surfaces Curved B. Exact	C. Approximate
$N = 1 + \dfrac{r_1}{f'}$	$N = \dfrac{-b - \sqrt{b^2 - 4ac}}{2a}$	$N = 1 + \dfrac{1}{\dfrac{f'1}{r_1} - \dfrac{1}{r_2}}$

N = refractive index

r_1 = radius of first survace $\Big\}$ + if centered to right,

r_2 = radius of second surface $}$ − if centered to left.

t = axial thickness

f' = paraxial efl

a = $f'(r_2 - r_1 + t)$

b = $f'(r_1 - r_2 - 2t) - r_1 r_2$

c = $f't$

The direction of propagation of the light is from left to right. By using formula A or B with good technique, N can be determined to ± 0.002 under very favorable conditions. For high accuracy, immersion and goniometric methods are employed, using costly apparatus such as monochromators, spectrometers, refractometers and/or precision hollow prisms. [3], [4], [5]

19. *Convex Sectors and Segments of Circular or Cylindrical Objects of Any Material:* Sector gears, cams, cylindrical and toroidal lenses, Woodruff keys, etc., can be measured for radius of curvature by mounting them on S at the level of the axis of V, lighting F and focusing the microscope on the surface of the specimen by sliding X. If, now, the specimen is oriented by sliding so that J can be swung around K without losing sharp definition of the specimen surface, the radius can be read directly on N if alignment operation IV, above, is first performed.

20. *Concave Sectors, etc.:* Change V to a telescope and mount a positive lens of greater efl than the radius to be measured on the objective of V, thus making it a long focus microscope. Focus it on the projected axis of K as in IV above. Read N. Mount the specimen level with

B.18.5. Testing Methods

V and with the surface to be measured facing V. Continue as in 19. Read N again. Difference is radius. For short radii, V as a microscope can be used by shifting N.

21. *Spherical, Unpolished Surfaces:* Use method 19 or 20, taking care to orient the specimen vertically so that the greatest bulge or depression is on the axis of V.

22. *Concave Spherical, Polished Surfaces:* Before measuring transparent material, smear the surface not to be measured with Vaseline to avoid confusing reflection. Arrange V as a microscope. Check IV above. Mount specimen in W on S. Set U and J to their respective zeros and light F. Slide S until dust, talcum or an ink dot on the center of the surface is in sharp focus. Slide X to the right until pinhole image is sharp. Read radius directly on N. If pinhole image is not centered in V, specimen should be reoriented and determination repeated.

23a. *Convex Spherical Polished Surfaces:* (Of radius shorter than working distance of microscope V.) Set U to zero on N after checking IV, P. Smear back surface of specimen with Vaseline and mount in W. Light F and slide S to sharp focus of F in V. Slide X to right until specimen surface is in sharp focus (dust, talcum or ink dot). Read radius directly on N.

23b. *Convex Spherical Polished Surfaces:* (Longer radius than working distance of microscope V.) Change V to a telescope. On W mount an achromatic lens of longer efl than radius of specimen. Mount specimen on L_2, facing V. Dust talcum on right surface, grease left surface and light F. Move specimen until powder is in sharp focus in V. Note position of specimen holder. Blow away talcum with syringe and move specimen to right until pinhole image is in focus. Distance through which specimen holder was moved equals radius of curvature.

B.18.5.7.1 Radius Measurement in General

By using objectives on V of large aperture and excellent correction (shallow depth of focus), it is possible to determine radii with an accuracy of ± 0.2 percent on the average with good technique and equipment. Such objectives are costly, however, and conventional methods serve as well as or better than bench methods in many cases. For example, long concave radii are best measured by Foucault's knife-edge method; concave and convex radii of moderate length are handled by spherometers of many types and interference methods, using master curves, are best for checking replicate surfaces, as in volume production.

Methods 22 and 23 are among the best available, however, for determining the lengths of short radii such as those used on microscope elements of small diameter and on elements of high-power eyepieces.

B.18.6 Summary

An optical bench, used with due regard for its mechanical nature, is a most useful and instructive instrument of wide application. Anyone pursuing general optics as a hobby should have one in his possession, though it be of the simplest type.

The instrument, as described in the above chapter, is only one of many types [6], [7], [8], [9], [11] and the methods outlined by no means cover the field of testing exhaustively.

For those wishing to construct a slightly more versatile optical bench than that described, photographs of the writer's equipment are reproduced. The apparatus accommodates objectives up to 8 inches in diameter, will allow the study of 120° fields by method 8a and of 30-inch diameter object and image fields by 8b and 8d. Focal lengths of from zero to 10 feet can be handled with little trouble. Also, objectives of 15-inch efl can be tested at unity magnification by method 8d.

The major differences to be noted are largely for convenience and for the saving of time. They are:

1. Separate microscope and telescope.
2. Metal angle-supported H and E elements, which slide into position and are automatically aligned by a music wire-pulley linkage.
3. Reverse illumination of target E, which is of engraved metal.
4. Double cross-slides.
5. Graduated scales at H and E.

Despite the fact that wood is employed, the unit is capable of giving good accuracy.

Although some of the methods presented are thought to be original, they may not be so. They are theoretically correct and practically satisfactory and have been intentionally made applicable to fairly inexpensive apparatus. The chromatic aberrations, for example, can be determined more accurately if a monochromator and an apochromatic or "orthokumatic" [10] objective are substituted for the reflecting collimator and the filters which are recommended. The cost, however, would not be justified for occasional use.

B.18.7 References

1. Catalogs of American Optical Co., Buffalo, N.Y. and Bausch & Lomb Optical Co., Rochester, N.Y.
2. Central Scientific Co., Chicago, Ill.
3. *Applied Optics,* Steinheil and Voit. Trans. French, Vol. II, 1919, pp. 45–68.
4. *Bur. Stds. Jour. Res.,* series of papers on refractometry by Tilton, Research Papers No. 262, 575, 776, 919, 934, 971, 1535 and 1572.
5. Faick and Fonoroff, "Precision Apparatus for the Rapid Determination of Indices of Refraction and Dispersion by Immersion," Bur. Stds. Research Paper No. 1575.
6. Kingslake, *Jour. Opt. Soc. Am.,* Vol. 22, pp. 207–222.
7. *Practical Optics,* B.K. Johnson, 1945, pp. 21–36.
8. *Dictionary of Applied Physics,* Glazebrook, 1923, Vol. IV, 5–7 and 164–167.
9. *Photographic Lenses,* Beck and Andrews, 2nd ed., pp. 273–288.
10. Hastings, *Jour. Opt. Soc. Am.,* Vol. II, p. 63.
11. *Bull 156-74,* Gaertner Scientific Corp., Chicago.

Chapter B.19

Interference of Light[†]

The performance of all optical instruments is influenced by the interference of light. Some, such as interferometers and grating spectrographs, could not exist if interference were impossible.

It is the object of this chapter to discuss interference and its relation to the construction, testing and use of astronomical telescopes, avoiding as far as possible theoretical controversy and mathematics. This attempt to explain interference phenomena, especially those caused by diffraction, without critical comparison of theories and without the aid of the calculus, is akin to trying to make whiskey without grain—something is bound to be lacking and the connoisseur will be disappointed.

The true nature of light is probably not known. Certainly no one simple theory is available which will satisfactorily account for all the things which light is known to do. However, the behavior of light has intrigued some of the most brilliant people of history who have, by countless experiments, established many facts. By conjecture and computation, these same men and women have advanced many general and special theories designed to explain the phenomena which they and others have observed. It is fortunate then that one simple theory appears adequate to rationalize all the phenomena associated with the subject of this chapter. This theory is called the wave theory of light and is due largely to Huygens,[1] Young,[2] Fresnel,[3] Fraunhofer,[4] and Sommerfeld.[5] The fact that no wave theory explains the facts associated with the emission and absorption of radiation while certain

---
[†]By Horace H. Selby, Received October 1948.
[1]Christiaan Huygens, "Traité de la Lumiere," Leiden, 1690.
[2]Thomas Young, "On the Theory of Light and Colours," *Phil. Trans. Roy. Soc. London*, Vol. XCII, pp. 12–48, 387–397, 1802. Vol. XCIV, pp. 1–16, 1804.
[3]Augustin Fresnel, "Oeuvres Completes," Vol. I, Paris, 1866.
[4]Joseph von Fraunhofer, "Collected Writings," Munich, 1888.
[5]A. J. W. Sommerfeld, *Math. Annalen*, Vol. 47, pp. 317– 1895.

corpuscular and quantum theories do is unfortunate, but the relative merits of the various theories fall outside the scope of the present discourse. By combining undulatory and corpuscular concepts, as is done in modern wave mechanics, a fairly satisfactory explanation of all optical phenomena is possible.[6]

In general, it is possible to transfer energy from point to point by three methods—a projectile traverses the intervening medium, the medium is forced to flow, or the particles of the medium execute rhythmic movements. In the third method, no particle crosses the space between the points in question. The particles of the medium move only infinitesimal, if any, distances in the direction of propagation of the energy being transferred. The particles of the medium do move to and fro across the axis of propagation at right angles to it, and in all azimuths, except when the light is polarized. This constitutes wave, or simple harmonic, motion. In Figure B.19.1a, a ray of light is represented as moving through the paper at right angles to it and its center is indicated by the dot. At b, the same ray is shown moving in the plane of the page from left to right and its center is represented by the line XX'. (Properly the curves depicting light waves should be sinusoids and a should be a small fraction of b. The curves are distorted for reasons of clarity and ease of preparation.)

For simplicity, only the vertical component of the ray is shown, but it should be clearly understood that other components are presented in ordinary light, vibrating in all azimuths, as indicated in Figure B.19.2, where one particle vibrates in plane AA', another in plane BB', etc.

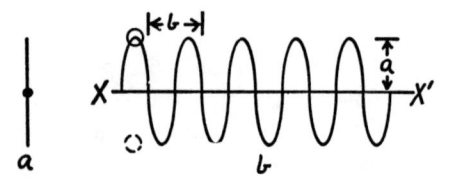

Fig. B.19.1a and b.

In Figure B.19.1b, a represents the maximum distance from the axis of propagation which a certain particle moves while vibrating. It is called the amplitude and is proportional to the square root of the intensity of the light. b represents the distance along the axis which the light moves while the disturbance executes one cycle. It is one wavelength (λ) and is frequently expressed in millimicrons (10^{-6} mm) or angstrom units (10^{-7} mm). A vibrating particle is shown as

---

[6]Louis de Broglie, "Matter and Light," trans. by W.H. Johnston, pp. 27–31 (1946).

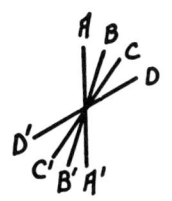

Fig. B.19.2.

a circle in Figure B.19.1b. It moves back and forth in a straight line from its indicated to the dotted position. Obviously, many particles representing many azimuths of vibration could not, as Figure B.19.2 suggests, collapse to one dimensionless point $\lambda/4$ later than the instant depicted. This absurdity is employed as an artifice to avoid discussion of vectorial resultants which would be difficult without resorting to mathematical complexity.

The number of cycles or wavelengths passing a point of reference in a unit of time is called the frequency. The frequency of light is a constant for any color and approximates $10^{14.8}$ cycles per second for visible radiations. Velocity, amplitude, and wavelength are not constants for they vary with the medium (vacuum, air, glass) through which the light is passing, all being less in denser media (Figure B.19.3).

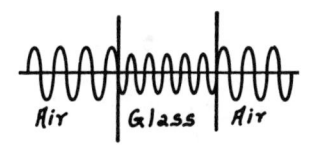

Fig. B.19.3.

Often it is more convenient to consider b of Figure B.19.1b as an angular quantity equal to 360° rather than as an interval of time or a length. Thus, two disturbances which are out of step by $\lambda/2$ are said to be "out of phase" by 180°.

As an illustration of the difference between the geometrical and physical representation of light, consider Figure B.19.4, which represents a beam splitter $ABCD$, composed of two 90° prisms with their hypotenuse faces cemented together. One of the cemented faces is aluminized so that the reflected light equals the transmitted light in intensity.

B.19.4a and b show, respectively, a ray and a fan of rays from x as treated in geometric optics by straight lines. B.19.4c and d represent the same, illustrated as wave phenomena.

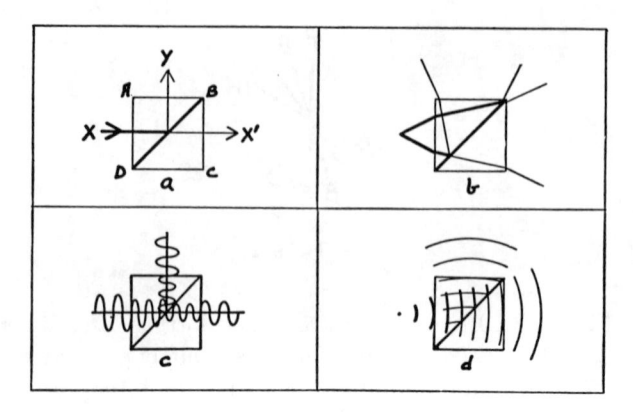

Fig. B.19.4.

Consideration of a fan of rays proceeding from a point source is facilitated by imagining an infinite number of rows of the curves of b, Figure B.19.1b, radiating from a common origin or by rotating one curve in a plane around the source. A corrugated surface is thereby generated about the source and a system of circular waves is the result. By rotating this surface about any of its rays, spherical waves can be generated. Cones of spherical waves are what we deal with in all optical instruments. In the case of the astronomical telescope, a cone of spherical waves of infinite radius of curvature reaches the instrument and is transformed into a cone of waves with a center of curvature at the focus, as in Figure B.19.5.

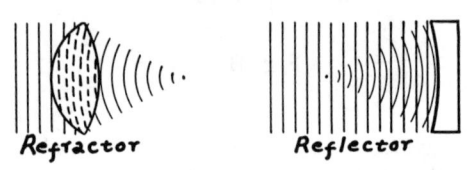

Fig. B.19.5.

In explaining optical phenomena in terms of the wave theory, Huygens evolved a brilliant hypothesis which states that every point on a wave front behaves as if it were a source itself, sending out spherical wavelets ahead of, but not to the rear of, itself. In this way, each wave forms a new one and the light is propagated, for the result of an infinite number of wavelets can be shown mathematically to be a new wave front, that from a point source being spherical, that from a line source, cylindrical. This propagation of light by wavelets from all points in each wave is called "Huygens' Principle."

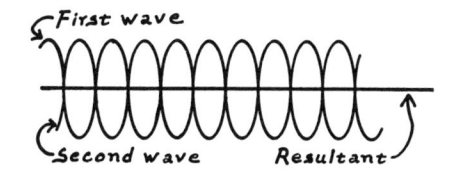

Fig. B.19.6.

It would appear that light could be made to destroy itself, if the wave theory were correct, by combining two rays which are precisely 180° out of phase. In other words, if the crests of one wave train were to coincide with the valleys of the other (Figure B.19.6), the sum of the motions would be a constant, no vibration of the medium would occur and therefore no light would be propagated. Experiment proves this to be the case.

L Monochromatic light
S Slit
M, M' First-surface mirrors
W Wedge of cigarette paper
B Backing
M₂ Magnifier
E Eye

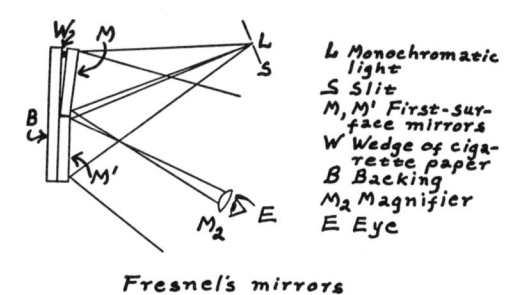

Fresnel's mirrors

Fig. B.19.7.

If two first-surface mirrors of good quality plate glass, say 2 by 4 inches, are arranged as in Figure B.19.7 and illuminated by a slit, accurately parallel to the line where the mirrors meet, dark areas can be found with a magnifier in the reflected beams, if sodium or other reasonably monochromatic light is used. Between the dark areas will be found bright areas which are much brighter than the same field similarly illuminated by one mirror of any size.

Fig. B.19.8.

The above shows that, at certain points, the wave trains have paths which differ by odd multiples of $\lambda/2$ and therefore annul each other, producing darkness, while at other points the paths differ by even multiples of $\lambda/2$ and reinforcement occurs, giving increased illumination, seen in Figure B.19.8.

This behavior of wave trains, giving annulment and/or reinforcement, is called interference.

By employing the wave theory, Huygens' Principle, and a knowledge of interference, which have just been discussed, the following phenomena relating to the optics of amateur telescope making can be explained:

1. "Newton's rings" or fringes of equal thickness, used in testing flats and lenses for surface contour.
2. Haidinger's fringes, which are fringes of equal angle used in testing plane-parallels for parallelism.
3. Low-reflection coatings on lens and prism surfaces.
4. Diffraction "spikes" radiating from star images.
5. Diffraction patterns at telescope foci.
6. "Edge diffraction," noted when using Foucault's knife-edge test.
7. Loss of definition due to obstacles such as Cassegrain secondaries, etc.

Each phenomenon will now be treated separately.

B.19.1 Newton's Fringes

When a transparent film is bounded by other media of different refractive index, the interfaces can be made to reflect wave trains which differ in path length and thereby produce interference.

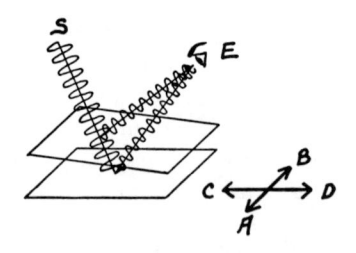

Fig. B.19.9.

In the case of a thin air film between glass surfaces, Newton's fringes are formed as shown in Figure B.19.9.

A ray of monochromatic light from any point in the source S, which may be of any size, passes through the first glass and is partly reflected at the first air-film surface to the eye at E. Since this reflection occurs

B.19.1. Newton's Fringes

between dense and rare media, there is no phase change. Later, the parent ray is again reflected when it reaches the second air film surface, sending a ray of different path length to E. This time, the reflection is at a rare-dense interface and a phase change of nearly 180° occurs.

As the film is scanned in the $C - D$ direction, film thicknesses of varying magnitude are encountered. Since the path difference between the two reflected wave trains is thereby changed, the phase difference also changes. The combined phase changes due to varying path length and to the two reflection sequences cause the two trains to interfere in such a way that alternating dark and light areas are seen in the plane of the air film. When the combined phase difference is an odd multiple of 180°, *i.e.*, $\lambda/2 \times$ 1, 3, 5, etc., annulment causes darkness. At the next 180°, or $\lambda/2$, reinforcement effects an increase in illumination.

Since there is no thickness change in the $A - B$ direction, constant path differences are found and scanning in this direction reveals no change in illumination. Therefore the interference fringes between two surfaces of identical curvature but of opposite sign will be straight lines parallel to AB, regardless of the magnitude or form of curvature (spheres, planes, conics, etc.). If the angle between the surfaces is reduced, the fringes become more widely separated because the eye must then scan a greater distance in the $C - D$ direction in order to reach the next region of half-wave path difference. Therefore, when the surfaces are parallel the fringes are infinitely far apart and the entire film will be uniformly illuminated. Coaxial spherical and aspheric surfaces of revolution, which differ in curvature, form circular fringe systems because the lines of uniform path difference are obviously circles themselves.

If the curves do not match each other, the fringes will be arcs or will have varying shapes, depending on the relative contours and curvatures of the surfaces. The fringes form accurate half-wavelength contour maps of the air-film thickness. (The reason is that the distance between any two adjacent bright fringes represents one wavelength of path difference but, since the longer path traverses the air film twice, the path difference is double the change in air-film thickness.)

In the mathematical treatment of interference in thin films, the following variables occur and greatly influence the appearance of the fringes and therefore the accuracy of any measurement made with their aid:

1. The wavelength of light.
2. The angle at which a fringe is examined.
3. The thickness of the air film.

When only moderate precision (±4 fringes for flats) is required, it is necessary to consider nothing but the illumination and the cleanliness of the

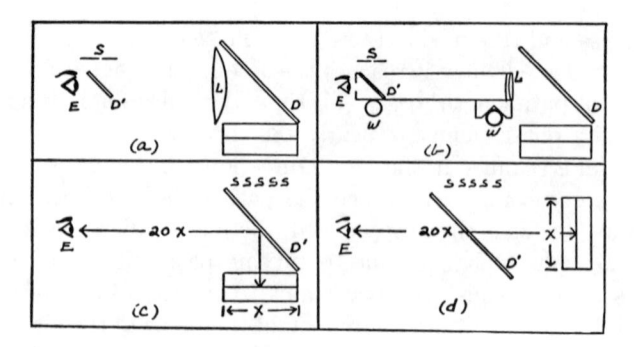

Fig. B.19.10.

surfaces. The light must contain but one fairly narrow band of wavelengths, which can be obtained by the use of filters, and the surfaces must be free from large particles which might keep them apart. Wiping with the palm of the hand is usually sufficient. The surface may be illuminated and viewed from any angle up to 20° from the normal and from any distance not closer than two diameters.

When high precision of the order of 0.1 fringe is necessary, extreme precautions must be taken. Each portion of each fringe must be illuminated and viewed from the normal to the surface because only along a normal or perpendicular will one wavelength of film thickness be indicated by two fringes. Arrangements for insuring this condition are shown in Figure B.19.10, where E is the eye, S a monochromatic light source, D' a semireflecting diagonal, L a spherically corrected lens, D a diagonal mirror and W parallel sliding ways. Only diametral fringes should be regarded, or diametral paths across fringe systems. B.19.10a and b have the advantage of yielding brighter fringes than B.19.10c or d.

If more than one color of light is present, the different fringe systems overlap, giving diffuse fringes which bear no exact relation to film thickness. Also, two or more wavelengths can resonate or "beat," giving maxima and minima which can be many wavelengths apart. Thus, surfaces viewed under mixed colors may appear to match within one fringe under certain conditions, whereas they may be dissimilar by several fringes as proved by examination under monochromatic light (Figure B.19.11).

The light need not be strictly monochromatic in the sense that but one wavelength is emitted. So long as only one dominant, fairly narrow band is used, accuracy can be attained. Other colors may be present if they are very weak in comparison with the dominant. If filters are used, they may be small and held before the eye or they may cover the source. If the fringes appear to have very high contrast over the entire surface (jet black and

B.19.1. Newton's Fringes

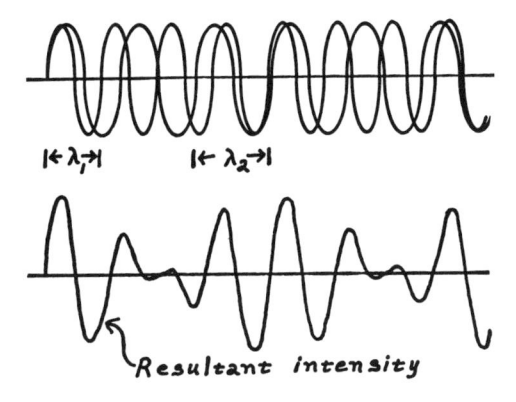

Fig. B.19.11.

brilliant, pure color) the light is probably satisfactory. One can be positive, however, only by examining the radiation with a spectroscope.

The thicker the air film, the greater will be the various errors due to obliquity, which can be serious. Also, the contrast of the fringes will be less for thick films. Therefore, separators should not be used. If the surfaces are scrupulously clean, they can be approximated closely enough for possibly all purposes. Cleaning with ethyl alcohol or pure isopropyl alcohol on a lintless cloth, followed by blowing with an infant's syringe, is recommended.

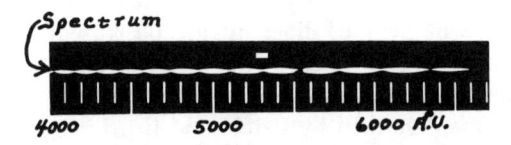

Fig. B.19.12.

If it is desired to measure the film thickness in order to apply corrections for its magnitude,[7] it can be done as follows: Arrange some form of spectroscope so that the distance between two lines of known wavelength is clearly indicated in the field of view. Illuminate the air film under consideration brilliantly with incandescent white light and examine the light reflected from it with the spectroscope. Across the spectrum will be seen a number of dark lines which are parallel to the slit. See Figure B.19.12. Since the lines will be narrow if numerous, a narrow slit should be used. The number of lines (n) present between the reference points representing

---

[7] ATM2, Chapter F.3, Footnote 13, "When observing a diametral ...

the known wavelengths can be counted and entered in the formula given below to find the thickness. α is the angle of incidence of the light.

$$t = \frac{n\lambda_1\lambda_2}{2\sin\alpha(\lambda_1 - \lambda_2)}$$

For example, assume the spectroscope reference marks to be set on $\lambda_1 = 6563\text{Å}$, the C-line of hydrogen and $\lambda_2 = 4861\text{Å}$, hydrogen F. Between these marks, if 28 dark lines are counted when the axis of the spectroscope collimator is coincident with the normal to the film at the illuminated point, the thickness of the film at this point will be in angstrom units

$$t = \frac{28 \times 6563 \times 4861}{2(6563 - 4861)} = 260,000.$$

If the result is wanted in millimeters

$$t = \frac{28 \times 0.0006563 \times 0.0004861}{2(0.0006563 - 0.0004861)} = 0.026 \text{ mm}.$$

(It will be noted that, for the same units, the thickness is a linear function of n and extremely easy to calculate.)

The reason that this method of thickness computation is valid is that any moderately thick film will be correct for the production of a maximum of illumination through interference for a number of different wavelengths and therefore will transmit only this number of wavelengths, holding back by annulment an equal number, which are therefore absent from the spectrum transmitted, appearing as dark lines in the spectroscope. This method is strictly correct only for thicknesses of several wavelengths, but is entirely adequate for the correction of fringes between flats. A more complicated but exact method is described by Peters and Boyd.[8]

B.19.2 Haidinger's Fringes

A different system of interference fringes—Haidinger's rings—is extremely useful for determining the degree of parallelism of the surfaces of plane-parallels.[9]

These fringes may be seen at infinity either by looking through the glass or preferably by reflection, as shown in Figure B.19.13. Some ray from every point x in the illuminant will strike the glass perpendicularly. Each such ray will be partially reflected back upon itself to M and will then be

---

[8] *Bur. Stds. Sci. Papers,* Vol. 17, pp. 693–704.
[9] ATM2, Section F.3.8.

B.19.2. Haidinger's Fringes

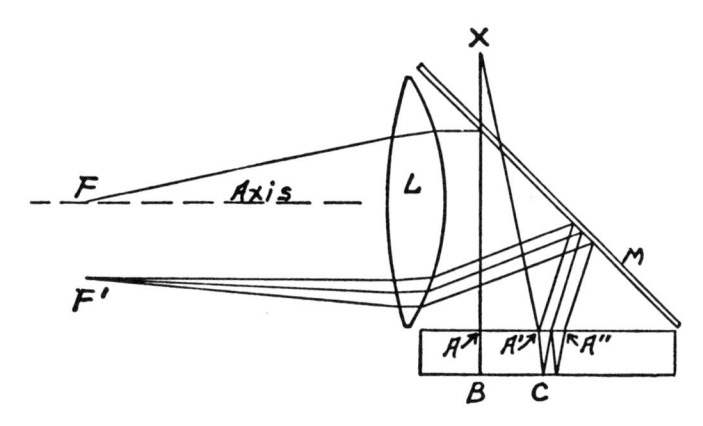

Fig. B.19.13: X = a point in the monochromatic source. L = Lens, focused on infinity. M = Semireflective mirror.

partially reflected through the lens L to its axial focus F. Some of the light will proceed simultaneously downward from A to B and be reflected upward from B to M through L to F. Due to the difference in path length of these two rays, they will reach F under conditions of interference, the out-of-phase relationship between them governing the amount of annulment or reinforcement at F. All such rays from all points, being perpendicular to the glass, are parallel and therefore contribute to the illumination at F.

Consider a ray from the first or any other point which strikes the glass at an angle to the normal. It, too, will undergo multiple reflection within the plate, sending many portions of itself (only three are shown) over different path lengths to F', where interference occurs again. By rotating the angular ray $XA'C$ around XAB, keeping angle AXA' constant, it can be shown that the point F' will form a circle about F, with the radius FF'. At greater angles, other, larger circles are formed around F. Therefore, a system of interference rings, centered in the field of view, is seen when looking through the plate with a telescope which is focused on infinity. The reason that the fringes are seen only at infinity is that the rays causing the interference (such as $A'M$ and $A''M$) leave the plate mutually parallel and therefore have the effect of coming from an infinitely distant source.

By transmitted light, the fringe system has very low contrast because wave trains of very small amplitude are interfering with directly transmitted ones of high amplitude. The minima are therefore nearly as bright as the maxima. By coating one or both surfaces with a semireflecting film of silver or aluminum, the contrast can be improved by making the interfering rays more nearly of equal brightness. With no treatment, the two rays have an approximate brightness ratio of 100:1, even with dense flint. By making one surface semireflective a 20:1 ratio is possible, while treatment of both surfaces yields 4:1. It is therefore wiser to employ the method using reflected light, for a ratio of 1.1:1 is obtained without special treatment. In fact, a semireflecting coating on one or both surfaces will give the following ratios in reflected light: Upper only, 80:1. Lower only, 8:1. Both, 11:1.

If, for some reason, transmitted illumination must be used, the phase changes of reflection at the high-index surfaces of the silver or aluminum coating need cause no concern. The only effect will be to change the fringe spacing, not the number appearing per unit of distance.

The formula expressing the path difference in films is $\Delta = 2Nt\cos(a-\theta)$ where N = ref. ind. of film, t = film thickness, α = angle between internal ray and normal to first surface and θ = angle between the surfaces of the film.

In good plane-parallels, N and t are practically constant and θ is insignificant in comparison with α. Therefore, Δ depends largely on α. Since α can only be constant in a circle around a normal, the fringes must be circular.

When t is not absolutely constant, the path differences will change as the plate is moved across the field of view, causing the central fringe to collapse and disappear as less thickness is encountered or to expand and make room for new ones, as greater thickness is found.

The difference in thickness between any two points is $n\lambda/2N$ where n is the number of fringes lost or gained, λ the wavelength of the monochromatic light used and N the refractive index of the glass for λ.

For example, assume that a disk of BSC-2 has been examined, using sodium light and that the greatest number of rings seem to disappear while moving the disk across many diameters was 11. Then the disk is lacking in parallelism by $(11 \times 0.000589)/(2 \times 1.517) = 0.0021$ mm.

B.19.3 Low-reflection Coatings

The third phenomenon of telescope interest caused by interference is that of reduction of reflection at glass-air surfaces by coatings.

The entrant and emergent faces of right angle prisms, the faces of the correcting lenses of Schmidt-type cameras and the lens surfaces of oculars and objectives all reflect some of the light which passes through them. This reflected light does not properly illuminate the image. The light that is reflected away from the image may be lost or it may, by other reflections, reach the image field with whatever light has been reflected toward the image and cause lowered contrast, flare, ghosts and other spurious images, especially when Venus or other bright objects are close to or within the field of view.

B.19.3. Low-reflection Coatings

By changing the refractive index of the glass surface, the amount of reflection can be altered, for the amount of reflected light is a function only of the refractive index of the glass and of the angle of incidence. If it were possible to coat a lens with a layer of a transparent material the refractive index of which changed continuously from that of air to that of glass where it touched the lens, no reflection could occur. Since no such substance is known, a compromise is adopted and a substance with an index intermediate between air and glass is employed. The best index would be $\sqrt{N}$, where N is the index of the glass.

One popular method is to evaporate magnesium or other fluoride in a high vacuum and permit it to condense on the glass, forming a film. Another method removes the components of the glass which raise its refractive index—such as lead, barium, sodium and potassium—by leaching with water, nitric acid, etc. Still another etches the surface with fluorine compounds in such a way that the pits formed are small in comparison with the wavelength of light. This has the effect of "diluting" the surface with air without impairing its ability to properly refract light. Organic compounds have been developed which can be applied to glass surfaces at atmospheric pressure and which can be made to decrease as well as increase reflectivity with negligible absorption.[10]

Without a knowledge of interference we should be able to reduce the reflecting power of glass but little by the above methods. By taking full advantage of this property of light, however, we can practically eliminate reflection of one wavelength at normal incidence and greatly reduce it over the remainder of the visible spectrum and over useful angles.

Consider the effects of two properly formed films, as shown in Figure B.19.14.

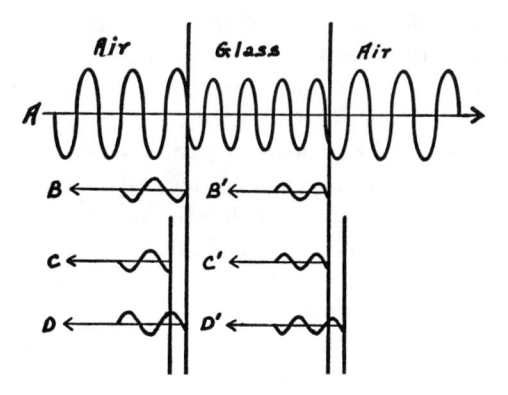

Fig. B.19.14.

---

[10] American Optical Co., Southbridge, Mass.

Imagine a wave train traversing a glass plate, as at A. Some will be reflected, as at B and B' and be lost or contribute to image faults. If, now, the glass is properly coated with a quarter-wave film, the first reflections at each face will occur as shown at C and C'. After $\lambda/4$, in terms of time, the second reflected waves reflected D and D' will start back, being propagated out of phase with C and C', respectively and annulling them. (Although all trains are coaxial, they have been separated for clarity.) After the first half-cycle, A will be reinforced by the amount of energy possessed by C, C', D and D' in their first cycle and the particles transmitting A will vibrate with increased amplitude. The fact that the reflections of C and D are at denser surfaces means that each has changed phase 180° and they are "in step" as soon as they reach the air, if their path lengths are equal. However, D has traversed $\lambda/4$ twice and is therefore $\lambda/2$ behind C or 180° out of phase, making annulment possible. C' and D', being reflected at rarer media, suffer no phase change. Otherwise, they do not differ from C and D. It has been argued that, when a ray passes from a dense to a rare medium, no coating is necessary because no phase change occurs and that the reflected ray would be annulled by a portion of the parent ray. That this is incorrect is shown by the fact that wave trains must be moving toward a common point, as are C and C', above, before interference can be utilized, quite apart from any phase changes.

It is obvious that a coating can be $\lambda/4$ thick for only one wavelength and that any odd multiple of $\lambda/4$ will do well for this color. However, the thinner a coating, the better it will perform with widely-separated colors and it is therefore wise to have a coating no thicker than one one quarter wavelength. This can be assured in the high-vacuum process by turning off the filament as soon as the first reflection minimum is reached for the chosen color. Another possible advantage of the $\lambda/4$ thickness is that it usually gives greater durability than does a thicker film. For visual instruments, it is best to coat all surfaces for the maximum transmission of 5500 to 5600Å unless the instrument is to be used for critical colorimetry. (In this case, color distortion can be avoided by coating each surface for maximum transmission of a different color, as recommended by Jacobs.[11]) When such a coating is examined in white light, it will appear to be faintly reddish-violet (purple). A brilliant, deep purple indicates a thickness of $3/4\lambda$, $5/4\lambda$, etc. A color differing from purple indicates a minimum reflectance differing from the visually brightest yellow-green.

Whether or not surfaces should be coated must be left to the owner's judgment. Properly done, coating improves instruments used under threshold illumination, especially if many glass-air surfaces exist. Some cases of

---

[11] *Fundamentals of Optical Engineering*, p. 121, 1943.

ghost and flare-spot can be completely cured. However, injudicious baking of the film may strain or break a large prism or lens (the durability of some fluoride and organic films is enhanced by heating to 200–450°F) and there are the usual hazards which accompany handling, transportation, etc. Certainly, an instrument which is perfectly satisfactory as it is should not be changed.

B.19.4 Diffraction

A particularly interesting type of interference occurs when a beam of light rays is limited in any way, as it is in telescopes by the edges of mirrors, lenses, supports, etc. This type of interference is called diffraction and is the cause of the spikes often seen extending radially from star images formed by reflectors. It also makes it impossible to form a true point image of an object; it causes the edge of a mirror under test by Foucault's method to appear brightly illuminated and it causes reflecting telescopes to give images of lower contrast and slightly poorer definition than those of high-quality refractors of equal aperture.

B.19.5 Image Spikes

Consider the train of waves shown in Figure B.19.15. The light is being propagated in the conventional left-to-right direction from a distant small source to the opaque knife-edge K, where a portion of the wave front is obstructed.

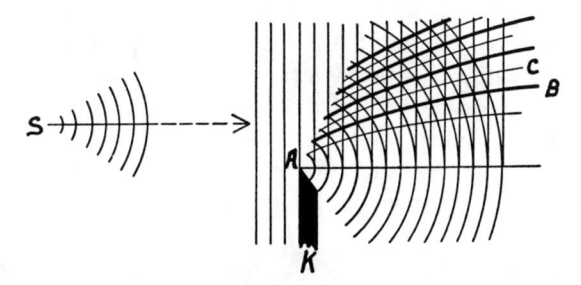

Fig. B.19.15.

According to Huygens' Principle, every point on every wave is sending out wavelets into the hemisphere ahead of it, the sum of all these wavelets at any instant constituting a new wave, the repetition of this wave formation resulting in a smooth flow of waves. When K is reached, this smooth flow is interrupted and the wave existing along the edge of K now radiates wavelets which cannot combine their energy with other wavelets on both sides to form the customary new wave front, since the medium on one side is devoid of wavelets, being shielded.

The wavelets along K therefore begin the propagation of an entirely new wave and the edge of K consequently behaves exactly as if it were an independent source of light of precisely the same character (frequency, constancy, etc.) as S, save that the wave front is cylindrical instead of spherical. As we have found earlier, waves from two sources of the same character proceeding toward a common point can and do cause interference where they meet, depending on their relative path differences. So it is with the waves from the edge of K and from S. Along the parabola AB, for example, the waves have path differences of even multiples of $\lambda/2$ and reinforcement causes a maximum of illumination while, along AC, the difference is an odd multiple of $\lambda/2$ and a minimum is encountered. Into the shadow of K, the obstructed wave front also sends light, but since there are no other waves in this area in proper condition for interference, the illumination merely decreases smoothly and rapidly, without exhibiting maxima and minima. With a very narrow monochromatic source, the fringes show good contrast, as in Figure B.19.16a, which was made with a slit source.

When a straightedge diffracts white light, a series of colored fringes are formed. They are, in fact, spectra. The lines forming these spectra will be parallel with the straightedge and the spectra themselves will extend at right angles to the straightedge into the unobstructed beam. If the light, after passing the obstruction, is brought to a focus, the spectra will be focused also and will be composed of lines which are short for small sources, such as pinholes, and long for large or extended sources.

Consider now, the support EA in Figure B.19.16b, which represents the diagonal-supporting mechanism of a typical Newtonian telescope. EA acts as a double straightedge. The diffraction pattern caused by one edge will form a series of spectra in the field of view extending horizontally to one side of the image, making one spike of the artificial star image of c. The other edge will form the spike on the opposite side.

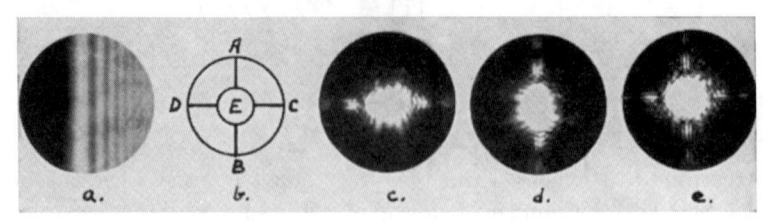

Fig. B.19.16.

B.19.5. Image Spikes

Simultaneously, the two edges of support ED will form d. Supports EB and EC will merely form patterns identical with those from EA and ED, which will merge precisely at the focus, intensifying the spike pattern e. Therefore, it is obvious why a vertical support forms a horizontal spike and why every arm of a support causes two opposing spikes.

This also shows why bending the supports[12] and the use of curved masks[13] causes elimination of the spike phenomenon—they multiply the spectra, forming them in many directions over such an area that they become too faint and too broad to see.

The image of a point formed by a telescope, as will be shown later, is a system of rings surrounding a central disk. These rings extend into the field, but rapidly diminish in brightness away from the center.

The energy contained in the spikes mentioned above and illustrated in Figure B.19.16.c partially combines with the feeble energy present in the rings of the image pattern to give reinforcement at an angle of 45° to the spikes, Figure B.19.17a.

As in the case of the prominent 90° primary spikes previously described, the secondary 45° spikes from the vertical supports of Figure B.19.16b are superimposed on those of the horizontal supports, giving the pattern shown in Figure B.19.17b. The secondary spikes are rarely found on astronomical negatives because of their faintness and because the mechanical constants of the telescope must be rather critically arranged in order to produce them most efficiently.

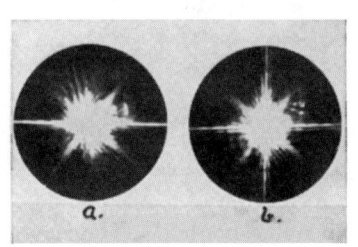

Fig. B.19.17.

By deliberately altering the dimensions of diffracting obstacles and the intervals between them, diffraction can be made to do certain things which demonstrate and prove the "bending" of light. Figure B.19.18a shows a slit image formed by a lens. By placing a second slit immediately in front of the lens the image is changed by diffraction into a band, as at B.19.18b. If a double slit of proper dimensions for the wavelength and distances con-

---
[12] $Sci.$ $Am.$, June, 1945, pp. 381, 382.
[13] ATM2, Chapter B.12.

cerned is substituted for the single lens slit, secondary interference, as at B.19.18c, is produced. If an opaque disk, which Editor Ingalls aptly calls a curled-up straightedge, is used to cast a shadow in a parallel beam, the bending of light can be conclusively proved, for, at the proper distance from the disk, the shadow will have a bright center, as in B.19.18d. Also, by photographing a disk, while backlighted, with a lens of smaller aperture than the disk diameter, so that no direct light enters the lens, the edge diffraction phenomenon seen when testing by Foucault's method can be demonstrated. See B.19.18e. Another refutation of the ancient idea that light travels only in straight lines can be obtained by photographing the light distribution within an illuminated slit and at various distances back of it, as in B.19.18f, g, and h.

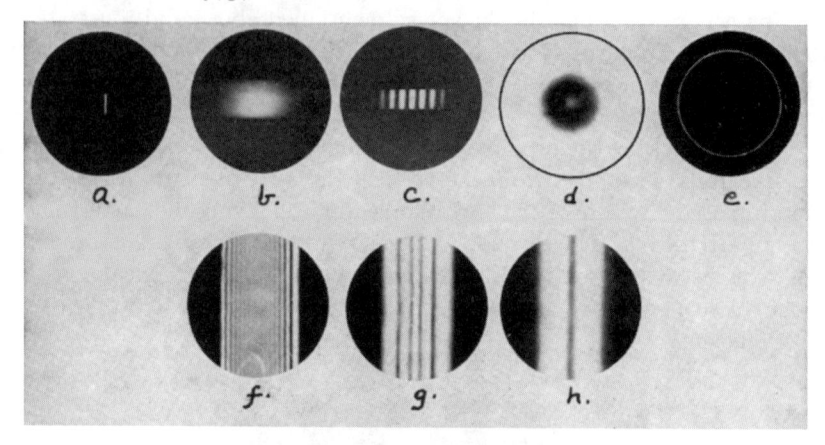

Fig. B.19.18.

The reason for the formation of a bright line at the edge of a mirror, when it is tested by Foucault's method at its center of curvature, is similar to that outlined above for image spike diffraction. (Mathematical treatment is different, however, for the former case is considered to belong to the class of Fraunhofer diffraction phenomena since this diffracting obstacle is in parallel light, while it is in nonparallel light in this instance and therefore is classed under Fresnel diffraction.) It is allied, too, with the Airy disk and rings which are formed at the focus of a telescope, instead of a geometric point. In keeping, then, with the time-tested policy of the editor of this compilation, let us treat the latter case first.

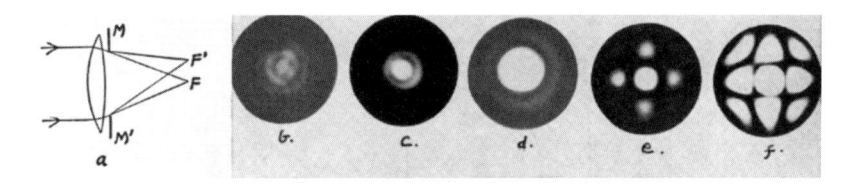

Fig. B.19.19.

B.19.6 Focal Diffraction

When a telescope forms an image of a point object, this image is never a geometric point. Instead, it is a small disk, containing some 80 percent of the illumination and surrounded by alternate circles of interference maxima and minima which contain the remaining 20 percent, approximately, of the light. The diameter of the central disk is an inverse function of the aperture ratio and approximates $2\lambda f/D$ for an aberrationless instrument. Imagine the aperture of a telescope, refractor or reflector, filled with parallel rays, which are brought to a focus, Figure B.19.19a.

Since the path lengths of all the refracted rays are equal in a perfect telescope, the wave trains which they represent all reach F in phase and F is always bright. From the distant axial point source, no refracted light reaches the area outside F, and it is black. The diffracted light from the stop MM' does reach the area surrounding F, however, producing minima at places where the path length difference between MF' and $M'F'$ is about $\lambda/2$ or odd multiples thereof and maxima at others where the difference is zero or an even multiple of $\lambda/2$. F is therefore surrounded by concentric circles of comparatively low luminosity, becoming fainter as the angle $MM'F'$ increases. At F, the light from all parts of MM' arrives along the same path length, which may differ from the path length of the refracted rays reaching F. The resulting interference effect is unimportant from the standpoint of the present paper. Figure B.19.19 shows the effects of focal diffraction. B.19.19b, c, and d were made at $f/1.5$, $f/3$ and $f/6$, respectively. B.19.19e and f show how the supports of Figure B.19.16b alter the diffraction pattern.

B.19.6.1 Edge Diffraction

When the objective of Figure B.19.19a is tested by autocollimation or when a concave mirror is tested at its center of curvature, using Foucault's method, and the surface is darkened by cutting off the image of the light source, a line of diffracted light can be seen on certain portions of the periphery and bordering some objects near the surface which obstruct parts

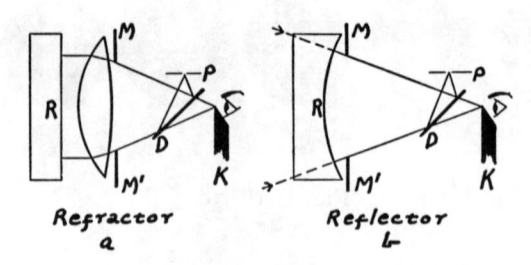

Fig. B.19.20.

Refractor a

Reflector b

of the beam, such as dust, scratches, supports and edges of many kinds.

The mechanism causing this edge diffraction can be explained by reference to Figure B.19.20 where R is a reflecting surface, MM' represents a circular mask, P indicates a pinhole and D is a semireflecting diagonal. K represents an obstruction, such as a knife-edge.

Light proceeding from P is limited by MM', which diffracts light according to Huygens' Principle. This diffracted light is not visible from the vicinity of K, however, until it has reached R and has been reflected back toward K; for, as was mentioned earlier, points on a wave front send out wavelets ahead of but not to the rear of themselves.

When the distances from P and K to R are equal, as well as when P is farther from R than is K, the above statement holds. When K is farther from R than is P, however, the situation changes and MM' does not limit the cone of rays until after reflection has occurred at R. In this case, the diffracted light from MM' is propagated in the direction of K and the luminosity is viewed directly.

When P is not on the axis, one side of MM' gives diffracted light which is seen directly while the illumination from the other is viewed by reflection. The reflected light will appear slightly dimmer and narrower than that directly viewed when the distance from MM' to R is great, due to physiological factors and greater path length.

When MM' coincides with R, *i.e.*, when no mask is used, the edge of the mirror itself functions precisely as a mask for the virtual rays (dashed lines, Figure B.19.20b). The edge of the mirror does in fact limit the cone of actual rays going from P to K and the *limitation* of waves, whether performed by a physical obstruction or by a discontinuity of surface is the fundamental cause of diffraction.

If a pinhole is used in the arrangement shown in Figure B.19.20b, and if an opaque, circular screen (such as an ink spot on a microscope cover glass) is employed instead of a knife-edge at the focus to mask the image of the pinhole, a line of light of uniform intensity will be

B.19.6. Focal Diffraction

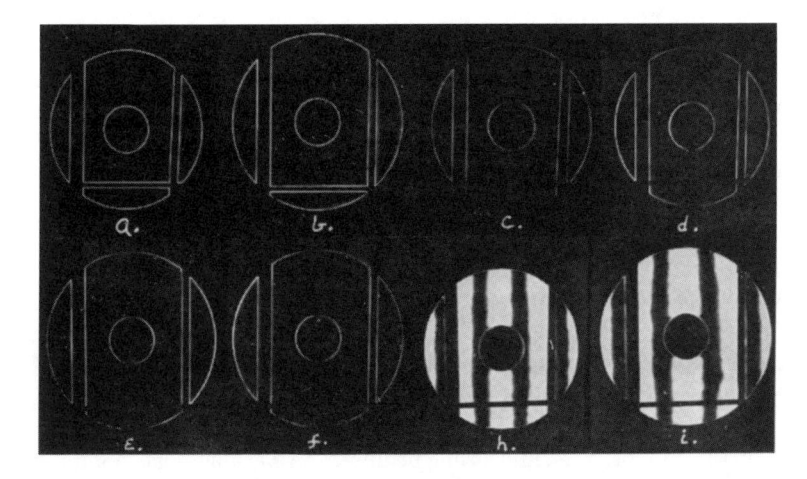

Fig. B.19.21.

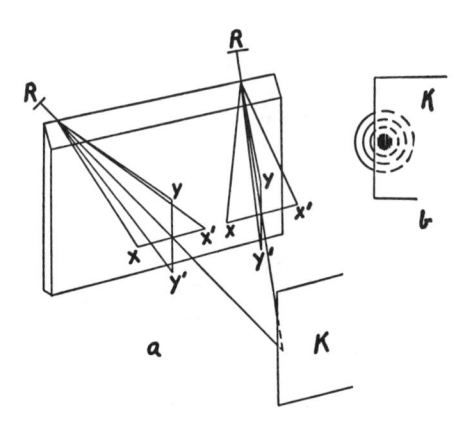

Fig. B.19.22.

found encircling the aperture. The light will be uniform (Figure B.19.21a) because path differences and amplitudes are equal in all azimuths around the axis and because all the RYY' planes of Figure B.19.22a are obstructed equally. Note also that both horizontal and vertical obstructions show uniform edge illumination.

When the symmetry of the path system is destroyed, as it is when the spot is replaced by a knife-edge or a wire or when the pinhole is replaced by a slit, the appearance of the mirror edge and of some obstructions is changed. Use of a vertical knife-edge or wire with a pinhole or a vertical slit causes the top and the bottom of the mirror and any horizontal obstacle to lose their illumination. (Figure B.19.21c).

The reason for this phenomenon is that light is diffracted by any given point on an edge in a two-dimensional, not a three-dimensional, angle, for the wave front originating at a line is cylindrical except at the ends. The plane of this angle contains the ray which just grazes but passes the point. The angle plane is also perpendicular to the diffracting edge. (See Figure B.19.22a, where the diffracted light from R is thrown forward only in the RYY' plane, not in other planes, such as RXX'.)

When the mirror is darkened by covering the center of the focal diffraction pattern, as in Figure B.19.22b, all the YY' planes from all horizontal obstructions are also covered and their diffracted light cannot be seen.

Further, more than half of the diffraction system at the image is occulted and, though the eye may still receive amounts of light directly from both sides of the mirror, the interfering beams from various edge points enter the relatively large aperture of the eye dissimilarly, causing one side, that opposite to the knife-edge, to be somewhat the brighter (Figure B.19.21b). If a large pinhole is used and the mirror edge is perfect, the diffracted light will be relatively fainter because each point in the pinhole produces its own diffraction system and appreciable angular separation of the systems tends to blur the visual picture. If the mirror edge is badly turned, however, a large pinhole causes the extreme edge to diffract more light to the eye.

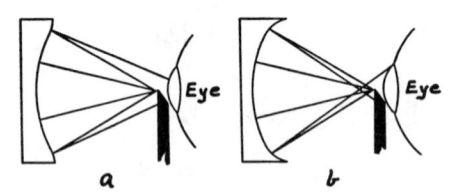

Fig. B.19.23.

The use of both a knife-edge and a slit will cause a further blurring in a direction parallel to the slit, for a slit represents a series of pinholes, each of which contributes its own little family of diffraction gremlins. The result—Figure B.19.21d—is a complete lack of light at the top and the bottom of the mirror with bright, though slightly unequal illumination at the sides. (Figure B.19.21 was made using a mirror having a perfect outer edge and a very slightly turned-down edge around the central perforation. See Ronchigrams B.19.21h and B.19.21i, photographed with and without a mask, respectively.)

If turned-down edge is present, the illumination from the edge of the

B.19.6. Focal Diffraction

mirror opposite to the knife-edge will be increased and that from the other edge decreased due to reflection, as in Figure B.19.23a. Turned-up edge will give the opposite effect, as in Figure B.19.23b.

Table B.19.1

Equipment	Perfect edge (A)	Turned-down (B)	Turned-up (C)
Pinhole and spot both on axis	(1) All azimuths uniform, very narrow and bright. (Fig. B.19.21a, mirror edge)	All azimuths uniform, broader and dimmer than 1A. (Fig. B.19.21a, edge of hole.)	Same as 1B.
Pinhole and spot separated off axis	(2) All azimuths bright and narrow. Side illumination very slightly unequal. (Fig. B.19.21b.)	Broader than 2A. Otherwise same. (Fig. B.19.21b) edge of hole)	Same as 2B.
Pinhole and knife on axis	(3) Top and bottom dark. Sides very slightly unequal, narrow. (Fig. B.19.21c, edge of mirror)	Knife-edge dimmer, more narrow. Top and bottom dark. (Fig. B.19.21c, edge of hole)	Knife side brighter, broader. Top and bottom dark.
Pinhole and knife off axis	(4) Ditto. Side inequality slightly less or greater—varies with right or left cut-off.	Same as 3B.	Same as 3C.
Slit and knife on axis	(5) Same as 3A with very slightly greater inequality. (Fig. B.19.21d, mirror edge.)	Same as 3B. (Fig. B.19.21d, edge of hole.)	Same as 3C.
Slit and knife off axis	(6) Same as 4A. (Fig. B.19.21e and f.	Same as 3B.	Same as 3C.

From the above it is clear that, when the focus of a mirror or lens is in the plane of the knife-edge or of the opaque spot and the aperture is darkened by proper cut-off, the extreme edge of the aperture will have the appearance indicated in Table B.19.1 when viewed with the eye precisely on the axis.

Caution concerning inequality of side illumination is advisable, for such illumination is a function of the position of the eye as well as of the factors just discussed. By attempting to peer around a knife-edge, instead of holding his eye precisely on the axis, a TN can unconsciously distort the appearance of the edge of his mirror. Practice and mental discipline probably make the warning unnecessary in the case of the experienced mirror maker.

B.19.7 Refractor vs. Reflector

For many years it has been asserted by some that the refractor is superior to the reflector as regards sharpness of definition and contrast. Pickering[14] in particular has stated the case against the reflector quite well from the practical standpoint.

A qualitative consideration of diffraction effects indicates that reflectors do suffer from more scattered light than do refractors, for, with the exception of Herschelian and other off-axis telescopes, all reflectors have supports, diagonals, plateholders or similar diffracting obstructions in the light beam.

The sum of the lengths of all diffracting edges in the light path of a telescope is an approximate measure of the amount of light which can improperly illuminate the image field to lower contrast and alter the size of the diffraction disk, while the area of the free aperture is a measure of the quantity of light which can properly form the image. In Table B.19.2 three telescopes are compared. In earlier parts of this chapter, it has been shown that diffraction greatly limits the performance even of refractors. It is probable, therefore, that the additional effects given in Table B.19.2 are significant, indicating over twice the diffraction effect in the case of the reflector.

Table B.19.2

Telescope	Area of free aperture	Circumference of free aperture	Support length (2 edges) each)	Circumference of center obstacle	Ratio of free area to diffracting edges
10 inch refractor	78.6 in.^{2}	31.4 in.	0.0 in.	0.0 in.	2.50
10 inch Newtonian, +2 inch diagonal, +4 supports.	75.5	31.4 in.	32.0	6.3	1.08
10 inch Cassegrain, +3 inch secondary +4 supports	71.5	31.4 in.	28.0	9.4	1.04

Table B.19.2 is valid only if the aperture ratio of the refractor is such that chromatic residuals do not limit its performance.

B.19.8 References

Mathematical treatment of some aspects of interference can be found in the references given below:

Cornu, *Jour. de Phys.*, Vol. 3, p. 1, 1874.

---

[14] ATM2, Chapter A.1.

Stokes, *Camb. Trans.*, Vol. IX, 1850.

Glazebrook, *Dictionary of Applied Physics*, Vol. IV, pp. 172, 181–188, 1923.

Page, *Introduction to Theoretical Physics*, pp. 515–531, 1929.

Rayleigh, *Encyclopedia Britannica*, Vol. VIII, pp. 238–255, 1910.

Nijboer, *Physica*, Vol. X, No. 8, pp. 679–692, 1943.

B.19.9 Notes Added in 1996

1. Some questions may arise about Figs. B.19.19a, b and c. The images keep increasing in size as focal ratios increase. This size increase does not imply superior resolution at fast aperture ratios, it merely indicates an increase in magnification at higher f-number. Angular resolution is the same.

2. Confusion can result from misuse of section B.19.7, an effect that Selby no doubt did not intend. Although readers may obtain an off-the-cuff estimate at the amount of diffraction by merely taking the ratio of clear area to diffracting edges, as is done here, diffraction is a great deal more complicated than that approximation would indicate. The shape of the diffracting obstacle has as much to do with the character of the scattered light as its perimeter. The comparison is not as clear-cut as Table B.19.2 taken alone seems to make it.

3. Selby gives a simplified discussion of the way diffraction spikes form at the edges of spider vanes. This treatment could be construed by those unfamiliar with diffraction effects to mean that the thickness of a spider vane is unimportant. In fact, thickness is very important. Thin vanes diffract less light and spread it out farther.

4. Since this article's original publication, a seminal and fundamentally important work on edge diffraction has appeared and should be listed among the bibliography. It is:

    Joseph B. Keller, "Geometrical Theory of Diffraction," *Jour. Optical Society Amer.*, Vol. 52, No. 2, pp. 116–130, Feb. 1962.

    Also, an excellent discussion of the effects of obstructions on the image appears in:

James E. Harvey and Christ Ftaclas, "Diffraction effects of telescope secondary spiders on various image-quality criteria," *Applied Optics*, Vol. 34, No. 28, Oct. 1995.

H.R.S.

Part C
Workshop Wisdom

Chapter C.1

Advice From TN's[†]

The following pages consist of notes that have been directly inspired by requests for help made by other amateurs; also by practical difficulties encountered by the Editor in his own work. Also included are abstracts and extracts from scattered and often obscure books and articles bearing on telescope making, and it is hoped that some of these will prove useful to the advanced amateur who wishes to explore beyond the beaten path.

It would be remarkable, indeed, if the beginner did not occasionally "strike a snag." Sometimes troubles arise over little points which seem obvious to the expert. This part of the book, then, is a miscellaneous collection of odds and ends, some of which it is hoped will prove useful.

C.1.1 General

C.1.1.1 Choosing a Mirror Size

If you want to make a ten or a twelve-inch mirror, at least make a six-inch first. Old hands have plenty of trouble with these larger sizes and they are therefore not the place to get one's first experience. Above a twelve-inch, special glass is necessary, that being a sort of critical or limiting size with regard to warping, distortion, and so on. A larger size than sixteen inches will require more elbow grease for the polishing than one man usually possesses.

C.1.1.2 Walking Around the Barrel

There is no need of making a slavish duty of "walking around the barrel" while grinding and polishing. One may safely stop in one place and take

---

[†]By Albert G. Ingalls, Associate Editor, *Scientific American*.

10 or 20 strokes. Because one does not then have to think of constantly shifting, more pressure can usually be brought to bear on the disk and excavation will be accomplished in proportion. There is no danger of getting the curve lopsided by this method. But when it comes to polishing, this method will usually give a surface which under knife-edge test appears lumpy, like that of a "dog biscuit." Therefore, just before getting ready to test, taper off by taking successive groups of, say, 8, 4, 2 and 1 strokes in a place. This will even up the surface. There is no objection to sitting down to work, provided the tool be turned a little now and then.

C.1.1.3 The Correct Level for Grinding and Polishing

This is exactly where you find it most comfortable. Most beginners work too low and their spinal column acquires a case of the washtub bends. J.W. Fecker states that in his shops, work on the lap is done at such a height that the forearms are horizontal.

C.1.1.4 Flexure

In one case a worker rested his tool on the heads of three nails, for a three-point, automatically-adjusted support, and the result was "a mirror that gave a triple image of the Sun at the focus, with all kinds of ligaments connecting the images." Similarly, a tool, especially if thin, may flex during cold pressing under heavy weight, as S.H. Sheib points out. "When the weight is removed, the glass tool returns to its original shape, and any work on it then cannot help but produce zones. It is well to watch the wooden board on which the tool is placed, since this may become wet and warp, bending down when cold pressing is done, but springing up when the heavy pressure is released. I use a flat steel plate. This permits me to use thin tools—half to three-quarter inch in thickness." To forestall temporary flexure after heavy cold pressing on pitch, unload mirror gradually.

C.1.1.5 Warping of Wood on which Tool is Mounted

"Saturating the wood with hot paraffin, in order to obviate trouble from warping when water enters its grain, is not entirely successful, due to difficulty of covering and sealing every pore in the wood. Binding and sticking of the mirror to the lap during polishing may often be traced to warping of the tool block. Shellac and paraffin help exclude moisture, but it is very difficult to seal every crack and joint, and moisture seeps in, destroying the figure of the mirror by flexure. Cement an iron or aluminum plate to the wooden block, and then cement the tool to the metal plate, using soft pitch of the consistency of chewing gum at mouth temperature.

C.1.1. General

The soft pitch will hold the plate and tool securely, but will yield under stress of warping, permitting the tool to retain its shape. The metal plate resists flexure and prevents distortion of the tool and mirror. The writer uses a $1/2''$ cast iron plate of suitable diameter. An ordinary stove-lid may be adapted in emergency. Metal plates may be readily cleaned of adhering pitch by playing the flame of a torch or Bunsen burner over the surface and then wiping off the melted pitch with an old rag. Even when using the iron plate between the block and the tool, it is well to paraffin the block nevertheless, to prevent the warping as much as possible. Also, it is best to use a very dense, hard wood such as birch (or teak, if the latter can be procured). Pine wood is utterly unsuitable; nor is oak, on account of its porosity, very much better. Time spent in properly preparing the support for the tool so that it will not be flexured will eliminate much trouble and exasperation."—*S.H. Sheib, Richmond, Virginia.*

Commenting on this, A.W. Everest writes: "There is a whole lot in this, if you are using thin tools like those supplied in the kits. I am a stickler for tools of the same thickness as the mirror. This overcomes most of the trouble with amateur sizes. I support the tool directly on the bench, with three generous gobs of pitch, or a ring just inside the edge, and don't seem to have any trouble from tool warpage."

C.1.1.6 Uniform Working Temperature

When the advice to work in the cellar is given it is not to be inferred that one who has no cellar cannot make a telescope, provided he can find a warm workshop that stays at fairly uniform temperature—say, within 5 degrees or so. As Ellison says, it is changing, not changed, temperature that plays havoc with the job. If forced to use a warm place one must, of course, experiment with harder laps of boiled-off pitch or some of the materials suitable for the tropics.

C.1.1.7 Keep a Log Book

Write down in it everything you do. It will prove invaluable on subsequent jobs, especially if they do not happen to follow closely on the heels of the last one. It is well to jot down various "don'ts," just when the situation arises that provides the don't. Then, some months later when you tackle the next mirror you will read this record, discover your mistakes, profit by them and get off to a flying start. There is still another reason for recording everything: amateurs of the future may come to treasure your records—should you evolve into a Herschel, a Ritchey.

C.1.1.8 Sagitta equals r²/2R

To find the sagitta of a spherical mirror, given r = radius of mirror and R equals radius of curvature: In the Figure C.1.1 (left), MAB is a right angle, being inscribed in a semicircle. MN over r equals r over $2R - MN$ (since the normal to the hypotenuse from the right angle is the mean proportional between segments). But when the sagitta is relatively small, as in a speculum, MN equals $r^2/2R$.

Another demonstration given by Pierce is as follows: To find the distance between centers of curvature of middle and edge zones of a paraboloid: Draw normals, as in Figure C.1.1 (right), from A and M. Then C is the center for zone A, and NC is the subnormal for the edge zone. O is the center for the central zone and MO is the subnormal for the same zone. MO equals NC (since all subnormals of a paraboloid are equal); therefore, OC equals MN. In a speculum, MN approximates the sagitta, and the centers shift an amount equal to the sagitta itself, or $r^2/2R$. If the knife-edge and lamp move together, the shift equals OC, these points being the centers of curvature of the edge and middle zones, or $r^2/2R$. But if the knife-edge alone is moved, the shift will be twice that amount, or r^2/R. (Q.E.D.)

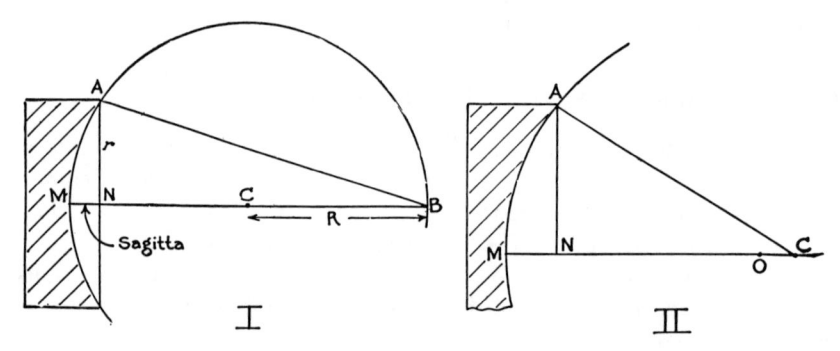

Fig. C.1.1: *Drawings by John M. Pierce*

C.1.1.9 When r²/2R is not exact enough

In John Pierce's figure he gives a hint that $r^2/2R$ is not precise, though it is precise enough for all but a small minority of mirrors. It is not precise enough for very short-focus mirrors, also for compound telescopes, and nearly all lenses. The armchair mathematician correctly argues that the telescope maker outrageously forces the formula $r^2/2R$, which is exact for the sagitta of a paraboloid, to serve for that of a sphere. The standard reply to his letters of thunderous protest is that ATM is a book on telescope making and must not be allowed to degenerate into one on mathematics.

C.1.1. General

Just where $r^2/2R$ leaves off being exact enough and the armchair boys' worries start having real practical significance to the dirty-handed shop optician has been worked out by the shop optician and amateur mathematician Allan Mackintosh. Derived algebraically the next formula for the sagitta S of the sphere is $S = R - \sqrt{R^2 - r^2}$. Derived geometrically it is the same. Derived from the Cartesian equation of the circle it is again the same. (Mackintosh has it surrounded on all three sides!) In *Sky and Telescope* (Cambridge, Mass.) 1957 February—the present note having been inserted in ATM later in the same year—Mackintosh provides an alternative formula for S, derived by expanding $r^2/2R$, thus: $S = r^2/2R + r^4/8R^3 + r^6/16R^5 \cdots$. This alternative formula has an advantage for those of us who have forgotten how to do square root, and its first two terms will give close enough approximation to exactness for any mirror.

In a table Mackintosh gives worked-out examples showing the discrepancies between the figures by the approximate $r^2/2R$ formula and the true formula for 15 different mirrors. Even for a 12″ $f/1$ the difference is only 0.0121″ (even then it is 3/8″ in focal length), while it is only 0.0001″ for a 6″ $f/4$. He summarizes his findings with some rules of thumb: For longer than $f/5.6$ the approximate formula gives answers within 0.1 percent of the true sagitta—that is, multiply by 1.001. If you are making a mirror between $f/5$ and $f/3.3$ add 0.1 percent to the figure you got from $r^2/2R$. Between $f/3.2$ and $f/2.5$ add 0.2 percent. Between $f/2.4$ and $f/2.2$ add 0.3 percent. Below $f/2.2$ work it out from the true formula.

Are these seemingly piddling amounts too small to bother with? Not on a compound telescope. Here, for a first-class job, the focal length should be accurate within 0.01″ and preferably 0.001″, says Mackintosh. He suggests that the use of the rough formula may explain why many short-focus mirrors have turned out inexplicably longer in R than anticipated. A very small difference in sagitta can make a big difference in R. In short, if you want focal lengths precise to 0.01″, abandon the rough formula for focal ratios shorter than $f/6$.

After standard algebraic juggling Mackintosh also gives the true sphere formula in terms of R instead of S, thus: $R = (S^2 + r^2)/2S$.

A poor spherometer may deceive by inaccuracy. How sensitive the radius of curvature is to slight differences in sagitta is shown by an example. An error in measurement of only 0.001″ on the sagitta will result in a difference of 1.1″ in the radius of curvature of a 6″ $f/6$ mirror. A more precise means for measuring S than a less than high-grade spherometer spanning the entire mirror is a straightedge and vernier depth gauge. Reverse the straightedge, top for bottom, and take the mean of the two readings.

Mackintosh makes no warning against a common source of inaccuracy—fictitious accuracy, or self-deception by spinning a calculation out to more decimal places than the mechanical means of measurement make significant; like measuring your sleevelength with a hammer handle eked out with a micrometer caliper measurement.

C.1.1.10 Scratches

Why can I not seem to avoid getting scratches on my mirror during fine grinding? Answer: Perhaps you have mixed a little coarse Carborundum with finer sizes. Don't run your finger into a coarse size before a fine size. If necessary to do this, do it afterwards. Keep the can closed when not in use, and put it away upside down, thus keeping dust and grit out of the crack around its cover. Tack a few newspapers on the ceiling over your work, otherwise, people walking about upstairs will keep jarring down grit from the joists above you. Again, you may have previously wiped coarse Carborundum on your working clothes and transferred it to your mirror several days later, during the finer stages of grinding. It is better to keep finer sizes—and rough also—in a clean wooden box, turned on its side. This keeps out the falling grit. It is a great temptation while polishing to wash the surface of the mirror from time to time. Whenever this is done, do not *rub* the surface dry, for this is a fruitful source of scratches; *blot* the surface with cloth or paper towel.

Scratches on the mirror do very little actual harm, since they cut off only a small fraction of the light, but they make the job look mussy. You will be more proud of your workmanship if your mirror is free from them. To insure this, one must be very painstaking, often to the point of seeming fussiness.

Newspapers are cheap and can be used to good advantage to prevent scratches on the mirror. One may lay or tack them on tables and all places likely to harbor coarse grains of abrasive, and then tear off a page or two with each change to finer abrasive. It is well to be fussy about not getting Carbo on tool handles, etc., from which a few coarse grains may easily be picked up later on. Any rag or towel that may be used as a general sop-up around the shop is a likely culprit. Ordinary cotton waste is safe for swabs only during coarse grinding but not for fine grinding, as it often contains grit. Here it is best to use absorbent cotton. For drying mirrors The Editor uses paper towels or a handkerchief. In favor of the latter is the fact that the same handkerchief is not likely to be in one's pocket long enough between trips to the tub to carry coarse grit over to the next finer stage of grinding. Worst of all habits is that of wiping one's hands on one's work clothes, for these are usually worn throughout several stages of finer and finer grinding. Old hands at mirror making appear careless about grit. This is an "optical illusion"—they know just when and just where they can

C.1.1. General

get away with it without risk of scratches. Until one has become wise it is better to play safe and be fussy.

If one has reached the last stages of fine grinding and is worried about scratches, it is just as well not to start off each new wet with a long stroke for if any coarse particles were present this would drag them clear across the mirror. But a group of very short strokes will at least localize such scratches and the offending coarse particle will generally be broken down quite quickly. Even this is, however, a botch or makeshift and contaminated Carbo should really be washed. Transfer it to a common drinking glass (clean), stir it up and observe how fast it settles. As the heavier particles are to be sorted out by taking advantage of the fact that they settle more rapidly, we strive to pour off or siphon off the water into a second glass after the right lapse of time, leaving behind any coarse particles. A bit of preliminary experimenting with this process will teach more than a ream of words.

C.1.1.11 Mirror Breakage

What if I drop my mirror and break it? Answer: Get another glass disk. One is enough. Turn the tool over, beginning again with the flat side, and place it on a ring of pasteboard, in order to insure steadiness. Your second job will go ahead altogether faster than the first one, thus giving you so much assurance that you will be almost glad you broke the first. This, at least, was the writer's experience with his first speculum, and it will also be the case if you make more than one telescope.

To keep well-meaning, but damaging, hands off the mirror and tool during your absence, buy a cheap galvanized water pail, remove its bail, bend the two "ears" out at right angles, invert the pail over the work, and "lock" it down by means of two short, thick screws. This excludes dust, as well as children and house-cleaning enthusiasts.

C.1.1.12 Workplace Cleanliness

Do not swab off the mirror with waste, for this is generally full of sharp grit. A two-ounce roll of absorbent cotton—a standard drug-store size—can be cut into a number of four-inch lengths, and if each length is rolled up separately in paper, and kept rolled up until needed, it will be free from grit. Paper towels used like blotters are also excellent for cleaning the mirror.

Instead of a shallow basin in which to rinse the cotton swab, try a tall container of some kind. The grit freed from the swab will then settle to the bottom, where it will stay. Thus you get practically clean rinse water each time. For example, the writer used the central container of the ice cream freezer, which is admirably shaped for this purpose, until found out and reprimanded! A clean, tall crock is also excellent.

When you begin polishing with rouge, you will need some kind of an "all-over" garment, covering the sleeves, of course. Otherwise, rouge will stain your clothing in spite of any and all efforts at cleanliness. A linen duster is excellent. It is well, however, not to wear it during the grinding with Carborundum—unless it is washed before the polishing begins. This is because grit may be easily transferred to a finely polished surface by means of the clothing.

C.1.1.13 Shipping a Mirror

Here a relatively small amount of packing material in the right place will serve better than wrapping it in a whole mattress. First, the face should be protected from rubbing against any loose packing, if any is used, while on the journey when the train will constantly jiggle the disk. Ellison received a mirror which had thus jiggled itself into a figure which required six hours of polishing to restore. Soft paper taped on the disk will move with it within the outer packing and prevent this kind of abrasion. A flat board laid against the face will bridge the concave surface, as will the cover of a round tin marshmallow box—an ideal cover, by the way, for protecting the mirror under some other circumstances. Outer packing need not be thick if it is properly bestowed. A glass disk will withstand a heavy blow on the middle of the back or at the middle of the edge, but a light blow near any of the edges will remove a big chip. Throwing or dropping the containing box ten feet, which must be expected in the mails, may deliver just such a blow on the edge, despite much packing of a loose kind through which the disk may have gradually burrowed its way toward the outside while en route. In a test, a mirror surrounded by a snugly fitting round wooden cut-out, or a square with a board screwed across the grain on either side, and no other packing, would take altogether more bumps without injury than another disk done up loosely in a whole featherbed; in fact it would be rather difficult to damage a mirror thoughtfully packed in the former manner, simple and compact as it is.

C.1.1.14 Take Courage

With regard to his mirrors Ellison is exceedingly particular, and it has often been suspected that the first paragraph of Section A.4.7 has been responsible for scaring off many a prospective telescope maker who did not know that a mirror may fail by quite a lot to come up to Ellison's high standard of perfection, yet function so well that the difference would not be worth the price of throwing up the sponge at the outset. Sir William

C.1.2. Mirror Substrates

Herschel's many mirrors were recently brought to light and given a knife-edge test. Some of them proved to be what a modern amateur would call "not so good." Yet they functioned passably well in use. Sir William did not have the advantage of the knife-edge test. If you are wavering, don't let Ellison's criterion of perfection scare you off.

C.1.1.15 Three in One

If one finds that one enjoys the first job of mirror making, the chances are that one may make several mirrors and learn the art. A good idea, therefore, is to provide for two or three sizes, when making the mounting. For example, in the simple mounting shown in ATM2, Figure B.1.9, the board in which the mirror rests can be recessed for 6-, 7- and 8-inch mirrors, each recess being turned out in the bottom of the next larger one. The size of diagonal should be calculated for the largest mirror.

C.1.2 Mirror Substrates

C.1.2.1 Fused Quartz

Fused quartz or fused silica, is a theoretically ideal material for telescope mirrors because its coefficient of expansion under temperature change is very low, only $1/18$ that of plate glass and $1/6$ that of Pyrex. Elihu Thomson of the General Electric Co., who developed its use for mirrors and optical flats, listed its qualities as follows:

1. Disks require but little annealing.

2. They can be rough ground by a Carborundum wheel without danger of fracture, an operation difficult with glass and rarely resorted to.

3. The disks can be made very thick and rigid more easily than glass.

4. The fine grinding is carried on with great facility, and the surface before polishing is usually of finer grain than with glass. The fused silica is considerably harder than glass, and not so easily scratched.

5. The polishing proceeds readily and can be carried on regardless of temperature changes. Incidentally, there is less liability of scratches forming in polishing.

6. In very accurate work, figured by polishing, as in high grade surfaces of astronomical mirrors, the polishing and testing need not be interrupted, as with glass, by long rest periods, with the mirror disk jacketed in felt for equalization of temperature. This is very important and involves great saving of time.

7. In service, none of the precautions against temperature variations and distortions arising then are needed, and even in solar work with full sunshine on the mirrors no evil result follows.

Russell Porter was the first to figure a fused quartz disk after Thomson's process of manufacture was developed, and found it true, as claimed, that the figure was entirely indifferent to temperature change whether the disk was worked in full sunlight or dipped in hot water immediately before knife-edge testing.

The much higher relative cost of fused quartz than Pyrex or plate glass puts it beyond the reach of most amateur telescope makers. However, unless the user is experienced and skillful enough to put on the quartz a virtually perfect figure, it offers no advantage, temperaturewise, over the less costly materials, since it would merely retain unchanged during temperature changes whatever figure was given it. A few of the most expert professional opticians are able to exploit its advantage, chiefly for ultra-precise optical flats of refinement far beyond that needed for telescope mirrors.

C.1.2.2 Pyrex

Next to quartz in low expansion and contraction qualities comes Pyrex, having about 6 times the linear expansion of quartz but still only one-third that of plate and optical glass. This explains why the familiar Pyrex baking ware does not break when suddenly cooled on the surface; common glass contracts so rapidly on the outside surface when chilled that the stresses set up are too great for it and it fractures. Pyrex is a borosilicate glass containing borax or boric acid or both, and is manufactured by the Corning Glass Works. Whether the beginner should choose this kind of glass, or plate glass, which has three times the expansion coefficient of Pyrex but is not so hard, and is therefore easier to grind and polish—requiring approximately half as much time—is perhaps a toss-up. The unanswerable question is, can the novice figure his first mirror accurately enough so that the gain in stability afforded by low-expansion characteristics of Pyrex disks will be exploited? If on his first job he cannot do this, the advantage will be illusory. Anyway, his demands on a mirror's excellence, and his skill, are not likely to be as high as they will be later on, after he has made two or three mirrors. On the other hand, the Pyrex disk is neater—perfectly round because it's formed in a turned mold, while plate disks are likely to vary a sixteenth inch or so from round. This does no harm but is not quite so pleasing, aesthetically. The hardness of Pyrex requires no added skill. A harder lap—less likely to cause turned edge—may be used without scratching. Best of all is comparative freedom from temperature effects in figuring.

C.1.2. Mirror Substrates 407

C.1.2.3 How Plate Glass is Made

Polished plate glass and pressed plate glass are apt to differ in qualities required by the telescope maker. Pressed plate is forced under pressure into a mold and is usually permitted to cool too rapidly to permit good annealing and freedom from internal strains. It can be told by its rough appearance. It is also less expensive—a poor economy, however, for the telescope maker. Polished plate is not necessarily well-annealed, at least it is not *necessary* to polish it in order to anneal it; but plate that is worth the cost of polishing actually is always annealed. This is done by slowly moving it through a tunnel 800 feet long called a "lehr," whose temperature decreases from one end to the other. This passage requires five hours. A visit to a plate glass factory would be of interest to the telescope maker, to see how the manufacturer goes at the job of accomplishing on a mass production scale the task which we amateurs do on a small scale and at such cost in physical effort. The molten glass at 2,500 to 3,000°F is poured upon a large steel casting table, a steel roller spreads it out just as a cook rolls out dough, and it quickly drops to a red heat. Then it is annealed in the lehr. Next it is placed on rotating tables and ground by big disks under heavy pressure, with sand and water. Finer and finer sands are followed by several sizes of emery. The polishing is done by means of a battery of 18-inch buffing disks of felt, using rouge. These buffing disks steam under the heat generated by friction. In all, about an eighth of an inch is removed from either side of the glass during these operations. The grinding of a sheet about 25 feet in diameter requires 500 horsepower. In the modern "continuous method" the glass is melted continuously in a tank and flows out continuously.

C.1.2.4 Optical Glass

Many beginners find it hard to believe that some special, aristocratic variety of optical glass, perhaps made in France, is not necessary for a good telescope. This belief is hard to put down. Poorly annealed plate glass is an abomination, but plate, *per se,* has not plagued the thousands of telescope makers who have worked from this book for the past years and used it on successful telescopes. The 72-inch mirror of the great Dominion Astrophysical Observatory's reflector is made of plate glass. Why, then, do some hunt all over the world for special glass? True, on sizes above 12 inches it is time to cast around for special materials—not so much with regard to their composition as to their preparation (annealing, etc.). But the man who has gotten that far doubtless knows his onions already and needs no such instruction, while this note is intended for that type of beginner whose name is "Doubting Thomas." Of course, there are quartz and Pyrex, and these are still better, but the point is that there is nothing wrong with commercial polished (but not pressed) plate. Both the 60-inch and 100-inch mirrors at Mt. Wilson Observatory are made of a kind of plate glass, and not "optical glass" as the term is generally understood. Plate is softer—easier to work—than Pyrex, which needs up to two hours per stage in grinding; it also polishes faster. Few on their first mirrors are skillful enough to exploit all the low-expansion advantages of Pyrex.

C.1.2.5 Glass Mirror Substitutes

Although the beginner will choose plate glass or Pyrex, and the average amateur will stick to it, there a number of other materials which may be regarded as possible substitutes. Some still require more research, some are a demonstrated "wash-out" and others, like quartz, are a grand success but are not at present as available as one could wish. Since many of the advanced amateurs will ultimately be likely to consider trying these substitutes and may have difficulty obtaining authentic information concerning them, they are described below. These descriptions are purposely inserted in some cases in order to call attention to the drawbacks of certain of them. This may possibly forestall futile efforts. Nevertheless, the resourceful amateur who loves to explore the byways is as likely as anyone to hit on some invaluable discovery, hence the consideration of the various materials is encouraged. Dr. F.G. Pease of Mt. Wilson Observatory, in canvasing possible substitutes for glass for very large reflecting telescopes, makes the following statements (in *Publications of the Astronomical Society of the Pacific*, August, 1926):

"Another method early proposed was to quarry a block of obsidian and fashion it into a mirror. Obsidian is a volcanic product occurring in large masses, usually fractured into small pieces, but Dr. F.E. Wright of the Geophysical Laboratory informs me there is a ridge in Iceland, from which blocks could be cut far larger than would be required for a 25-foot disk. Obsidian is easily ground and polished, and silvers well. It contains, however, quantities of fibrous and crystalline material which cause defects in the polished surface.

"It has been proposed to build mirrors of concrete or cement and face them with silver. It is almost certain that such mirrors would be failures for the class of instrument which we are discussing. Concrete changes its form as it ages and no amount of grinding can remove its surface grain. With a coating of silver thick enough to cover this grain and permit a uniform surface, the differential expansion of the silver and the concrete would cause large distortions of figure, and the chances are the silver would buckle or peel.

C.1.2. Mirror Substrates

"Various marbles, resins, waxes and grainless cements have been tried as mirrors, but fault can be found with all of them when considered from the standpoint of a large, permanent, precision mirror.

"Attempts have been made to coat various materials with alloys or metals by spraying with an air gun. Microscopic examination shows that such surfaces are grained and fibrous, and unfitted for fine astronomical purposes.

"There is a promising field for research in the investigation of metal alloys suitable for mirrors. When one considers the enormous number of possible combinations of metals, he has hopes of finding an alloy which would be light in weight, which could be cast either solid or as a ribbed plate, and which could be easily silvered, if not in itself possessing excellent permanent reflecting properties.

"Alloys are known which possess some of these desirable properties, and it may be that the addition of other metals, or new combinations of them, would yield the desired material.

"Metals very quickly adjust themselves to variations in temperature, and, consequently, would have a good figure most of the time. Their coefficient of expansion is large, but this property is not inherent in all alloys.

"Invar, for example, is an alloy of 64 percent steel and 36 percent nickel, which has a coefficient of expansion practically equal to zero. Its reflecting power is not high, but it can be silvered. If it were possible to cast a large ribbed plate of *Invar,* or to build up a mirror from sheets and separators, it might serve our purpose.

"Speculum metal is a bronze composed of 68 parts of copper and 32 parts of tin. Most of the early reflectors, including the 6-foot mirror of Lord Rosse, were made of this material. Its reflecting power gradually drops from 70 percent in the red to 50 percent in the violet. It tarnishes in the open air and must be repolished and figured in the optical shop.

"Stellite has been used for mirrors in small sizes. Its reflecting power varies from 64 percent at 6500 to 54 percent at 4000. (6500 refers to wavelength in Angstrom units and would be in the red; 4000 would be in the violet.—Ed.)

"Magnalium mirrors, made of 31 parts of magnesium and 69 parts of aluminum, possess a reflecting power of about 83 percent in the visual region, which gradually drops to 67 percent at 2510 (in the ultra-violet.— Ed.). Early mirrors made of this material were poor, but the art of casting aluminum alloys has since improved greatly and it is possible that good light castings could be made today.

"Stainless steel mirrors, containing 11 to 14 percent chromium, now enjoy considerable use in small sizes and possess the advantage of retaining their brightness over long periods of time under circumstances which would ruin a silver coating. Measurements of their reflecting power yield values from 60 to 80 percent in the visual region of the spectrum. The coefficient of expansion is higher than that of steel.

"The alloy which the astronomer looks forward to might be called 'Mirrorite,' and the time may arrive when metallurgists, by careful research, will so combine metals as to produce this remarkable material, having the reflecting power of silver, the zero coefficient of expansion of *Invar,* the freedom from tarnishing of stainless steel, and the lightness of magnalium.

"Telescopes have been proposed in which many independent paraboloidal mirrors all point to the same spot. The mechanical mounting of such a system of mirrors would not be difficult to build. The light-gathering power of such a telescope would equal the sum of that gathered by the individual mirrors, but the resolving power would be only that of a single mirror. Owing to the fact that most of the mirrors would be used in oblique positions, the quality of the image would be poor.

"The construction of a large mirror composed of separate pieces, whose surfaces are parts of a single paraboloid would be more difficult to make than a single disk. To grind and polish the parts simultaneously would require a large solid backing. To grind and polish each part individually would involve an enormous amount of 'local' work. The resolution of such a compound mirror would equal that of a single mirror of the same aperture, provided the pieces were in the shape of sectors whose adjoining edges were covered by the diaphragms supporting the secondary mirror. If many small pieces were used each image would be accompanied by surrounding spectra, just as though the mirror were covered with a grating."

Obsidian or volcanic glass is usually black. The Editor purchased a chunk of this mineral, obtained originally in Utah, from Ward's Natural Science Establishment, and Porter sawed out of it a $2\,1/2$-inch disk, ground it, polished it and figured it. He states that "it sliced as readily as glass. The only difference I could note between working obsidian and plate glass," he continues, "was that the obsidian would not take quite so fine ground a surface as that of glass, and a little longer time was correspondingly required in polishing. The resulting surface was a lustrous black, giving an admirable background for showing Newton's rings by ordinary daylight, when brought into contact with a glass flat." Obsidian in some cases is pure glass, in others a mixture of glass and crystals, depending on the original rate of cooling from lava.

Invar, an alloy used for tapes, etc., because altered but little by temperature and having the smallest coefficient of expansion known, has been tried out by at least one amateur, G.H. Lutz. He has devoted years to research on various alloys regarded as mirror candidates, and has made seven mirrors of *Invar.* He states that, in common with all alloys that he has

C.1.2. Mirror Substrates

tried, it reveals its crystalline structure when polished, the crystals of the softer ingredients not behaving the same as the harder parts. *Invar* is a trademark; some refer to this alloy as "36 percent nickel-steel" or "thermostatic nickel-steel." The alloy is made by the Crucible Steel Company of America; the Holcomb Steel Co.; and the Simonds Saw and Steel Co. The finished *Invar* mirror must be plated to prevent tarnish, and, says Lutz, "the plating has its own problems that call for a lot of patience. Chromium and many other materials have been made use of for plating, and the end is not yet. I am still experimenting."

Stellite was tried by Lutz on several mirrors, one of which, a 10 $1/2$-inch, he showed to The Editor. He says it cost him many hours to master, as the outstanding quality of this alloy is its extreme hardness and resistance to abrasion (also corrosion). This is why it is used for high-speed machine tools, knives, oil well tools, dredge dipper cutting edges, etc. The relative resistance to abrasion of Haynes Stellite is 4 to 9 times that of steel. Lutz states that he used up 8 inches of brass tubing and 4 pounds of Carbo in drilling a $1\,3/4$-inch hole through his mirror, made of this remarkably hard alloy, and, he writes, "I will not wantonly advise anyone to start to make a mirror from *Stellite*, as I do not wish to make any enemies. However, if one has a machine, plenty of time and unlimited patience there is a chance." The alloy, which is said to be one of chromium, cobalt and tungsten, is not cheap. Lutz had trouble due to the crystals of tungsten remaining above the surface, but the later results were gratifying when he obtained a form of the *Stellite* in which the tungsten was reduced to the lowest possible limit. He did not have to pay any attention to temperature while working it, the figure not being thus affected; cold water could be run on the hot mirror (this is also true of *Invar*). Air temperature changes during observation did not alter the figure, as with glass. The coefficient of expansion is about half again that of glass, but heat is conducted very much more quickly through all metals, hence the point is largely academic. The manufacturer states that *Haynes Stellite* reflects from 83 percent of the incident light in the red to 68 percent in the violet.

Stainless steel mirrors are made by W. Ottway and Co., Ltd., London, but not in paraboloidal surfaces. Ernest Brookings, metallurgist, Jones and Lamson Machine Co., has made recent experiments with chrome steel for mirrors.

Copper, electrolytically deposited, reflects 48 percent in the blue to 90 in the red; commercially pure copper, 32 and 83, respectively. *Gold,* electrolytically deposited, 29 and 92. *Silver,* 86 and 95 (when untarnished, of course). Figures given are from the *Smithsonian Physical Tables* and are for perpendicular incidence and reflection.

C.1.2.6 Rotating Mercury Mirror

Dr. R.W. Wood, Professor of Experimental Physics at Johns Hopkins University, attempted in 1908 to make an automatically paraboloidal mirror of variable focal length by the theoretically practicable method of rotating on a central, vertical axis a round, shallow pan of mercury. Under centrifugal action the mercury takes on the figure of a true paraboloid. Using a 20-inch pan, a rubber thread transmission and a magnetic clutch, Dr. Wood obtained interesting results, the focal length being varied with ease by changing the speed. Minute irregular disturbances injured the perfection of the mirror's surface, despite the velvety transmission or drive.

In the *Scientific American,* March 27, 1909, Prof. Wood stated that the mirror was set up on a massive concrete foundation at the bottom of an old well 15 feet deep. To afford approach and room for the driving motor a second well was dug 6 feet away and the bottoms interconnected. A building with a sky hatch was erected over the wells. On a vertical axis a round, flat-bottomed basin was centrally mounted, its bottom everywhere absolutely perpendicular to the axis (necessary to prevent ripples) and filled half an inch deep with mercury. Around it, but carried on a support not at any place in contact with it, was a collar on ball bearings, driven from a motor by a rubber thread belt, and this was coupled with the basin by rings of horseshoe magnets on each. At 12 r.p.m. the f.l. was 15 feet, but at 20 r.p.m. it was only 3 feet. "On the whole," Prof. Wood wrote, "the definition was found to be surprisingly good when one considers the difficulties." In the *Astrophysical Journal,* March, 1909, Prof. Wood stated that surface ripples were at first caused by jars from the driving mechanism, but the addition of the magnetic clutch eliminated these; by jars from the bearings, but these were eliminated; by imperfect leveling, which set up a wave which was eliminated through very accurately plumbing the axis by centering the image during rotation; and, finally, by variations in velocity, which was the worst source and was not eliminated before the experiments were put aside. Damping the ripples with glycerine gave much better definition, and 5-second double stars were separated—a fair beginning. Prof. Wood then wrote: "It may be necessary in the end to use a motor, the speed of which is controlled by a clock." Today such clocks are everywhere—the work awaits the worker.

In a private communication Prof. Wood, in 1928, stated that the same experiments were continued the next year, and that he "got it to work much better. I put a 20-inch flat over it," he continued, "and had excellent views of the Moon. The final conclusion was that constant speed of drive would eliminate the slight tidal wave, which was all that remained. I did not even have a synchronous motor."

C.1.2. Mirror Substrates

Prof. Wood also hoped to find some substance that could be fused, rotated and allowed to solidify when rotating, which would indeed be a quick way to make paraboloidal mirrors. To save mercury he also suggested using a copper vessel, only roughly paraboloidal. A thin film of mercury would climb up its slopes from the bottom when rotation began.

Fig. C.1.2: Ritchey Cellular Mirror: The whole structure, a 30-inch flat, is propped up on edge. The dark grid is the partition work of ribs and plates seen edgewise and through the thin mirror disk itself. The evenly spaced ventilating holes in the ribs may clearly be made out. Professor Ritchey states (Journal Royal Ast. Soc. Canada, July–August, 1928) that this mirror "has retained its optical figure for three years without change which can be detected by the most sensitive optical tests."

C.1.2.7 Cellular Mirror

Professor George Willis Ritchey, who figured the 60-inch and 100-inch mirrors at Mt. Wilson Observatory, and also the 24-inch reflector mirrors at Yerkes, some years ago worked on a new cellular type of glass disk for optical mirrors of large telescopes. Each mirror disk of the new type consists of a front and a back circular glass plate, with a deep rib system also made of glass plates, between, and separating, the two. For mirrors from 60 inches in diameter up to the largest sizes ever considered the glass plates are all about one inch thick. They are made by the celebrated St. Gobain Glass Company, of a special, low-expansion glass perfected for this purpose. All plates for a given mirror are carefully selected for uniformity of thickness, of coefficient of expansion, and of flexure index.

The various plates are fitted together by fine grinding, and are then cemented together with a thin layer of Bakelite cement much less than one micron [practically 1/25,000 inch.—Ed.] in thickness, thus forming a comparatively light and very rigid cellular structure. For a concave or a convex optical mirror the front glass is curved to the proper degree, but is of *uniform thickness* throughout, and is of the *same* thickness as all other plates composing the cellular disk.

In order to permit the air to circulate freely within the multi-partitioned interior, round holes are cut through the sides of each rib or partition. A positive or forced circulation of air is used. Thus, adaptation to temperature changes takes place rapidly, especially as no glass thicker than an inch enters into the structure. The whole mirror has only one-fifth the weight of a solid disk, and holds its figure so perfectly under widely varied working conditions that no change of form can be detected with the most sensitive optical tests. For descriptions, see *L'Astronomie*, Feb., 1926; *L'Illustration*, April 24, 1926; *La Science et La Vie*, August, 1926; *Popular Astronomy*, May 27, 1927, p. 258.

Thus far Professor Ritchey has made mirrors of this kind (cellular) up to 60 inches in diameter which, he states, have remained optically plane under temperature changes such as those which occur in the open dome at night.

Fig. C.1.3: *Ritchey Cellular Mirror: The ribs and plates show plainly. This is Ritchey's second cellular experiment and is a concave mirror about 60 inches (1.50 meters) in diameter. Made in 1925–1926.*

C.1.2. Mirror Substrates   415

Fig. C.1.4: *Alan R. Kirkham and a 16-inch cemented disk made of two disks of 1-inch plate glass separated by 20 blocks of glass, 1 at the center, 7 in the inner ring and 12 in the outer ring. For a time the disk seemed promising, but later gradually failed.*

C.1.2.8 Cemented, Built-Up Disks

It is possible, by cementing together two relatively thin disks of glass, either with or without spacers between them, to build up rigid disks of the desired thickness-to-diameter ratio, but will they stay rigid? Evidently yes—for a time: weeks, months, perhaps longer. Alan R. Kirkham, also others, after numerous experiments at first very promising, later disappointing, comments as follows. "There are plenty of cements which would be tough enough, hard and permanent enough, but they have all failed, not because of inherent deficiency in the cement, or any lack of adhesive qualities, but rather because of the difference in thermal expansion of the cement and the glass. A conceivable explanation is that, when the temperature changes, the glass may try to expand faster than the cement, or vice versa, with the result that stresses, perhaps of several tons, are set up within the disk, and the constant small changes of temperature eventually force the cement free from the glass, and these disks come apart within a few months. Many have tried soft cements, like pitches and balsams, but it is fairly evident that several disks piled up, as is often done, will deform as much or more so." Some day, however, this problem may be solved.

C.1.2.9 Suction Mirror

Another unusual but possibly worthwhile effort would be the conversion of a flat mirror into a paraboloid by atmospheric pressure. By producing a partial and variable vacuum behind the mirror the latter will be caused to sag inward and the curve should theoretically be a paraboloid. One worker is known to be experimenting with this method at present, while Porter states that he tried it out in a preliminary way in 1910, getting far enough with it to conclude that it was worth further effort.

C.1.2.10 Magnesium Oxychloride Mirror

This was described by F. Le Coultre in *Bulletin de la Société Astronomique de France*, 1925, page 484; see also *Zeitscriften für Instrumentkunde*, 1926, page 588. The following is an abstract published in *Journal of the Society of Glass Technology*. "Several attempts were made to find a satisfactory substitute for glass for large telescope mirrors. The most successful results were obtained by mixing 100 parts of magnesium oxide with 24 parts of a one-half percent aqueous solution of magnesium chloride which, on being thoroughly mixed, set readily into a hard, white mass. For the production of a concave mirror, the viscous mixture was poured into a mold, which was slowly rotated until the mixture became stiff. The surface was then polished by means of emery and rouge. This mixture was not impervious to water, and it had the further disadvantage that it was attacked by CO_2, and the surface when silvered was not so good as a silvered glass surface. To overcome these defects the surface to be silvered was dipped into a 40 percent solution of formalin. Afterwards the block was immersed, with its reflecting surface upwards, in a 2 percent silver nitrate solution saturated with formalin, when the magnesium oxychloride immediately became covered with a brown deposit of silver oxide. The mirror was then taken from the bath and dried. The surface became black through the decomposition of the protoxide into sesquoxide and metallic silver. When it was quite dry the mirror was polished with rouge. The finely divided sesquoxide was rubbed off and the metallic silver formed a layer on the surface, as hard and as highly reflecting as if on a glass surface."

While glass is still king, after many attempts to discover a better material, this does not prove that nothing better will ever be found. And

C.1.3. Cutting Circular Disks and Holes

amateur experimenters are as likely to make a valuable discovery as professionals.

Coefficients of Expansion and Relative Thermal Conductivities, both of which are likely to prove useful to the telescope maker, are quoted below, from the Smithsonian Physical Tables. These are for degrees Celsius.

	Expansion	Conductivity, Heat
Plate glass	0.00000891	
Crown glass	0.00000954	0.003
Flint glass	0.00000788	0.0018
Quartz, fused	0.00000057	0.0023
Speculum metal	0.00001933	
Cast iron	0.00001061	
Steel	0.00001322	0.107
Silver	0.00001921	1.006
Stellite No. 6, cast	0.0000165	
Stellite No. 6, forged	0.0000146	
Invar	0.000000374 to 0.00000044*	
Stainless (chromium) iron	0.0000010*	

*Source: Bureau of Standards

The decimal fractions quoted relate to that portion of the total dimension of the piece of material used, which the material will be increased or decreased by each increase or decrease, respectively, of 1 degree Celsius, in temperature. A degree Celsius is equal to 1.8 degrees Fahrenheit.

C.1.3 Cutting Circular Disks and Holes

Here is a way to cut out a circular disk of glass from a slab. Assuming that we have access to a drill press, we proceed as follows: A cutter is made by fastening a tin can to a bolt, as shown in Figure C.1.5 (left). The can A is cut down, leaving walls about twice as long as the thickness of the glass to be cut. A hole large enough to take the bolt is punched through the back of the can, washers and nuts are added loosely. The bolt end is then inserted in the chuck C of the drill press. Shift the can on the washers until it runs smoothly, then screw the nuts home.

The glass slab B is pitched to the board and clamped to the table. Carborundum (about No. 200) and water are fed under the cutter, and in about half-an-hour the slab will be cut through. The cutting edge of the can, though not perfectly circular, will however cut out a perfect cylinder from the glass. If the cutting edge is notched as shown, it will cut faster. The drawing shows the slab nearly cut through.

Fig. C.1.5.

The old fashioned bow drill in Figure C.1.5 (right) will cut small disks if a drill press is not available. The setup required needs no explanation. Of course the bolt must be longer than the one used with the press.

C.1.4 Tools and Fixtures

C.1.4.1 Inverting Tool and Mirror

In grinding, Ellison gives instructions to allot to certain sizes of Carbo certain reductions in length of radius. An alternate method is to employ a rig which will permit quick inversion of mirror and tool—*i.e.,* tool either on top or on bottom, at will—and then excavate all the way to ultimate radius with the first size of abrasive. From then on one may hover more or less around that radius by alternately inverting and reinverting. This also brings plenty of work on the outside zone of the mirror, which ordinarily polishes out last and, therefore, needs it. With this method, however, one must be on the watch for turned edge, due often to the application of force too high up on the handle. As finer sizes are used, make the simple pencil mark test for sphericity, by carefully drying and cleaning the tool and mirror drawing pencil marks across each, working them dry with a few very short strokes and observing where they are in contact by noting where the marks are rubbed thin or entirely off. To get rid of the turned edge, which probably amounts, when noted at this early stage, to what would be considered gross and incurably turned if permitted to pass on to the polishing stage, remove the handle and grind a few wets with strokes less than one inch long. It is well, if this more elastic method of reaching the desired radius by inverting tool as mirror is employed, to arrange to leave an inch or less of radius at the end of polishing, and to do most of the polishing right side up, thereby avoiding possible turned edge from that source and perhaps facilitating the observation of contact with lap by permitting the worker to look through the glass while polishing is in progress. (The above-described pencil test may cause scratches.)

C.1.4. Tools and Fixtures

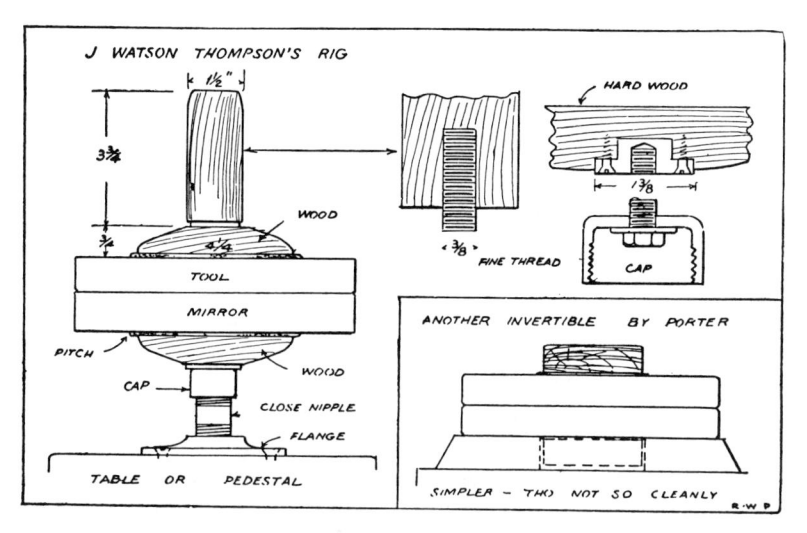

Fig. C.1.6: *Inverting Devices: It is a marked convenience to be able to invert tool and mirror, and without time-consuming bother.*

C.1.4.2 Inverting Device

For those who really enjoy elaborating the few simple tools needed for telescope making, a rig for hand grinding and polishing, made by J. Watson Thompson and presented to The Editor, may be constructed—provided one has a lathe. This has proved entirely satisfactory and most convenient when it was desired to invert the normal positions of the tool and mirror. The sketch explains it. It also has the additional advantage of keeping the work up out of harm due to grit from the usual bench or pedestal. A common pan with a hole cut through the bottom for the stem of the stand is usually placed under it. This catches all drippings and can be cleaned up in a jiffy. Another advantage this rig embodies is variable height—obtainable by inserting pipe nips of various lengths in the upright stem. The wooden flanges shown in the drawing are a slight variation of Thompson's original job, in which disks only about 2 inches in diameter were used. These do not "blindfold" the mirror while it is being worked (you can see the contact through it), but they did not, at least in one man's hands, provide enough area to keep the disk from breaking off under stress of hard work.

C.1.4.3 Limiting Devices

Every now and then someone devised an automatic limiting device for insuring strokes of uniform length, or one for turning the tool by some even part of the circle. These ingenious devices are worse than useless, for the very merit of hand work is that the combination of stroke length and direction is never twice repeated, thus preventing the development of zones, etc. On the other hand, no slavish effort should be made to vary the length of strokes. This will take care of itself.

C.1.4.4 Making Templates

The following is from a letter written by Prof. Elihu Thomson: "My method of making templates for the curves of lenses or mirror surfaces is very simple and effective as well as accurate. I make them of sheet glass (thin photo glass is best) by using a radius bar as usual on the floor with a glass cutter (glazier's diamond) set in at radius distance from pivot or center with diamond edge projecting at proper angle. A glass plate is thus marked by the diamond and if the cut is a real cut and not a scratch, the plate will break along the line quite nicely. To finish the pair it is only necessary to fine grind the edge of the cut by laying them down on a flat surface (a board) and moving one on the other lengthwise, using fine Carborundum and water. Very rapidly the fit becomes perfect. Even greater accuracy is attained by reversing one of them so as to exchange ends. The edge, being now a fine ground surface, can be marked or blackened by a lead pencil before applying it to the disk of the lens or mirror and it will mark (by slight end movement) the places of contact with the disk. I have even used this method with short curves, using very thin glass, with success. Of course for the best results near the finish I use a spherometer of usual three-legged pattern with micrometer screw in center."

C.1.4.5 Metal Template

"It is very difficult," J.V. McAdam of Hastings-on-Hudson writes, "for even a good mechanic to scribe a curve and file a *square* edge to within half a frog hair of the line. My method is to file a piece of thin brass to the approximate curvature, put a hole in either end and screw it down to the end of a rough T-square or board. Drive a nail through the other end at desired radius and into the work bench or a mahogany table. Clamp a file to the bench so that the edge contacts the brass and push the T back and forth over the file, gradually rapping the file into closer contact as it cuts away the high spots, the last few strokes having very light contact." Metal is not, however, the only suitable material for templates. Cut single strength window glass with a glass cutter attached to the radius bar described. After breaking apart, slightly grind the curves with Carbo.

C.1.4.6 Removable Handle for Mirror Disks

Several have discovered that rubber-cup sink clean-outs make excellent handles. When wet they are said to stick like a bulldog to a root (performance not positively guaranteed, however). These are the kind of rubber cups, attached to the end of a stick, which the plumber brings—or, rather, goes back for—when your sink traps become plugged. The smaller size, about three inches in diameter, is most suitable for the mirror worker. The handle may be sawed off. To attach, wet, apply to disk, and press hand on one side. To detach, press on opposite side.

C.1.4.7 Grinding Stands

A convenient grinding stand which permits sitting down to work consists of a base A of heavy plank, with a center pin (an old bolt end) B, a wooden ring C, nails for ease in turning the tool, stops E if needed, and a wooden wedge F. This was designed by James C. Critchett. It permits quick removal of the tool and quick inversion of tool and mirror but, because of the occasional need to invert, it must be used in connection with the rubber clean-out force-cup mentioned elsewhere as a handle for the mirror disk, or else with no handle. It will require careful cleansing after each grade of abrasive.

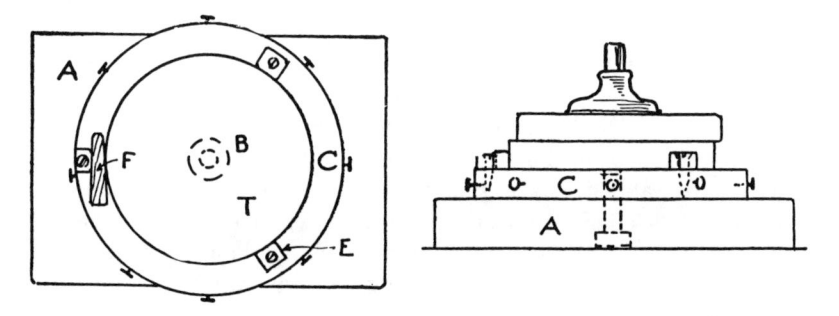

Fig. C.1.7: *Drawings by Russell W. Porter, after James C. Critchett*

C.1.5 Grinding Abrasives

C.1.5.1 Designations for Carbo Grain Sizes

Sizing Carbo grains is mainly nominal and in most cases do not express the rate of fall through water, as formerly. The following statement by Henry R. Power of The Carborundum Company explains them: "Abrasive grain sizes No. 220 and coarser are defined clearly in the Simplified Practice Recommendation R118-30 of the Department of Commerce. Grit

Numbers 240, 280, 320 and 400, which are the powder sizes, are defined by standard samples of the Abrasive Paper and Cloth Association. These may be placed in a sedimentation tube containing a fluid of known viscosity, and the amount collected in a unit time measured. The size in microns is plotted against the percentage, giving a curve which should be matched in the commercial preparation of the powders. The symbols are nominal. The symbols '1 min.,' '15 min.,' '60 min.,' etc., are now obsolete with us, though I presume they once referred to the time required to settle the powder in water. The symbols F, 2F, 3F and 4F are used for powders less accurately graded than the numerical gradings 240, etc. above referred to. These symbols are nominal."

C.1.5.2 Crushed Steel and Pyrex

Crushed steel and Pyrex are said by Napoleon Carreau, optician, of Wichita, Kansas, to be a logical combination. Pyrex is about twice as hard as plate glass. "I can do in one day with Crushed Steel what I cannot do with Carborundum in four days, when grinding Pyrex," Mr. Carreau writes. "Crushed Steel can be used a great many times over without apparent loss in grinding quality. It scores the grinding tool and wears it out very fast, but saves time. When one throws a teaspoonful of crushed steel between the grinding tool and the Pyrex, one hears a sharp noise that keeps on until the steel is all thrown out. The same crushed steel may be thrown back between the tool and Pyrex and the noise is again about the same as it was before. Those who do not notice very much difference between the grinding qualities of the two abrasives on Pyrex may not be using enough pressure."

For the benefit of beginners it may be pointed out that, while crushed steel is a gross abrasive, the gain will scarcely pay when making small mirrors, while the coarse grains in relatively inexperienced hands may leave pits which will cause much trouble later. When working on large disks, or on smaller hard ones such as Pyrex, this material will probably effect a saving in time, as it will hog out the bulk of the glass in rapid fashion.

C.1.6 Grinding

1. The sharp edge produced in case the disk becomes ground down excessively is liable to chip off and scratch the mirror. Hone or chamfer this edge with a spare piece of glass and 600 Carborundum, washing the mirror thoroughly afterwards.

2. If the grinding stand is not quite level, no harm will result; you will always get a perfect figure of revolution, because you keep turning the tool and the mirror around from time to time by varying amounts.

C.1.6. Grinding

3. It is convenient to keep the coarser sizes of Carborundum in a large cleaned salt cellar; the finer sizes may be made up into a cream before using.

4. While Porter advises the use of sizes 80, 120, 280, 400 and 600 Carborundum, Ellison recommends sizes 80, 200 FFF, 400, 500 and 600. While these seem contradictory, what we have in both cases is simply a series of grains regularly diminishing in size. Either group is equally suitable.

5. Bubbles that form between mirror and tool may be pushed out if one wishes, simply by sliding the top disk away out over the lower one and then carefully drawing it back; then the same on opposite side.

6. Grinding out of doors will do no harm, regardless of temperature. In addition, it provides assurance that the coarser abrasives never will be brought near the indoor place where rouge is later to be used, and is therefore all in favor of freedom from scratches.

C.1.6.1 Who Discovered the Method of Concaving a Glass Disk?

The common method in which two disks of equal diameter are ground together was discovered by Professor Elihu Thomson and described by him in a communication to the *Journal of the Franklin Institute,* 1878, pages 117–121, entitled "A New Method of Grinding Glass Specula." The same paper was reprinted in the *Scientific American Supplement,* Sept. 14, 1878. However, there is a question whether the same method was not previously discovered and used by Short, more than a century previous. Even if so, it would make no difference in this case, because, as Ellison has stated in a letter, "If Short, in 1750, found that he could make specula quicker and better with an equal disk of the same metal as a tool, he would certainly not tell the world, but would keep the secret for his own profit." By a sacred tradition in science he who makes a discovery and thus withholds it for his own use receives no credit for it; not even if he bobs up later, when someone else independently makes the same discovery and unselfishly publishes it, and attempts then to claim credit for it. Short was excessively secretive (see Bell, *The Telescope,* page 27).

Previous to this discovery the accepted method was to form the tool to the desired curve first.

Ellison states that "Wassell had the disk-on-disk method a good many years before 1878." He refers to some articles by Francis, describing it, in Volumes 7 and 8 of *Amateur Work,* which antedate the Thomson articles cited above. He believes Professor Thomson discovered the method independently.

C.1.6.2 Nature of Grinding

The nature of the grinding operation is not generally believed to be the more obvious one. Without taking particular thought one might assume that the grains of abrasive act simply as plows, like the tool on a planer or shaper. It is thought, however, that this sort of thing occurs only when a grain drops into a depression in the one glass disk or the other and becomes lodged there. Mainly, the operation of grinding is something like what we would have on a large scale (visible) if we were to place a number of large steel balls on a cake of ice, with a second cake on top of the balls, and move the upper cake over the lower. Behind each ball, as it rolled, there would be a path, not planed or gouged out but *fragmented by pressure*. On glass we get something of the same thing with a glass cutter, which is not a cutter at all, but a local splinterer. With abrasives, the pressure directly under the particles is transferred downward and outward, and conchoidal chips of glass are pushed and sheared out laterally. And the greater the pressure the greater the depth of shearing involved. Grinding takes place in direct ratio to pressure per unit area, a point which answers the frequent query "How hard should I bear down?" The answer is: "Do you feel stronger than you feel patient, or more patient than you feel strong?"

C.1.6.3 Wets

How long is a wet? In grinding, the best test for the proper length of a wet is when the grit stops sounding gritty. If one is of an investigative turn of mind and has a medium-powered microscope it is interesting to start with a charge of fresh Carbo, grind 20 or 30 strokes and examine a dab of the charge under magnification, noting how the grains have broken down and how many particles of glass are commingled with the abrasive. Carry this out until the Carbo is all broken down and glass predominates. It will prove instructive and worthwhile.

C.1.6.4 One-third Strokes

What is meant by a one-third stroke? Answer: When the upper disk overhangs the lower by (roughly) one-sixth of its diameter at either end of the stroke. Thus, for a six-inch mirror, come at each end to an overhang of about an inch—not two inches. But in rough grinding, as Ellison says in Section A.4.4, the stroke may be much longer. Too long a stroke will probably hyperbolize the mirror.

C.1.6. Grinding

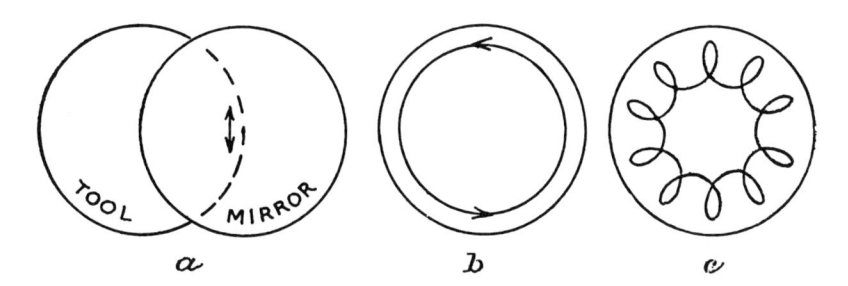

Fig. C.1.8: *Drawings by Russell W. Porter, after the author.*

C.1.6.5 Strokes in Grinding

In addition to the more conventional 'cross-and-back stroke, several others are possible and are used more or less in various combinations by professional opticians, mainly for hogging out of the rough curve without great loss of thickness at the edge. In Figure C.1.8a the center of the upper disk moves over the edge or near the edge of the lower one, as shown by the arrow, which is its path during one stroke cycle. This represents only one position with regard to the pedestal; of course, the worker will shift gradually round the pedestal, just as in the more familiar method of grinding. In *b* the upper disk is swept round and round with its center always near the edge of the tool, or perhaps three-fourths way out. The arrows represent the path of the center of the mirror. Both of these strokes differ from the 'cross-and-back stroke, in that they bring practically no abrasion to bear on the edge of the mirror and the center of the tool, almost the whole weight of the upper disk being balanced on or near the edge of the lower. In the more conventional 'cross-and-back stroke there is considerable wear all over both disks and the concavity is gained only because there is more wear—perhaps 50 or 75 percent more—in the one area than in the other. This wastes glass, abrasive and elbow grease. However in the various off-center strokes much abrasive is lost by being pushed off the disk; also these strokes are not as comfortable to perform as the 'cross-and-back stroke.

A third variation is the epicycloid, as in *c*, where the curve, rather idealized, represents the path of the center of the mirror. This has its uses; sometimes for evening up a curve that has become irregular (sometimes, too, if care is not taken, it will make a regular curve irregular); sometimes just to vary the monotony of grinding. Stroke *c* will act about the same, as far as distribution of wear is concerned, as the 'cross-and-back stroke.

All of these strokes are capable of playing bad tricks, but if the worker who has already made a mirror by the 'cross-and-back stroke will experiment with them, or with combinations of them or other modifications, he doubtless will acquire more general skill than he previously had. In different hands they exhibit different idiosyncrasies; hence the individual worker may experiment for himself. In general, they will tend to deepen the center unduly and should be used with circumspection when employed in the fine grinding stages—if used then at all.

C.1.6.6 Why Disks Grind Concave

This is explained in Chapter A.1. The abrasion is increased when the upper disk overhangs, due to increased pressure on a given area. A more common explanation is that the center of the upper disk is under abrasion all of the time, while the edges are under abrasion only a part of the time. This at first seems quite logical, but if it is the true explanation, why is the center of the lower disk not also abraded more than the edge?

C.1.6.7 Refusal of the Mirror to Become Concave

This is a rather frequent difficulty among beginners. The cause is generally too rapid strokes. Workers who attempt to combine sitting-up exercises or fat reduction with grinding will get exercise and fat reduction, but will discover a stubborn mirror disk, which will either become concave very slowly or not at all. Sixty cycles per minute—that is, across and back once a second—should be the speed limit, but even this is rather frantic. The habit of taking mirror making in low gear is in general a good mental discipline to acquire. It conduces to cautious, thoughtful workmanship.

There is a reason why rapid strokes annul much or all of the normal concaving tendency of a mirror. At opposite ends of the stroke a mirror must each time be brought to rest and its motion reversed in direction. This is done by force applied on the handle, above the center of mass of the mirror. But the momentum of the mirror tends to keep it moving in the same direction in which it has been moving. The resultant is that, under these two offset forces, it tends to ride up on edge. (It actually would, if the strokes were rapid enough.) This throws a share of its weight on the edge of the mirror and center of the tool, at the exact time when it is desired to accomplish just the opposite, *i.e.*, to abrade the center of the mirror and edge of the tool more than the other portions. Therefore little or no positive progress is made. This tendency increases directly as the height of application of effort, and as the square of the velocity of motion. When about one stroke-cycle per second is exceeded it begins to mount up rapidly enough to annul some of the desired concaving tendency, and at two stroke-cycles per second it will almost entirely offset that tendency.

Ideally, the effort should be applied at the center of mass of the disk, and then this tendency could not occur. Practically, a short handle (or

C.1.6. Grinding

none) will help. So will a reasonably deliberate attack on the job. The attachment, when a machine is used, may be a pin-and-socket or ball-and-socket low down, close to the face of the mirror.

C.1.6.8 Amount of Glass Removed for Each Stage

While we are plugging away with the various stages of grinding we shall want something to think about. And here is something: How much actual depth of glass does each stage of grinding ordinarily remove? And especially, how much removal of glass does polishing involve? If we know these amounts, even in round numbers, it will frequently be of use to us, improving our judgment when it comes to making certain decisions.

To arrive at these figures by calculation is easy. Knowing the radius for each stage of grinding and polishing, we simply substitute in the formula $r^2/2R$. This is none other than our old playmate, the r^2/R formula, altered a bit because here we are no longer dealing with a beam of light whose aberration is doubled by reflection (which accounts for the R, instead of 2R) but determining the actual depth of the curve with regard to a straightedge placed across the surface of the mirror. This depth is called the "sagitta," the Latin word for arrow (the line of the sagitta perpendicular to the center of the glass resembles an arrow held on a bow).

Let us assume that we are working with a 6-inch mirror having a radius of 96 inches. Assume, also, that we reduce the radius of curvature to 132 inches with No. 60 or 80 Carbo; 22 inches more with No. 100; 10 inches more with No. 220; and about an inch more with each successive size; finally alloting a generous inch to polishing. We substitute in the formula and obtain the following figures:

Table C.1.1

Carbo. No.	Radius (inches)	Substitution	Sagitta (inches)	Difference (inches)	Or Roughly (inches)
60	flat to 132	9/264	0.03407	0.03407	1/30
100	132 to 110	9/220	0.04090	0.00683	1/100
220	110 to 100	9/200	0.04500	0.00410	1/250
280	100 to 99	9/198	0.04545	0.00045	1/3000
400	99 to 98	9/196	0.04591	0.00046	1/3000
600	98 to 97	9/194	0.04638	0.00047	1/3000
Rouge	97 to 96	9/192	0.04682	0.00044	1/3000

Thus we see that we have worked about as hard to excavate $1/3000$ inch with one size Carbo as we did to remove $1/100$ inch with another; and about five times as hard to polish off the last three-thousandth of an inch with rouge as we did to remove a whole $1/100$ of an inch with Carbo. As a matter of fact, these are not, strictly speaking, the actual depths of glass worn away, but rather the differences between the amounts worn away at edge and center of disk. (Probably half as much glass is worn away at the edge as at the center.) But no matter—we now have in mind a rough idea of the amounts we are dealing with for each stage of the job, and these amounts will apply in a general way to most of the mirrors we are likely to undertake.

Finally, in altering our final sphere to a paraboloid, how much do we actually scoop it out? A mere 87,000th of an inch! For the inside zone of the parabolized mirror has about one-tenth of an inch shorter radius than the outside zone; and we may thus calculate the amount of deepening by the same formula as above.

F.W. Preston of the Research Laboratory of Messrs. Taylor, Taylor and Hobson, Ltd., states that in their workshops it is known that there is commonly removed in the polishing process from $^6/_{100,000}$ to $^{12}/_{100,000}$ of an inch in thickness. It is found, he says (*Trans. Opt. Soc.*, No. 3, 1921–22, p. 159), that with automatic polishing, and when the polisher is working at its best, this amount is removed in 50 or 60 wets. Consequently, the average amount removed per wet is in the neighborhood of one or two millionths of an inch.

C.1.6.9 Chamfering Disks

Sometimes the disks when received are not chamfered on the edges, and sometimes one will quite wear away the chamfer, leaving a sharp, delicate edge. This edge is liable to chip off and the chips are more than likely to scratch the surface. Accordingly, the business edges of both disks should be honed with a Carborundum stone held at an angle. In the average case The Editor finds that a bevel carried one-sixteenth inch back from the edge will nearly but not quite grind out—for we wish to leave a small width of bevel on the finished job. It is not so well, however, to put off chamfering until later, for tiny round chips are likely to be flaked off from the surface near the bevel, and polishing or even fine grinding will not excavate enough glass at the edge of the disk to work them out. [As later revised: It was found that the stone causes tiny fractures that spread. Glass detaches and scratches. Substitute No. 600 Carbo grains, wet, rubbed with a scrap of glass.]

C.1.6.10 Getting Overall Contact while Grinding

There is one place where Ellison might well have been more explicit. In the third paragraph of Section A.4.4 he recommends the use of the long (full) stroke for roughing out. This is good, not merely because it saves much work and time, but because it leaves the outer zone of the disk comparatively untouched while a shorter, roughing-out stroke would

C.1.6. Grinding

bring nearly as much work on the outside zones as on the center and thus take quite an appreciable total thickness off the mirror—a point which may become important if the disk is none too thick to begin with. Now, the important point is this: after the long strokes of rough grinding have scooped out the center irregularly, as Ellison says, leaving it a pronounced hyperboloid, the two disks *must* be brought back to *full* spherical contact all over before polishing begins. Preferably they should be brought nearly so before fine grinding. It has been noted that, since the first edition of the present work was published, several mirrors, when finished, proved to be very seriously overcorrected. In two cases the knife-edge test showed that the radius of the outer zone was a full inch greater than that of the inner zone, instead of about one-tenth of an inch (in the case of a 6-inch mirror of f/8). Here the assumption is fairly safe that the hyperbolization came about, not in polishing, but before fine grinding was finished, and was probably due in part to Ellison's failure to stress the point so urgently stressed above.

Watch the liquid sludge through the mirror disk while grinding. This film of sludge looks thicker than it really is. To measure it roughly, and follow its diminution in thickness from time to time while working, wash both tool and mirror, dry them, brush them absolutely clean with the hand and place a tiny scrap of paper in the center of the tool. Usually the mirror will teeter on the paper to a noticeable extent when it, with the tool, is grasped on either side and between forefingers and thumbs. By choosing thinner paper from a book it is possible to watch the discrepancy between tool and mirror narrow down as the work continues. While a single sheet of India paper is only about one-thousandth of an inch thick, even that is quite a large departure from the spherical if one foolishly leaves it to deal with by means of rouge, whose cutting powers, quantitatively reckoned, are comparatively slight; or for that matter, even with No. 600 Carbo.

C.1.6.11 Bad Central Contact in Final Grinding

Problems with central contact in final grinding, with the outer zone refusing to wear down and allow the inner zones to make spherical contact, may be due to the use of too thin and watery a paste of Carbo. Capillary attraction draws the abrasive off the outside zone and down over the edge, and the central zones receive more abrasion than the outer zones. When this situation arises, a rather stiff paste of Carbo will usually deal with it and permit the two disks to be brought to concentricity. To this, Porter adds the suggestion that side pressure be applied to the edge of the disk (as in Figure A.1.16).

This note partly repeats statements made in other notes. It might, however, be inserted **at ten different places and capital letters,** to the advantage of the beginner, who very often falls heir to trouble in polishing or figuring perhaps because he does not sense these realities until he has come hard up against them. This is one important matter which Ellison omits to emphasize.

C.1.6.12   Estimating Fineness of Grinding

Dr. James Wier French rates Carbo as four or five times the cutting power of a corresponding grade of emery (*Transactions of the Optical Society*, Jan.–Feb., 1917). It will do no harm to try emery for a stage or two of grinding, just to see what the old-timers were up against before the days of faster abrasives. (The Editor tried it and came away with a profound respect for their patience.) Other notes follow, chosen from Dr. French's article. In working with Carbo, emery or sand the abrasive effect, determined by weighing the quantity of glass removed, was found to be directly proportional to the speed. Dr. French made interesting quantitative determinations on fine ground surfaces, refining the method described in Subsection C.1.6.18—"When is fine grinding finished?" .... He placed his glass on edge upon a table graduated in degrees around its periphery. Near it he placed an electric lamp, permanently fixed. He then placed his eye so as to catch the image of the filament in the glass. "If the plate (*i.e.,* the glass) is rotated," he says, "so as to reduce the angle of incidence, the eye being moved so as to keep the image central, there will be observed at one point an abrupt change in the whiteness of the image, due probably to the scattering of the more intense short wave rays.

"In the first instance some difficulty may be experienced in observing the change, but once it has been observed, no difficulty should be experienced in locating the point of change to within half a degree without the use of any special apparatus other than the graduated circle.

"If now the angle of incidence is still further reduced, the image will quickly become red and then rapidly fade away. The point of maximum redness can be recorded to within three-quarters degree. ...These two points—where the white image changes and the red is a maximum—afford a definite record of the surface that is independent of the personal element or the intensity of the illumination. Repeated measurements, accurate to within half a degree, can be made of the point at which the image disappears, but as these measurements are not independent of the illumination and the individual, it is necessary to standardize these factors for closely accurate work."

Dr. French determined microscopically that the particles removed in pitch and rouge polishing are about $1/25,000$ inch in diameter.

C.1.6. Grinding

C.1.6.13 Streak of Rouge Test for Contact in Fine Grinding

"Use rouge instead of a pencil, on the tool only. Rub the mirror at right angles to the line of rouge, which should be applied very thinly with the finger tip. Examine the mirror for red stain. Put no pressure on the mirror in testing for contact—its own weight is sufficient. Pressure often results in serious scratches."—*Cyril G. Wates, Edmonton, Alberta.*

C.1.6.14 Water Drop Test for Contact in Fine Grinding

Kirkham's idea. Suppose the head of a pencil is dipped into water and the adherent drop transferred to the mirror. The volume of this hemisphere is approximately $1/2 \times 4/3 \pi r^3 = 0.003$ cu. inch, assuming $1/4''$ diameter of the drop. For an $8''$ mirror, with approximate area 48 square inches, $0.003 \div 50 = 0.00006''$. This gives some idea, at least, of the thickness of the water film, assuming that it covers the entire area of tool and mirror and has uniform thickness—which it probably would not have in the average case. But the order of size of thickness indicated may be about $0.0001''$, which is considerably finer than we can measure by the insertion of scraps of tissue paper. How much of the water goes into the pits is another factor. Doubtlessly, instead of trying to cut the calculation too fine, the better way is to make the test on a few surfaces and empirically derive an idea of its degree of precision—that is by later observation of the figure of the mirror after polishing has been started.

Some think that the scratches which are sometimes the result of the lead pencil test described in Section C.1.4.1 are from grit in the graphite. Others believe some sort of molecular cohesion between the surfaces pulls out pieces of the glass.—*Ed.*

C.1.6.15 Sticking Mirror

Occasionally a mirror will stick and suck, even during fine grinding. If the disks are spherical and have an equal radius they will not stick (although any mirror ought to push a little harder as it passes over the other, for more surface is in contact at that part of the stroke). If the disks stick they may be badly hyperbolic. If the latter, they cannot make good contact at ends of strokes, there is a tendency to create a vacuum between them, and this will account for the sticking. A very short stroke—sometimes less than an inch long—will usually doctor any such condition in a few minutes. "Or," Porter states, "one may press down locally at the *edge* of the mirror as it revolves under his hand thus wearing away the edges of mirror and tool and bringing their center portions back again into contact."

C.1.6.16   Mirror Sticks to Tool

"When one is fine grinding a mirror on a glass tool, and the amount of abrasive is reduced to a minimum, the tool and mirror will sometimes seize and stick to one another as tight as the proverbial 'tail in a mule.' This is most disconcerting, especially since it happens suddenly without warning.

"Do not attempt to wedge them apart—the results may prove to be disastrous. Rather, submerge tool and mirror, (everything, in fact except the barrel) in a basin of cold water, the water to cover the joint well between tool and mirror. Then heat slowly until you cannot bear your hand in it. Keep water hot for a couple of hours or so, and then set it aside to cool over night. In the morning the water will have been drawn in between the two surfaces and they will separate easily. Of course, the pitch that fastens the tool to the block and the mirror to the handle will need attention, but then, what of that? This method has worked 100 percent for the writer."—*Rev. J.G. Crawford, Saunderstown, R.I.*

"When my mirror froze to the tool during fine grinding I took the contrary disks out in the backyard and played a fairly strong jet of water from a garden hose on the edge of the glass at the line of contact, and after a little while had the satisfaction of seeing the water creep in between the two until I was able to lift off the mirror without any resistance whatever. It would be well to be careful about the temperature of the water."—*Dr. John W. Straight, Santa Ana, California.*

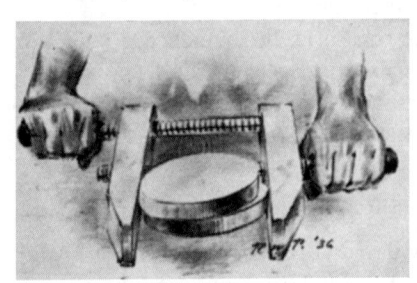

Fig. C.1.9: *Porter's method of unsticking stuck disks.*

"If the two disks are not quite coincident, but slightly overlapping—as usually happens—I have placed them between the wooden jaws of a large cabinet maker's hand screw (Figure C.1.9) and left them for a time. Generally they were found to have come apart."—*R.W. Porter, Pasadena, California.*

"When my two disks stuck together I wiped the surplus water off the edge and applied alcohol. The alcohol penetrates between the disks and

C.1.6. Grinding 433

heats a little as it mixes with the water."—*W.B. Hiner, San Jose, California.*

"Your suggestion, also those made to me by the Alvan Clark Co. and the American Telescope Co., failed to separate the two disks which I reported to you a year ago as being stuck together. I present a statement of my efforts to separate the two, which became stuck together while grinding with No. 600 Carborundum:

July	9	Placed them in warm water	10 minutes
	10	Placed them in turpentine	10 minutes
	15	In carbon tetrachloride	23 hours
	18	Denatured alcohol	96 hours
	24	In hot water	15 minutes
	24	On ice	24 hours
	25	In ice water	15 minutes
	26	In hot water	15 minutes
	27	Carbon tetrachloride	72 hours
Total useless effort			240 hours

Then I placed the two disks on edge on the work bench and, using a Durham safety razor blade and a piece of two-by-four, separated them in 10 seconds."—*R.A. Bell, Cleveland, Ohio.*

"Stand midway between the rails on a convenient railroad, holding the two stuck disks exactly opposite the pit of the stomach (precision is important here) and wait till something arrives. This has separated the disks in most attempts, but if it fails the first time try again."—*Casey Jones, at the round-house.*

C.1.6.17 Get a Sphere before Beginning to Polish

When a worker states, as many have, that after polishing his mirror he has a difference of 3 or 4 inches between the radii of different zones, it is unlikely that he has derived so gross a discrepancy as this merely from polishing, no matter how poor a lap he has used. Doubtless it is an earlier inheritance from grinding. When the pencil test indicates in the finest stage of grinding that the mirror and tool are out of sphericity (contact) by more than the thickness of a pencil mark at any point it is almost futile to go ahead with polishing in the vague hope that, somehow, something will "happen" to straighten it out later on. It will sooner or later become necessary to return to fine grinding because, short of almost a lifetime of hard work, the thickness of glass involved cannot be removed with so fine an abrasive as rouge. If the pencil mark is rubbed off in some places along its length, but only touched in others, it is safe to begin polishing, for rouge will handle that discrepancy, since it is of the order of size of $1/5{,}000$ of an inch—about the amount normally shaved off the glass by the rouge during the polishing operation. Sometimes it rubs everywhere except at the edge. This, if permitted to remain, will inevitably mean a grossly turned-down edge later on. (But see warning at the bottom of subsection C.1.4.1.)

C.1.6.18 When is Fine Grinding Finished?

How can I tell when the fine grinding is finished? Answer: Whittle a wedge, three inches long and $5/8$ inch thick at its blunt end. Lay it on the mirror, hold the latter in line between the eye and an electric light placed a foot or two farther away. Then slowly lower the mirror, keeping it horizontal. If the red image of the filament remains visible on the mirror until you can sight down the slant of the wedge (that is, at about a 12-degree angle), you can begin polishing—provided no larger pits than the average size remain.

C.1.6.19 Finishing With Emery

Is it worthwhile to finish fine grinding with emery? Answer: The writer finds it decidedly so. The polishing is done in much less time than the rules called for. Emery, like, Carborundum, makes pits, but they are shallower. The emery is used exactly as the Carborundum is used. Every minute's work with emery pays fine dividends in time saved during polishing, later on.

By means of abrasives finer than No. 600 Carbo, used before polishing proper is begun, the polishing period may be shortened. The American Optical Company's M 301, M 302, M $302 1/2$, M 303, and M $303 1/2$; the Norton Company's E 108 (super-finishing); The Bausch and Lomb Optical Company's $902 1/2$, 903, 904 and 906—all are suitable for the purpose. They may be used either on glass, pitch (Section A.4.4) or HCF, but preferably on pitch or HCF. On glass they will grind, on pitch or HCF polish, and it is a question whether the final goal will be sooner reached if thus preceded by very fine grinding or by coarse polishing. An underlying principle is this: On glass—that is, an unyielding backing—abrasives, coarse or fine, will grind, that is, crush or splinter (Section C.1.6.2); even rouge will not polish between glass and glass. On pitch or HCF—that is, a yielding backing—abrasives, coarse or fine, will polish, that is, minutely scour or plane (Section C.1.12.1); a piece of wood charged with Carbo, even if coarse, will give a kind of polish but will not grind. Hence, to obtain a prepolish, either charge a pitch lap with finishing abrasive or rub a scrap of HCF over the glass tool, so that the lap will not skid, lay on a sheet of HCF, then the mirror, cut around the HCF, and add one or two tablespoonsful of thin

C.1.6. Grinding

cream (thick will accomplish less) made from some of the fine abrasives—for example, M 303 1/2. This should last through the one hour of work which is recommended. Later, polish with rouge on another pitch or HCF lap. A further refinement is to follow the 303 1/2 with a half-hour or so of Levigated Alumina (made by the Norton Company and nicknamed "Levigal" by Uncle Ephram, because Levigated Alumina is a mouthful) or to use Levigal alone as the prepolishing agent. This abrasive, which does not stain, is not white "rouge," as some have called it, and is composed of Al_2O_3. It is much finer than the others—only 2 to 4 times the size of ordinary rouge particles, according to Dr. S.H. Sheib, a Richmond amateur who measured its grains.

C.1.6.20 Searching for Tiniest Pits

Pits which can be allowed entirely to escape an ordinary examination, even with a lens, may be detected if the lighting is correctly arranged. The disk should be placed within one or two feet of a strong light (*e.g.*, a 40- or 60-watt bulb), and tilted so that the reflection of the bulb appears on it. With a lens the areas at either edge of this reflection are then studied critically. Lighting at such an angle will reveal tiny pits and defects through a common reading glass which will get by when a strong lens is used with incorrect lighting. Another way is by transmitted light—through the disk from behind.

C.1.6.21 Futility of Attacking Big Pits with Rouge

"Although I have polished, to date, something like 30 hours, there seems to be no diminution of pits.

I took what I thought to be the utmost care in fine grinding, and although the rouge polishing went quickly enough, getting rid of the remaining pits is the stumbling block I have encountered. I have tried several new laps, thinking that might be the source of the trouble, but nothing seems to help."

So writes one worker. Here is a note from another: "There seems to be something rotten in Denmark, for I have been polishing my mirror for 46 hours. It looks perfect to the naked eye, but a lens shows up many pits."

These letters are typical of many. There *was* "something rotten in Denmark." Quite a few beginners (and some others) get into this fix. Pits—even large ones—are real adepts at the art of camouflage, often "hiding out" despite inspection with a lens.

We may look at it thus: Each size of abrasive in the series used has its characteristic depth of pit, this depth being roughly proportional to the size of the individual grains of abrasive. Let us assume that we have a surface

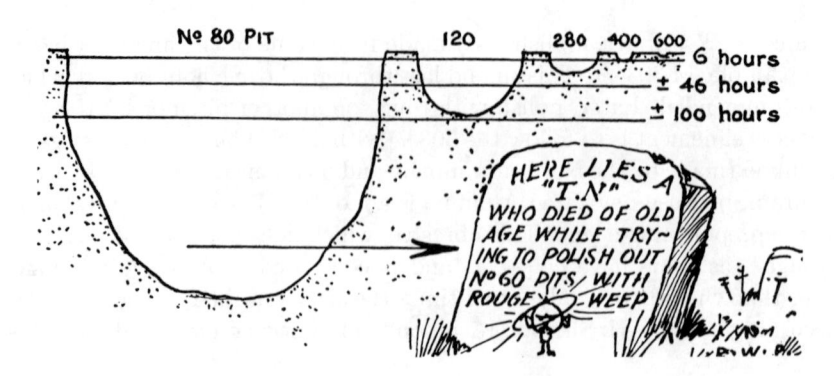

Fig. C.1.10: *Drawing by Russell W. Porter, after the author.*

on which there is a mixed collection of pits due to all of the sizes of abrasive (see Fig. C.1.10).

First, the flat level areas between the pits will come to a polish within a few minutes, and if a way could be found to grind a glass and leave no pits of any kind, the polishing job would be complete then and there. Or, if we had a series of abrasives logically filling in the big gap between the smallest available size of Carbo, and the far tinier particles of rouge (about $1/25{,}000$-inch), we doubtless could shorten the polishing operation quite considerably. But we have no such ideal series. The finest grade of Carborundum is of the order of 100 times the size of the particles of rouge. Therefore we must dedicate some six or more hours of work with rouge to planing the glass down to the level of the bottom of the pits even for those due to the finest abrasive used in grinding, not to speak of the coarser ones. The latter would not, however, be there at all if the medium-sized abrasives had been used *long enough at the right time.*

Now suppose we discover that we have some No. 400 pits but no larger ones. With rouge we could also wipe these out—probably by about twice as much work as required normally for No. 600 pits. But suppose we had pits inherited from much coarser abrasives, No. 80 or 120. The diagram shows how brave a man would have to be to expect to abolish these by polishing with the comparatively slow cutting abrasive called rouge. This is what the workers whose plaints are quoted above were doubtless trying to do.

The moral is obvious: It is more costly to let go the clearing up of pits due to *coarse* grinding than those due to fine grinding. All the subsequent fine grinding will probably leave them scarcely at all reduced in depth. And at that—and worse yet, too—for a long period of hours most of them will be playing hide and seek among smaller sizes, making the worker think they

C.1.6. Grinding 437

have gone. But they will turn up, like the cat that came back, to plague him and make faces at him after hours of polishing removes all the pits due to finer abrasives, thus causing the mirror actually to look *worse and worse,* after more hours of polishing.

Each size of abrasive requires at least an hour. An extra half-hour on the second, third, and fourth sizes may prove to be a paying investment, even if the surface does not look as if this were necessary. Remember Benjamin Franklin's story of the man who shirked the pits in his ax: "The man who, in buying an ax of a smith, my neighbor, desired to have the whole of its surface as bright as the edge. The smith consented to grind it bright for him, if he would turn the wheel; he turned, while the smith pressed the broad face of the ax hard and heavily upon the stone, which made the turning of it very fatiguing. The man came every now and then from the wheel to see how the work went on; at length would take the ax as it was, without further grinding. 'No,' said the smith, 'turn on, turn on, we shall have it bright by and by; as yet it is only speckled.' 'Yes,' said the man, 'but I think I like a speckled ax best.' "

C.1.6.22 Test for Center of Curvature During Grinding

Many have passed up Ellison's test for center of curvature (Section A.4.4) because it is "hard to perform," "awkward," and so on. If we will persist it will soon become simple enough, just as is the case in learning to do anything new. Used in coarse grinding, it will give the center of curvature within 3 inches, and in fine grinding, down to an inch or sometimes less. The test is a valuable aid. An electric flash lamp is splendid for this test, but is more accurate when both the bull's-eye and the reflector are removed. Another method is to take the mirror, wet, out into the sunlight and see where the reflected disk of light (at focal plane) is least indistinct. This test, of course, gives focal length, not radius of curvature.

C.1.6.23 Keeping Track of the Radius of Curvature While Grinding

This may be done as described by James C. Critchett. The mirror is centered on the tool and a sight is taken along its top before grinding is begun. Mark the point on some convenient wall or vertical fixture at a distance from the center of the tool equal to the radius of curvature desired. Next, measure and mark a distance below this reference mark equal to one half the diameter of the mirror. Whenever it is desired to measure the amount of concavity of the mirror as grinding progresses, the mirror is slid toward the wall one half its diameter, so that its edge rests on the center of the tool. It is then weighted so that it will not tip off, and a sight is again taken along its top. When the line of sight reaches the lower mark the desired radius has been reached. If nothing on which to make the marks happens to be at the right distance from the tool one may choose some more distant surface and increase the vertical distance-to-go in the same proportion. Further to refine this method, Critchett made a simple sighting telescope consisting of an inch lens of a few inches focal length and a small lens of shorter focus, mounted in notches on the edge of a straightedge. He fastened a needle with its point at the focus of the eyepiece and with this, which is simply a small telescope without a tube, obtained still more accurate readings. With a little planning beforehand the same lenses may later be used as a finder for the telescope.

C.1.6.24 Very Exactly Finding the Radius of Curvature

This may be done by giving the surface a five or ten minute pseudo-polish with an old lap and rouge, and then measuring by the knife-edge test. The lap need not make good contact. This method also facilitates greatly the study of pits. It works poorly at the 120 stage but better following the 280 stage.

C.1.6.25 Bubbles Between Mirror and Tool

Bubbles that form between mirror and tool may be pushed out if one wishes, simply by sliding the top disk away out over the lower one and then carefully drawing it back, then the same on the opposite side.

C.1.7 Pitch Laps

C.1.7.1 Pitch

Wegener says in *The Origin of Continents and Oceans,*

> Pitch behaves as an absolutely solid body when subjected to blows and percussions, but, given time, it begins to flow under the influence of gravity; a piece of cork cannot be forced through a sheet of pitch, but after a lengthy period its slight buoyancy is sufficient to allow it to rise slowly through the pitch from the bottom of a vessel.

He goes on to observe that pitch is harder than a candle, yet if one lays a stick of pitch and a candle horizontally between two supports, the pitch will bend of its own weight while the candle will not. According to definitions given in Maxwell's *Theory of Heat* (1872), the candle is, therefore, classed as a "soft solid," while the pitch is a very "viscous fluid." For an interesting

C.1.7. Pitch Laps

article on pitch, by F.W. Preston, see *Transactions of the Optical Society*, No. 3, 1922–1923.

Porter says:

> While many amateurs advocate coating their laps with beeswax or paraffin, or mixing these with the pitch, it is a significant fact, that Brashear, whose establishment has produced most of the finest optical surfaces in this country, used nothing but plain (strained) pitch.
>
> Cloth laps are sometimes used. Felt or broadcloth is cemented to the glass tool with hot pitch and is pressed into shape with the speculum. Polishing is done with rouge and water as usual. But the surface produced in this manner is never as good as that produced by pitch; it is slightly wavy. The poorest work of this kind is called by the mirror working fraternity, 'a lemon peel finish.' Pitch starts in work by *shearing* off the tops of the irregular elevations left by the fine grinding; the valleys between are not touched until, at the last, they—that is, the few remaining pits—vanish as if by magic. It is estimated that any effort to bring a surface nearer to perfection than a quarter of a wavelength of light, viz., $1/200{,}000$ of an inch, is time wasted. This amount—five millionths of an inch—is just detectable with the knife-edge test.

C.1.7.2 Making the Pitch Lap

Next to silvering, no part of the telescope making art has proved so difficult for the beginner as making a lap of pitch. Some have reported failure after five, ten, or as many as fifteen attempts. It is suspected that the majority of laps made in such cases would have performed well if they had been saved and used, but the available instructions have been too sketchy to enable the beginner, who may never have even seen a lap, to judge whether the one he has made is actually good, bad or indifferent. The common pitch lap may start in life like many youths—an untidy, unlicked object of pity, yet may come around in good shape after all. Or it may really require remaking.

At the risk, therefore, of insulting the intelligence and gumption of those who are gifted with more natural aptitude than the rest of us, a rather detailed outline of one way—there are many—of making a pitch lap will be presented.

Place the tool in a pail, blocking it up off the bottom with some metallic object which is fairly open, so that the glass will heat evenly—a heavy wire suitably bent will serve well. On top of it, place a second spacer. On this place the mirror. Unless thus exposed to the water nearly all over their surface, the disks when heated may expand unevenly and crack, rendering them worthless. Fill the pail with cold water just barely above the level of the top of the upper disk and heat its contents slowly. It should be given as much as 20 minutes to reach about 110° F. This corresponds to the temperature of hot bath water. Do not hurry the heating or you may crack the disks. If disks at higher temperature, or even at 110°, are exposed to icy air or drops of cold water they may crack.

Place some pitch in an old can and heat it rather slowly—if heated rapidly the fumes may take fire. Pitch does not explode, but it burns rather vigorously and it is, therefore, well to have on hand an old coat to throw over it in case of eventualities. If the heating is well managed, the disks and the pitch will reach their respective required temperatures together. Do not let the pitch boil, as this favors the development of bubbles in it. You will have some bubbles anyway, but there is no point in getting more of them than is necessary.

While the disks and pitch are heating, assemble within reach a dry towel or cloth, a bit of absorbent cotton dampened but not soppy with turpentine, and a strip of heavy paper a few inches longer than the circumference of the tool and about $5/16$ inch wider than its thickness at the edge. It should have perfectly straight edges. Provide some heavy twine or several good strong rubber bands and a little rouge mixed with water, or else some very soapy water, preferably lukewarm since even a drop of cold water may crack a warm plate glass disk.

Find a level place to rest the tool on later when pouring the pitch. It is well to use a level if your eye is not pretty good.

Remove the warmed mirror from the water and set it aside. Remove the tool and replace the mirror in the warm water, but do not permit it to go on becoming warmer. Dry the tool and be sure to dry both hands, for drops will otherwise fall on the glass and prevent good adhesion of the pitch. Take plenty of time—there is no rush. Many laps are botched because of hurrying the job.

Wrap the strip of paper around the tool, keeping its lower edge level with the bottom of the disk, and tie it on with string or snap the rubber bands around it. Another way to attach its ends is with a dab of hot pitch. The joint need not be very tight, as even melted pitch is thick. Make sure that the paper extends a uniform distance above the glass—say, within $1/16$-inch of the $5/16$-inch thickness previously suggested. Moisten the tool with the turpentined cotton.

Pour the pitch on the tool, distributing it reasonably well. If it is not finally level no very particular harm will result, but a wedge-shaped lap will

C.1.7. Pitch Laps

always be bothersome to work with. Melted pitch will shift some and partly level itself but not as water will, so it should be fairly well distributed.

Now sit down and have a smoke and cool off.

After a few minutes—perhaps five or ten or even more—the pitch, judged by a touch of a wetted finger tip (Ecclesiasticus XIII, 1), will be firm enough to stand up without the paper. Strip off the paper, or as much of it as you can easily get off, smear rouge and water or soapy water liberally on the lap and mirror so that the latter will not stick, and let the mirror lightly down. Do not slap it down, for this causes bubble-holes in the lap. It will make bad contact at first, but you still have 15 or 20 minutes in which to make it conform reasonably well, and the lap will not really harden for another hour or two.

At first supporting a part of the weight of the mirror, in order to prevent it from pressing in too deeply wherever it rests off-center and thus leaving a ridge in the pitch at its edge which will be hard to manage, move it an inch or less in different directions. This will tend to make the pitch conform. As it cools this may be done with more pressure but only for a brief time in any one exact position. If possible, do not at any time short of an hour or two lift the mirror entirely off, as its return to the soft lap may trap air under it and introduce unnecessary bubbles. If you must lift it off, replace it by sliding it on from the side, lightening its weight at the same time, so that its edge will not plow up a ridge.

During all this conforming process some pitch will continue to creep over the edge of the tool. This will do no harm.

Remain near by for an hour or so longer, occasionally moving the mirror. Even after that there will still be time to repair any big bubble holes you may have seen through the glass. Either pour melted pitch into these with a spoon, never quite filling them since an excess will be bothersome, or stuff into them little chunks of pitch softened in the warm hand, replacing the mirror and working them down level. They will not look pretty, but never mind. The pitch may not appear to give promise of joining, on account of the rouge or water already in these holes, but it will stay there quite well enough to do business. Remember that one way to make a lap is actually to dig holes in it, here and there, instead of channeling it. Hence holes, in themselves, cannot ruin it—unless the lap is practically *all* hole. It is believed that hundreds if not thousands, of newly made laps have been needlessly sacrificed by beginners because they resembled a piece of Swiss cheese. Examine the grouping of the holes *with regard to zones,* not with regard to sectors, and size up the situation. For example, holes fairly evenly distributed throughout all zones, even if all on one side of the lap, will probably do no harm, since the mirror is to be rotated during polishing and this will prevent any sectorial effects on the mirror. Judge all this by a blow of the eye and try to plug up the larger holes where needed or, on a pinch, even make compensatory holes in certain zones, which should do no harm. It is true, the holes will reduce the total area of abrading surface, but the pressure per unit area will be increased correspondingly. No matter how inexpertly you perform this preliminary balancing of noncontact spots in zonal areas, the results will show up very soon under the knife-edge test and, long, long before it matters much, you will know through that test where to alter the lap further. Softened pitch can be worked into holes at any future time. So it is not yet time to condemn the ugly-looking lap.

This wet-nursing job on an infant lap will probably demand your intermittent attention for two hours the first time you make one. Do not, even then, leave the mirror on the lap and go off fishing, as it is still partly warm (feel of under side of glass tool). Warm pitch crawls—faster too, when you are not looking—and it may in some way bring about disaster; for example, depositing mirror and all on the floor.

In addition to large cheeseholes myriads of little tiny holes often appear on a lap. These are due to bubbles in the pitch. They will not do any particular harm, but enough of them in one zone may cause unequal abrasion. As they are too small to plug up with pitch they may be compensated by making a few shallow holes elsewhere, and they will gradually fill in under pressure as the polishing proceeds. Do not be in too much of a hurry to compensate them, however, but first polish twenty minutes, or even more, and then try the knife-edge test. Regardless of the appearances of the lap, that test is what will tell what, if anything, is wrong with it, and where. If the lap is polishing evenly, leave it alone, no matter how bad it looks.

Of course, a nice smooth lap without any holes is altogether desirable, and after a little experience the worker will learn to make such a lap.

If the lap turns out to be rather thick, more than one half-inch, no particular harm will result unless the working place is hot, above about 80°, when it may give considerable trouble by slumping. It will also require more subsequent trimming up. If thinner than one eighth inch it may be slow in the matter of establishing contact, but it possibly will serve. Ellison recommends one eighth inch, which is thinner than most other workers use. A quarter inch or $5/16$ inch is a fair average. It is impracticable to specify too closely, because many of the factors—local conditions, temperatures, quality and hardness of pitch, personal working idiosyncrasies of the worker, etc.—vary. For the same reason there is little use in attempting to work out a standard of hardness of pitch for the amateur. Suffice it to say that laps having quite a wide variety of characteristics have performed well. It is best to aim at the norm but, within limits, variations need not spell disaster.

An alternative way to make a lap is not to warm the two disks at all. This is far less trouble and consumes comparatively little time—if you

C.1.7. Pitch Laps

can make a good lap in that manner. After trying it, many may vote to warm the disks. The hot pitch strikes the cold glass and hardens almost immediately, allowing the inexperienced worker little time to nurse the lap into shape.

The methods described above state only one way of making a lap. There is no law or set rule about the job. However, this way has worked well many times. After you have made a few laps you may find ways to short-circuit much of the detail, do it all in a few minutes and no longer make a botch lap requiring "pointing up." An old hand may make a lap and be working on it within 30 minutes.

The tool and lap now being entirely cold after perhaps a couple of hours (next day is better, if convenient), channeling is in order. The overhanging pitch at the edge may be trimmed off, because it may cause a turned-down edge. It is better on the same account to reduce the diameter of the lap to a quarter or eighth inch less than that of the mirror by chamfering the pitch at an angle. This may cause turned-up edge, but the latter is easier to deal with than the former. Use a sharp knife and shave with little short strokes, instead of plowing boldly ahead before you have learned how peculiarly pitch behaves. This will tend to prevent taking big chips out of the face of the lap.

For Everest's test and scale of hardness, press one pound with thumbnail (practice on scales beforehand). If a $1/4$-inch dent is made in 2 seconds the lap is very soft; if in 10 seconds it is medium (normal); if in 20 seconds it is hard, and if in 40 seconds very hard. Judge hardness only after several hours of cooling: regarding this see opposite page.

Four purposes for the channels in laps are:

1. To distribute the rouge—local pick-up sources.

2. To prevent suction. This will be demonstrated the minute they are allowed to close in—the mirror behavior then becomes cantankerous.

3. To facilitate perfect contact, by providing the pitch, under pressure, with an escape; otherwise it must travel all the way out to the edge of the lap. Hence the deeper the channels (taller the pitch columns) the quicker the response, on pressing, and so it is generally well to cut the channels clear to the tool. Beware of chipping the tool; the chips may scratch the mirror.

4. Perhaps the most important purpose is to overcome the lubricating effect of the fluid film between an ungrooved lap and the work. When two surfaces of nearly the same radius are brought together in a fluid medium, and relative motion occurs, unless the surfaces are held rigidly parallel a pressure film of the fluid is formed between

them, which tends to force the surfaces apart. The film will be wedge-shaped and thickest where the lubricant enters the film. For example, in a well-designed oil lubricated bearing, the minimum film thickness may be of the order of 0.001 inches, increasing to a maximum of perhaps 0.002 inches. Should we divide our bearing surface into facets by cutting grooves the lubricant will have open paths of escape when the load is applied. This destroys the film and permits the surfaces to come into contact.

With a straightedge and everything wet, to minimize sticking, draw a pair of pencil marks for each channel, starting with the one just off-center, as explained by Porter and Ellison. Then cut them out. Do not bear heavily on the straightedge, as this will indent the lap; and do not bear heavily on the knife, as this will be likely to press out large chips of pitch where it is not desired to remove them. Little short strokes, especially before the "feel" or knack of cutting pitch has been learned, may prove best. Do not try to cut far down at one cut; the better method is to keep enlarging the cut by increments, working on one side of the channel, then on the other, etc. Slope the sides of the channels about 60° more or less, as the pitch will not then chip so readily from the sides of the facets.

It will probably pay to practice a bit by cutting some pitch poured on a board, and thus learn something about pitch, which is rather "cussèd."

All this is a mussy job. Clean up with rags and turpentine.

In case you have to boil the pitch because it is too soft, a short boiling will have little effect. A lot of boiling is required to alter its hardness more than a little. Ellison tests the pitch from time to time while thus boiling it, by dropping a sample into cold water to cool. If this is done it should be given "several minutes" (Ritchey) to cool, else the apparent new hardness may be deceptive. The water should be "at the temperature of the polishing room" (Ritchey). There seems to be a sort of time lag in the hardening of pitch—perhaps this is an illusion of some kind—whereby it often seems to have cooled to the temperature of its surroundings, yet will become harder some hours or a day afterward. The probable answer is that it has not actually cooled to the temperature of its surroundings when one thinks it has. When Ritchey makes a lap he allows the pitch to cool "for six or eight hours."

C.1.7.3 Modifications of the Plain Pitch Lap.

These are numerous. There are almost as many as there are lap makers. Each worker, after following the rules of Hoyle in making a lap of two, usually introduces his own variations. If the variations are good this expression of individual inventiveness is good, but where variations are frowned on it may be just possible that the variation is intrinsically bad, and not that the frowner is stuck in a rut of orthodoxy.

John H. Hindle's pet lap is made of asphalt (pure Trinidad bitumen) thinned down by turpentine or one of its substitutes until it is quite soft. After the lap has been channeled he brushes on the surface one even, thin coat of beeswax, using a warmed brush of good width. The coat of beeswax should be scarcely any thicker than a sheet of paper. Ritchey also uses beeswax. In the famous treatise *On The Modern Reflecting Telescope*, etc., which is now out of print, Ritchey describes the manner of application: He strains the beeswax, then brushes it on by a single stroke of the brush. "The wax should be very hot," he says, "otherwise the layer will be too thick." His brush is made "by tying several thicknesses of cheese cloth around the end of a thin blade of wood $1\,1/4$ inches wide." Incidentally, Ritchey makes his own pitch by melting rosin, and after removing it from the stove, as turpentine is inflammable, adding turpentine to the extent of about $1/25$ of the weight of the rosin, depending on the amount of turpentine already in the lap, and stirring thoroughly. A very little turpentine will effect a big change in the hardness of the mixture.

Others have similarly painted on the plain pitch lap a coating of paraffin wax, grafting wax, floor wax and almost every conceivable variety of wax except possibly ear wax. The intended purpose of using wax is to give the glass a smoother polish, which some claim it does, and to reduce risk of scratching. Note that Ellison mixes the beeswax with the pitch, largely for another purpose (to render it more tractable), while these other workers merely paint it on the surface at the top.

Some have also painted wax or pitch in a thin layer directly on the glass tool. This will polish the mirror and may enable the worker to figure it if he is lucky, but the adaptive qualities of the thick pitch lap are then entirely abandoned, for wax will not flow as pitch will. In fact, a thin layer of pitch will flow relatively little. With wax on pitch the valuable qualities of both materials are exploited.

C.1.7.4 Straining Pitch

Melted pitch will not strain through muslin. Use cheese cloth, and double it once or twice. This removes any grit that may be in the pitch.

C.1.7.5 Speckling of lap

Why is my pitch lap all speckled over with little bubble holes? Answer: The pitch was probably boiled and the bubbles that arose from the bottom did not escape from it. Do not heat it so fast.

C.1.7.6 Pitch Flammability

Is pitch very inflammable? Answer: Take a match and try a small piece. During the melting of pitch, it is well to have a pot cover handy, to clap on in case of fire.

C.1.7.7 Channels in Laps

What is the purpose of cutting channels in the pitch lap? We often hear that they are for facilitating the spread of the rouge. This is true, but there are two other and at least equally important reasons. The channels admit air, and thus break up sucking or adhesion of the mirror. But, most important of all, they permit contact to be established by cold pressure, for they provide the pitch with a place to escape to when it is slowly deformed under pressure. Without them the pitch would have to flow clear out to the edges of the tool. With this in mind, it should be apparent that shallow, halfhearted channels do not greatly facilitate good contact. Cut them down to the glass itself—they will fill in rapidly enough by slow flow, while one is working.

C.1.8 Nonpitch Laps

C.1.8.1 The HCF Lap

The HCF Lap has been widely used since Chapter A.3 of the present work was published in 1928. Many workers swear by it; some swear at it. When the correct conditions are obtained it gives an excellent polish, as there is much less tendency to cause zones than there is with a channeled pitch lap, probably because the pattern of the facets is finer. The more enthusiastic early claims for much more rapid polishing than pitch affords have not been altogether borne out. Russell W. Porter reported in *Scientific American,* July, 1932, page 53, that "comparison tests have been made here (*i.e.,* at the optical shops of the California Institute of Technology, Pasadena, California) on a number of lenses, half of them polished with a cast iron tool covered with the comb foundation and half with the ordinary tool of pitch facets painted with beeswax. Aside from the differences in the tools these lenses received identical treatment on the machine." The results were that the length of time required to produce a complete polish was about 10 hours. The rapidity of polishing with the two tools was about the same. These tests were made on a group of $7^1/_4$-inch lenses. At first the HCF lap was made of three layers of HCF on the cast iron tool, but the polish came up unevenly. Three more layers of HCF were added, and the six then gave a sufficient cushioning effect to permit good contact to be

C.1.8. Nonpitch Laps

secured after cold pressing. This is an unusual way to use HCF—a sort of glorified club sandwich of HCF sheets.

The tests mentioned above are really tests of beeswax in the form of HCF, against beeswax painted on a pitch lap. But the time consumed was nearly as long as that required in plain polishing. Others have claimed faster results when polishing on HCF than on plain pitch.

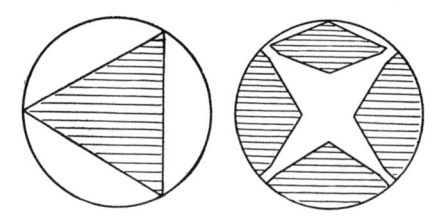

Fig. C.1.11.

HCF may be used in several ways, the one in Chapter A.3 having been recommended by Everest because, after experiments with many kinds of HCF laps, he believes that one superior. Some of these ways are:

1. HCF applied directly to the tool. (*a*) Rub beeswax (a scrap of HCF) over the glass, to charge it, lay on a piece of HCF, add mirror and cut around it, rub soap on its face and go to work. The HCF will not skid sidewise but may be lifted off at any time. (*b*) Another way, also cleanly to make, is: Heat tool to temperature of a hot potato, lay a sheet of HCF on top, then mirror, and add 3 to 5 pounds weight. Let cool. Trim HCF and go to work. If potato was not too hot or too cold, HCF will not be harmed, yet will adhere strongly and permanently.— B.R. McCrary's lap. D. Everett Taylor attaches HCF to the tool by means of what he calls "pitch paint"—pitch cut in acetone.

2. HCF cold pressed deeply into a pitch lap of ordinary thickness and left there. (*a*) On unchanneled pitch. With this lap good contact will be good luck, since the pitch must crawl a long way sidewise to give contact on top. Note: If the pitch and under side of HCF are heavily soaped beforehand, the HCF may be stripped off later, if desired; or channels may be cut through both, leaving HCF on facets. (*b*) On channeled pitch. The HCF bridges the channels. Fine contact until channels flow full.

3. Whole HCF pattern transferred to a pitch lap. Soap pitch and under side of HCF, cold press HCF in deeply. Strip off. In channeling, follow the pattern on the pitch, entirely removing 1 in 7 or 8 rows.

Begin one row off center. Cross channels will be diagonal, following the HCF pattern but beginning two rows off center. A third diagonal set of channels is hardly necessary, hence facets will be diamond-shaped. This lap will drag hard and even screech out loud. Combines qualities of pitch on top parts and the HCF "wedge" at forward edges of depressions. Is a pretty lap to look at but is sometimes tricky. Produces too much heat for final fine work.

All of these laps, if used at all, will bear close watching.

Joining HCF sheets edge to edge is possible though difficult. They come in two widths, 8 inches and $10\,1/8$ inches, and if a larger mirror is to be polished two sheets must be joined. If they are merely butted without being welded the pitch will gradually work up between their edges, spread out near by and, because of the added abrasion, cause a depressed zone to form. Two sheets of HCF are therefore slightly overlapped and the two are cut at one stroke (to insure an accurate fit). They are then pressed against the respective sides of a piece of warm metal, the metal is quickly withdrawn and the HCF sheets are instantly slid together before cooling. To do this skillfully one must almost be a prestidigitator but it can be done, and neatly, too, after some practice.

Special HCF laps for figuring are described as follows by James C. Critchett:

> While experimenting with superimposed pieces of HCF, I hit upon the following principle, which has so far worked out nicely in securing spherical surfaces, preliminary to parabolizing: When a polygon of HCF is placed upon an HCF-covered tool, so that the edges of the polygon form chords of the tool circle, the abrasion delivered by the polygon is eased out to the edge of the mirror without leaving narrow zones or any zones, if ordinary care is taken. Thus a high center can be rapidly and evenly worked down with a triangular secondary lap, as shown in the first sketch (Figure C.1.11). Following this, if so justified by the resulting figure, the mirror is worked with a square or rectangular piece of HCF of which the edges are chords of the tool. Low edges can be nicely brought up with pentagons or hexagons, etc. A partial chord will have the same abrasion-distribution effect, more locally applied. In case of a low center one can make a secondary lap by cutting a circle of HCF to fit the top of the tool, quartering it and trimming it as shown in the second sketch, thus getting the chord action. By cutting segments from the edge zone one can also use these pieces for bringing up a low edge, thus securing hyperboloid rectifiers.

C.1.8. Nonpitch Laps

The second sketch is thus a modification of Ellison's star lap. The ends of the pieces should be cut off enough to allow some shifting between edge and center of the tool.

C.1.8.2 The Paper Polishing Lap

This lap turns up, every now and then, as a substitute for the pitch lap. Some defend it stoutly but the majority of fine workers condemn it roundly. What Ellison thinks of it is reprinted from *English Mechanics*, Feb. 1, 1929. A certain inquirer, he states, "need not regret the absence of any notice of the paper polisher from my book and my letters on speculum working. He evidently has no experience of it. It is just one of those things which appeal to the beginner and especially to the lazy one, who eagerly seizes on the idea, exclaiming, 'How lovely and simple.' It is such a contrast to the nasty messy pitch.

"In actual fact the paper polisher is everything that a polisher should *not* be. It has no yielding surface to embed and swallow up particles of grit. Consequently there are very few samples of rouge which will not leave a mass of ugly scratches on the mirror, if used with a paper polisher. Another objection is even more serious, and is fundamental. We know, I suppose, that the conclusion of fine grinding, if carried out with proper care, leaves the mirror and tool with absolutely coincident surfaces; in other words, each is a segment of a sphere of the same radius. Now, if we paste a sheet of stout paper on the tool, it *is no longer of the same radius* as the mirror, but has a radius longer by the thickness of the paper. 'Oh, but what does the thickness of a sheet of paper matter?' It matters everything. We are dealing with quantities of the order of a millionth of an inch, when we figure a mirror, and therefore 1/100 inch, or the thickness of an average paper, is a gross error. The net result is that the curves of mirror and tool are no longer coincident, and we infallibly get a turned edge. The paper, saturated with paste, is a hard and unyielding layer, which no pressure will persuade into coincidence with the mirror. The value of pitch lies just in the fact that it will yield to pressure, but slowly and regularly. A blow breaks it; but it flows under long continued pressure, and takes any required form that the operator has patience to give it. The first lesson which a beginner must learn, before he ever can hope to be a successful mirror worker is, **do not shrink from the pitch polisher. Master it.**"

Bell, *The Telescope*, page 71, states that "cheap lenses are commonly worked on cloth or paper laps but that they leave microscopic inequalities which scatter light." He adds that all first class objectives and mirrors are polished on pitch. All good opticians know this fact.

C.1.8.3   Laps in Hot Places

In the tropics, pitch may refuse to perform well, due to the heat. Vard B. Wallace used rosin tempered with beeswax while making a telescope in Guatemala. Professor George H. Hamilton of Jamaica, British West Indies, author of *Mars at Its Nearest* and the maker of two telescopes described in the *Scientific American,* January, 1928, has had considerable experience with laps in warm climates.

C.1.8.4   Substitutes for Pitch

As soon as the first edition of the present work appeared, several amateurs experimented with other substances than pitch for laps. H.L. Rogers and a co-worker coated a pitch lap with ordinary Johnson's Liquid Floor Wax, giving it two coats (very even and very thin). Rogers states that a fine polish was obtained. A variation of this was to dry the liquid wax over a gas stove. The lap and mirror were kept in a bowl, just covered with water, when not in use, thus maintaining the lap in perfect condition.

C.1.9   Polishing Agents

C.1.9.1   Rouge Size

If rouge is left in water 10 to 15 minutes the particles remaining in suspension will have an average diameter, according to Beilby, of $1/50{,}000$ to $1/30{,}000$ inch. This compares roughly with the average wavelength of light. Beilby believed these particles were actually aggregates of cohering units which would probably be reduced under work to finer form. The microscope will not resolve scratches made by an abrasion finer than $1/100{,}000$ of an inch in width, unless strong dark-field illumination is employed, and a wavelength of violet light—the shortest of the visible rays—is just about $1/100{,}000$ of an inch.

C.1.9.2   How to Make Rouge[†]

There are but three iron oxides which interest us—ferrous oxide, FeO, which is an intermediate oxidizing in air to ferric oxide, Fe_2O_3, which is red rouge. Fe_2O_3 can be changed to ferrous ferrite, $FeOFe_2O_3$, which is black "rouge."

Next, common rouge and how it is made:

[†]By Horace H. Selby.

C.1.9. Polishing Agents

1. Ferrous sulfate, copperas, $FeSO_4$, can be ignited in air to form ferric oxide. Product bad. Reaction:

$$4FeSO_4 + O_2 \rightarrow 2Fe_2O_3 + 4SO_3$$

2. Ferrous hydroxide, $FE(OH)_2$, can be made from a filtered solution of ferrous sulfate, $FeSO_4$, and ammonium hydroxide, ammonia water, NH_4OH. The $Fe(OH)_2$ is filtered off, washed, dried and ignited as above. Product good. Reactions:

$$FeSO_4 + 2NH_4OH \rightarrow Fe(OH)_2 + (NH_4)_2SO_4$$

$$4Fe(OH)_2 + O_2 \rightarrow 2Fe_2O_3 + 4H_2O$$

3. Ferrous oxalate, FeC_2O_4, can be made from ferrous sulfate and oxalic acid, $H_2C_2O_4$. The two solutions are filtered and mixed hot. The precipitated yellow oxalate is washed, dried, and ignited as above. Product good. Reactions:

$$FeSO_4 + H_2C_2O_4 \rightarrow FeC_2O_4 + H_2SO_4$$

$$4FeC_2O_4 + 3O_2 \rightarrow 2Fe_2O_3 + 8CO_2$$

If ignited out of contact with air, ferrous oxide, FeO, results:

$$FeC_2O_4 \rightarrow FeO + CO_2 + CO$$

There are other methods, but they are possible only in a laboratory.

Now for ferrous ferrite, $FeOFe_2O_3$, black "rouge." This can be made by heating the Fe_2O_3, obtained as above, in an atmosphere of hydrogen or carbon monoxide, H_2 or CO.

Reaction:
$$3Fe_2O_3 + H_2 \rightarrow 2(FeOFe_2O_3) + H_2O$$

or

$$3Fe_2O_3 + CO \rightarrow 2(FeOFe_2O_3) + CO_2$$

At 400° C the product is soft, polishes slowly, leaves no sleeks with a soft lap and has a density of 4.86.

At high temperatures—1000° C—it is harder, polishes more quickly, has density of 5.05.

Both are octahedral crystals, black and magnetic. This is the most stable iron oxide.

Now to return to the ferrous oxide, FeO, which is unstable in air. If it is heated to 1570° C (approx.) immediately after manufacture, in contact with oxygen, it will melt and upon cooling, will form red rhombohedral plates of Fe_2O_3, rouge, which have a density of 5.19. This must be powdered.

In the first methods of making red rouge, Fe_2O_3, low ignition temperatures give bright, fine rouges which powder well and polish slowly. High temperatures (600–1100° C) yield dark, dense, coarser rouges which are less friable but which polish faster. Such products are not called rouge, but crocus and are used in the foundry for metal polishing.

C.1.9.2.1   Editor's Note

As the attentive reader will have noted, there are several points in the note on the preceding page at which additional instructions might have been inserted. For example, just how to "ignite out of contact with air," also how to handle the low and high ignition temperatures mentioned.

Accordingly, it was suggested to the author, who is a chemist, that he expand somewhat on these points. His reply was that most of "those who do not find it possible to make rouge from the data given would only bungle the job if told how to do it. Probably," he continued, "a certain background of laboratory experience is needed, so why not head the story 'For those chem. hounds alone who want to experiment with rouge making, we give this dope. For others, if it isn't clear, better not try—it is cheaper to buy rouge, anyway.' "

No doubt this is true, but it still is not entirely in accordance with the nature of human nature. Past editorial experience with the human race in general indicates that many will not take such a no for an answer, especially where the situation contains something of a challenge. Hence the likelihood was foreseen that, down through future years, many inquiries would be addressed to author and publisher, each asking for "full details" on rouge making laboratory technique, and so the matter was referred to Dr. S.H. Sheib of Richmond, another chemist-telescope maker, for opinion. He replied, "I think Selby is exactly right. Give them warning that rouge making is not as easy as it looks and that, unless they have a year of lab. work, preferably two, they had better not tackle it. If then they insist on doing it, let them prepare for it by studying in the lab. This sort of work must be done in a properly equipped laboratory. Take, for example, igniting out of contact with air. That means heating in a metal bomb with a tightly fitting screw top; or else, if every trace of air must be removed, some sort of refractory container provided with air inlet and outlet tube by means of which the air may be swept out by passing some inert gas through the container. It is not a job for a beginner. This thing isn't like silvering, for which one requires only a pan and a couple of tumblers. Here you need pretty complete and expensive equipment—gas generators, combustion furnace, etc., and a fundamental working knowledge of chemistry."

Doubtless by this time some are sure they want to try to make rouge. Hence, if you are one of these, good luck, and may no hard-hearted cynic label your product road ballast.

C.1.10 Another Method for Making Rouge

"Dissolve steel wool in nitric acid, filter and mix with a filtered solution of oxalic acid. Precipitate oxalate of iron by heating. Filter, dry, and heat to igniting point. Result, a very fine, dark red rouge."—*H.A. Lower, San Diego, California.*

C.1.11 Breaking-up Rouge

"Boiling lump rouge vigorously for an hour or so will reduce it to fine particles."—*Winston Juengst, Rochester, New York.*

C.1.12 Polishing

C.1.12.1 Polishing, Theory

"The process by which a ground or smoothed surface is turned into a polished one has been the matter of a good deal of debate," says F.W. Preston in *Glass Industry*, (Feb., 1928). "Some contend that the ground surface is liquified by the drag of the polisher and, as it were, smeared about like butter on bread. Others contend that polishing is really just an exceedingly fine grinding operation."

Amateurs who are of a more or less investigative turn of mind and who have access to a large city or university library will find in the files of the *Transactions of the Optical Society* (London) some interesting food for thought on this much-debated question: Exactly what actually takes place during the polishing operation? Is it simply a case of ordinary abrasion on a finer scale, or is it something else—perhaps molecular flow? Discussions of this question are to be found in several of the earlier numbers of the journal mentioned, most of which The Editor found to be available from the Optical Society. These journals came rather high, averaging about two dollars an issue, with a 25 percent duty to pay the postmaster at this end. However, if one wishes to obtain them, inquiry should first be made direct to London, regarding the availability and price, postpaid, of each individual issue, for the prices are not uniform. Most readers will, however, be satisfied with a summary of these several theories of polish, which is quoted from a lecture by Dr. L.C. Martin of the Imperial College of Science and Technology and which was published in the *Journal of the Royal Society of Arts*, August 12 and 19, 1927, under the title "Recent Progress in Optics." The Editor has taken the liberty to insert in Dr. Martin's statement, within parentheses, the exact references to the *Transactions* wherever he mentions papers previously published in them.

"In 1907 G.J. Beilby described researches on the nature of the polishing process (*Transactions Opt. Soc.*, Volume 9, 1907, page 22); he distinguished this as very different from grinding. Herschel had beleived that the final polishing as not merely a very fine grinding, the direct continuation of the process by which, using finer and finer grades of emery, the outstanding irregularities of a surface are worn away by the actual removal of material. Beilby showed, however, more especially with regard to the polishing of speculum metal, that the polishing action consisted, in part, of a flowing action which would cover over the irregularities of the surface. The skin of the material, he concluded, owed its nature and stability in all cases to the surface tension, a conclusion markedly different from that of Lord Rayleigh, whose view in 1901 was that the polishing of glass consisted of an almost molecular wearing away of the highest parts of the surface, and who had supported this view by proving that a thickness of glass equivalent to six wavelengths of light (about $1/10{,}000$ inch.—Ed.) was removed during the polishing, as distinct from the fine grinding process.

"J.W. French, writing in 1916 (*Transactions Opt. Soc.*, Vol. 17, page 24), suggested that the surface layer of a polished piece of glass consisted of a portion which had been caused to flow, more or less, under the strong surface forces of the polishing operation. This layer, which he termed the 'Beta layer,' was considered, from evidence based on the study of fire cracks, to be about eight wavelengths deep.

"In more recent work by the British Scientific Research Association the depth of the deepest pits in a finely ground surface was examined microscopically. The results showed that irregularities of the order of six to eight wavelengths or more are present in such surfaces, thus indicating again that polish is only complete when the surface is removed to the depth of these deepest pits, when Lord Rayleigh's observation is borne in mind.

"F.W. Preston, in 1921 (*Transactions Opt. Soc.*, Vol. 23, page 141), expressed the view that the surface of finely ground glass is of the nature of a 'flaw-and-fissure complex'; in other words, the surface skin is a layer of finely 'cracked' material. The polishing action would then consist in the removal of the fissure complex. In such a 'grey' surface there are probably (so it is suggested) a certain proportion of fine cracks slightly wedged open

C.1.12. Polishing

by the displacement of parts of the material. This is a possible cause of the strain first observed by Twyman (*Proceedings of the Optical Convention of 1905,* page 78—same address as Opt. Society; this is a book, not a periodical.—Ed). The action of polishing allows these fissures to close up; when their width becomes much smaller than a wavelength of light they cease to be visible."

In contradistinction to the various molecular flow theories of the majority of British theorists, Dr. Elihu Thomson states below his theory of the nature of the polishing operation. Dr. Thomson is Director of the Thomson Research Laboratory of the General Electric Company at Lynn, Mass., where fused quartz was developed and is made; he has been an amateur worker in glass for 60 years, has also made optical surfaces of diamond, and has made objective lenses up to 10 inches in diameter. His theory is reprinted, by permission, from *The Journal of the Optical Society of America and Review of Scientific Instruments,* Vol. 6, No. 8, October, 1922. He states:

"The problem of how it is that, for example, a glass surface which has been smoothed or finely ground can, by proper means, be polished not only so as to be invisible ordinarily, but so that under the severest tests it shows no diffusion of light (as of the Sun's rays falling on it) has at times engaged the attention of the ablest physicists. The late Lord Rayleigh studied the matter and his paper (Lord Rayleigh, *Proc. Roy. Inst. Gr. Britain,* March, 1901; *Trans. Opt. Soc.,* 19, Oct. 1917) on the subject is well known. He properly explains the polishing process on the principle of removal, by a process similar to grinding, of the high points of the surface, and progressively so until the whole ground surface has been cut away, but the cutting is by an action so fine that the grain produced is beyond the power of resolution by a microscope or other powerful optical means.

"It is the purpose here to show that while this view is measurably correct it does not go far enough, and that the polishing is a unique mechanical process; a self-regulated planing down of the surface to a real level without even the finest scratches or other character which would lead to diffusion of any light falling on the surface.

"Some have most erroneously tried to explain the result of the process, by assuming that the glass has, during the polishing, actually flowed; or that there was some peculiar plastic condition brought about which allowed the glass surface being polished to take on the characteristics of a liquid surface. There is no need for such hypotheses and no validity in such assumptions. This will be made clear.

"In burnishing of plastic metals by a hard burnisher there is, of course, such flow, but with hard, brittle, nonmalleable materials like glass the process is decidedly not like burnishing.

"Glass may receive an optical polish in either the wet or dry way. Other materials of a brittle, nonmalleable nature are dealt with similarly; such are quartz, agate, calcspar (Iceland), and many jewels and minerals.

"In the manufacture of plate glass the ground surfaces (the last, or smoothing stage, being often called *mud ground*) are not worked by grinding to so fine a grain of surface as in the better class of accurate optical work, and the polishing is done by runners of felt charged with rouge (crocus) and water moved over the plate by machinery. The result is that the surface obtained is not an optical one; it has a smoothness and polish similar thereto, but is wavy throughout, as can easily be discerned by a skilled eye in regarding the reflection of an edge from such surface; and, of course, by other simple tests. It is neither optical in the large or small elements of surface. The yielding felt runners have swept out indiscriminately the hollows, small and large, and have not held the surface to a definite figure. Similar yielding polishers are used in finishing the very irregular surfaces of cut glass. The cheaper kind of lenses, where accuracy of figure is not needed, are often cloth polished, a process which, if carefully conducted, gives a result intermediate between the plate glass surface and the true optical surface, such as is obtained by a pitch polisher with rouge and water. The considerations as to the true nature, the mechanics, of the polishing process are applicable to all such cases, but will be given in connection with the pitch polishing, most usual in good work. They apply, too, to the case of dry or paper polishing with paper-faced tools charged with tripoli (diatomaceous earth), a method of polishing which has been used to some extent in France for medium grade lenses.

"In rouge polishing with pitch for a carrier, as is usual, the surface of the pitch is molded to fit the glass and is divided (usually) into small square facets by grooving. It is worked over the glass, or the glass worked upon it, by movement in all directions on such innumerable paths are given that no definite course is repeated. This is essential to the best result.

"The conditions as found, in successful work, are as follows: The rouge, though very hard, is friable and breaks down to a very fine powder. Too hard (nonfriable) rouge will tend to fine scratches. These scratches are not like grinding or crushing, but are smooth-bottomed grooves, discoverable by a magnifier.

"The pitch is at all times yielding. It is made so by tempering and testing. If too hard, it tends to cause fine scratches all over the surface which is being polished. These, with very hard pitch, may resemble grinding, but ordinarily they show no crushing, but are smooth cuts.

"In grinding, on the contrary, the surface is crushed, while in polishing it is clean cut. Smooth cutting is the rule. The polishing is indeed a kind of planing process; the particles of rouge set themselves into the pitch surface

C.1.12. Polishing

and cut smoothly; they do not roll or grind. There are millions of fine planing or cutting edges at work fixed in position by becoming, at least temporarily, embedded in the pitch surface, which readily yields to receive them. They make smooth cuts as can readily be seen by examination of the scratches when the pitch is overhard or the rouge too hard and nonfriable. Good rouge is friable without apparent limit, and rouge washed out of a used polisher may be so fine as to float for days in colloidal solution.

"All the above considerations are fairly well-known and recognized, but there is one additional condition or circumstance which, so far as the author knows, is worthy of recording, no attention having been hitherto drawn to it.

"It is this: by the very nature of the case the particles which are doing the cutting in polishing are all *automatically adjusted*, in successful work, to cut to the *same depth* during any stroke. The yielding nature of the pitch surface not only ensures this, but makes it a necessary consequence, for any particle of rouge riding higher than another is at once depressed to the proper level by sinking into the pitch surface. The innumerable cutting edges of all the particles reach a common level, and with motion of the polisher in all directions, and cutting smooth (no crushing or grinding) the result cannot fail to be what it is, an optical surface without grain or irregularity. The rouge is friable without limit, so that the polishing particles may, in the process, become finer and finer. With felt, cloth or paper as a carrier for the polishing powder, the effect is much the same; the particles are held to position when cutting, as planing tools. Even fine washed Carborundum will polish glass if held in the surface of soft wood or cork, and the author has even produced a fair polish on a glass lens by a soft metal tool charged with fine Carborundum. In such case, the polishing takes place in a few seconds, but the technical difficulties are very great. In dry polishing, a sheet of paper is pasted down on the surface of the polishing tool, and a special high-grade pure paper, rather heavy and uncalendered, is used. This is charged by gently rubbing its surface with a lump of fine tripoli selected for the purpose, the fine silicious skeletons composing which constitute the polishing powder. The first application to the smoothed surface, as of a lens, which surface has the fine grain usual in such a case, is to show innumerable fine scratches, crisscrossing in every direction. They are, however, smooth scratches. As the work goes on, the tripoli works down to finer and finer conditions, while the polishing comes up gradually, no new application of the powder being required after the start. It is manifest that here, too, is the condition of smooth cutting and particularly a self-adjustment of cutting depth, owing to the yielding character of the paper surface, so that at the end all the cutting is done in one surface of movement. It is believed that this dry paper process is much less used than formerly. It cannot be expected to yield the high accuracy that may be obtained with the wet pitch.

"It is thought that in pointing out the mechanics of the polishing process, and more especially the smooth cutting and self-adjustment of cutting particles above described, the interesting process of the production of an optical surface may be relieved of something of the mystery which has been its accompaniment.

"The author has drawn upon an experience of more than fifty years in occasional working of optical surfaces on glass of many kinds and on media, such as crystal quartz and fused quartz, Iceland spar and others.

"The amount of material removed from the surface under treatment is, of course, seen to be almost infinitesimally small per stroke, and it is only by the long continuance of this action that at last there is a sufficient removal to secure an optical surface. Time is saved by carrying the fine grinding or smoothing as far as possible before applying the polisher. As Rayleigh has stated, and it is, of course, the common experience, polishing begins on the highest or most elevated parts of the surface, seen only under a magnifier, and these are removed while the polished spots widen out, and, if the surface has been well-prepared, or *bottomed,* as it is termed, spread to include the whole surface. If the surface has not been well-bottomed there will remain pits which the slow planing action of the polisher is incompetent to remove in reasonable time, and if the polishing is continued too long the surface is more than likely to have lost its truth, or has been seriously deformed. This, however, depends on the polisher itself keeping its form. Too soft pitch is a guard against polisher scratches from particles of grit, but not conducive to accuracy. Accuracy can be helped by remolding the polisher at intervals by slight warming of its surface and application to a true surface of the same character as that being produced, while moistening the said surface to prevent adhesion.

"No matter what degree of smoothness has been attained in polishing, the continued smooth removal of the glass surface goes on as long as there is rouge, pitch and water applied; a fact which is, of course, taken advantage of in parabolizing a concave astronomical glass mirror."

In addition to the references to the back files of the *Transactions of the Optical Society,* quoted above, the scientifically inclined amateur will find valuable material in the same periodical, No. 1 of Volume 18 (1917), "More Notes on Glass Grinding and Polishing," by James Wier French; also in Volume 19 (1917), No. 1 (Oct.), "Polish," by Lord Rayleigh; and in No. 3 for 1925–26, pages 181–189, Preston on "Nature of the Polishing Operation." See letter in *Nature,* London, Sept. 4, 1926, page 339.

H. Dennis Taylor, of Taylor, Taylor and Hobson, Ltd., states in *Transactions of the Optical Society,* Volume 21, page 82, "If the rouge imbeds

C.1.12. Polishing

itself thoroughly in the pitch surface where it touches the glass, so that the said surface appears of a rich orange red, *not* glazy when dry, but of a matte surface, then we always know that both the polishing and figuring are going on satisfactorily. In hand polishing the sweetness and evenness of the frictional resistance is then most noticeable, whereas if the pitch surfaces in contact with the glass appear to be glazy in appearance when the polisher is dry, we know that the polishing and figuring are not going on satisfactorily; we then expect trouble and generally get it. In the case of hand polishing, a polisher in this condition is apt to suck and only move in jerks.

"Still worse is it when the pitch surface in contact with the glass refuses the rouge and shows a black, glazy appearance; then we usually get the glass covered with fine scrubs, while it refuses to take a good figure. This sort of polisher clings hard to the glass and can only be moved by hard jerks."

To The Editor, at least, this statement by Taylor would seem to bear out Dr. Elihu Thomson's theory of polish. While it is possible to polish an optical, glass surface on a pitch lap from which the rouge has been washed, an attempt to do this with a lap which has never been armed with rouge will possibly prove illuminating.

C.1.12.2 Grabbing

For the first few minutes after I begin my daily polishing, the pitch lap acts as if possessed of pure cussedness, gripping the mirror suddenly and letting go unexpectedly. No two strokes are alike, nor are they even. Answer: Have you cold pressed the lap, as explained in Section A.4.6? If so, and if the "acting up" persists after a few strokes, stop and cold press again. Pitch laps usually act this way at first while still cool, even if they are made right. They may soon settle down to business, if the pitch is not too hard. In all cases, however, the "bad acting" may indicate poor contact. This is a point about which it is hard to be definite. Whatever you do, avoid haste, and think the matter out. In cold pressing, use 10 or 15 pounds weight.

Mr. Porter adds to the above.

> Keeping good contact is the secret of avoiding zones. Whenever the glass is removed from the lap, evaporation takes place, lowering the temperature of the lap and altering its shape. Moral: Have patience to polish for long intervals, and *always* cold press after exposing the lap. I have often safely left my glass on the lap, between polishings, for a day or so, simply by swathing it in wet compresses and covering with an inverted pan to retard

evaporation, watching it from time to time to see that the rags do not go dry. After these intervals, the glass is always found in perfect contact, giving the operator by the feel of the stroke a sense of assurance that every part of the lap is doing its work.

C.1.12.3 Sleeks

"Sleeks" is a term applied by glass workers when, for some unknown reason, a number of very fine scratches appear on a glass surface during the polishing. They are so fine as to be indistinguishable unless the glass is held up to the light and its surface viewed by reflection. It is said by English opticians that if the pitch lap is thoroughly dried, a few moments of vigorous dry polishing will remove them. My experience has been that they appear only during the last stage of figuring. Another method is to start polishing on them with short strokes and scarcely any rouge, and gradually increase stroke and amount of rouge.

C.1.12.4 Scratches From Rouge

A large part of the rouge on the market is so inferior in quality that experienced amateurs are following Ellison's advice (Section A.4.6) and washing their rouge before using it. There are various ways. S.H. Sheib writes: "I lay down two pieces of cotton flannel, each 8 inches square, fluffy side up, and on this put about 4 ounces of moist (paste) rouge, tie it up in bag shape and work it under water, à-la-Ellison. Another way is to stir up 4 ounces of rouge in 3 pints of water, allow it to settle 1 minute, then carefully pour the liquid part into a two-quart jar. Next let this settle five minutes and pour off the liquid, using the residue at the bottom." This last liquid may also be saved and its content of very fine rouge, after a few days' settling, used as finishing rouge. Some workers do coarse polishing with American Optical Co. rouge, and finish with Bausch and Lomb's best. Harold Lower is partial to black "rouge," which he finds very free from scratchiness.

C.1.12.5 Stuck Disks

Should the disks become stuck together, try a wooden mallet. A hardwood lever with chisel-shaped end, inserted at edge notch, and pried carefully over a fulcrum, may lift the mirror. If necessary, melt out the pitch in water warmed and heated very slowly and make a new lap.

C.1.12. Polishing

C.1.12.6 Test for Complete Polish

"We note that not more than five percent of mirrors sent to us for silvering are fully polished. The following test, used by us on our own work, might be a revelation to many.

"Take a small reading glass or other simple lens and, using it as a burning glass, throw a bright spot upon the supposed polished surface. Either an electric light or the Sun may be used, though the latter is much the better. An insufficiently polished surface which seems quite fair using a magnifying glass in the usual manner will appear as a gravel bed even to the naked eye. If the test seems too drastic, try it out on a professionally polished surface, such as a B. and L. prism. If the prism surface is clean and free from grease, even the most intense 'burning glass' spot can scarcely be detected by front face reflection." —*Tinsley Laboratories.*

C.1.12.7 Detecting the Most Minute Pits and Scratches

"Place a lamp at the right of center of curvature, where the pinhole normally goes in the knife-edge test. Then place the eye where the knife-edge would go. Move the head so that just the edge of the cone of reflected rays is seen." —*J.H. Hindle.*

"I have found when looking for small pits or scratches without the aid of a lens, that these are more easily seen if one looks through the glass from the back. I have found fine scratches from the back that I have never been able to detect from the front until they showed up in the polishing process. Apparently light is more readily caught by a pit or scratch and sent into the glass than reflected out of the pit. If I am not mistaken this is the reason why the inside frosted light bulbs allow more light to pass out than did the old outside frosted ones." —*John Tom Hurt, Rice Institute, Houston, Texas.*

C.1.12.8 Cold Pressing

Ritchey says, "When it is sufficiently pressed the surface appears uniformly smooth and bright." Incidentally, this points out the value of keeping the back of the mirror relatively free from blindfolding obstructions, so that one can see what goes on at all stages of the polishing.

C.1.12.9 Prolonged Cold Pressing

"The mirror may safely be left in contact with a pitch lap for several days if the rouge is moistened with a 50 percent solution of glycerine and water. The glycerine prevents the evaporation of the moisture, and the local humidity will govern the amount required. This suggestion is particularly valuable if one is able to polish or figure the mirror rather infrequently. While the glycerine keeps the surface of the pitch moist, the glass, by its own weight alone, will gradually make perfect contact, without unduly squeezing down the pitch and springing the glass caused by the use of weights."—B.L. Souther.

C.1.13 Figuring

C.1.13.1 Correcting Turned-up Edge

This is easy to deal with. Invert tool and mirror and work on the edge of the mirror with the tool, but do so with discretion; that is, do not apply the work at the very edge, for this probably would "bevel" it at the very outside without producing a uniform job. The best way is to make one or possibly two rather rapid revolutions around the pedestal, taking elliptical strokes in each of which the tool is drawn toward you at about three-fourths of the distance from center to the edge, and pushed back along a path halfway or less toward the edge. A very little of this treatment will go a long way, and only a bit more will convert the turned-up edge into a badly turned-down edge.

C.1.13.2 Correcting a Hole

Ellison gives practical directions (Section A.4.7) for doctoring a hill, but does not tell how to get out of a hole. One way is to remove the lap opposite the hole. It is necessary to scrape off only the thinnest possible depth of pitch. Then, the amount scraped off will fill in automatically by the time the hole is cured; but if not, the lap should be cold pressed before going on. Shaving a lap just opposite any kind of depression is both a logical treatment and one that is usually effective.

Fig. C.1.12: *How a hole looks with knife to the left. Invert the book and you have the appearance of a hill. Drawing by Russell W. Porter*

C.1.13. Figuring

C.1.13.3 Correcting a Hyperbola

Working with a pitch lap, H.L. Rogers states that he has successfully reduced a hyperbola, simply by painting a ring of rouge on the outside inch or so of the tool, keeping the remainder of the tool just wet enough to slide nicely, and using short strokes.

C.1.13.4 Forestalling Turned-down Edge

Ritchey forestalls turned-down edge by "diminishing the area of the squares around the edge of the tool, by trimming their edges." On this, Porter comments, "It is O.K. Have done it frequently to advantage."

C.1.13.5 Turned-down Edge

If turned-down edge has been inherited from fine grinding the chances are that it is too gross to cope with by means of as fine an abrasive as rouge, short of vast labor, and a sad return to No. 400 or 500 Carbo is indicated. The pencil test, if carefully applied, should have brought the disks close enough to sphericity to permit rouge to deal with the remaining discrepancies within reasonable time.

One makeshift way to avoid turned-down edge is to court its opposite, by reducing one or two of the edge facets—though a little of this usually goes a long way.

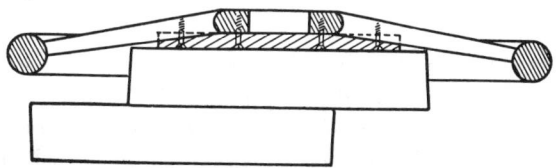

Fig. C.1.13: *A steering wheel handle used by J.V. McAdam, Hastings-on-Hudson, N.Y., for mirrors of medium size, and shown in his elevation drawing, applies the effort nearer the plane of the work, though the desire was also to provide a more firm, secure grasp than a knob or handle permits, and to facilitate handling a 12½-inch, 21-pound disk when off the tool.*

Too rapid strokes will often cause turned-down edge. The cause is the same as explained in an earlier note (subsection C.1.6.7).

C.1.13.6 Turned Edge

Both in grinding and polishing, turned edge may result if the push-and-pull are applied too high above the plane of the work, which is the bottom of the mirror disk; likewise from too rapid strokes. Both sources have a similar cause. Source 1: If the push-and-pull are too high, as with a long handle like the one shown in Figure C.1.6 when used carelessly—that is, when the force is applied far above its base or if it is actually grasped in the whole fist instead of acting merely as a centering convenience for the outspread hands—the tendency will be to lighten the load on the trailing side of the mirror and cause the leading side to plow in, rounding the edge. Source 2: If the strokes are too rapid the momentum of the moving disk will act as explained in subsection C.1.6.7 and turn the edge, even if the force is applied low. Remember, even if the force could be applied no higher than the top of the disk, the point of application would still be considerably above the actual plane of the work. Sometimes we are warned that long strokes will turn the edge, and therefore when we discover a turned edge we may resort to short, timid, mincing strokes, in trying to cure it. But, on hard laps at least, firm, fearless one-third strokes will often provide the cure when we finally "get mad at it."

Strictly speaking, no handle or knob is needed in mirror making. Professionals use none. The bare disk is held in the hands. Many amateurs who have accustomed themselves to working in this way wonder whether the handles or knobs were not used only because it became customary.

The Ronchi test will make very evident a slightly turned-down edge, in fact few mirrors are made on which this test will not reveal at least some sign of it, even if no more than $1/32$ inch wide.

C.1.13.7 Test for Slight Turned Edge

A.W. Everest contributes the following note: "A rigorous method of testing the perfection of the edge of the mirror is by diffraction, in connection with the knife-edge test. Lay a straightedge against the mirror, in a vertical position. Bring the knife-edge across until the mirror completely darkens. At this point both edges of the straightedge will be brightly illuminated by diffraction. At this point also, the right hand edge of the mirror will always be illuminated, either by turned edge or by diffraction, even if the edge is good. If the latter is the case, the left-hand edge of the mirror will also be illuminated by diffraction. Now continue with the knife-edge until the illumination at the straightedge just disappears and, if the illumination at the edge of the mirror disappears at this point also, the mirror's edge cannot be improved. If the illumination at the right hand edge of the mirror persists beyond this point, the mirror's edge is turned and will scatter light around bright star images (knife cuts beam from the left).

"It should be understood that the foregoing refers to that bright semi-circle of light which, with amateur workmanship, is usually seen under the knife-edge test on the side of the mirror away from the knife after the rest of the surface has completely darkened. In extreme cases, this illumination of the edge persists during *several inches* of knife-edge travel, and only most careful workmanship by an experienced workman will eliminate it entirely. But the beginner need not worry. If his mirror is mounted in a cell, the retaining flange will always cover the defect. If it is to be mounted in the open it may be fine ground for a few seconds on a piece of flat plate glass, using the finest washed grit obtainable, after which, lo and behold, it stands the diffraction test. If not, give it a few seconds more and test again. In bad cases it may be necessary to grind in $1/16$ inch from the edge. This grinding is, of course, done after the final figuring operations and the writer's experience has been that it never upsets the figure."

C.1.13.8 Diffraction Effects

John H. Hindle contributes the following. "Diffraction is the cause of the bright rim at the extreme edge of the mirror disk, which is often confused with turned-down edge. By applying a *cleanly cut* mask, say, one inch or more smaller in diameter than the mirror, we can observe what a legitimate diffraction ring should be like. It is extremely difficult to get the edge of the mirror to conform to this standard, but that is what we have to aim at."

"What is diffraction, and why is there more diffraction with four-legged prism supports than with three? When light passes near an object some of it is bent slightly around behind. In the telescope light bends around both edges of a support and is spread out in a streak at right angles to that support. Two legs are just like one which reaches straight across, and the effect is to make the streak twice as bright. Four legs obviously have two directions and cause two diffraction streaks, which cross over the image. Three legs have six sides and produce six-legged stars. Diffraction can be spread around and made very weak in all directions by making the supports wavy in profile, as per Everest, Couder and others, or by curving them so that no part goes very far in one direction."—From a note by Alan R. Kirkham. A clear understanding of diffraction is fundamental in telescoptics.

C.1.14 Testers

C.1.14.1 A Cool Pinhole

A pinhole which will not toast the eye and face may be secured by reflecting the image of an ordinary pinhole by means of a small prism, as shown in Figure A.1.18. Another method is that used by Hindle. Round and burnish the end of a piece of copper wire, about No. 20, and amalgamate the spherical end with weak sulphuric acid and mercury. The image of a distant lamp on the spherical surface forms an artificial star, which may be placed directly in the optical axis, with the knife-edge and eye immediately behind. Haviland uses a drop of mercury on the end of a stick.

C.1.14.2 Permissible Distance Differential of Pinhole and Knife-edge along the Axis

If the pinhole and knife-edge are not maintained at the same distance from the mirror, will an error be introduced thereby? Apparently yes, but in practice no. Prof. F.L. Wadsworth analyzed this question in an article on the mathematics of mirrors (*Popular Astronomy*, August-September, 1902, page 345). He takes the case of a mirror of 50 inches focal length and shows mathematically that the knife-edge may be moved a whole inch nearer it or farther away from it than the pinhole before an error as great as the error of measurement (assumed to be 2 percent, though the uncertainty is likely to be greater than this). Hence in a mirror of any considerable focal length, an error of several inches in the relative placing of the lamp and screen will not affect the measurements by as great an amount as the average error of measurement itself. However, high and wide separation of pinhole and pinhole image will give a false monad shadow.

C.1.14.3 Electric Lamp for Knife-edge Test

Figure A.1.18 gives a hint that an electric lamp will prove suitable in place of the hot, smoky, cumbersome old-fashioned kerosene lamp for this test. Several years ago a practicable method was found—and a most simple one at that—by J. Watson Thompson, a lawyer. He simply frosted the lamp bulb with Carbo. The resultant illumination through the pinhole was perfectly uniform when seen on the mirror. It would be well, however, to see to it that the pinhole does not come quite opposite the actual filament of the lamp. A 110-volt, cylindrical candelabrum lamp bulb about an inch in diameter was employed, having straight sides and a miniature base. The lamp is easily frosted by daubing it with No. 220 Carbo, wrapping around it a strip of thin sheet metal and working it in the hand a few minutes. By purchasing an "adapter" the miniature base may be made to fit standard lamp sockets. One may fill in the "throat" left around the base of the lamp with plaster of Paris, in order to hold the bulb more firmly. This bulb is simply introduced from above into a piece of tubing perforated for the pinhole. It is a great convenience thus to do away with the mussy oil lamp. In localities where the voltage is a bit high the lamps may burn out very quickly. This is because they become too hot in their prison, electric lamps

C.1.14. Testers 467

being designed for a normal amount of air cooling which they do not get when inside of the tube. Here it may become necessary to insert a small resistance in series into the line, to ease off the voltage a little bit. The new inside frosted electric lamp bulbs can be used to almost equal advantage.

C.1.14.4 Tiny Pinhole for Advanced Workers

Prof. Ritchey says: "When the knife-edge test is used with an extremely small pinhole of between $1/250$th and $1/500$th of an inch in diameter, illuminated by acetylene, or what is much better, oxy-hydrogen or electric arc light, minute zonal irregularities are strongly and brilliantly shown, which are entirely invisible with large pinhole or insufficient illumination."

C.1.14.5 Slow Motion Devices for Testing

Fig. C.1.14: *The slow motion device in the left-hand photograph was made by Henry H. Mason, from a description by Rev. C.D.P. Davies, in Monthly Notices, Royal Astronomical Society, March, 1909. Hinges and thumbscrews permit motion in two planes and there is a fixed magnifying glass with which to read the scale. The lamp is also shown. The apparatus in the right hand photograph is described in Section C.1.14.3. The annular ring takes a standard eyepiece when the eyepiece test is to be made. The knife-edge is attached to it merely by wax and may then be removed. Apparatus of the kind shown in these photographs is not required, but some enjoy constructing it and it is useful.*

C.1.14.6 Reversing Knife-edge

If the knife-edge is made to cut the cone of rays from the direction opposite the normal—for example, with the pinhole on right and the knife-edge to left of it but moving into the rays from the right to left—the normal shadows will be reversed, those of a paraboloid becoming those of an oblate spheroid. This method is helpful in more closely finding the "crest of the doughnut," etc. A special knife-edge made like a broad slit, will permit quick change from normal to reverse direction and back.—*Ed.*

C.1.14.7 Testing Tunnel

A testing tunnel is any kind of structure—tube, box or what-not—surrounding the path of the rays to and from the mirror in testing. Especially in winter when there is likely to be a large temperature gradient between parts of the house, or on windy days in summer, such provision usually will render unsteady wavering shadows instantly fixed and steady, and thus on the whole it is very likely to lead to much better results in mirror making. For testing large mirrors professional workers employ well-built structures, virtually long houses which are often double-walled, but the average small job may be tested in any single-walled structure of paper, cardboard, cloth or what-have-you, tacked over a few sticks, and this will often work wonders. It should, however, be quite tight, having no cracks larger than, say, $1/32$-inch wide. It should be tightly closed at the mirror end, where some kind of door can be rigged up to give access to the mirror. The other end must be left open.

C.1.15 Testing

1. Soon after rouge polishing begins, the Foucault shadows may be seen.

2. Attach a long stick of wood loosely to one end of the rack on which you place the mirror during the knife-edge tests. This will enable you to control its position from the testing position several feet away, an invaluable aid.

3. If an electric lamp is used for the knife-edge test it must either be one with a frosted bulb, or the bulb must be frosted by the user. If an attempt is made to use it without first frosting it the result will be, not an evenly distributed diffused illumination, but a sharp inverted image of the filament thrown on the mirror, much as in the case of a pinhole camera.

4. Sometime, while testing the speculum, lightly place the finger tips on its face for a few seconds. When quickly viewed from the knife-edge position the "hills" due to expansion under your warm finger tips will all stand up sharply. This trick always impresses visitors with the delicacy of the Foucault test. Also, have someone hold his hand below and near the speculum while it is in the testing rack: the waving currents of air due to the heat of the hand are a revelation!

C.1.15. Testing

5. Experts advise keeping the eye very close to the knife-edge during testing even removing one's eyeglasses—provided it is possible to see without them. To discover why the eye must be kept close, first place it close, then move it slowly back; the mirror apparently becomes covered with black radial streaks. This is an unavoidable condition existing in the human eye itself.

Fig. C.1.15: *Jean Bernard Leon Foucault (1819–1868).*

C.1.15.1   First Announcement of Foucault Test

True devotees to the ancient and honorable art of telescope making may discover real interest—though no new information—in a holy pilgrimage to some large library to look up the first publication, in 1859, of information concerning the famous Foucault test with the knife-edge. They will find it in a long paper by Foucault himself, safely ensconced in Volume 5 of the *Annales de L'Observatoire Impérial de Paris,* pages 197 to 237, and entitled "Mémoire sur la Construction des Télescopes en Verre Argenté." Do not miss the incomparable engravings in the back of the volume, showing the characteristic shadow of ellipse and paraboloid and the theory of the test. After studying these engravings and reading Foucault's lucid account it will become evident that he did not rush into print until he had both theory and practice of the new test well-worked out. A rather rare old book entitled *A Compleat System of Opticks,* Cambridge, England, 1738, by Robert Smith, describes on page 310 a test which is really the eyepiece test but which bears certain superficial resemblances to Foucault's test. The test

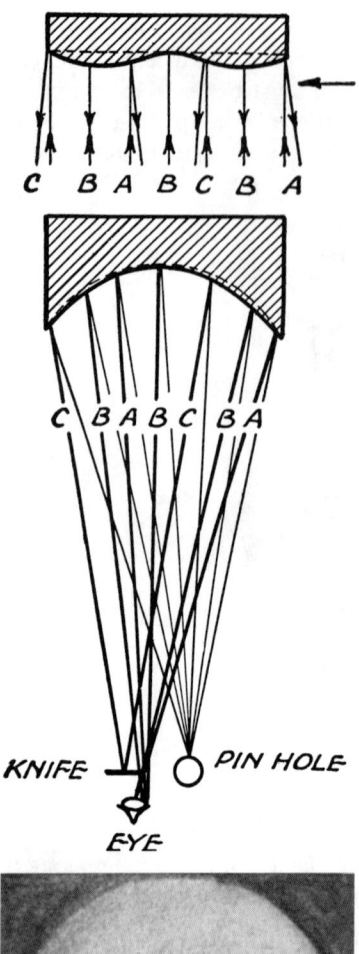

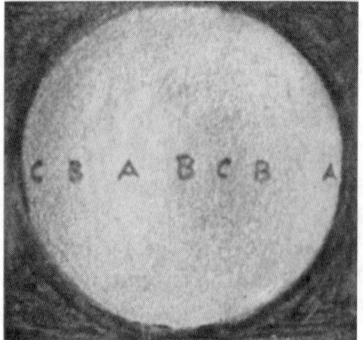

Fig. C.1.16: The Paraboloidal Shadow. The accompanying illustration may explain the peculiar distribution of lights and shadows that is characteristic of a paraboloidal mirror with the knife-edge at its mean center of curvature (when the lights and shadows on either half of the mirror balance in respective areas). Under these conditions the mirror has the appearance of being illuminated from the side (see arrow), and of having a ring-like bulge. At bottom is a shadow of this sort, with letters corresponding to those of the apparent cross section at top. In testing a spherical mirror we saw that the reflected rays all converged as radii to the eye and made the concave disk look flat. Such a spherical curve is dotted in behind the parabola in the central figure. The parabola touches this curve at only three places. In between, and between the curves, are two narrow strips and it is to the effect produced by these strips that we can credit our shadows. As the concavity of the mirror is actually very slight, we shall straighten out these strips, and then their analogy with the bulge of the top figure will be obvious. Thin as they are on the mirror itself, these strips alter the direction of the reflected rays by a few thousandths of an inch along the axis of the mirror, and thus they may be thought of as long "pointers." For example, at eight feet the minute changes in curvature are multiplied, in effect, many times. Therefore, rays A,A enter the eye, and the areas from which they are reflected are bright. Rays marked C are swung slightly to the left, striking the knife-edge, hence, their areas are dark. But rays marked B graze the knife-edge and the areas from which they are reflected appear gray. In practice we make use of this parabolic shadow only in order to make sure that the curve is an evenly flowing one, for the amount of parabolizing is determined by diaphragming out all but the center and margin, as explained by Ellison and Porter.

C.1.16. Conic Sections

was devised by John Hadley, inventor of the sextant. After reading Robert Smith's description of it, one is struck by the fact that Hadley, had he gone on experimenting, might have blundered on the knife-edge test 125 years before it was actually discovered by Foucault. How Herschel would have thanked him!

C.1.16 Conic Sections

Those who are not familiar with the conic sections (See Figure A.1.15) may find the accompanying drawing a help in visualizing some of the curves. Here they are purposely exaggerated. A correctly figured speculum has the form of a paraboloid, the surface which is generated when a parabola is revolved about its axis through 360°. This is the only existing curve which will reflect the parallel rays of light arriving from a distant object to a single point, called the focus. As we saw in Chapters A.1 and A.4, the proper way to bring the surface of a speculum to a paraboloid is first to bring it to a sphere, and then very slightly to deepen the central portions of this sphere into a paraboloid.

We might even remain satisfied to leave it as a sphere, or still less useful, as a mere approximation of a sphere, and this is the method (described in ATM3, Chapter A.1) for those who cannot master the more exacting but intensely interesting process of parabolizing. A sphere will not, however, reflect the rays to a perfect focus. This is shown clearly, although in an extremely exaggerated form, in the lower half of the Figure C.1.17. Here, as in all similar cases, the rays that strike the mirror at different parts of its curve are reflected away again at the same angle, just as a billiard ball without spin bounds away from the cushion at the same angle as that of its approach.

Now our six-inch speculum, having an eight-foot radius or center of curvature, must be thought of as a very small part of a great hemisphere of that radius, and if it were possible to make such a mammoth hemispherical speculum (see lower half of cut), then the ray B, striking near its outer *edge*, would be reflected to that point on the axis of this hemisphere shown by the arrow; the ray A, striking near the *center*, would be reflected so as to intercept the axis about three feet forward of this point, while the other rays would fall between these extremes. Therefore our focus would not be at a single point, as is desirable, but along the axis.

Since, however, our speculum (indicated to exact scale, in solid black) is only six inches in diameter, it occupies only a small fraction of this hemisphere, and that part is close to the axis. Therefore the rays reflected from this shallow curve will be found to cross the axis so closely together (note rapidly decreasing spaces between reflected arrow of B, and that of A) that

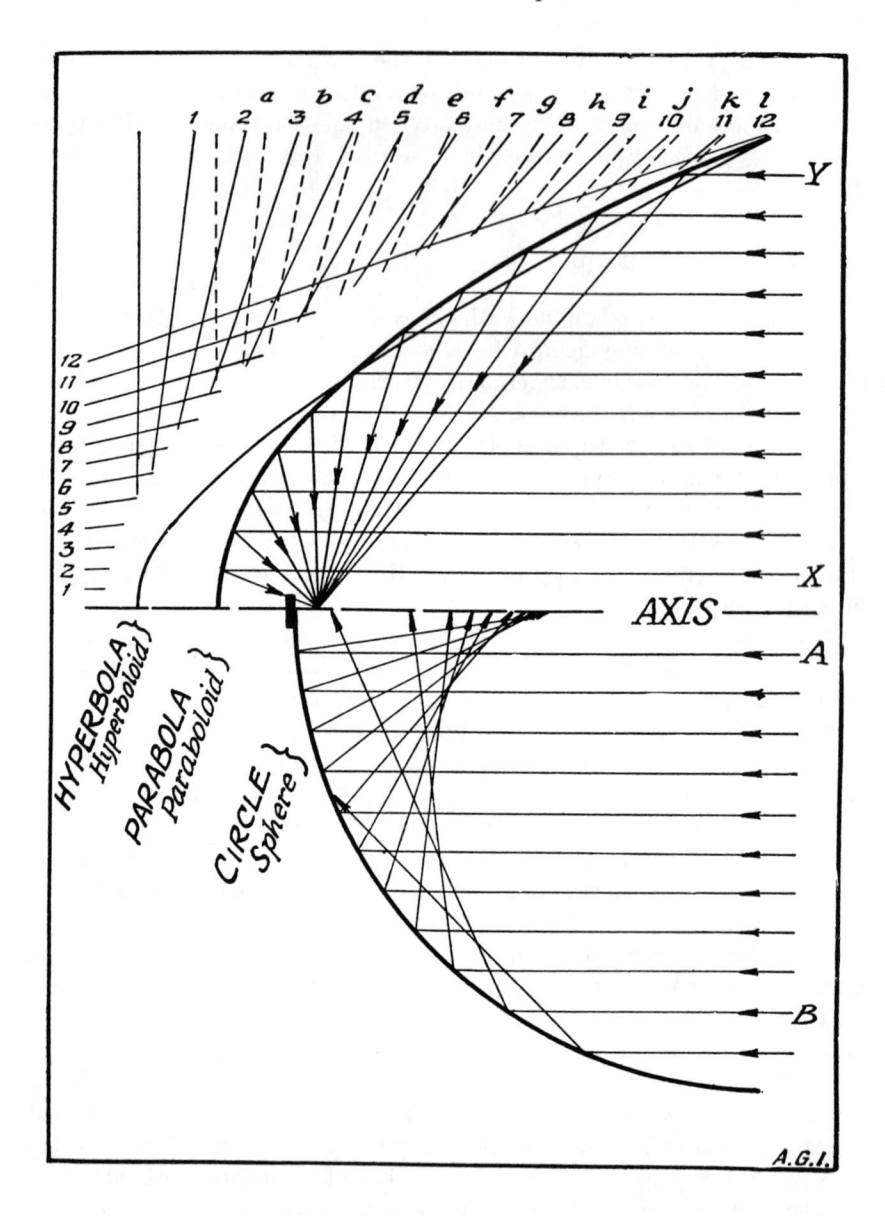

Fig. C.1.17: Three common curves. See Section C.1.16 for explanation.

the definition of the telescope will not appear to be seriously injured—at least until the beginner, after some hours or evening of observing, develops a taste for better things. Hence some have recommended the omission of parabolizing, calling it quits when a sphere has been attained. However, the worker who can make a good sphere (not easy) can surely make a parabola.

The upper half of the diagram referred to, shows how parallel arriving rays are reflected from a paraboloid, which is a curve whose radius, unlike that of a sphere, shortens continually as the axis is approached. Here again the angle of reflection of each ray is the same as the angle of incidence, but the changing radius of curvature of a paraboloid compensates for this and brings all of the rays exactly to a single point—the focus.

The hyperboloid is still deeper than the paraboloid, having a shorter radius at the vertex. It does not bring the reflected rays to a single point, although for the sake of avoiding confusion the reflected rays were omitted from the drawing.

Regarded from the aspect of the making of specula, these three curves should be shaped so that they all start together at the vertex and cross each other at the edge of the speculum. To have done this here would, however, have added confusion to the drawing. If one has time, it is well worthwhile to hunt up one's dusty school books and lay out a curve of each type. Enough of the working lines were purposely left or indicated, so that the student can reconstruct the method, by prolonging them until they cross. There is a set for each curve, based on equal divisions of unequal spaces, horizontal and vertical.

The two arms of a parabola are more nearly parallel as the distance to the right increases, and approach parallelism as the distance to the right approaches infinity. If the arms of the parabola are opened out so that they are no longer ultimately parallel, we get the hyperbola.—*Moulton.*

C.1.17 Record Keeping

It is advisable to keep a systematic record of one's spells of grinding, polishing, and especially of figuring, for such a record will prove invaluable if a larger mirror is attempted later, as is likely. A typical record showing how a hyperbolized mirror was remedied, is shown on the next page, not only as a sample but in order to show how an expert "doctored" a mirror which was brought to him with a hyperbolic figure and a turned-down edge, both of which had resulted from the use of too long a stroke in fine grinding.

The first line of the record shows an apparent section like that described in Chapter A.1 (hyperbola). The center was now shaved away from a pitch lap, leaving a ring, and an effort was made with this tool to move the ridge of the apparent cross section farther toward the center, using an elliptical

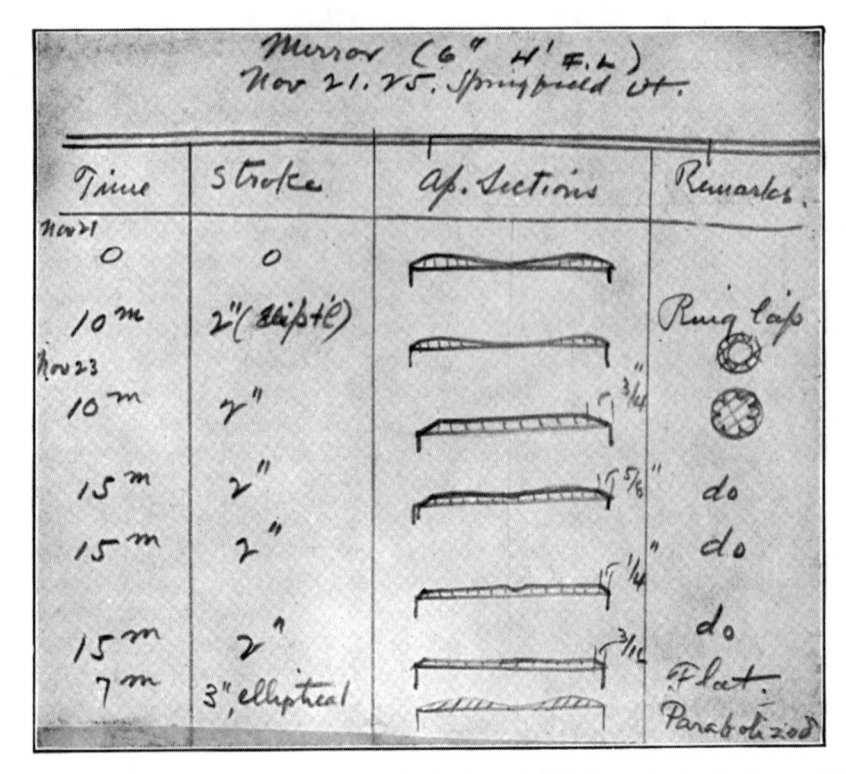

Fig. C.1.18: *A sample record of work. It shows how a hyperbolized mirror with turned-down edge, due to use of too long a stroke, was treated and brought to rights. Such a record should be kept from the very beginning. It will prove invaluable in making the next and larger mirror.*

stroke. This proved unsatisfactory, so it was decided to bring the mirror to a sphere and parabolize from that. With a pitch tool pared like the one shown in Figure A.4.3 and using a two-inch, elliptical stroke, the surface was gradually planed down to an apparently flat plateau (sphere), leaving, however, the turned-down edge. More planing narrowed this edge until it, too, disappeared. (Note that a central pit developed and disappeared again during this process.)

Finally when the mirror was spherical practically all over, it was parabolized in only seven minutes, using a three-inch, elliptical stroke. (Here the stroke was lengthened for it was *desired* to deepen the curve.) Only an old hand should attempt to go ahead and parabolize in so short a time, or in a single spell, without frequent testing.

C.1.18. Correct Paraboloidal Shadow 475

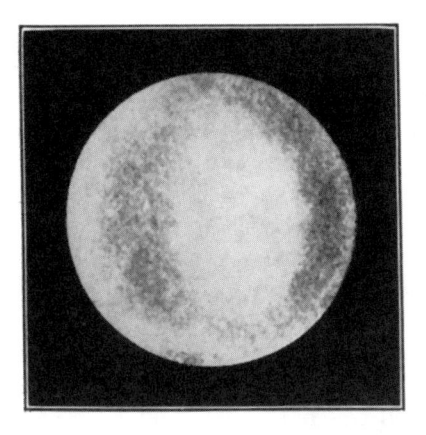

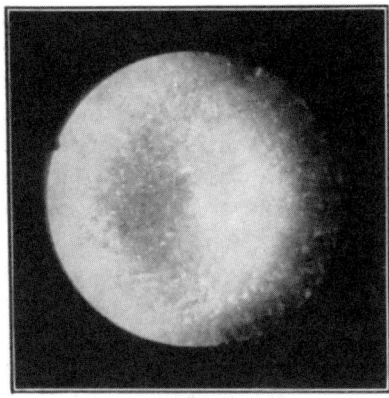

Fig. C.1.19: *Paraboloidal shadows. The one on the left is reproduced from the plate which accompanied Foucault's original announcement of the discovery of the knife-edge test. Foucault placed the knife-edge on the right, which will account for the location of the shadows on the disk. The picture on the right is a direct focogram made by Dr. Nakamura. See the text.*

C.1.18 Correct Paraboloidal Shadow

Astrophysical Journal, June, 1918, contains an article by Porter, entitled "Knife-Edge Shadows—Photography as an Aid in Testing Mirrors." A camera was placed behind the knife-edge, at the center of curvature of a mirror, obtaining thereby beautiful focograms comparable to what one sees with the eye when making the knife-edge test, but more revealing. A similar focogram made from a $12\,1/2$-inch Calver mirror of f/8 is reproduced. This was sent to The Editor by Dr. K. Nakamura, of the Astronomical Observatory of the Kyoto Imperial University, Kyoto, Japan. Dr. Nakamura, a pupil of Ellison, has produced over 45 parabolic and 80 plane mirrors of high quality. He states that when measured on 10 zones this mirror showed no aberration of more than one-hundredth of an inch. He points out what some may not wholly realize, that the judgment of a surface from the shadow requires delicate estimation. (He is speaking, of course, of really first-class mirrors—let not the doubting beginner become discouraged, for even a poor mirror will provide plenty of thrills when directed on the heavenly bodies, and a better one may be produced either the first time or after further efforts. But even the word "good" does scant justice to Dr. Nakamura's mirrors.) "The paraboloidal shadow," Dr. Nakamura continues, "cannot be judged from its position, but one must study its shape and tones of shadow."

As reproduced, the Nakamura focogram is too dark in shade, but apparently this cannot be helped. Any reader who is familiar with photo-engraving and printing may appreciate the difficulty of carrying a reproduction involving delicate, thin, wispy, gray lights and shadows through the numerous processes without deviating from the original. The distribution of lights and shadows—the facial *map* of the shadows—in this focogram is excellent, but workers who are satisfied merely to copy the tone or depth of these shadows without measuring zonal radii will almost surely overcorrect.

We also reproduce the famous drawing by Foucault, which, while it makes poor pretense of depicting the subtlety of the actual shadows, does show very clearly one thing of importance, namely, that the one half of the mirror's shading is precisely the converse of that of the other (with the vertical diameter as an axis of symmetry): *i.e.,* where there is shade on the one half there is light on the other, and in exactly corresponding positions. The late Dr. Calver of England pointed this out very definitely (*English Mechanics,* Aug. 28, 1925, page 96), as does Dr. Nakamura in his letter. If one squints at the Nakamura picture with nearly closed eyelashes this complementary nature of the correct shadows will be better brought out. The manner in which the respective top and bottom horns of the respective dark and bright shadows cross one another should be got clearly in mind.

Finally, and by way of further emphasis, the following is quoted from the Reverend C.D.P. Davies (*Monthly Notices Royal Astronomical Society,* March, 1909): "The one point to be noted above all others is the exceeding delicacy of the shadow. It is impossible to insist on this feature too forcibly. In spite of what Wassell and Blacklock have written, it seems almost hopeless to impress this on the mind of the average worker, who seems to think that because the shadows come on right, he has therefore got a parabolic mirror; the inevitable result of this fallacy being that the tone of shading is in reality far too deep, and the mirror markedly, often profoundly, overcorrected. In most cases the crux of the question lies in the temptation to shirk the trouble of making a proper and effective zonal measurement apparatus and cutting out diaphragms."

C.1.18.1 Measuring Zones

Guarding against self-deception when measuring the radius of zones demands rigorous honesty. It is no more than natural that the worker should wish the mirror on which he has labored so hard to measure up as it ought, but it is virtually certain that if he knows beforehand where he hopes the marks will fall he will tend in some measure to give the mirror the benefit of the doubt. We measure in the dark but we usually make the marks in the light and, of course, we cannot help but remember where we put the last one. Unconsciously obeying a kind of wish-fulfillment, we then tend to crowd the knife-edge toward the last marks when making the next measurement of the series. Instead, after each measurement of the three which usually are taken on each zone the knife-edge should be lifted some distance off the test-stand and put back in the dark, or in some other way thoroughly shifted, so that the location of the former marks will be lost and forgotten. One man wrote rather frankly, "I have tried this method but given it up, as I cannot make the marks come in the same place again"(!).

It is also best to measure the zones of shorter radius first. Then, when measuring the others, the base of the knife-edge will conceal the previous marks.

There are two ways of placing the knife-edge at the exact center of curvature of a zone. One, and possibly the better of the two, is to "bracket the target," as artillerists say. Shell No. 1 lands some distance behind you, then No. 2 lands some distance in front of you. You then know that No. 3 will land behind you again but somewhat nearer. No. 4 lands just in front of you, while No. 5 lands on your center of curvature—the entertainment being that you know, after No. 2, what is about to happen and can say your prayers. Bracketing thus, over and under, over and under, with the knife-edge, may be done quite quickly. No mark need be drawn, of course, until the center of curvature is found.

The second method is to attempt to sneak up on the exact center of curvature entirely from one side—*i.e.*, moving the knife-edge either away from you or toward you. Some can do this well, but if it is given a series of blind tests from both sides the respective marks are likely to show quite a spread; for it is not easy to say, when working entirely from one side, exactly when to "call it." However, the *average* of several tests by this method will generally fall just where the bracketing method may previously have placed the center of curvature.

Russell W. Porter was observed while testing a mirror; he took a long time—perhaps a full minute—for each measurement. But all three marks fell precisely on top of one another. Do not attempt to hurry the process.

C.1.18.2 Testing without Masks

In routine tests of a paraboloid which are not final and critical, experienced workers generally do not bother with masks. At first it is somewhat confusing to test in this manner, but one soon becomes familiar with it. To place the knife-edge at the radius of the outside zone, find the place along the axis at which a large half moon of shadow fills all except the outside zone in the right hand half of the mirror. It then will be observed that the entire left-hand half of the mirror is illuminated. Of course we must ignore all except the outside zone—not seeing it, though it is there. The part of the outside zone on the right lies next to a dark shadow, hence it is well-demarcated, while on the left it lies next to a high light (actually the reversed counterpart of the shadow), with no intervening demarcation, and on this account it appears to be a part of the adjacent high light. This will perhaps be confusing at first. We are now approximately at the radius of the outside zone and by a little closer adjustment, so that the whole zone all the way around darkens simultaneously and evenly, we can arrive at its radius.

All this is precisely what happens when we find the same point with the aid of a mask, though only a couple of patches of light are then actually visible and we may think of it in a different manner. Therefore an easy and instructive way to make the mental transition between the two methods of testing, and become familiar with the appearance without masks, is to find the radius by the more familiar means of a mask and then, without disturbing the knife-edge, remove the mask and study the mirror until the real (*i.e.*, whole) appearance is familiar.

Having found the radius of the outside zone, again the simplest way to find that of the inside zone without a mask, until the method has been learned, is to find it by means of a mask and then remove the mask. We now note that, when we have the knife-edge at the correct place along the axis, the shadow starts at the left, as a slender new moon, spreads or widens at first slowly, then more rapidly to the right and, as the central zone is approached, it takes a comparatively sudden leap clear across. The same thing of course occurs when a mask is used, but most of the appearance is hidden.

When the worker reaches the point where he can test in this manner, and understand the reasons for the several phenomena observed, he may promote himself out of the rookie squad.

Masks for final tests are, however, advisable no matter how expert the worker has become in isolating zones without them, because no one ever becomes perfectly capable of ignoring the subtle influences of appearances adjacent to the zone being examined. Even in making a spherical mirror for some special use, if it is to be a really precise one, Hindle advises the use of masks. If there happens to be a very gradual, uniform transition in shade between two widely separated areas on a sphere this may "get by" unnoticed without a mask, whereas the same two areas would show their difference in shade by contrast if they were isolated by this means.

C.1.18.3 Interpreting Shadows

Many beginners (others, too) are stumped by shadows, often complicated, whose interpretation is equivocal but important. If the interpretation is really important enough it may be worthwhile to make special masks, isolating each doubtful area and actually measuring its radius. If the measurements themselves are right this method will give findings which admit of only one interpretation. Performing these measurements is something of a chore, but perhaps less of an annoyance than writing to someone a few thousand miles away and sitting down to wait a week or two for an answer.

C.1.18.4 Precision in Reading Knife-edge Shadows

"Very few persons find it possible to read Foucault shadows to closer than $1/100''$, and indeed it requires considerable skill and practice to read accurately to $1/50''$. Claims made by those who profess to read more closely are generally based upon fallacy. However, it is possible with patience and care to obtain results of considerably greater exactness.

"In reading photometers, colorimeters and similar instruments, the fields of which present the same gradual and uncertain transitions from light to shade that we observe in the Foucault test, it is recommended to make many readings, throwing the instrument well-off the end point after each reading, and making the next reading by approaching from the opposite side. Thus in the Foucault test, if one reading is made by approaching the focal point from within focus the next should be made by approach from without focus. After obtaining eight or ten readings in this manner, the results are averaged, and any individual readings which depart widely from the average are rejected, after which the average is taken of the remaining results. After resting the eye, and then making a similar set of observations, it will be found that the averages of the two series agree very closely.

"The procedure consumes considerable time, and it is very necessary to rest the eye at intervals, but the greater precision obtained in this way is worth the extra effort. The method is unnecessarily exact for any except the very last stages of figuring, or in checking the figure of a completed mirror, where the utmost precision is desired. It is particularly to be recommended in making zonal tests, especially on mirrors of short focus, where the central shadows are deep and indefinite.

"The observer is fully justified in rejecting any 'wild' results which obviously depart too far from the averages. If one obtains eight or ten readings that are fairly close together, errors in one direction will be at least partially balance errors in the opposite direction, and the average will be very close to the truth. For greatest accuracy, several series of readings should be taken, and the average of all the series used as the final result.

"To satisfy one's self of the degree of precision of which the method is capable it is recommended that, after having measured a mirror as outlined, it be rechecked after the lapse of a sufficient period to remove any lingering recollections of earlier readings which might insensibly bias the observer. Still better it would be to have the mirror checked by a disinterested person. If the mirror is tested in zones, the use of a second mask, exposing zones of different radii from those of the first mask, is recommended, particularly if the results are plotted on graph paper. The results obtained by both masks should fall along the same curve."—*S.H. Sheib, Richmond, Virginia.*

C.1.18.5 The Inside-and-outside Test

"Some years ago while making zonal measurements on a 6″ mirror, which, under the Foucault knife-edge test, had all the characteristics of a correctly parabolized mirror, a zone was found about 1 1/4″ in from the outer edge that was slightly higher than the rest of the curve. With the pinhole on the right and knife-edge on the left, the high zone happened to be so placed that its shadow was thrown to the left edge of the mirror, and was consequently lost in the crescent-shaped shadow normally characteristic of a parabolic surface of revolution.

"Previous to the zonal measurements, described above, the mirror was checked for turned edge with the eyepiece test. The outer edge of the expanded disk, seen with the eyepiece inside of radius of curvature, appeared as it should, with the central area uniformly bright and the outer edge shading delicately into the gray. On close inspection, however, a ring, just a bit darker than the adjacent area of the image, was noticed immediately inside and concentric with the shaded area on the outer edge. This darker ring corresponded exactly with the high zone found in the zonal measurements. Drawing the eyepiece outside of R. of C., the ring which appeared darker inside of R.C. reversed and appeared brighter than the adjacent area.

"Other mirrors on hand at the moment, which were in various stages of figuring, were scrutinized and in each case it was found that high areas appeared dark and low areas appeared bright inside of R.C. and conversely, outside of R.C., dark areas denoted low spots and bright areas were indicative of high spots.

"Applying a bit of reasoning it was readily seen why this should be so. With a perfectly spherical mirror, rays of light from a divergent source placed at the radius of curvature, are reflected to form an image at the source, and the expanded image seen with an eyepiece inside of R.C. is uniformly illuminated. Any departure from a perfectly spherical surface changes the radius of curvature of the area affected, causing the reflected rays to come to a focus slightly ahead or back of R.C. Consequently, in viewing the expanded image inside of R.C. of a mirror having a high zone, its light at the eyepiece will not be quite as close to focus, and therefore will be slightly more diffused than the light reflected from the remainder

C.1.18. Correct Paraboloidal Shadow

of the surface and is slightly darker than the rest of the image. Outside of R.C., the converse is true, *i.e.*, the high area, having the longer radius of curvature, the reflected light passes through focus slightly farther back of R.C. and, being less diffused than the light in the main beam, would therefore appear brighter. The same reasoning may be applied in the case of a low zone, except that the reverse is true. A low zone, having a shorter radius of curvature than the rest of the surface, its reflected light comes to a focus ahead of the main beam, is therefore less diffused and appears brighter in the expanded image inside of R.C. Outside of R.C., it is more diffused and appears darker.

"This simple but very effective test has been found to be a great time saver in testing and figuring the hundred odd mirrors which have passed through the writer's hands; in fact, with experience, one can very accurately judge the correction applied to a mirror after a few moments' study of the expanded image inside of focus. Beginners have found it exceptionally valuable in helping them interpret the Foucault shadows."—*L.E. Armfield, Milwaukee, Wisconsin, in Amateur Astronomy, Sept. 1935.*

C.1.18.5.1   Editor's Note

The germ of the above test was briefly described by Leon Foucault, in his celebrated "Mémoire sur la Construction des Télescopes en Verre Argenté," published in Vol. V of the Annals of the Observatory of Paris, 1859. Mr. Armfield developed it independently and considerably further than Foucault. It is interesting to note in the same memoir, that Foucault similarly described the germ of the Ronchi test; in other respects he was far ahead of his time. He was a physicist and combined with this a flair for inventing apparatus for his experiments—a practical man, no armchair theorist. His picture (Figure C.1.15) deserves to hang in every amateur's workshop—chief patron saint of the telescope making hobby.

Note how lucidly Foucault described his great test, in the memoir named above—the first account of the then new test. "We place a point of light in the vicinity of the center of curvature, in such a way as not to obstruct the returning rays; after crossing one another these rays form a divergent cone in which the eye is placed and moved up toward the focus until the surface of the mirror appears entirely illuminated; then, with the aid of a vertical screen, we intercept the image to the point at which it disappears entirely. This maneuver produces for the observing eye a progressive extinction of the brilliance of the mirror which, in the case of exact sphericity, remains until the last moment and with a uniform intensity over the entire surface. In the contrary case, the extinction does not take place simultaneously at all points, and some contrast of lights and shadows gives the observer, with an impression of exaggerated relief, the perception in black and white of the prominences and depressions which mar the spherical figure."

Many writers have described the same phenomena, but few so clearly as this.

The germ of the Ronchi test is suggested in the following, by Foucault: "If we wish to inspect the mirror as a whole, by a blow of the eye, it is necessary to take for the object a piece of square-meshed netting whose image becomes very sensitive to deformations at whatever part they display themselves. Suppose, as often happens, the mirror, exactly spherical in its central portions, is opened out toward its edges by a progressive lengthening of radius of curvature ... such a mirror will give an image in which all the lines are curved as in Figure ...." How easily, from this point, Foucault might have hit on our Ronchi test.

C.1.18.6 Learning to Understand the Knife-Edge Test

It requires years to learn all about the knife-edge test—although enough can be learned about it on a single occasion to use it. There is almost no end to the fresh insights about shadows, their significance and fine interpretation, which more and more use and general familiarity with them, on more and more mirrors, will keep right on giving.

In explaining things that are not at first easy to grasp, the exact turn of a phrase, like that of an ankle, has everything to do with its power to reveal niceties, and therefore the mirror makers who are closer to their maiden efforts often describe the test (not the ankle) more graspably, and with a fresher point of view, than the old-timer. For example, Robert Hurley's comments, written when their author was a student at the University of Cincinnati, are to the point: "If the pencil of rays," he says, "is cut with the knife-edge at the point where they focus, until half of the image has been observed, the light that passes the knife-edge will have come in equal amounts from all parts of the mirror, provided the mirror is a perfect sphere. When the observer places his eye so that it catches the light that passes the edge, he will see an evenly lighted disk. Suppose there is a mound in the center of the mirror, that is slightly elevated from the surface that would be a perfect sphere. The light coming from one slope would be thrown more into the obscured side of the image, while that from the other would be thrown in the free side. The observer would then see an exaggerated light and shadow relief of the errors on the surface." (This is very clearly stated.) He continues: "The paraboloid can be divided up into a number of annular zones, each of which has a definite radius of curvature. If the radius of curvature of the central zone is designated as R, then the radius of any zone will be $R + r^2/R$, where r is the radius of the zone." We do not recall anyone

C.1.18. Correct Paraboloidal Shadow

previously explaining r^2/R in just this way, the usual way being to state that the difference between foci of inner and outer zones is r^2/R. While this is the same thing, the way quoted above seems to be a clearer statement for the beginner. "Besides this," Hurley continues, "the paraboloid can be recognized by its characteristic pie-shaped appearance when tested at a radius intermediate between that of its center and outermost zone." A paraboloidal mirror is usually tested, not at its "center of curvature," as is commonly stated in print, for it has no such thing, but at its mean center of curvature (c of c).

Another amateur who has a flair for stating things clearly, is Hugh Hazelrigg of Evansville, Indiana. In a letter he explains the behavior of the characteristic paraboloidal shadows at mean c of c, thus: "Keep in mind that when the knife-edge cuts into the rays reflected from a spherical mirror from outside the center of curvature, the shadow moves in a direction opposite to that of the knife-edge. Now the outer part of a paraboloidal mirror has a longer radius of curvature than the inner part. Consequently when we place the knife-edge between the center of curvature of the inner part, and inside the center of curvature of the outer part, the shadow will move, in the outer part, in the same direction as the knife-edge and in the inner part in the opposite direction." We might look at the same thing as if we had two separate mirrors of slightly different focal lengths, one inside of the other.

C.1.18.7 Avoiding Fatigue in the Knife-edge Test

Beginners often complain of eye strain from this test, but nerve strain—tension—is the usual cause of the headaches. When you learn how, you will be able to test for hours without any fatigue at all. First, arrange the setup for the test so that your whole body is relaxed and sprawled out limp; have the knife-edge at such a height that the rays come into the eye just where it is when you are so deposited—no scrooching or stretching. Do not squint the eye or tense any facial muscle but compose your face blankly and then convey it that way to the knife-edge, looking as easily and naturally as if taking in the scenery on the other side of a keyhole. Do not close, half close or even squint the other eye; this is most fatiguing. Learn to ignore it, as all astronomers and microscopists do when looking through eyepieces. Finally, don't use up nervous energy over self-accusation because you cannot interpret the shadows right off. It takes a long while for this to become automatic. Probably no one reaches that stage very soon. Have faith, however, that it will come in time, and that fatigue will disappear.

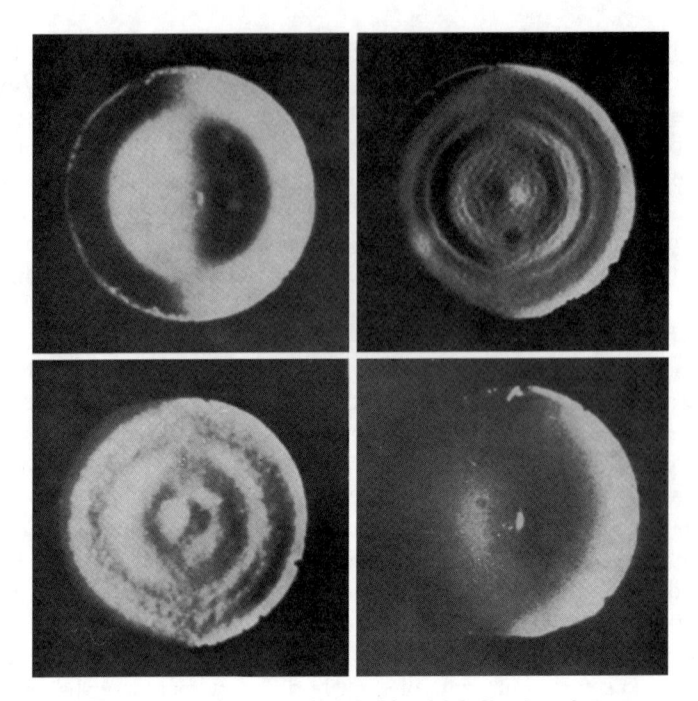

Fig. C.1.20: *Four focograms made by E. Lloyd McCarthy, showing progress on a recalcitrant mirror which was finally brought to book. Mirrors, like youths, usually pass through a few of these "unlicked" formative phases.*

C.1.18.8 Making Focograms

Of these, Dr. C.P. Custer, 155 East Sonoma Street, Stockton, California, has made many. For a camera he uses a cigarbox, film taped inside one end, in the other a $1/2''$ hole, no lens, the hole placed about where the eye was. The source of illumination for the pinhole lamp is a circular, frosted refrigerator bulb, $1 1/2''$ in diameter, $7 1/2$ watts, which can be screwed into a standard sized socket.

First he adjusts the knife-edge. He then places the camera on its narrower edge, on top of another cigarbox which has three nails for stops along the sides and front of the camera, so that the hole is in proper position.

Now the room light is turned off, the camera is loaded by taping a $2 1/4 \times 3 1/4$ Eastman Super Press Type B film, concave emulsion side facing the hole, to its end inside, and a ten minute exposure is tried using a 1.25 mm ($1/20''$) pinhole. With the smallest pinhole, about an hour's exposure is needed; for paraboloids about two hours. A very sharply turned-down edge may require a narrow mask on the mirror, probably due to light reflected from its back surface.

C.1.18. Correct Paraboloidal Shadow

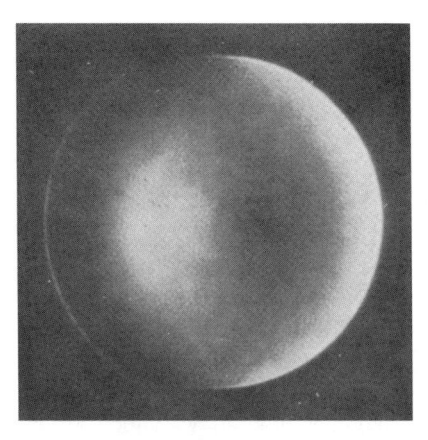

Fig. C.1.21: *Focogram of a 12-inch mirror (95 percent of r^2/R) made by Harold A. Lower. The heavy contrast in shadows is due partly to the short focal ratio (f/6) of the mirror, and partly to the difficulty of preserving the original delicacy of shadow through the various engraving and printing processes.*

The films are developed by using the 10-cent packages of Eastman developer and fixer according to directions. The $1/2''$ wide image enlarges well. One good focogram in three tries is a good average.

Time can be saved if the camera is carefully removed after the exposure is finished, and the image examined by leaning over from the side. If it has turned black or brilliant white, a movement has occurred, and development of that film is useless. Correct the knife-edge and reexpose. The vibration from people walking around the house will ruin the exposure.

The late Harold A. Lower similarly omitted the lens, placing the camera a foot or so back of the knife-edge and shielding the pinhole to assure that no light would leak directly to the film. He used panchromatic film and started with three-minute exposures. He wrote:

"One is not likely to have difficulty, except with mirrors of short focal ratio, when the knife-edge and pinhole must be close together or the mirror will appear warped in the focogram. However, this is true of visual testing, so photography does not add any new problems. Panchromatic film or plates are used because they are faster."

Lower's method of making focograms is essentially the one which was described by Porter, though in less detail, in June, 1918 (*Astrophysical Journal*). The lens is removed from the camera.

In the first focograms the lens was retained in the camera. (See Hartmann, in *Zeitschrift fur Instrumentkunde,* Vol. 29, 1909, page 217, and in *Astrophysical Journal,* Vol. 27, 1908, pages 237 and 258.) This is the method used by E. Lloyd McCarthy, who makes the following comments: "I have found it necessary to mount the camera on a bench entirely separate from that supporting the knife-edge and pinhole, as the most minute jarring causes the shadows to disappear. I use regular Verichrome film and find that the exposure for the unsilvered mirror may vary from 10 seconds to 20 minutes. The former is sufficient for a bad figure but, as the figure approaches the sphere, up to 20 minutes is required. A ground glass at the back of the camera is a help in telling beforehand whether one may expect to find anything on the developed negative, but a puff of smoke near the camera will reveal the path of light and tell you whether or not it is all entering the lens. See that no light escapes from the lamp except through the pinhole, or else there will be reflections from the back of the mirror."

Focograms provide an excellent method of comparing mirrors when at a distance. It is fervently hoped that focograms will be substituted for some of the freehand sketches of shadows which are submitted by mail for diagnosis of mirror trouble. Focograms are admittedly more troublesome to make than hand sketches but they tell far more than the latter.

C.1.18.9 Diffraction Ring (Star) Test

The diffraction ring test is made on a star, by observation of the extrafocal rings. This has nothing to do with the knife-edge test, though that test too may be performed on a star by removing the eyepiece and placing a knife-edge across the end of the adapter tube, when the appearances will be the same as those of a mirror tested at the focus (Chapter A.1).

If the telescope, with an eyepiece giving it a magnification of 30 or 40 diameters for each inch of aperture, is focused on some moderately bright star, preferably of 2nd, 3rd or 4th magnitude, and if the eyepiece is then pulled out or pushed in an eighth or a quarter of an inch, the image of the star will be seen expanded into an area of light. If the atmosphere is fairly steady, and especially if the observer persists for a few minutes in order to train his eye to a new and unfamiliar appearance, this area of light may resolve or tend to resolve itself into several concentric rings, perhaps perfectly round, perhaps distorted and, unless the air is steady, constantly shifting and changing and clearly visible only by occasional short glimpses. If the mirror is perfect and in perfect adjustment, and if the air is very steady, these rings can be made to look exactly like Fig. 23 in Figure C.1.22; except that, in the case of a reflector, the shadow of the diagonal will cover up the central rings, leaving in their place merely a black area. Note that in Fig. 23 each ring is a little wider than its immediate neighbor; also that the dark interval between the rings likewise is progressively wider from center to edge.

But, lest the worker suffer undue sorrow concerning his mirror, it can be said that such good fortune as to behold a perfect set of rings is rela-

C.1.18. Correct Paraboloidal Shadow 487

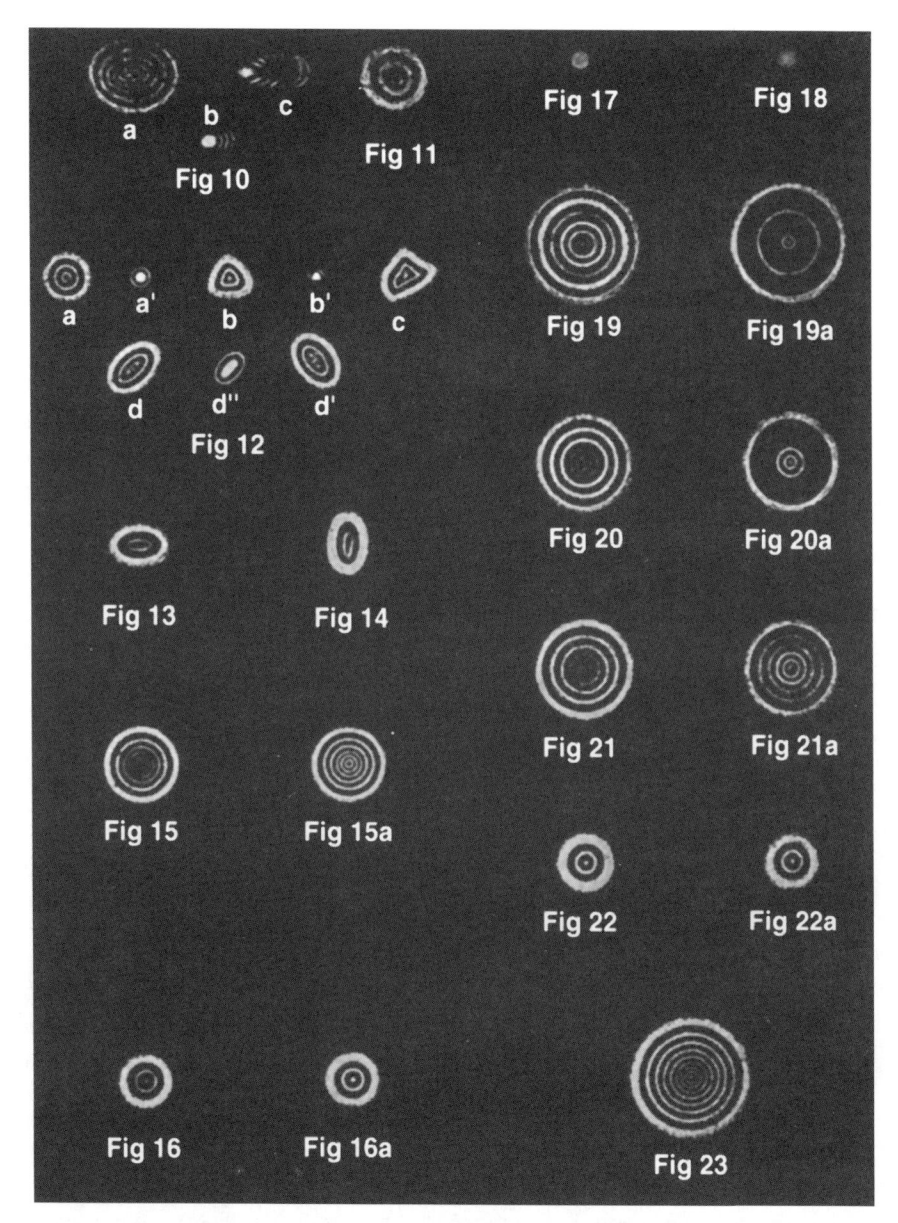

Fig. C.1.22: *From* The Adjustment and Testing of Telescope Objectives, *by Dennis Taylor, courtesy Messrs. Cooke, Troughton and Simms, Ltd.*

tively rare, and that "perfect" mirrors showing none of the bad symptoms indicated elsewhere on the same plate, at least in some degree, are made chiefly in Heaven. Figure C.1.22 is reproduced from the book *The Adjustment and Testing of Telescope Objectives*, by H. Dennis Taylor published by Cooke, Troughton and Simms, Ltd., of York, England, well-known manufacturers of engineering and scientific instruments. The various appearances indicated on it are treated in greater detail in that book than they can be treated in the following briefer comments.

For good telescopic seeing a night when the stars are sharp and "twinkly" is likely to be much inferior to a night when they stand out less distinctly; in fact some of the best seeing is available on nights when the air is slightly hazy. The same will usually be true of conditions, and for the same reasons, when testing the mirror by means of its extra-focal rings. However, if the rings show ideally for even an occasional tenth of a second on a night when the atmosphere is unsteady, this and not the sadly distorted rings seen during the rest of the time, may be taken as a gauge of the mirror's probable worth; and if one persists long enough a "perfect" night may come.

Fig. 10 of Figure C.1.22 includes three appearances. In a, which is lopsided with the smaller end the brighter, the mirror (or objective lens, for the same characterizations apply to both) is not square with the telescope. Move one side of the mirror or the other by means of its adjusting screw, and note whether the rings become less lopsided or more so; if the latter, move it thereafter in the opposite direction while adjusting, and continue until the rings are round.

In b, somewhat below, and c, top row, the eyepiece has been placed as close to the focus as possible ("best focus"). In b the rings show on one side of the star image and are fanned or tailed out; in c they are farther fanned out into a distorted pattern. This too shows that the mirror is not properly squared-on.

In Fig. 11 of Figure C.1.22 the previous fault has been partly remedied, a high-powered eyepiece has been put on and moved up closer to the focus. This reveals fewer rings. More delicate movement of the mirror screws will be required at this later stage of adjustment.

Fig. 12 of Figure C.1.22 includes eight different appearances: top row, left to right, a, a', b, b', c; bottom row, d, d'', d'.

a is related to Fig. 23 of Figure C.1.22, indicating a perfect mirror, but again the observer should remember that the central rings will be covered by the black shadow of the diagonal or prism; also that each support for the diagonal may reveal itself by a delicate streak traversing the appearance from center to edge (diffraction).

a' is the star image *at the focus*, showing one "spurious ring." This spu-

C.1.18. Correct Paraboloidal Shadow

rious or outer disk is normal, not in the sense that something is necessarily wrong when it cannot be seen, but in the sense that theoretically it should always be seen. If one or more spurious rings are not seen the fault may be the state of the air and not the telescope. Compare this sketch with Fig. 17 of Figure C.1.22, which also shows the star disk or image of star exhibited by a perfect objective or mirror but with higher power. Both of these are *at the focus;* they are not extra-focal images.

b is extra-focal and suggests the triangular, while b' to its right is the same mirror at the focus, the central image being similarly distorted and the spurious disk broken into three equilateral segments, not all of which show clearly in the reproduction. These suggest flexure. Doubtless a flotation system for the mirror would change these segments into a full circle.

c is simply one more of the 1001 or 1,000,001 possible forms which the images may show on different telescopes due to various peculiarities. It suggests flexure.

The three appearances in the lower row denote astigmatism. If a mirror has a certain radius of curvature across one diameter, and another radius of curvature across a different diameter, the reflected rays obviously will not come to a focus at equal distances from it. d shows the effect on an otherwise good image of a mirror which has a longer focus in a "northeast-southwest" direction (speaking mapwise) than at right angles to that direction. If we now pull the eyepiece out farther we shall then be past the focus of the first rays and, since they meet and cross over, the ovals will be turned through 90° as in d', the last of these eight figures; between these extremes the oval will be turned at intermediate angles. d'', between these two, is the appearance caused by astigmatism on the central image and one spurious disk at the best focus.

Before condemning the mirror see whether the astigmatism, if any, is in your own eye or in the eyepiece. Twist the head through an angle of 90°, more or less; if it is the eye which is astigmatic the axis of the oval will follow around with the head. Rotate the eyepiece; if the astigmatism is in this, the oval will rotate. (One interesting combination is when the oval is long at one eye position but round or more nearly round at right angles to it. Here both the mirror and the eye are astigmatic.) It may require a higher power to detect a smaller degree of astigmatism in the mirror, but a low power reveals the same fault, if in the eye, and to better advantage.

If there is astigmatism in the eye, spectacles will correct it but these are often a nuisance when using a telescope, for they may prevent the eye from coming up closely enough to the eyepiece to take in the whole emergent pupil. It may in some cases be worthwhile to add a correcting lens to the eyepiece and remove the spectacles. A small cap may be designed and fitted over the eye end, and one of the observer's individual spectacle lenses trimmed down and placed in it. Since the position of the tube of the telescope will vary, this auxiliary device should be arranged so that it will rotate, and preferably a single detachable lens cap should be made to fit all of the observer's eyepieces.

Fig. 13 and 14 of Figure C.1.22 are sections respectively a very little way inside focus and a very little way outside focus, under high power, showing astigmatism.

Fig. 15 of Figure C.1.22 exhibits undue progressive strengthening or brightening, or both, of the outer rings, and weakening of the inner rings. If seen inside focus it denotes undercorrection; if outside focus, overcorrection. Fig. 15a of Figure C.1.22 is the reverse of the last. If seen inside focus this denotes overcorrection; if outside focus, undercorrection.

The distances from focus chosen for this test are such that about the number of rings shown, and no more, will be seen. The outer ring is similar in appearance to that of Fig. 23 of Figure C.1.22, but the adjacent rings are proportionately more strongly stressed; the curve of change is steeper.

If no undercorrection or overcorrection is found at some distance from focus the eyepiece should be shifted closer. Here there will be fewer rings, as in Figs. 16 and 16a of Figure C.1.22, which are analogous to Figs. 15 and 15a. The reason for shifting closer to the focus may be understood if rays from each zone of an undercorrected and an overcorrected mirror are sketched on paper. If there is much undercorrection or much overcorrection the two sets of rings will cross at a considerable distance apart along the axis; if little, they will cross nearer together. Some of the rings will be unduly enhanced because rays which cross one another pass on to combine with rays from other zones; others will be weakened because rays which have thus jumped the track are absent from their rightful place.[1] When this is clearly understood, the reason for testing with different numbers of rings will be obvious: it is not any given number of rings, in itself, that is desired, but the number used is a measure of the point along the axis which the various rays will be intersected by the eyepiece.

Undercorrection refers to a mirror which is ellipsoidal (also oblate); the marginal rays come to a focus nearer the mirror than the central rays. Overcorrection refers to a mirror which is hyperboloidal; the marginal rays come to a focus farther away from the mirror than the central rays.

---

[1] See Figure C.1.23. As constructed, rays start from zones on the undercorrected (ellipsoidal) mirror at equal intervals across diameter and cross the axis at intervals which increase toward the mirror. The cross section marked "inside focus" clearly reveals greater crowding at the outer edge, corresponding to Fig. 15 from the Cooke, Troughton and Simms book. The outer cross section, marked "outside focus," reveals the opposite, corresponding to Fig. 15a of Figure C.1.22. For an overcorrected mirror (hyperboloid) the rays should be started from the same places on the mirror but the intervals along the axis should decrease toward the mirror.—*Ed.*

C.1.18. Correct Paraboloidal Shadow

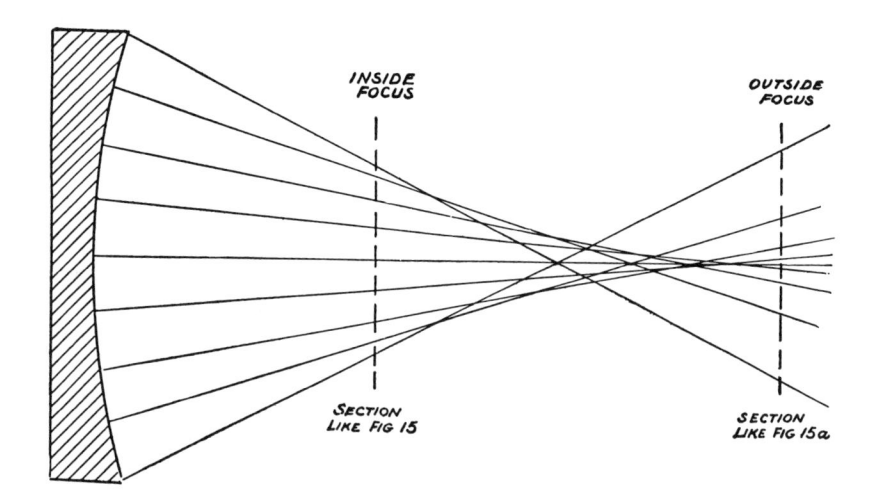

Fig. C.1.23: *Drawing by Russell W. Porter, after the author.*

In both cases parallel incident rays are assumed.

The extra-focal tests reveal the condition of a mirror at the time they are made, and allowance should be made for changes in figure due to changing temperature. But, which way should allowance be made? This, in a sense, begs the question, which is: what really *are* the effects of changing temperature? Not all are agreed concerning them.

Fig. 17 of Figure C.1.22, at the focus, is "the spurious disk or image of a star yielded by a perfect objective and viewed under a very high magnifying power." The ring will be visible only on rare nights. Very rarely two rings may be visible—perhaps even three. The larger the relative aperture the smaller the disk and its system of rings.

Fig. 18 of Figure C.1.22, like Fig. 12b, has a triangular suggestion and in the Cooke, Troughton and Simms book from which the plate is reproduced this is ascribed to an objective lens flexured by being mounted on three supports. (Figs. 17 and 18 are rather faint in the reproduction.) Doubtless the analogous effect from a mirror too large for so few supports (see ATM2, Chapter B.8) would not be altogether dissimilar; at least it will usually show some "threed" appearance.

Fig. 19 of Figure C.1.22 with 19a is another pair; so are Figs. 20 and 20a, and Figs. 21 and 21a of Figure C.1.22—all denoting raised or depressed zones. Zones cause the rays to focus at different distances along the axis and to pass on and across one another, meeting other rays from other parts of the mirror, and not only to aggrandize these places but rob their own.

For detecting zones the eyepiece should be placed a long way from focus, revealing about a dozen rings, and for their proper illumination more light is needed—a brighter star may be chosen.

Fig. 19 of Figure C.1.22, inside focus, and Fig. 19a, outside focus, represent a bad case of zones, more easily interpreted for position (as the zones all are, for that matter) by means of the knife-edge test than by the present one.

Fig. 20 of Figure C.1.22, inside focus, or Fig. 20a, outside focus, denotes an intermediate zone of shorter radius than the ideal curve of the rest of the mirror, and rather nearer the edge than the center.

Fig. 21 of Figure C.1.22, inside focus, or Fig. 21a, outside focus, denotes a raised center. The drawings are for a refractor but the diagonal on a reflector will rob the members of this pair of most of their contrasts and render these effects difficult to detect.

Figs. 22 and 22a of Figure C.1.22 represent a perfect mirror, respectively outside and an equal distance, but not far, inside focus, under high power. Both appearances should be exactly alike.

Fig. 23 of Figure C.1.22 is the appearance of a perfect mirror when the eyepiece is far enough from focus to take in eight rings. Note the perfect roundness of all rings, the perfect progression in width of rings and of dark intervals, except that the outer ring is legitimately broader and brighter than the same progression calls for. The same appearance should be had outside focus as inside, and at equal distance. It speaks loudly in praise of a mirror if it can pass this test. However, as Captain Ainslie has pointed out, most eyepieces have spherical aberration of their own, low-powered ones the most, hence if the mirror seems perfect with a high-powered eyepiece but under corrected with a low-powered eyepiece this may be due to the eyepiece. On a rare night the appearance of a perfect mirror should be as good or better than this figure, which is not idealized. But, rare is a perfect mirror, and rare is a perfect night!

As in the case of the knife-edge test, the observer's judgment and discernment will become very much more critical after a few weeks of practice in observing the various appearances just described.

Elementary discussion of the optics of diffraction will be found in any college physics textbook, though not, of course, as applied specifically to the telescope. The latter application is explained in *Telescope Objectives,* the book from which most of the data just presented were abstracted. Briefly, these rings are due, as Captain Ainslie states in *The Splendour of the Heavens,* to "the mutual interference between light waves from the various parts of the aperture on their way to the focus."

Why cannot diffraction rings similar to those just dealt with be seen by means of an artificial star—that is, a pinhole? They can. There is no

C.1.19. Abnormalities

essential difference in theory between the two sets of conditions. However, some special refinements must be introduced. J.R. Haviland of New York has experimented with this interesting problem. First, an actual star subtends an extremely small angle, so the artificial star used must be very small. Haviland makes such pinholes by stacking a number of small sheets of tinfoil (in practice, one long strip folded zigzag) on a hard surface and pressing a fine needle through a number of them. The last hole thus made will be the smallest and may happen to be very small. The others are discarded. This tiny pinhole demands extremely powerful illumination if the rings are to be seen at focus, but ordinary filament illumination will reveal the extra-focal rings. Finally, it must be remembered that the real stars used in extra-focal tests are at "infinite" distance and that they therefore send parallel rays; while artificial stars are usually placed at the average focus of the mirror and send convergent rays to it. Therefore, in order to obtain normal rings, the mirror must either be a sphere or, if a paraboloid, the artificial star must be at something comparable to "infinite" distance. Here Haviland employs the bright spot on a Christmas tree ball placed in the sunlight at several hundred feet distance. The rings actually will show at the average focus of a paraboloidal mirror but will not have a normal appearance.

C.1.19 Abnormalities

C.1.19.1 Warped Mirror

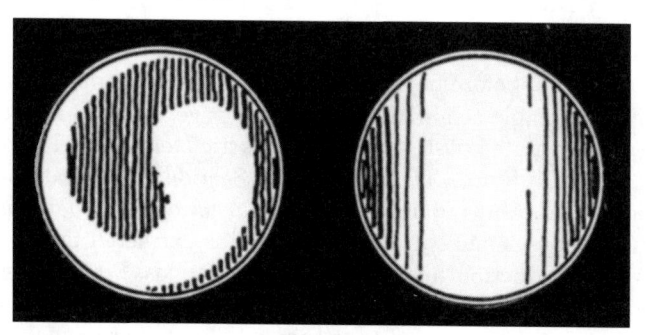

Fig. C.1.24.

The shadowgraph characteristic of a warped or "twisted" paraboloidal surface is similar to the emblem of the Northern Pacific Railroad. A mirror showing this figure persistently should, according to Porter, be treated as follows: Seek out a good hard, solid hydrant. Hurl the mirror as fiercely as possible at said hydrant. Walk home. Such a mirror is strained and is hopeless; the glass was not well-annealed.

A high and wide separation of knife-edge from pinhole will also produce a this shadow.

C.1.19.2 Astigmatized Mirror

Suppose the shadow comes in from both sides at once, making two half moons of deep shadow, one on either side. This indicates that the mirror is no longer a figure of revolution but an oval; it has astigmatism. One must go back to fine grinding again, then polish a few minutes and test. If it still shows oval, the chances are that there is a weak diameter, the glass was badly annealed and the disk is hopeless. Only one case of this kind of tragedy is known to The Editor, so far.

C.1.19.3 Striae

Striae in optical glass are veins or irregularities whose refractive index is a little different from that of the surrounding glass. Their understanding, detection and so on are described in Scientific Papers of the Bureau of Standards No. 373, "Characteristics of Striae in Optical Glass," to be had only from the Government Printing Office, Washington, D. C., price 5 cents. (Postage stamps in payment are not accepted by the Government.)

C.1.19.4 Strains in Glass and Their Detection

Poorly annealed glass often refuses to hold its figure and causes trouble. Advanced workers, before beginning operations on large disks, may wish to test them for possible strain. Bell devotes a page to this subject, and the following extract may assist. It is from an article entitled "The Annealing of Glass," by A. N. Finn, Chief of the Glass Section at the National Bureau of Standards. This was published in the *Journal of the American Ceramic Society* (Columbus, Ohio), August 1926, and is extracted by permission:

"The rapid reduction in the temperature of glass which is incident to practically all commercial molding almost invariably produces a condition in the glass commonly known as 'strain.' (This is permanent strain and should not be confused with temporary strain resulting from the application of external force or temperature differences.) This condition must, in general, be removed before the ware is satisfactory for commercial purposes and its removal is called annealing or tempering. (Tempering sometimes means the internal introduction of strain, as, *e.g.*, in the heat treatment of glass to produce special physical properties.)

C.1.19. Abnormalities

"Strain is due to an inelastic yielding under the stresses developed within the body of the glass while cooling, because at any moment during this time the different parts of the body are not contracting at the same rate. During molding the surface of the hot glass, coming in contact with the relatively cold mold, shrinks or contracts very rapidly and is quickly cooled to a temperature at which the glass is rigid. The interior loses its heat more slowly and, consequently, contracts less rapidly than the surface which must, therefore, if the stresses become large enough, either crack or yield inelastically.

"After the surface becomes rigid or at least practically so, it does not yield readily to any force applied to it and tends to resist the contraction of the hotter and relatively soft interior as it continues to cool and contract. This produces stresses in the glass, the surface usually being compressed while the interior is dilated. If these stresses exceed the strength of the glass, it will break. This may occur either during the initial cooling or subsequently whenever the glass is subjected to sudden temperature changes, such as those resulting from the action of hot or cold water. In order to reduce the liability of breakage of ordinary ware to a minimum, glass must be annealed. "The annealing of glass consists of two fundamental steps. The first is to heat it to such a temperature that the entire body of the glass becomes sufficiently soft to permit it to yield. Then, if sufficient time be allowed, the stresses developed during molding will disappear. The second is to cool it so slowly that the temperature differences between the interior of the glass and its surfaces, with the consequent differences in contraction, are not great enough to produce objectionable strain.

"Strain in glass is easily detected by means of polarized light. A beam of light (white light) is either reflected from or passed through a 'polarizer' which in the first case is usually a plate of black glass or, in the second, a number (pile) of thin transparent plates, and in both cases the surfaces of these plates must make a definite angle with the beam. The reflected or transmitted beam then passes through the sample being tested and is again reflected at the same angle as before from another plate of black glass or is passed through a second pile of plates set at the proper angle. In either case these are called the 'analyzer' and they are sometimes replaced by a nicol prism. From the character of the light finally passing the analyzer the degree of annealing is determined. If the sample is free from strain, no change in the intensity of the field (as compared with the field when no sample is used) or in the appearance of the sample will be observed; if it is moderately strained, the light will be brighter in certain areas; and if it is highly strained, it will appear colored, and the color will vary over the field of view.

"If the analyzer is a nicol prism, a sensitive tint plate may be advantageously used because this, in a way, intensifies the effects obtained with moderately strained glass by producing a series of colors which can be more easily interpreted and compared with those obtained in testing other (standard) samples.

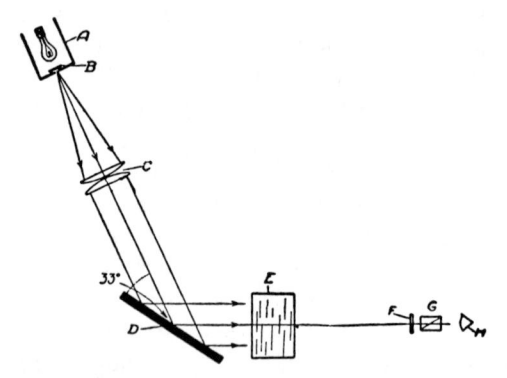

Fig. C.1.25: *Sectional view of polariscope for testing glass for strain. A, source of illumination; B, ground glass; C, condensing lens; D, black glass plate (polarizer); E, sample to be tested; F, retarding plate; G, nicol prism (analyzer); H, observer's eye.*

"The apparatus used at the Bureau of Standards for determining the 'amount of strain' or the degree of annealing is so simple in construction and yet so eminently satisfactory that a detailed description of it will be given. (See Fig. C.1.25.) A 200-watt electric light bulb is enclosed in a box made of bright sheet metal. A $1/2$-inch hole is cut in one side of the box and covered with a piece of glass whose surface is 'fine' ground. This small piece of glass, illuminated by the lamp, is the actual source of light.

"A condensing lens, 7 inches in diameter and having a 15-inch focal length, is so placed that its focus is at the 'source of light.' The light after passing through the lens is practically a parallel beam, but not entirely so since the source is not a 'point.'

"The beam of light then strikes a piece of polished black glass at an angle of approximately 33° (between the beam of light and the glass) or the angle which is necessary with the particular glass for obtaining the best polarization possible. (An angle of 33° can easily be obtained as follows: Draw a right triangle with one side $15\,1/4$ inches and the other side 10.0 inches long. The angle between the long side and the hypotenuse will be about 33°.) Although the beam of light reflected from the black glass is apparently not affected, it is, nevertheless, polarized.

"At some distance in front of the black glass polarizer a nicol prism is placed in the polarized beam, so that the enlarged image of the 'source' is visible through it. By turning the nicol, the intensity of the light will

C.1.19. Abnormalities

change progressively from bright to dark. The nicol is properly set when the field is as dark as possible.

"The tint plate, which should be a 'First Order Purple' (retardation, about 575nμ), is mounted directly in front of the nicol prism and is set by turning it until the color of the light passing through the prism is as marked as possible. Nicol prisms and tint plates can be obtained from practically any dealer in optical instruments.

"If, now, a piece of strained glass is placed in the beam of light between the polarizing plate and the nicol, parts of the glass will apparently be colored differently than the original field. If well-annealed glass is examined, practically no change in color will be seen and if a highly strained piece of glass is used, the colors will be vivid. If the tint plate is removed and the same pieces of glass are examined, the first (moderate strain) will appear brighter in certain areas, the second (no strain) will produce no change, and the third (high strain) will be very bright or even colored in some places.

"Glass being tested should be absolutely uniform in temperature. Ware taken from a room warmer or colder than the room in which the testing is done will show different amounts of strain than if allowed to come to thermal equilibrium. Even the effect of the heat from an inspector's hand will be pronounced, especially if the ware is thick."

Fig. C.1.26.

C.1.19.5 Editor's Note

Above item was inserted in 1933. By 1938, however, Polaroid had become available, superseding costly nicols, and tester shown was made by F.M. Garland, Pittsburgh amateur, after design by Dall. Starting at lower left: Tilted wooden box with about 1″ hole covered by half of a pair of ordinary Polaroid goggles, horizontal dimension vertical; 25-watt lamp inside. Easel with window glass frosted on one side with Carbo. The piece under test, held in gloved hands to obviate heating effects. Pair of Polaroid goggles worn by tester. Test in dark or subdued light. Dark and light streaks indicate bad strains: reject glass. Faint shadings probably are inconsequential.

C.1.20 Notes on the Eyepiece

Nominally, in a Huygenian eyepiece the focal length of the field lens is three times that of the eye lens. In practice this varies somewhat from the prototype, for there are many modifications. In general, the ratio will vary with the power. With low powers it may be 1:1.5 or 1:2. The 1:3 ratio pertains more to the higher powers. As stated by Bell (*The Telescope*), the exact figure will also vary with the amount of overcorrection in the objective (or mirror) and undercorrection in the observer's eye. The separation may be adjusted by trial on the telescope, at the point where the best color correction is found.

Bell's entire chapter on eyepieces should be read by anyone who aspires to make more than one or two very simple eyepieces and, of course, the late Professor Hastings' contributions, Chapter B.1 of ATM3, should not be missed. The article on "Eyepieces," in Volume 5 of Glazebrook's *Dictionary of Applied Physics* contains 5 pages of eyepiece theory (mathematical). John M. Pierce's "Hobbygraphs," No. 2 on "Eyepiece Making" and No. 1 on "Amateur Lens Making," contain much practical help, the former giving working formulas for the design of Huygens and Ramsden eyepieces of different e.f.l. Orford's *Lens Work for Amateurs* contains some practical shop suggestions.

The following is from Ellison's *The Amateur's Telescope*:

> "The Huygenian eyepiece consists of two plano-convex lenses, having their plane sides both toward the eye. The focal lengths of these should theoretically be as 1:3, the eyelens being the shorter, and their distance apart should be half the sum of their focal lengths. These dimensions are, however, rarely adhered to in practice. The German form of Huygenian has a meniscus field lens with the concave side toward the eye. The Ramsden eyepiece is simply a pair of equal plano-convexes, with their convex sides next to each other and separated by a distance slightly less than the focal length of either. The Kellner construction, sometimes used for low powers, has a double convex field lens with radii as 2:3, the deeper curve being toward the object glass, and

C.1.20. Notes on the Eyepiece

a plano-convex eyelens, placed as in the Huygenian, and distant its own focal length from the field lens. Browning's achromatic eyepieces are Kellners, with an achromatic combination for eyelens, the side next to the eye being concave.

"To ascertain what are the powers of a set of eyepieces, a dynamometer is necessary. The most usual and cheapest form of this useful little instrument we owe to the Rev. E. L. Berthon. It consists of a pair of metal straightedges, fixed at a very acute angle to each other, one edge being provided with a scale of $1/100$ inch divisions, showing the distance between the edges at each division. It can be used as a wire gauge, or to measure the thickness of sheet metal, etc. As a telescope dynamometer it is used by measuring the diameter of the 'Ramsden disk.' If the telescope is directed to the open sky, a sheet of white paper, a whitewashed wall, etc., the eye placed about 12 inches behind the eyepiece, sees in it the image of the object glass as a sharp bright disk. This is the 'Ramsden disk.' Being a real image, and in front of the eyelens, it can be focused with a magnifying glass or pocket lens, and a very sharp view of it obtained. The dynamometer is now placed in position and adjusted till each edge is tangent to the disk of light. The scale is read at the point of contact. The aperture of the telescope, divided by the diameter of the Ramsden disk, gives the power of the eyepiece on that particular telescope or any other of the same focal length. A telescope is not even necessary for the measurement. The image of a windowpane, clock dial, or any other sharply-defined object can be used, and the result will be the power of the eyepiece on a telescope whose focal length is the distance of the eyepiece from the object used. Then, taking this power as a divisor and the focal length as a dividend, the quotient is the equivalent focus of the eyepiece. For example, on a 5-inch telescope the Ramsden disk was $1/10$th inch diameter. Power was therefore 5 inches $\div$ $1/10$th = 50. The focal length of the o.g. was 75 inches : $75 \div 50 = 1.5$. Equivalent focus of eyepiece was $1 1/2$ inches. Again, with another eyepiece on same telescope, the Ramsden disk measured 0.03 inch. Power was therefore $5 \div 0.03 = 166$. Equivalent focus of eyepiece was $75 \div 166 = 0.45$ inch.

"It is necessary, before making these measures, to make sure that the aperture of the o.g. is 'clear'—*i.e.*, that none of the stops in the tube are narrow enough to stop any portion of the cone light converging from it. Any other instrument capable of giving an accurate measurement of the diameter of the

Ramsden disk can be used as a dynamometer, an ordinary micrometer calipers being an excellent substitute, especially if its two contact faces are lightly smoked in a candle flame to prevent reflection from them. It is sometimes stated that the equivalent focus of a Huygenian eyepiece equals half the focal length of its field lens. If this were true it would be easy to measure powers, as the field lens, being the larger of the two, has a fairly long focus and easily measured directly. But, unfortunately, it is only true if the proper proportions for a Huygenian (foci as 1:3 and distance apart = 2) are strictly adhered to by the makers, which they rarely, if ever, are.

"It may be well to state, in conclusion, that the equivalent focus of any two-lens eyepiece can be obtained from the formula:

$$\frac{f_1 \times f_2}{f_1 + f_2 - d}$$

where f_1 and f_2 are the focal lengths of the components, and d their distance apart. The distance apart of plano-convex lenses should be measured from their *convex* faces, not from the plane ones, as might at first be supposed. The only drawback to this formula as a means of measuring powers is the difficulty of obtaining an accurate measure of the focus of very small lenses, such as the eyelenses of high powers always are."

The following statements were chosen from an article by M.A. Ainslie, in the *Journal of the British Astronomical Association*, Nov. 1930: "The amount of the spherical aberration of a given eyepiece varies very nearly as the square of the aperture ratio of the object glass or mirror with which it is used. The same eyepiece used with a mirror of ratio $f/5$ will have 9 times as much aberration as with an object glass of $f/15$, and it may be very serious. In short, the Huygenian does very well for a refractor for all powers, but it is not good enough for a reflector. If used for a speculum of aperture ratio $f/5$ a 1-inch Huygenian eyepiece would require the speculum to be *overcorrected* to the extent of about one-twelfth of an inch."

In a private communication, H.E. Dall of England, an engineering instrument designer who has made a number of interesting telescopes of amateur size, and who makes Tolles eyepieces (ATM3, Chapter B.2), points out the interesting fact that "when using a standard form of Huygenian giving a power of, say, 20 per inch (a fair average) the spherical aberration of the eyepiece gives as much error in the image as the difference between a

C.1.20. Notes on the Eyepiece

sphere and a paraboloid for about 6-inch aperture, this being independent of angular aperture. Under these conditions," he states, "it is rather futile to figure to a precise paraboloid. The remedy is of course to use eyepieces of small aberrations."

To make the solid metal type of lap which has been in use ever since the year 1 for grinding and polishing small spherical lenses such as those used in eyepieces has always required much labor and patience, and to keep it spherical has required further labor and patience. Within relatively recent years large balls for bearings have become readily and widely available. In 1926 John M. Pierce made lead laps by hammering such balls into chunks of lead. George Croston of Tacoma has improved this method, in a similar way making accurate laps of sheet metal. His method is essentially as follows: "Lay a sheet of copper, lead, soft steel or other soft metal on a ring; place a steel ball on this and, with a vise or hammer, force the ball into the sheet metal. The latter may be about $1/16$ inch thick, and the ring should have a hole slightly smaller than the full diameter of the ball. For extreme accuracy press both ball and sheet metal into a block of lead. As the lap wears out of shape during use it may be restored in the same way it was formed.

"The cup-shaped lap thus formed is pitched to a rotating spindle, similar to the lens in ATM3, Fig. B.5.7, at a. Use a different lap for each grade of abrasive. In grinding with the last two grades of abrasive hold the lap in the palm and rotate the lens with the other hand, using pressure. If the motor drive is used on these final sizes the excessive speed will deprive the center of abrasive by centrifugal action and the edges of the lens will be ground too much. On the final stage a lead lap will give a finish which will polish in 10 minutes with rouge.

"To make the pitch polishing lap turn the spindle vertically, pour melted pitch into the concavity of a spare lap (a spindle of wood with a concavity concentric with the lap turned in one end may be substituted for the tubular spindle, either here or at the outset) and, while the pitch is still melted, lay on it a patch of silk or cotton. Before the pitch entirely sets, dip the lens in cold water and start the spindle rotating. Press the lens into the cavity, making it run true."

Alan R. Kirkham of Tacoma furnishes the following note: "Serious defects in small lenses for eyepieces may be revealed by holding the lens at such a distance from the eye that a distant street lamp appears completely to illuminate the lens, and then introducing a Ronchi test grating very near to the eye. The figure of the lens will be indicated, just as in Chapter B.4, by the bands then observed."

C.1.21 Telescope Mechanics

C.1.21.1 Tubeless Telescope

If one is forced to observe near strong street lamps, or if one wishes to use his telescope as a terrestrial instrument, the tubeless mounting described in ATM2, Section B.1.1 and shown in Figure B.1.9 will not function any too well, the view being somewhat fogged by the outside sources of light. It is better to employ a tube for these purposes.

C.1.21.2 Hardening Brass

The little brass ears at 1, on the prism tube C, in ATM2, Fig. B.1.9, may be made springy simply by tapping them with a hammer. This hardens brass. To soften brass, heat it in a flame.

C.1.21.3 Blackening Brass

For blackening brass the following formula has been used with success. Photographer's hypo, 8 oz., lead acetate, 4 oz., water, 2 quarts. Let the brass soak in this solution. To obtain a blue opalescent surface substitute 4 oz. of alum for the lead acetate and boil the brass in this.

C.1.21.4 Paint for Inside of Tube

A dull or flat black paint is best for the inside of the tube. Ordinary black house paint well-doped with boiled oil will do fairly well. Better than anything is "coach black," which comes in cans and is to be diluted with turpentine, for use. This gives the dull black which one sees on the inside of cameras and other optical instruments.

C.1.21.5 An Adapter Tube

An adapter for eyepieces is often a decided convenience, as the length of the eyepiece may not always permit handy focusing. The adapter is a piece of tubing 3 or 4 inches long, into which the eyepiece fits, and which in turn slides within a larger tube—usually the one that holds the prism on its opposite end. Brass tubing that fits other tubing, inside and out, may be purchased from Patterson Brothers or Chas. H. Besley and Co., but one should specify that it must telescope smoothly.[2]

---

[2] If 1 1/4 eyepieces are used, you can find a 1 1/4-inch bathroom drain extension at a hardware store. Saw off any flare on the ends. Often, these are made of chrome plated brass.

C.1.21.6 Finders

For a finder one may rig up some kind of a gunsight, perhaps employing radium paint, visible at night (The Radium Luminous Material Corp.). In any case, a finder is a decided convenience. One which was purchased from the Gaertner Scientific Corporation for a moderate price proved applicable to almost any telescope and well worth the price paid. A good one will include a field of view at least 3 degrees (six Moons) in diameter. H.L. Rogers, a real estate broker, made his own finder and equipped it with a small total reflection prism. By placing the finder near the eyepiece he was able to use both eyes at the same time in locating a star, an ingenious wrinkle. The Editor asked Mr. Rogers to describe his finder, which he has done as follows: "I took one lens of an old pair of opera glasses of about $1\,3/4$-inch aperture, and bought at a hardware store a tube to fit. The lens was seated against a piece of the same tube, split, shortened, faced on the lens side, sprung inside the tube and soldered in. There is an outside retaining ring also, made like the other. Before deciding on the length of the tube the focal length of the lens must be ascertained (by focusing in the parallel Sun's rays, not those of a local light source, and then measuring distance from lens to clearest obtainable image—Ed.) The eyepiece holder for the eyepiece is soldered to the main tube from the inside. Two lengthwise hacksaw cuts divide the eyepiece holder into four prongs which may be sprung inwards to hold the eyepiece. The latter was a low-powered microscope eyepiece smaller in diameter than the $1\,1/4$-inch American standard. The sketch shows the prism which is held in a sheet brass clip soldered to a support of stout soft brass. This may be adjusted by the screws and by bending. The end fixture for the main tube is merely a piece of sheet brass with a $1/2$-inch strip soldered across the inside face, and bent at right angles to take a small screw at either end through the main tube. Finally, the finder was attached to the telescope tube at a point so that I could use both eyes at one time in finding an object, a convenience whose unusual value will instantly become apparent on using it. The finder must, of course, be adjusted so that its field of view coincides with that of the telescope. No crosshairs were used, as it is easy to place the star in the center of the field of the finder, or sufficiently near the center to bring its image somewhere on the main mirror."

To this Porter adds: "It would, however, be easy to add crosshairs, or at least a kind of sight, simply by bending a piece of spring brass wire into suitable form. If the eyepiece used on the finder is positive, the sight will be snapped inside the tube just beyond the field lens; if a negative eyepiece is used, remove the field lens and place the sight between the two lenses of the eyepiece, against the diaphragm, where it will be in sharp focus. The idea in either case is to place the sight in the focal plane of the eyepiece used."

In one of his little brochures or "Hobbygraphs" John M. Pierce tells how to make the objective lens for a finder; in another how to make the eyepiece.

C.1.21.7 More about Finders

The simplest finder—some say the best of all—is a pair of homemade gunsights, one member being a ring and the other a point. Before putting anything into finished form a few outdoor tests on visibility at night, using wires variously bent, may save surprise and disappointment later on. Luminescent paint may prove useful on the sights.

Two simple lenses costing about a dollar can be made into a finder of a kind. Let the front lens be an inch or so in diameter, with a focal length of 8 or 10 inches, the eye lens an inch or more in focal length and whatever diameter is readily available—say, half an inch. Many kinds of cheap pick-me-up magnifying glasses will be found to contain suitable lenses for this purpose. Lenses may be purchased from Bausch and Lomb, or even homemade, following instructions in John M. Pierce's "Hobbygraph" No. 1, on "Amateur Lens Making." Such a finder will scarcely be achromatic, but it will find—and at low cost. Separate the two lenses by a distance equal to the combined focal lengths of the two lenses. The focal lengths may be ascertained by measuring the distance from each lens to the smallest, sharpest image of the Sun it will give. Place the curved side of the eye lens next to the eye if it is a plano-convex, as it will then show less spherical aberration. Two spectacle lenses may also be used. No tube is needed, the lenses being merely attached on supports of some kind on the main telescope.

Crosshairs, discussed earlier, if used at all—and they are not really needed since almost anyone can center the finder on a given star closely enough to place the star in the field of the main telescope, provided it has a low-powered eyepiece—should be placed in the focus of the eyepiece. For night use the crosshairs, if any are provided, should not be spider-webs but No. 30 to No. 36 wire, which is large enough to be visible in the dark. Instead of crosshairs a small ball on the end of a wire may be placed in the focus and bent about until it is in line.

John M. Pierce's "Hobbygraph" No. 3 tells how to design and make a very good finder having a one-inch objective lens. The objective lens is achromatic but the eyepiece is a simple lens. Such a combination will give far less color effect than the reverse—that is, than a simple objective lens and a compound eyepiece, but as a finder rather requires a broad field, it

C.1.22. Telescope Designs

will be improved if provided with a two-lens eyepiece.

A finder should give a field at least three to five degrees in angular diameter—$15\frac{1}{2}$ to 26 feet wide—at 100 yards. For example, Carl Zeiss, Inc., lists three typical finders: The first has a 1-inch objective lens of about 8 inches focal length and a 1-inch eyepiece. It magnifies 8 diameters and has a field 5° 13' in angular diameter. The second has a $1\frac{3}{16}$-inch objective of about 12 inches focus, $1\frac{3}{16}$-inch eyepiece, magnifies 10 diameters and gives a field 4°10' in diameter. The third has a $1\frac{3}{4}$-inch objective of 14 inches focus, a 1-inch eyepiece, magnifies 15 diameters and has a field 2°47' in diameter. A finder designed about like the first of these three would probably be the best for general use. Comparing such a telescope, made as a finder, with binocular specifications—for example, those in a Bausch and Lomb catalog—we have for the latter: 1-inch objective, 8 diameters, field 6°30' in angular diameter or 34 feet at 100 yards. Thus the binocular and the regular finder are not very different. A prism binocular is simply two telescopes, jackknifed by means of prisms in order to gain compactness. Half of a binocular or even a whole one makes a splendid finder.

C.1.22 Telescope Designs

C.1.22.1 Ritchey-Chrétien

In a communication to the editor, Professor Ritchey wrote from *l'Observatoire de Paris,* where he was conducting many of his researches: "I am also developing two new types of photographic telescopes—the Schwarzschild and the Ritchey-Chrétien. The first of these has a large concave and a small concave (secondary) mirror; the second has a large concave and a small convex (secondary) mirror. Both have *new curves* of the mirror surfaces (other than paraboloid and hyperboloid). Both give much larger fields and much smaller, more concentrated, images of out-of-axis stars, than the paraboloid gives.

"The Ritchey-Chrétien type allows the shortest tube and smallest dome of any reflecting telescope type, about one-half those required for the type of the 60-inch and 100-inch at Mt. Wilson. The out-of-axis images are so small, round and concentrated that, on an average, for a field of 40 minutes of arc diameter, we will photograph out-of-axis stars which are at least two magnitudes fainter than those photographed with a Newtonian of equal aperture and focal ratio.

"This is a revolutionary result, and has been fully demonstrated in this laboratory. These small images demand refinements of guiding and focusing, of convenience and consequent efficiency of the observer, and of protection of the telescope mounting and tube from temperature changes and from flexure, *far beyond those attained at Mt. Wilson by the writer;* these refinements have all been fully worked out in this laboratory.

"The Schwarzschild type also gives a very large field of small, round images, together with very small focal ratio—as short as 2½ or 3 to 1—so that the light concentration for nebulae and faint stars is very great. But it is a most inconvenient type in use, unless it be used as a fixed telescope with a coelostat. When so used it becomes as important and revolutionary as the Ritchey-Chrétien. The two new tyes of mirror curves, used interchangeably, with various focal ratios, in conjunction with the fixed, universal telescope, together with the cellular, ventilated mirror disks made of extremely strong, rigid plates of low-expansion glass, will inaugurate a new epoch in telescope efficiency, in astronomical photography, and in accuracy of astronomical measurement."

Fig. C.1.27: *Professor George Willis Ritchey. From a photograph taken in his optical laboratory at the Paris Observatory, by James Stokley, 1927. (Science Service Photo.)*

These new conceptions are described in *L'Astronomie* in a series of six monthly articles beginning December 1927; also in *Journal Royal Ast. Soc. Canada*, beginning with the May-June issue, 1928.

In the first article Professor Ritchey outlines his plans for a modern observatory with a great, vertical, fixed telescope having a coelostat and quickly interchangeable, cellular mirrors of the two new types, Schwarzschild and Ritchey-Chrétien.

Commenting on the first edition of *Amateur Telescope Making,* Professor

C.1.22. Telescope Designs

Ritchey writes: "I am *most* heartily in sympathy with your efforts in regard to amateur optical work such as you describe in the little book which you so kindly sent me. Henry Draper's remark that 'the future hopes of astronomy lie in the multitude of observers, and in the concentration of action of many minds' is true. The greatest single need of astronomy today is the thorough popularization of it. This is easily possible by means of the finest attainable astronomical photographs. Astronomy could have, and soon *will have*, a thousand devoted friends assisting in every way in its development, where it now has one such friend."

The amateur should read well Ellison's comment in Chapter A.4. The professional, he points out, has nearly always started as an amateur. No professional has risen to higher attainments than Professor Ritchey.

C.1.22.2 Cassegrainian Notes

A diagonal may be used to reflect the cone of rays to one side of the tube, instead of allowing them to pass through a perforation in the primary. This kind of construction is used on the 100-inch telescope and others of importance. As pointed out by Horace D. Dall (*Journal of the British Astronomical Association*, March 1931), it permits the eyepiece to be located near the declination axis where the observer will find it more comfortably accessible. It also removes the body heat of the observer farther from the mirror. However, it reverses the image. A pentagonal prism will keep the otherwise reversed image straight but would be expensive.

Ritchey, in his work *On the Modern Reflecting Telescope and the Making and Testing of Optical Mirrors*, treats the Cassegrainian as the special type mentioned above. Since there have been frequent requests to know just what Ritchey did say in this rare work, the *whole* of his comment on the Cassegrainian is quoted from it herewith. As will be seen, the comment is rather limited:

"The writer has recently made two convex mirrors of different curvature, for use with the 2-foot reflector. These give equivalent focal lengths of 27 and 38 feet respectively.

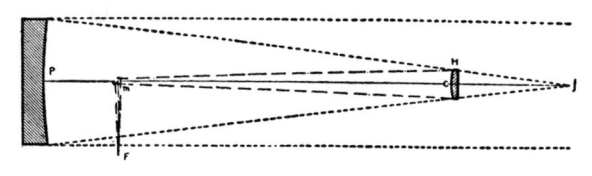

Fig. C.1.28.

"Figure C.1.28 shows the arrangement of mirrors employed in the 2-

Fig. C.1.29: *A 4-inch Cassegrainian built a number of years ago by John M. Pierce. Primary f/5; overall f/20.*

foot reflector when used as a Cassegrain; a small diagonal plane mirror is used at m, to avoid the necessity of a hole through the center of the large concave mirror. P is the paraboloidal mirror, with its focus at f; H is the hyperboloidal mirror, the secondary focus or magnified image produced by the combination being at F; the point c is the center of the hyperboloidal surface. Calling the distance $fc = p$ and the distance $cm + mF = p'$, then p'/p represents the amount of amplification introduced by the convex mirror. The radius of curvature R of the spherical surface to which the convex mirror is ground and polished preparatory to hyperbolizing is found with sufficient accuracy for all practical purposes by the formula

$$\frac{1}{p} - \frac{1}{p'} = \frac{2}{R}$$

whence

$$R = \frac{2pp'}{p' - p}$$

"For example, let the focal length of the paraboloidal mirror P, Fig. C.1.28, be ten feet; let $fc = p = 2$ feet and $cm + mF = p' = 8$ feet. Here

C.1.22. Telescope Designs

Fig. C.1.30: *A modified Cassegrainian (spherical secondary and ellipsoidal primary) made by Horace E. Dall, Luton, Bedfordshire, England, and exhibited in 1932 before the British Astronomical Association. The calculations for designing a spherical secondary Cassegrainian are stated in Scientific American, June, 1938. The telescope shown above is very compact and light: length, 19", weight complete with finder, 5½ pounds. Primary is 6", f.l. 19¼". Secondary, 1½". Overall f ratio, 13.*

$p'/p = 4$; the image of the Moon or the other celestial object produced at f is therefore four times larger in diameter than it would be at f, the focus of the paraboloid; and

$$R = \frac{2pp'}{p' - p} = 64 \text{ inches.}$$

In ATM3, Chapter E.7, Hindle recommends an eye stop for the Cassegrainian, to delimit the Ramsden disk, and this is the most common method of cutting out the direct light of the sky. An alternative dodge is a short conical tube with its small end fixed permanently through the perforation of the primary and into the cell. This delimits the light entering the eyepiece to the secondary cone. This tube extends forward from the mirror.

A point not often mentioned in favor of the Cassegrainian is the fact that the long e.f.l. and slender cone of light favors the eyepiece. This is explained by Captain Ainslie, in the note on *"Eyepieces."* The effect, in a Cassegrainian of $f/20$, would be the same as in the case of a refractor, or

Fig. C.1.31: *A 12-inch Cassegrainian-Newtonian combination made by Harold A. Lower of San Diego, Calif., with a housing which rolls off on metal tracks.*

in a Newtonian reflector, of the same focal ratio.

The Cassegrainian is essentially for the observation of fine lunar and planetary detail; it has a small field and for general observation, rather than particular observation, it will be likely to prove disappointing. There is no difference in this respect, between a Cassegrainian of $f/20$ and a Newtonian of $f/20$.

Alan R. Kirkham contributes the following note. "When adjusting the optical train of a compound telescope, the first approximation may be made as follows: Direct the telescope toward a bright planet or a distant street lamp and replace the eyepiece with a grating of the kind used in the Ronchi test, placing it a trifle inside focus. The characteristic bands will be seen, appearing as if they were on the secondary mirror. These bands will show astigmatic shapes with the slightest misalignment of prism, secondary, primary or adapter tube, spreading more at one side than at the other—that is, diverging. If the bands do not appear almost straight (temperature changes sometimes bend them one way or the other), there is a serious error in the figuring of the mirrors, primary or secondary or both. As will quickly become apparent when performing this test, stiffness in the various parts of the mounting is essential."

The following note was contributed by Harold A. Lower. "It is possible to test convex mirrors from the back, just as if the mirror were concave. The back surface of the mirror must be fairly flat and the glass well-annealed, as any irregularities in the plane surface, or any strains in the glass, will

C.1.22. Telescope Designs

Fig. C.1.32: *One of several 12-inch Cassegrainians designed by Russell W. Porter and used in the investigation for selecting the site of the 200-inch telescope. Primary focus 5 feet; e.f.l. of primary and secondary 20 feet. Duralumin mounting. Built ruggedly for mountain transportation on mule back or man back. Note rugged yoke, which is hollow. The cell is threaded and screws on. Clock drive in pier. Made by Fred G. Henson of Pasadena, California, for the California Institute of Technology.*

appear as black lines. If not too bad, the streaks can be ignored, as they do not change with changes in the figure of the mirror. This test does not give the radius of curvature, as refraction at the plane surface shortens the apparent radius. Chromatic aberration due to refraction produces objectionable color effects, but a deep red filter (ruby glass from an old darkroom lantern) in front of the pinhole will, to a large extent, prevent this trouble. While this method of testing the secondary of a Cassegrainian does not take the place of the usual method of testing with the speculum and a large flat, it is very useful during polishing, for keeping track of the figure and preventing any very large departure from a spherical curve. The final figure on the secondary mirror of a Cassegrainian, when tested from the back, will present the familiar appearance of a long-focus conic, and as the figure can be seen from center to edge, it is easy to see whether it has a uniform curve. This method of testing was discovered while making a 12-inch Cassegrainian, but I have since heard that it has been reported

Fig. C.1.33: *A combination Cassegrainian-Gregorian telescope designed and made by Ben L. Nicholson of Tacoma.*

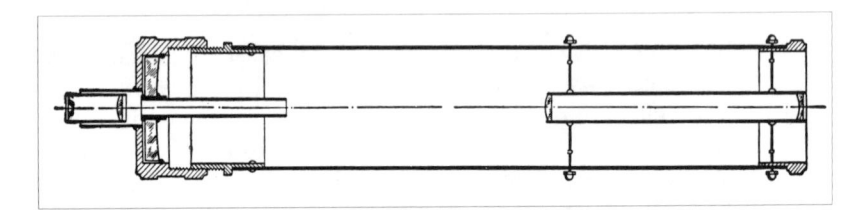

Fig. C.1.34: *Diagram of Ben L. Nicholson's Cassegrainian-Gregorian combination. The Cassegrain secondary mirror is removable. Drawing by J.F. Odenbach, after Ben L. Nicholson.*

by Ellison a number of years ago."

A note received years ago from Ellison may prove of interest. He says: "There used to be a man named Whittle, in Liverpool, who made Gregorians on what he called the 'mirror principle,' the mirrors being glass, silvered on the back and with the front surfaces worked to a curve; that of the small mirror correcting the chromatic aberration of the large one. The

C.1.22. Telescope Designs

Fig. C.1.35: *A compact, stubby Cassegrainian of f/12, with 10" × 20" tube, made by Alan R. Kirkham of Tacoma, Wash. Primary, 8", unperforated, a diagonal turning the secondary cone upward to a comfortable downward-looking eye position which never moves more than 4" and obviates the usual neck-wringing contortions of the perforated Cassegrainian. The primary, of f/2.8, was roughed out in 4 hours, while inverted over a 6" tool, mainly with 4" strokes; fine ground in 4 hours by working a small tool over it; polished in 7 hours with a 7" tool on top; and figured in 2 hours with 3" and 5" tools on top (Draper's method, which obviates turned edge). Secondary mirror $1\frac{1}{2}$" diameter, 4" inside focus.*

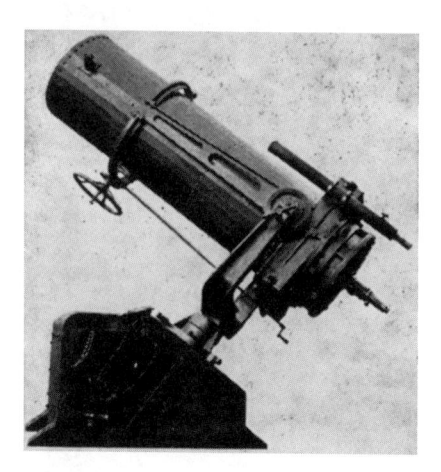

Fig. C.1.36: *A $20\frac{1}{2}$-inch Newtonian-Cassegrainian combination made by John H. Hindle and now in use in London by Dr. W.H. Steavenson, F.R.A.S., a well-known English variable star observer. The cast iron cell contains a mechanical flotation system with 18 equally loaded points. The tube is of wood. The polar axis rotates on ball bearings. The mirror, of $20\frac{1}{2}$ inches aperture, has a 5-inch perforation. Interchangeable spiders carry the diagonal for the Newtonian and the convex for the Cassegrainian. The e.f.l., when used as a Cassegrainian, is 30 feet.*

Fig. C.1.37: *A 15-inch Cassegrainian made by Dr. H. Page Bailey of Riverside, Calif. The primary is not perforated and the secondary cone is reflected out to the eyepiece by means of a diagonal. One head of the double yoke of the mounting is depressed, enabling the tube to reach the pole of the heavens. This depressed yoke head rolls on ball bearings and is counterweighted. The framework was made from parts of a motor truck chassis. The telescope is driven by a synchronous motor.*

principle is sound. I believe he got Professor C.V. Boys to work out the curves for him. But it was too complicated. I saw one once. It might be a good stunt to set some of the 'C.P.R.' boys, who are longing for fresh worlds to conquer." ("C.P.R.," that is, Carborundum, pitch and rouge, Ellison's term for the amateur telescope maker.)

C.1.22.3 The Herschelian Telescope

This telescope is favored by few. The inclination of the mirror causes astigmatism and distortion and the body heat of the observer has a maximum bad effect on the seeing, because of his position next to the incoming rays of light.

Theoretically we could dispose of that part of the distortion which is

Fig. C.1.38: *A simple, straight line observatory. The roof rolls off the walls on tracks at the plate level. A little practical experience in building the conventional hemispherical dome for an observatory will speak loudly in favor of the straight line type which is free from the "fusswork" involved in fitting materials to curved surfaces.*

caused by the inclination of the mirror, by figuring the mirror as if it were a portion of a larger paraboloid chosen at the side, as hinted at in Wood's *Physical Optics*. There are mirrors at Mount Wilson Observatory which are Herschelian in principle but their figuring was laborious. Such figuring might be done by local polishing with a subdiameter rose tool similar to the one described in ATM3, Figure E.6.2. But would the game be worth the candle?

Captain Ainslie of England has stated that "Herschel always used single biconvex lenses as eyepieces and with these a very small displacement from the center of the field, in the proper direction, would go a long way toward correcting the image."

C.1.23 Literature of Interest to the TN

C.1.23.1 Observatories

Literature on observatories is scarce. See Bell, *The Telescope*, which describes several. In *Popular Astronomy*, May, 1922, J. Ernest G. Yalden described an observatory for a 9 1/2-inch reflector, with pyramidal revolving roof. He also described another small refractor dome in *Popular Astronomy*, October, 1920. William Braid White described a revolving dome (tin) observatory of small size, in *Popular Astronomy*, October, 1916. Charles Early described in *Popular Astronomy*, October, 1914, a tin-domed house-top observatory. A small, brick-walled observatory with dome was described in *Popular Astronomy*, issue for June-July, 1913. A small turret reflector of unique design, combining in one structure the telescope mounting and observatory was described in *Popular Astronomy*, Aug.-Sept., 1927. In general, it is best to avoid complicated curved domes and stick to straight lines. One of the best of all types of "observatories" (actually it is only a housing) consists of an ordinary barn-like structure mounted on rollers on a track. During observation this "barn" is rolled away entirely, leaving the telescope in the open. Better yet, one may place the rollers and track at the plate level of the structure and merely slide aside the roof. These types require plenty of room.

C.1.23.2 Herschel's Mirrors

In *Transactions Optical Society*, No. 4 for 1924–25, Dr. W.H. Steavenson, F.R.A.S., outlines a "Peep into Herschel's Workshop." Sir William Herschel left four complete volumes in manuscript, relating to his various processes and experiments, in which he sums up the results of 40 years of experience in the art of telescope making. These manuscripts are now in the hands of the Royal Astronomical Society, and, says Dr. Steavenson, "it is greatly to be desired that means should some day be found for publishing them." (Here is an opportunity for some generous amateur enthusiast or group of enthusiasts.—Editor.) The same issue describes and illustrates Herschel's many mirrors, eyepieces, etc., which are still in possession of his granddaughter, Miss Francisca Herschel. Herschel did not use the knife-edge test, because Foucault had not yet hit upon it. He worked by feeling and by a remarkable sense of intuition. Some of his mirrors had been soldered down in cans prior to his death in 1822. Opened in 1924, a century later, they were found to be splendidly polished and without a trace of tarnish! Most of them were found to be overcorrected and to have two or three zones, yet on the whole the figure of all the mirrors was up to a very fair standard, and it is probable, says Dr. Steavenson, that they performed quite well in actual use. The complete report is extremely interesting. Think of the thrill of opening up cans containing mirrors sealed up 102 years previously by Herschel and testing the work of this great master, for the first time, under the knife-edge!

C.1.23.3 Properties of Pitch

Transactions of the Optical Society, No. 3 for year 1922-23, contains an article on the properties of pitch, by F.W. Preston. The Editor has already made numerous references to the *Transactions,* which are not, however, mainly devoted to telescope making, as might logically be inferred from these references. The articles mentioned are the few of that nature which were found while going systematically through all the back files.

C.1.23.4 Wassell and Blacklock Letters

Lest they be lost sight of, mention is made of the long series of letters on mirror making published many years ago in *English Mechanics.* Ellison in Section A.4.2 refers to those of Wassell, though they continued until 1886, not merely until 1883, as stated by him. The dates follow: For 1881: Sept. 23; Nov. 11; Dec. 9; Dec. 30. For 1882: Jan. 13; Mar. 24; Apr. 14; May 12; June 16; Aug. 11. For 1883: Mar. 30; Apr. 27; May 4; June 1; June 8; July 27; Aug. 31; Nov. 9. For 1884: Feb. 8, May 30; June 6; July 4; Aug. 22; Oct. 24; Dec. 12. For 1885: Feb. 6; Apr. 3; May 29; Nov. 30. For 1886: Feb. 5; Mar. 26; June 4; Sept. 17; Nov. 12. A second series, by Dr. Blacklock, may be found in the volume for 1895, pages 403, 449, 495, 543; and in the volume of 1896, page 26. Much of this matter is out of date, but the dyed-in-the-wool enthusiast who enjoys sherlocking around public libraries and poring over ancient, dusty tomes in search of hints will enjoy hunting up this series. (Purchasing the back numbers would be a most difficult task. A few large libraries contain them; for example, the Library of the Associated Engineering Societies, 13th Floor, 25 West 39th Street, New York City.) The advanced amateur should not look upon the methods of the old-timers as necessarily sacred, nor consider in the light of a sacrilege a wide deviation from them. This, in fact, is the way progress is attained, for a few such wild stabs in the dark, out of a multitude attempted, will likely land on something new and wholly worthwhile. One great trouble with mirror making in the past has been that there always has been an "established way", to deviate from which constituted a desecration of a holy of holies. "Try anything once" has proved a fruitful source of discovery in all fields of science. With several thousand workers doing that, there is no telling what may be hit upon.

C.1.24 Graduating Setting Circles

On equatorial mountings having setting circles there are two graduated circles—one in degrees for declination settings, the other in hours and frac-

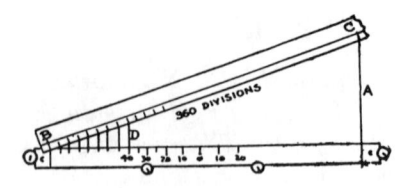

Fig. C.1.39.

tions thereof—for setting off the hour angles. If desired these graduations may be cut on the castings in a machine shop, but his is an expensive operation and the cost can be circumvented by going about it in this way: Prepare two strips of thin sheet brass half an inch wide and long enough to go around the castings and overlap a little. Wrap them tightly and make a scratch where they overlap. Then fasten on of the strips to a drawing board with thumb tacks , and in Figure C.1.38 and draw the perpendicular A. Place a scale on the board at such an angle that there will be 360 divisions between B and C. These divisions are now to be transferred to the strip by dropping the perpendiculars D and scratching them into the brass with a sharp tool like a knife blade. Every tenth division is made longer than the others and numbered, starting with zero at the middle and increasing to 90 either way, and then back to zero at the ends: 0. 90, 0, 90, 0. This is the declination circle.

The strip for the hour circle is similarly treated, except that here there are but 144 divisions divided up into 24 hour divisions, each hour subdivided six times, giving a least count of 10 minutes of time.

The precision obtained by making the circles by hand in this manner is enough for setting purposes, for all that is needed is an accuracy sufficient to bring any star desired well into the view of a low power eyepiece.

Part D
Observatory Buildings

Chapter D.1

Telescope Housings

D.1.1 The Warmed Observing Room

The time-honored manner of enclosing a telescope permanently mounted on its pier is by means of the hemispherical dome revolving on a circular track. A slit in the dome allows any part of the heavens to be reached with a minimum of exposure of the observer to the weather. All large telescopes are housed in this way, but these domes are expensive and difficult to make and are hardly needed for the relatively small instrument of the amateur.

It may be of interest to know what has been accomplished in the way of providing the stargazer with a closed and warmed observing room entirely independent of outside temperature. So far, in order to accomplish this end, an additional reflection has been needed in order to bring the image into the room and still preserve the principle of the equatorial mounting.

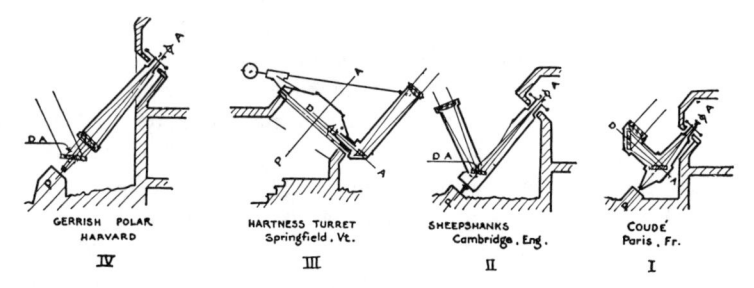

Fig. D.1.1.

In Figure D.1.1 are several schematic diagrams showing how the problem has been solved for refractors. In each case the mountings are meridional

Chapter D.1. Telescope Housings

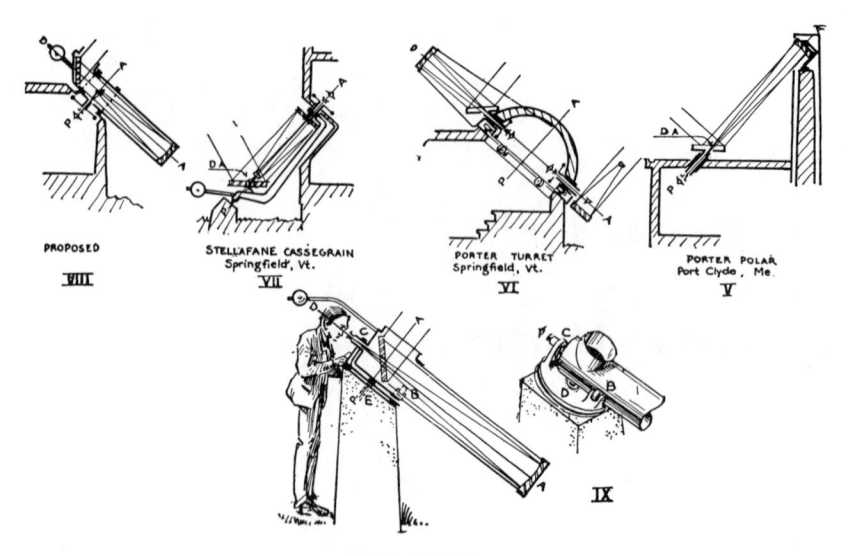

Fig D.1.2.

sections in the plane of the paper, and the polar and declination axes are lettered PA and DA, respectively. All of these types have been built and are in daily use. In I, II, and IV the observer looks down the polar axis through a fixed eyepiece. In III one looks down the declination axis, and the eyepiece describes a small arc of 180° in covering the heavens. I and III take in the entire heavens. II and IV are cut off from part of the northern sky.

In reflectors the relations of the primary and eyepiece, with respect to the object viewed, are reversed, thus changing the problem. So far as is known, only one or two attempts to adopt the Newtonian silvered glass telescope to an enclosed observing room have been made. I used the arrangement shown at V, Figure D.1.2, in my former home on the Maine coast for several years, but it is now dismantled. It is not recommended on account of the inconvenience of looking up the polar axis. This becomes very trying to the neck muscles after long periods of observing. This mounting was described in *Popular Astronomy,* in 1917, and the description was reprinted in the *Scientific American Supplement* for August 4, 1917.

Mounting VI, a photograph of which is shown elsewhere, is at *Stellafane,* Springfield, Vermont. The turret carries two telescopes, a 16-inch Newtonian of 17-foot focal length, and a Cassegrainian 12-inch of 16-foot equivalent focal length, so that two observers may study the same object simultaneously.

Mounting VII might be called a polar Cassegrainian. It was in use for

D.1.2. Turret Telescopes

some years at *Stellafane*. And mounting VIII, so far as I know, has never been built.

Mountings V and VII leave parts of the northern heavens obscured; VI and VIII are universal.

The upper telescope of VI demonstrates that it is possible to bring the focus of a mirror into a room without additional reflection. I have shown the arrangement again in IX in another position and adapted to a pier outdoors. The prism is replaced by a large flat mirror with a hole at its center. The tube carrying the two mirrors revolves in declination on the two rolls B and hollow bearing C. These supports are carried on the circular plate D which in turn revolves in right ascension on the stud E. The difficulty involved in perfecting this train of optical parts is in the perforated flat; cutting out the central hole after the mirror has been figured, results in a raised rim about the hole, due to released strains in the glass. If it is attempted at all, the central core should first be cut out and then replaced, cementing it to the glass with plaster of Paris. The whole is then fine ground, polished, and figured, and the central core is removed afterward. The perforated mirror must be very flat, otherwise any slight departures from a plane will seriously affect the image, due to the fact that it is outside of the concave mirror.

D.1.2 Turret Telescopes

The Hartness Turret telescope (refractor) was fully and technically described, with scale drawings, in the *Journal of the American Society of Mechanical Engineers,* December 1911, pages 1511–1537; also in the *Transactions* of the same society, same year. These are on file in some large libraries. A similar turret telescope (reflector) designed and constructed by Russell W. Porter and erected by him and the Amateur Telescope Makers of Springfield stands in front of "Stellafane" near that community in Vermont, and is shown in one of the illustrations. A turret similar in general design to the Hartness telescope is owned by J. Milo Webster, a Wyomissing, Pennsylvania, optician. A still more compact turret telescope is shown in the accompanying drawing made by the owner of the telescope, P.R. Allen, who describes it as follows:

"If the cone of rays from the flat of an ordinary Newtonian reflector is directed down the right ascension axis, instead of out of the end of the tube as usual, and the R.A. axis is expanded until it is big enough to accommodate an observer, the result would be the working principle of my turret reflecting telescope. I obtained my inspiration and ideas for this from a description of the Hartness refracting turret telescope.

Fig. D.1.3: *The Porter turret telescope at Stellafane, atop Mount Porter, near Springfield, Vermont. This is the instrument shown in Figure D.1.2, at VI, but at the time the photograph shown above was taken the Cassegrainian had not been added. The turret is made of concrete.*

"Briefly, my scheme consists of one cast iron ring about 30 inches in diameter, mounted on a cylindrical wooden observatory, with room inside for the observer to stand. A similar ring is mounted on the first, with ball bearings between, so that the upper ring is free to rotate with its axis toward the true north. The upper ring is prevented from sliding off from the lower by a pair of ball races supported by a casting attached to the lower ring. The R.A. scale is mounted on the inside of the upper ring. It is graduated down to intervals of four minutes. The declination circle has gear teeth cut on its periphery and these mesh with a pinion with handle attached, for ease in turning on this axis. Small electric lights illuminate the dials and my star maps.

"Entrance is from below, up ladder steps nailed to the inside of the turret. The observer stands on a little floor which can be dropped into

D.1.2. Turret Telescopes

Fig. D.1.4: *Bonneview Observatory, constructed by J. Milo Webster of Wyomissing, Pa. The eight-foot turret carries a refracting telescope mounted in a manner similar to the Hartness turret telescope.*

Fig. D.1.5: *Inside the turret of the Webster turret telescope.*

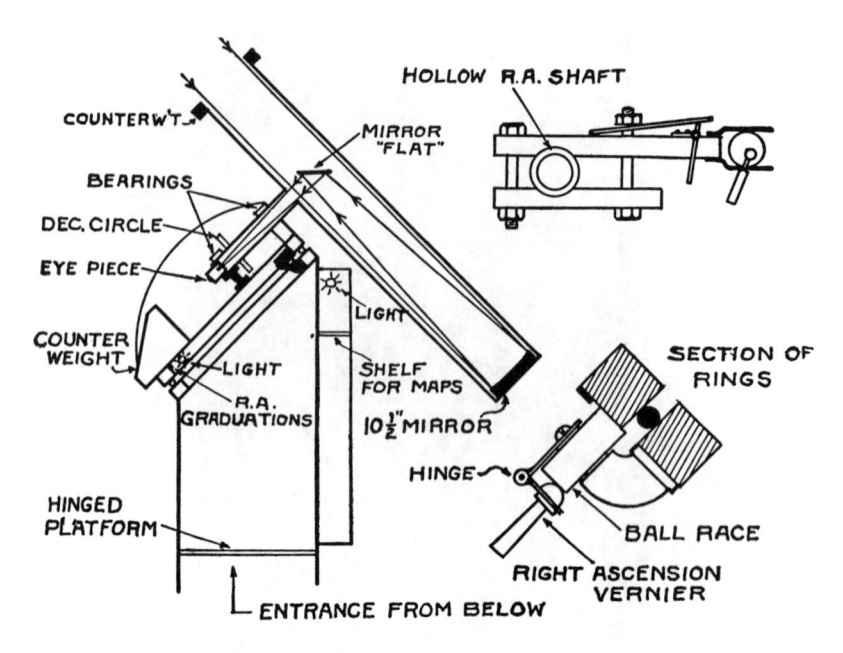

Fig. D.1.6: *The Allen turret telescope. (Drawing by P. R. Allen)*

place. Most of the work done with this telescope is variable star work using powers of 50 to 200, the latter rarely. Rocking of the telescope is noticeable with the high powers, due to the instability of the house and turret."

One special precaution might well be observed in connection with this type of telescope: the heavy ring should be secured in some positive manner against jumping the track, otherwise it might on some occasion slide off and do as neat a job of decapitation as a guillotine.

Chapter D.2

The Amateur's Observatory[†]

What is probably one of the best arguments in favor of an amateur observatory is contained in a letter from Alan Kirkham of Tacoma, which reads in part as follows: "The speed and ease with which some amateurs assemble their portable telescopes is amazing; the procedure is usually like this. First, the maker calls in half-a-dozen of the huskiest members of the group, and together they move the junk pile from the attic or garage to the back lot and, fortified with monkey wrenches, screw drivers and sledge hammers, the 'delikut' adjustments are completed with a speed and accuracy that leaves one breathless. By this time, however, everyone is so tired, and it's so late, that they are unable to observe, and immediately begin dismantling."

Undoubtedly you have encountered this condition in some degree in your own circle of amateur friends; but a permanently mounted instrument, protected by an observatory of whatever type, is ready for use at all times at a moment's notice; it is almost axiomatic that one cannot become fully acquainted with the whims and possibilities of a portable telescope in less than a year's service. This time could be greatly reduced in an observatory, and consequently greater appreciation of the telescope would ensue and a longer useful observing life result.

[†]By Leo J. Scanlon, Pittsburgh, Pennsylvania.

Fig. D.2.1: *Las Estrellas Observatory of the Texas Observers, Fort Worth, Texas. Type A1. Stone structure 12 by 14 feet. Roll-off roof permits view of Polaris. Cost: for materials, $125; for labor, nothing (Robert Brown, chief planner, stone mason, mechanic). Described in* Journal of the Royal Astronomical Society of Canada, *April 1934. The Patton Observatory, also of this type (A1), was described in* Popular Astronomy, *Dec. 1930.*

Fig. D.2.2: *The Bouton Observatory, St. Petersburg, Florida. Type A2. Built by T.C.H. Bouton and described in* Popular Astronomy, *Feb. 1930. It is 14 feet square, with two stories (shop below). Walls are cement stucco on wood. Raised projection on one roof section overlaps other section.*

Three additional reasons for building an observatory, fully as important as the foregoing, are: it protects the instrument when not in use; cuts off wind and stray light; and, last but not least, keeps the feet dry while observing, which is of great importance.

Fig. D.2.3: *An observatory built by Harold B. Webb, Jamaica, Long Island, N.Y. Type A3. Described in* Popular Astronomy, *March 1931. Inside dimensions 10 by 10 feet.*

Fig. D.2.4: *Waldo Observatory, built by J. Ernest G. Yalden, Leonia, N.J. Type B1. Described in* Popular Astronomy, *May 1922.*

These advantages permit one to carry on worthwhile observations from a fixed location, from which the appearance of the sky quickly becomes familiar, saving much time in locating celestial objects. Under such favorable conditions the telescope is more likely to be placed in accurate alignment and remain so, setting circles become more than theoretically desirable, thus widening the field of observation, and with a clock drive, which is a foregone conclusion in an observatory, work can be carried on with a certain amount of comfort and freedom from interruption. To complete the ideal picture, we need only set up a well-figured mirror in a Springfield mounting, to have all that any amateur could desire. This mounting, with its fixed observer's position, comes into its own in an observatory.

Fig. D.2.5: *The Wright Observatory, Berkeley, Calif. Type B2. Lift-off shutter, with hinged zenith hatch. Same dome is shown with picture of Mr. Wright, in ATM2, Fig. E.2.5.*

D.2.1 Observatory Size

Unless your surroundings are such as to make it an impossibility—in which case you had better look for a more favorable location—don't build an observatory less than ten feet square. Make it even larger if possible; anyone who has constructed an observatory of this size or smaller will be the first to advise you against cramping yourself for space. It requires no more effort to build one of proper size than to make one too small; it makes little difference, when you are on the north end of a south-bound saw, whether the plank you are cutting is ten feet, or ten miles, long—the work is the same. The slight additional cost of materials will be more than paid for by the extra space and convenience.

Avoid building the observatory on top of an existing building, especially if the lower one is of frame construction. If it is of brick or masonry, however, it offers a good base for the observatory, but presents the problem of installing a suitable pier. If at all possible the observatory building should be separate from all others, and as far removed as conditions permit. The base of the observatory (which is that part from the ground up to the track or rollers) should preferably be rectangular in outline; this shape is easier to construct and its corners offer ample space for desk, chairs, chart cases, etc. Climatic conditions often play an important part in determining the shape of this base; if it is likely to be piled high with snow for several months

D.2.2. Weatherproofing

Fig. D.2.6: *Left: Dome of the 24″ Clark Refractor on Mars Hill at the Lowell Observatory, Flagstaff, Arizona. From a photograph by O.S. Marshall of Springfield, Vt. and Pasadena, Calif. Aesthetically, the angles of side and top are pleasing. Right: Dome of the 61″ Wyeth Reflector at the Oak Ridge Station of Harvard College Observatory, near Harvard, Mass. The square base of brick surrounds a concrete cylinder, on top of which the dodecagonal turret revolves. The walls of the turret are faced with corrugated asbestos board over Celotex; the roof is of similar material, water-proofed with five-ply tar gravel roofing. The internal structural steel girders are painted with aluminum. To provide the shutter, one vertical section of the turret opens, giving access to very low altitudes, and a section of the gently sloping roof slides back—a simple arrangement.*

of the year, sufficient slope must be given the projecting areas, or they should be eliminated entirely, as in the case of a circular dome.

D.2.2 Weatherproofing

One question that arises in the mind as soon as an observatory is mentioned is, "Will it be weathertight?" This should give little trouble. After investigating this point, in connection with at least a dozen amateur observatories, only one case was found in which the builder claimed that water entered during rains. Further correspondence revealed that the shelter in question was not a permanent structure and admittedly was a makeshift.

D.2.3 Rollers

No matter what type of observatory is finally decided upon, use as few rollers as possible to achieve the motion desired, whether it be upon a straight or a curved back.

These rollers should be about 4″ in diameter and not more than four to six in number, for a revolving dome. A roll-off roof would require four to eight, depending upon the size and whether it was a one-piece or two-piece

Fig. D.2.7: *Observatory similar in type to Mars Hill, built by A.R. Leuchinger of Wantagh, N.Y. Type C1 which the author favors as combining the advantages of the hemispherical dome with flat construction. Note simple hinged vertical shutters and sliding zenith hatch.*

roof. If these wheels can be secured with ball-bearing centers, so much the better. Warehouse truck wheels serve admirably for this purpose. Ball-bearing roller skate wheels have frequently been used but, due to their smaller axle diameter, more of them must be used, in order to take the weight. In a roll-off roof observatory, the part of the track which extends beyond the observatory should be well-supported, to prevent sagging when the weight of the roof is on this track. The track itself should preferably be a "T" of angle iron, with the stem of the "T" vertical; in this case grooved wheels would be required. The use of this type of wheel and track eliminates the necessity of removing snow and sleet from the track in the winter.

D.2.4 The Sliding Roof Observatory

For a practical, inexpensive amateur observatory, one with a sliding roof has much to recommend it. In one plan, type A1 (see Table D.2.1 at end of this chapter), the entire roof rolls off as a unit (Figure D.2.1). In another (Figure D.2.2) the roof is divided either transversely or (Figure D.2.3) longitudinally. This point can be settled by the builder—all are practical. This type of observatory offers an unobstructed view of the skies, permitting quick setting on objects. Since there are no shutters to operate, it is easy to open, even when heavily laden with snow.

When the sliding roof is opened, the interior of the observatory quickly adapts itself to the difference in temperature, which is a great advantage

D.2.4. The Sliding Roof Observatory

Fig. D.2.8: *Observatory of Adrian Williamson, Monticello, Ark. Type C2. Brick Walls. Entrance room (study and library) at front, hexagonal 16 foot observing room in rear. The revolving dome is of galvanized iron and has a shutter opening with a lift-off sector for winter use. In milder weather, the entire dome may be lifted off by means of the jib and jack shown in the two pictures.*

with a reflector. If it is desired, one may paint the outside of the observatory a light color (aluminum paint is best), so that the heat absorbed during the summer day will be minimized; conversely, the interior of the observatory should be painted dead black or at least a dark color, so that the absorbed heat will be quickly radiated. A dark dome also reduces the glare of moonlight, and permits the eye to remain at maximum sensitivity for work at the telescope.

When the observatory is in use, one side of a split roof can remain closed or the one-piece roof can be rolled off part way. This helps to protect the instrument and observer from sudden gusts of wind and stray light, against which this type of observatory offers but slight protection. Many well-known variable star observers, both in our own country and abroad, favor the sliding roof observatory. This indicates that an unobstructed sky is of great practical value.

Fig. D.2.9: *Neat interior of an Observatory built by Robert E. Millard, of Portland, Oregon. Type Wc. Note the inter-rib bracing.*

D.2.5 Domes

While the average amateur hesitates but little before planning to build a flat, roll-off observatory, he often labors under the impression that a revolving dome is a job only for the professional mechanic. This is not really the case—but a knowledge of the use of tools helps a lot. The only difference is that, instead of working to a straight line in making the dome or shutter or in lining up the track, you work to a curve, and must be a little more exact.

The connecting link between the roll-off roof observatory and the revolving hemispherical dome is the pyramidal or cylindrical structure. It is in these B and C types (see Table D.2.1) that we find the most practical "in-between" observatories. The simplest, easiest, and least expensive is the low pyramid, embodying four inclined triangles for the roof area, and a simple lift-off or hinged shutter, with or without a separate zenith shutter. The latter is well-illustrated in the Yalden Observatory (Figure D.2.4), designed

D.2.5. Domes

Fig. D.2.10: *Exterior of the Millard Observatory, shown in Fig. D.2.9, attractively photographed by W. Boychuk. Diameter 10'. Height from floor to top of dome 11'6". Panels of fixed base part are of Celotex. The light canvas dome turns on twelve ball-bearing roller skate wheels, with eight similar wheels acting as thrust bearings on the inside of dome ring. Described in* Popular Astronomy, *May 1930.*

many years ago and still in service. The entire roof revolves on a track and roller system, with guide and hold-down brackets to prevent lateral or vertical motion at all times. An observatory similar to this has been constructed by Woods and Watson, of Baltimore, in which the entire building, including side walls, revolves on a track system placed on the floor of the building; Croston of Tacoma went a step further and placed the track and rollers under the floor, so that the entire building revolves by motor. The latter observatory consists of an 8 foot cube, the entrance door serving as part of the observing slit, while part of the cornice and a section of the roof are hinged in order to give vertical vision.

A further development of the pyramid is to increase the number of sides of the dome structure, and bend the triangular areas of each side at one or more points between base and apex, giving a more pleasing structure (Figure D.2.5). Such an observatory, by Wright, illustrates the facility with which hinged shutters can be used on this structure, which is composed entirely of straight lines and flat planes, except for the track and roller system.

Fig. D.2.11: *Observatory built by Fred L. Farmer, Yakima, Wash. The dome ring is made of a thickness of 1″ × 4″ superposed on a thickness of 1″ × 3″, each cut to 5 foot radius. Frame of the shutter is ½″ angle iron, on which wooden strips are fastened to support the slats. The dome is covered with heavy canvas. The walls of the fixed base are made of Beaver Board. Cost of material (1932) $60.*

Fig. D.2.12: *Observatory and 12″ reflector of K.F. Davidson, Marshfield, Wis. type Wc. Dome is of Masonite, and is made in spherical sections bolted to wood ribs. Slot 30″ wide, concrete floor 7″ thick. Shutter one piece of Masonite screwed to wooden framework, sliding in grooved oak track*

D.2.5. Domes

Fig. D.2.13: *Pacific Union College Observatory, Angwin, Calif., type Wm. Dome housing 22 foot diameter, 14" open fork telescope on low pier, flush with floor. In this example the conventional fixed walls have been eliminated and head room at the side wall of the dome is gained by providing a vertical skirt integral with the hemispherical movable dome. The entire building rotates on a series of roller skate trucks bolted to the circular concrete foundation below, the track being formed by the dome ring, a section of which is removable in the entrance doorway for oiling and adjusting rollers. In the basement is a workshop. A story placed below ground level will usually be cool in summer and free from vagrant drafts in winter. Some such hideout connected with an observatory provides the owner with a place he can call entirely his own, where peace and quiet may be obtained.*

Fig. D.2.14: *A 16 foot observatory by R.W. Steber, Warren Pa., type Mc. Housing 16½" reflector. The thin ribs are arcs of ¾-inch angle iron (1½" around shutter opening) to which wooden slats are screwed and canvas is laid over these. Shutter consists of four hinged sections, overlapping.*

If we do not attempt to follow too closely the hemispherical idea in constructing our dome, a structure that is at once pleasing to the eye, easy to construct, and highly practical, can be secured by building more on the cylindrical side. The fact that two of the large observatories in America, the Oak Ridge Station of the Harvard College Observatory and Mars Hill at Flagstaff, Arizona (Figure D.2.6), are constructed along these lines, should encourage the amateur to duplicate them. Retaining the straight-line features for ease of building, if we construct a slightly tapered cylinder, topped with a flat or sloping roof, we secure such an observatory as built by A.R. Leuchinger, of Wantagh, N.Y., (Figure D.2.7), which carries us over into the C types. It is the writer's opinion that this is the most practical observatory between the roll-off roof type and the true dome. It is simple to construct, and has few faults. Note that Leuchinger elevated the zenith hatch track high enough above the general roof level to clear the usual depth of winter snow. The hinged shutters are easily installed and do not depend on pulleys and cable for their operation. Note the hinged board in the zenith hatch, to form a water-tight joint at the top of the shutter doors.

A further development in cylindrical observatories, more costly but not more practical, is achieved by making a short section of round cylinder and topping this with a flat or cone roof. A good example of the latter is shown in the Williamson observatory (Figure D.2.8), in which the builder achieved not only a pleasing and practical observatory, but embodied a unique feature—winter-and-summer adaptability. Note that the entire dome structure can easily be hoisted by means of a screw jack operated from inside, and then swung clear of the observatory. Needless to say, this feature has its advantages during the summer, when it is often unpleasant to observe indoors.

The hemispherical dome that represents the least investment for the amateur is probably one of ribbed canvas. Here all the work can be done with hammer and saw. The usual procedure in making a dome of this type is to lay out and build up two circles or rings of wood at least two ply thick, of $3/4''$ lumber, from $3''$ to $4''$ in width, and with the correct diameter to suit the finished dome. One of these rings is set stationary on the fixed base of the observatory, and is known as the base ring, while the other is used as the foundation of the revolving dome. If we attach the rollers permanently to the fixed wall of the observatory, instead of building this second ring, we can use the dome ring itself as an inverted track, and eliminate one of these circles.

The "dome ring" is used as a foundation to which ribs, and possibly guide rollers, are attached. Usually the ribs are built up two ply, of $3/4''$ material about $3''$ wide, similar to the construction of the dome and base rings. Two main ribs are run uninterrupted across the center of the dome,

D.2.5. Domes

Fig. D.2.15: *Valley View Observatory, Pittsburgh, Pa. Constructed by Leo J. and Larry Scanlon. Type Om. First of several all-aluminum self-supporting observatory domes, 12 foot diameter, with weathertight reinforcing seams of great strength. Simple and effective construction requiring only hand tools and no previous experience in sheet metal work. Described in* Scientific American, *July 1931.*

and their distance apart establishes the width of the slot opening. **Make the width of this opening at least one-fourth the diameter of the dome.**

To these main ribs are attached the shorter remaining ribs. Inter-rib bracing is desirable, at least at one point about halfway along the arc of the ribs. These braces may be light, and will serve as additional supports for the canvas (Figures D.2.9, D.2.10). The rollers are usually attached to the underside of the dome ring, and roll on the track, but as suggested heretofore, if we mount the rollers permanently on the fixed base of the observatory, we can eliminate one ring and cause the dome ring to act as an inverted track. If this method is used, it will be well to provide about four "guide" rollers acting against this revolving ring to prevent lateral movement of the dome.

The rib construction is clearly shown in Figure D.2.11. Either canvas, composition board or sheet metal sections may without difficulty be fastened to the wooden ribs so formed. Davidson's dome (Figure D.2.12) represents type Wc.

A fine example of wooden-ribbed dome, metal covered (type Wm) was constructed under the direction of Prof. Newton, at Pacific Union College, and embodies the revolving observatory feature (Figure D.2.13). This dome, 22 feet in diameter, houses a 14" reflector constructed by the students, under the direction of their instructor. It is mounted in an open fork, close to the floor of the observatory, and to take advantage of the full sweep of clear sky near the horizon it was found necessary to extend the slit opening downward, or else to mount the telescope on a higher pier, requiring

540   Chapter D.2. The Amateur's Observatory

Fig. D.2.16: *Left: How the gores of the dome at Valley View Observatory were jointed—making the double-turned standing seam. Right: After the right-angle bend (stage 3, Fig. D.2.17, left) is bent with the tongs it is finished with a mallet backed by an iron dolly.*

the observer to perch atop a ladder. Extending the slit opening was considered more practical than raising the pier, and, to eliminate the necessity of installing an undersized entrance door (which has frequently been done in amateur observatories), the track and roller system was moved down to the base of the observatory walls, so that the entire building revolves. The entrance door to this observatory is directly opposite the slit, and through it objects on the horizon are within range of the telescope. The basement of this observatory, reached through a trap door in the floor, contains a dark room and workshop. Gangs of ball-bearing roller skate wheels mounted on small tracks are placed at frequent intervals around the foundation of the observatory, the track being attached to the dome and revolving with it. The wheel trucks are so mounted that they can be adjusted individually for height.

If, instead of wooden ribs, we are able to substitute arcs of flat or angle iron, to which the canvas, wood or sheet metal is attached, we may get a more substantial job, but one which is a little more difficult to construct (Figure D.2.14). Similar construction, but on a more elaborate scale, was employed in a 12-foot observatory in Morrisville, Vermont (Figure D.2.19).

D.2.5. Domes

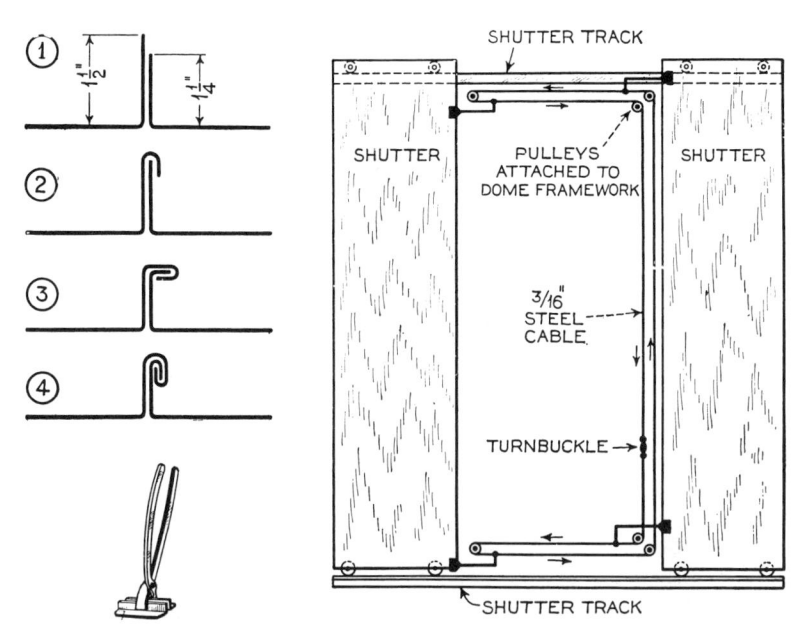

Fig. D.2.17: Left: Four stages in joining the gores of the seam, also the turning tool for bending the metal, improvised from pieces of one-inch angle iron and blacksmith's tongs welded together. Right: Diagram of arrangement by David L. Brown, Laurel Gardens, Pittsburgh, Pa., for opening and closing shutters on his observatory, adopted at Valley View. A cable of heavy picture wire is guided around the shutter opening by rollers 2″ in diameter at right-angle turns and made continuous by a turnbuckle. Shutters are attached at points indicated, by metal arms which are so formed as to pass each other when moving. Shutters can be opened manually by a pull on the cable, or by motor if the wire is passed around a drum at some point.

In this case, "T" ribs of steel were rolled with the vertical rib outward, flooring boards were used to fill in the space between, and the whole was sheathed with copper. The cost of this observatory was well over $1000, the fixed walls being of brick construction.

Why not entirely eliminate these cumbersome, expensive ribs? A practical way to accomplish this is to use a material which will be self-supporting. Sheet metal is probably the only material that answers this purpose; one of the early examples of a sheet metal dome was described by Bell, in *The Telescope*, but since the construction required professional assistance, is not entirely suited to amateurs. Later examples of sheet metal, self-supporting domes indicate that others have preferred to dispense with these ribs. Dr. Camilli of Pittsfield, Mass. (*Scientific American*, March 1934), riveted and soldered the

Fig. D.2.18: *An observatory 12 feet in diameter, built by H.B. Ross, Benton Harbor, Mich., to house a 12″ reflector. Type Mm. The observatory is at the edge of an 85 foot bluff above Lake Michigan, and one side stands on steel columns. Shutter slides over the top of dome.*

gores of his observatory together, requiring about 50 pounds of solder, 3,000 rivets, and considerable labor, to achieve this end. David Brown of Pittsburgh completed a 14 foot dome having 36 gores of sheet metal, bolted and riveted together, forming a very substantial, smooth-appearing dome. A slight lap was required for each sheet, and the joints were painted with a waterproofing material after fastening.

Both of these methods of joining the sheets of metal, and several others, were considered before building the Valley View Observatory (Figure D.2.15), but both were discarded in favor of a method which would require no solder, bolts, or rivets, yet which, when completed, would be more rigid than either. Since such a saving of time and money could be effected, it was thought desirable to invest the money thus saved in a better material for the dome itself. Accordingly, 20-gauge, half-hard aluminum sheet was chosen because of its light weight, rigidity, and resistance to corrosion. A 12 foot hemisphere was constructed (*Scientific American,* July 1931) by interlocking 12 sections, each 3 feet across at the base of the triangle and approximately 9 feet long. The joint between the sheets is technically known as a "double-turned standing seam" (Figures D.2.16 and D.2.17), and gives the same effect as an external rib of aluminum, five ply in cross section. This stiffens the entire structure to such an extent that the completed dome easily supports the weight of two men without perceptible distortion. Plans for this observatory have been lent to amateurs in several states. These domes weigh less than 250 pounds. The cost of the complete observatory at Valley View, exclusive of amateur labor used throughout its construction, was $250 (1930).

D.2.5. Domes

Fig. D.2.19: *Grout Observatory, 12 foot diameter, Morrisville, Vt. Type Mw. Brick walls, concrete floor, steel-ribbed dome covered with wood sheathing and copper flashing. Dome revolved by hand rack-and-pinion.*

D.2.5.1 Shutters

The problem of moving the shutter on a dome is basically no different from opening a roll-off observatory—in fact, since the shutter is lighter and its travel shorter, this is really a simpler job.

In general, the problem can be solved in a number of ways. A shutter that slides laterally in a grooved track is shown in Figure D.2.12. The builder declared that an improvement would have consisted of providing grooved wheels (like sash-weight pulleys) running along a narrow iron track or rod, in order to prevent the accumulation of sleet which forms in the present wooden track. This dome is constructed of semiflexible building board, nailed to wooden ribs, the joints being covered with additional strips. The shutter is made of the same material. It all was given three coats of white paint, and has been found satisfactory after some years of use.

Probably the simplest practical method of making a shutter is to use a single sheet of building board, metal, or even a curved framework covered with canvas, made to fit over the arcs around the slot opening and sliding, flexed, on these arcs across the top of the dome and down over the back of it, out of the way—like the sliding part of a roll-top desk (Figure D.2.18). Snow accumulating on the top of the dome in heavy weather probably will offer some resistance to opening this type of shutter. The same type obscures the zenith, unless it is made in two parts or short arcs which telescope into each other, with complicated construction.

If the shutter is made to move laterally, it is possible to make the opening and shutter extend past the zenith, so that full advantage can be taken of this freedom from obstruction in photographing an object in that most

Fig. D.2.20: *Left: Self-explanatory method for turning a dome, used by Harry Footer, of Cumberland, Md., in his 12 foot aluminum observatory. The sprocket chain completely encircles the dome, and pulls against 12 hook-shaped brackets as shown. $1/4$ h.p. motor, 1725 r.p.m., speed reduced 20:1. Right: Friction drives on Laurel Garden Observatory Dome, Pittsburgh, Pa., built by David L. Brown. Reversible, motor, $1/6$ h.p., reduced 20:1 by worm gear, drives the rubber-faced truck wheel running on the upper side of the track through a 2:1 sprocket gear reduction, causing one revolution of the dome in 35 seconds. Tension on friction drive is adjustable—note tension spring and nut. Motor controlled from pier of 10" Springfield type telescope.*

favorable location.

If the shutter is so large that it is desired to distribute its weight on the open dome, it may be divided meridianally, and the same arrangement of rollers and track applied to each half. A single rope-and-pulley arrangement serves to open both halves simultaneously. This plan was adopted in the two Pittsburgh amateur domes (Figure D.2.17, right).

No matter what pattern of shutter you select for your observatory, **make the width of the opening about one-fourth of the diameter of the dome itself.**

D.2.5.2 Revolving the Dome

While it is usually a simple and easy matter to revolve an amateur observatory dome by grasping one of a number of convenient handles and giving a pull, various methods have been devised to do this work by motor (Figure D.2.20).

If, instead of attaching our dome rollers to the upper or revolving ring of the pair mentioned in a previous paragraph, we attach the rollers permanently to the fixed base of the observatory itself, we can at once eliminate one entire laminated ring and provide a suitable arrangement for revolving

D.2.5. Domes

Fig. D.2.21: *The author in the basement workshop of his Valley View observatory at Pittsburgh.*

the dome by motor. The only requirement is that we gear down a motor (such as a second-hand washing machine motor) and, by using a friction pulley pressing against the revolving ring—preferably downward against and directly over one of the dome rollers—we can revolve the dome at any desired rate, depending upon the speed reduction used. A reversing switch may be attached to the motor, thus permitting rotation of the dome in either direction. Motors can be procured with these reversing features built in. With a 10 foot dome an 1800 r.p.m. motor would require about a 30-to-1 reduction, in order to cause the dome to rotate in approximately one minute—which is fast enough for all purposes. The motor may be controlled, of course, from the pier of the telescope, and during photographic exposures this is a great convenience, since the observer, probably following his guide star through the telescopic eyepiece, can move the dome as required without leaving position.

To sum up, if you wish your observatory to look professional and indicate by its appearance the purpose for which it was built; if you wish to satisfy your own artistic desires and live up to the expectations of your friends; or if you wish to engage extensively in a photographic program—by all means

Fig. D.2.22: *An enclosed observatory and telescope designed and constructed by John H. Hindle. Left: With the flat and diagonal covered. Right: Uncovered and ready for use.*

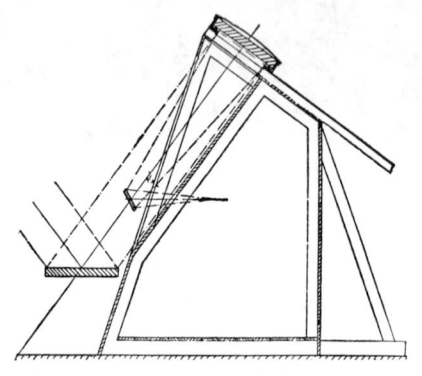

Fig. D.2.23: *Diagram of the optical train of the Hindle enclosed observatory. The flat of the coelostat has a diameter of $25\frac{1}{2}$ inches. The primary disk is $25\frac{1}{2}$ inches in diameter but only 20 inches are excavated, so that the telescope actually has a 20-inch paraboloidal primary. Its focal length is 96 inches. The tip of the cone is turned into the house by means of a diagonal. The only moving part is the coelostat. The extension below the diagonal contains a flat for a finder. The building faces south and the latitude largely determines its slope. It is very rigidly braced. Drawing by John H. Hindle.*

build a real dome.

If, on the other hand, you are removed from the interference of lights, and wish to construct an observatory that will consume a minimum of labor and be least expensive; if the appearance is of secondary consideration; or if you are going to engage in the observation of meteors or variable stars—by all means build a sliding roof observatory.

D.2.5. Domes

Table D.2.1 Types of Observatories and Domes
Classified according to difficulty and cost of construction.
STRAIGHT-LINE OBSERVATORIES

Type A: Includes all flat or low-angle sloping one- or two-piece roof, which rolls or slides aside, or is hinged.

    A1: Rectangular base, one-piece, roll-off roof on tracks.

    A2: Rectangular base, two-piece transverse division, roll-off roof.

    A3: Rectangular base, two-piece longitudinal division, roll-off roof.

Type B: Includes all flat pyramids with lift-off, roll, slide, or hinged shutters. (Roofs revolve).

    B1: Pyramidal roof, each section in one plane.

    B2: Pyramidal roof, each section in two planes.

    B3: Pyramidal roof, each section in three or more planes.

Type C: (All cylindrical observatories with flat or slope top.)

    C1: Cylindrical polygons, flat or sloping roof.

    C2: True cylinders, flat or conical roof.

HEMISPHERICAL DOMES
Classified according to rib construction, covering, and total cost.
Rib designations:
W = wood, O = none, and M = metal.
Cover designations:
c = canvas or composition board, w = wood, and m = metal.

Type Wc: Wood ribbed dome, canvas covering (or compo board).

Type Wm: Wood ribbed dome, metal covering.

Type Ww: Wood ribbed dome, wood sheathing and shingles.

Type Mc: Metal ribs, canvas covered.

Type Om: No ribs, metal covering.

Type Mm: Metal ribs, metallic covering.

Type Mw: Metal ribs, wood sheathing, metallic flashing.

SHUTTER CONSTRUCTION FOR HEMISPHERICAL DOMES
Classified according to simplicity and least cost.

1. One piece canvas flap; tied down, roll down.
2. Single wood framework, canvas covered; lift off, slide vertically or horizontally, roll aside, or over the top.
3. Single wood framework, composition board, moved as above.
4. Single wood framework, metallic cover, moved as above.
5. Single metallic framework, metallic cover, moved as above.
6. Double metallic framework, metallic cover, moved as above.

Fig. D.2.24: *Mr. Hindle at the eyepiece of the enclosed observatory telescope, which is fixed, horizontal and comfortable to look through.*

Chapter D.3

Thermal Effects of Observatory Paints[†]

D.3.1 Introduction

Little information appears to exist on the subject of the most suitable surface treatment of observatories for the purpose of reducing undesirable thermal effects. From general considerations it seemed that the choice was practically limited to two paints, white and aluminum, but it appeared to be worthwhile to carry out some simple tests to determine which of the two was the better. The results of these tests proved to be quite interesting, and were in fact somewhat surprising to me until the reasons were studied more closely.

D.3.2 Test Procedures

Two stout cardboard boxes of identical size were used for the first test. Both were ventilated to a small extent and comparable with an observatory. Box A was painted with aluminum paint and box W with white-lead paint. These were placed on pedestals in exposed positions in the open air, and a yard or so apart. Each contained an identical maximum and minimum thermometer, and readings were taken daily over a period of some eight weeks, commencing in late spring. Occasionally the thermometers were interchanged to eliminate systematic errors. In addition, two large metal boxes were coated with the same high-grade paint and similar tests made with and without ventilation.

[†]By H.E. Dall, From *The Journal of the British Astronomical Association*, 48 (1938) by permission.

Table D.3.1
TYPICAL TEMPERATURE READINGS (°F) IN BOXES W AND A

Date DD/MM	White (W) Max.	White (W) Min.	Aluminum (A) Max.	Aluminum (A) Min.	Remarks Weather and State of Boxes at 11 P.M.
17/6	63	40	69	42	Mixed weather. (A) quite dry. (W) heavily dewed.
20/6	69	44	75	47	Mixed, showery. Both wet after shower.
27/6	79	47	82	47	Metal boxes Cloudy night. Fitful sunshine. Both dry.
23/6	72	44	76	47	Metal boxes. Clear—cool winds. (A) quite dry. (W) heavily dewed.
30/6	65	38	66	39	Clear at night, but dry. Cold and very windy. Both dry.
21/6	74	36	79	41	Fairly clear day and night. (A) dry. (W) wet.
7 days from 8/8	84	51	92	56	Variable week—clear on several days. (A) dry. (W) twice wet.

The character of all the tests was identical—showing conclusively that the aluminum-painted boxes maintained a higher temperature *both by night and by day* than the white boxes.[1] Table D.3.1 shows a typical series of readings, including notes on the weather. As was expected, the greatest difference between *A* and *W* occurred on days and nights with least cloud and least wind. Strong winds, even with clear skies, result in convection effects swamping those of radiation, and both boxes attain closely atmospheric temperature.

The difference is quite sufficient in amount to justify the tests and to warrant careful consideration as to the most suitable paint for the purpose in view. If the observatory is to be used primarily *for solar or other daylight observations, the benefit of white paint will be most marked,* and convection-current troubles will be reduced to a minimum. If, on the other hand, night observations are principally intended, then aluminum paint is far superior to white. All temperature gradients within the observatory will be lessened;

---

[1] For the results of other experiments on the effects of paint in sunshine, see letters by H.R. Beckett and H. Spencer Jones, *The Observatory* (Oxford), 1936, 59, 14 and 17.

D.3.2. Test Procedures

performance should be quite noticeably improved. Dewing of the optical surfaces will probably be eliminated and external dewing much reduced.

My first impression before the tests was that white paint, having a visibly higher reflection coefficient that aluminum paint, would be cooler by day and warmer by night. This reasoning was fallacious, because it ignored the large variation of reflectivity and emissivity with the wavelengths of the radiation.

The temperature attained by an object in an exposed position is that at which equilibrium is reached between the heat lost and the heat received. Apart from the effects of conduction and convection, the heat radiated from the object depends on the emissivity of its surface at the particular wavelengths corresponding to its temperature. Similarly the heat received depends on the absorptive power at the wavelengths of the arriving radiations.

The emissivity of white paint at the wavelengths corresponding to atmospheric temperature is indistinguishable from that of black paint, i.e., *approximately 95 percent (reflective power = 5 percent).* A good grade of leafed aluminum paint has the relatively low emissivity of 45–50 percent under similar conditions.

The absorptive powers of the two paints for radiations from the high temperature sources of daylight are approximately 15 percent and 30 percent respectively. It is thus easy to see why the A box becomes higher in temperature than the W box when exposed to solar radiation. At night, *if the sky is clear,* the effective temperature of the sky will be considerably lower than that of the exposed object, therefore if it is coated with the highly emissive white paint it will attain a lower equalizing temperature than the slowly emitting aluminum-painted object. Metallic aluminum emits considerably more slowly still, and should in consequence be even warmer.

Condensation occurs on the object if its temperature falls below the dewpoint, and, as the falling gradient of air temperature at night often results in its being saturated, the white-painted object falling *below* air temperature becomes wet with dew, while the aluminum-painted object remains dry. The notes in the table indicate that this condition was observed frequently, and would imply that the surface of box A did not fall appreciably below atmospheric temperature. From the point of view of reduced convection currents this is very advantageous for an observatory used at night. If the surface is hygroscopic (and this appeared to be the case with the white-painted surface) there is the further objection to the formation of dew that additional heat is abstracted by subsequent evaporation, and the danger of dew on the optical surfaces is increased.

Aluminum paint should, for best results, be mixed just prior to use, as true "leafing" or interlacing of the scale-like powder does not occur more than 48 hours after mixing, and emissivity would increase due to the greater thickness of medium above the metal.

D.3.3 Interior painting

Radiation escaping from the aperture in the dome is practically blackbody radiation corresponding to inside wall temperature, even if the emissivity of the walls is as low as 50 percent. Similarly the color of the interior walls has practically no effect when sky or solar radiation enters the aperture. The interior walls may therefore be painted without regard to thermal effects. An observer highly sensitive to feeble extraneous light would prefer a black interior, but otherwise a light color would be preferable to assist illumination. It is, however, desirable to coat with white paint any part of the instrument subject to direct solar rays through the aperture, or, if night observation is more important, with aluminum paint.

Fig. D.3.1: *Horace E. Dall*

Horace E. Dall was born in Chelmsford, England, in 1901. He lived during his formative years and had his schooling in London and in Luton, Bedfordshire, 30 miles north of London. He trained for engineering and specialized eventually in instruments and methods of flow measurement, and has sat on National and International Committees on this subject.

His astronomical interests developed at the age of 16 and he made spectacle lens and other rigged up telescopes. Three years later he acquired an $8\,1/2$-inch Calver reflector. His telescopics proper started a few years later from following Ellison's articles in *English Mechanics*. Since then he has

D.3.3. Interior painting

made many mirrors, object glasses and complete instruments, with hundreds of eyepieces now distributed in various parts of the world. His other hobbies include microscopy (making, designing and repairing their optics) and photography. He was married in '34. Has no children.

Prior to marriage he made a hobby of cycling in many countries including Iraq, a "first ever" crossing of Iceland's interior wilderness via Askja, and another first ever of the African High Atlas in southern Morocco bordering the Sahara. "When, after dodging the French," he once wrote, "I rolled down the mule trail into Sous it was the first bike ever seen there. I spent a night in a wonderful old caid's castle and had to demonstrate on the bicycle in the courtyard to the great glee of many Negro slaves." While snooping through a keyhole Zeph once overheard a neighbor of Dall's say that on these cycling adventures he stripped all nonessentials from his kit, then stripped off nonessential parts from these essentials, and carried it all in his pockets. In later years he built the $3\frac{1}{2}$-inch, × 80 Cassegrainian shown in Fig. D.3.1 (and in detail in *Scientific American* '47 Dec.). Stripped to a bare 8 oz. it fits the weskit pocket.

Index

This is a cumulative Index for *Amateur Telescope Making 1, 2, and 3*. The number (1, 2, or 3) preceding a dash indicates volume number.

A

Abbe's sine theorem 2-147
Abbe, Ernst 3-155
aberration
    appearances of common forms 1-289
    axial chromatic 1-159
    axial spherical 1-158
    extra-axial 1-161
abrasives 3-62
    action, rolling and stationary 1-111
    crushed steel 1-422
    how much to use 1-35
achromatic refracting autocollimator 3-120
adjustments
    binoculars 2-379
    indoors of an equatorial 2-367
    lens cell 2-109
    polar axis 2-366
    telescope by equatorial star 2-375
advantages of refractor 2-46
Ainslie, Maurice A. 1-10, 2-352
air currents in the tube 2-7
Airy, Sir George 3-155
Allen turret telescope 1-526
Allen, P.R. 1-523
aluminum oxide 3-65
aluminum, reflectivity 3-302
Alvan Clark Co. 1-433
Amici prism 3-94
amount of glass removed 1-427
    during figuring 1-174
Ampliplane 3-155
anastigmatic 3-155

Anderson, J.A. 3-521
aperture
    critical 3-155
    ratio 3-155
aplanatic 3-155
apochromatic 3-155
Armagh Observatory 1-104
Armfield, L.E. 1-481
Arnold, G.P. 1-297
Arnulf, Albert 1-61, 3-247
astigmatic 3-155
    curves 1-82
    eyes 3-358
astigmatism 3-153
    causes of 1-339
    eyepiece test for 3-22
    in refractors 2-35
    infallible test for 1-338
    optical bench, testing for 1-358
    shop test 1-338
astronomical oculars 3-139
atmosphere, currents 3-362
autocollimation null test 1-315, 2-33, 2-89
    detection of a turned edge 2-35
    Ellison's method 1-213
axial chromatic aberration 1-159, 3-149
axial spherical aberration 1-158, 3-147
axial zonal aberration 3-148

B

Baade, Walter 2-507
Babendreer, A.B. 2-288
backlash, removing 2-307
Bailey, H. Page 1-514, 2-250, 2-428, 2-444
Bailey, Karl S. 3-112
Baker, James G. 1-61, 3-447, 3-447, 3-481
Bakker, Harold 2-334
balsam cementing methods 3-69

Barkelew, James 2-227
Barlow lens 3-177
    design of 2-134
    testing 3-181
Barlow, Peter 3-177
Barnes, R. Bowling 1-325
Barnesite 3-64
Barr's scale 1-243
barrel distortion 3-152
Barrett, S.B. 3-8
Barry, E.H. 3-271
Beardsley, J.D. 3-443
Beardsley, R.L. 3-34
beeswax in lap making 1-83
    Ritchey's usage 1-445
Beilby, G.J. 1-454
Bell, R.A. 1-433
Bergstrom, H.O. 2-310
Berthon, E.L. 1-499
bi-concave lens 3-213
binocular
    adjustment 2-379
    body 2-383
    centering and cementing lenses 2-403
    central focusing devices 2-414
    collimation 2-406
    coverings and finishes 2-392
    evaluation 2-417
    Galilean 2-416
    hinge 2-384
    lubricants 2-386
    objective assembly 2-385
    ocular assembly 2-388
    overhaul 2-379
    prism
        assembly 2-389
        checking 2-399
        collimating rules 2-401
        matching 2-399
        pyramidal errors 2-398
        right angle errors 2-398
    semimodern 2-412
    tolerances 2-393
    types 2-380
    waxes 2-387
black oxide finish, applying 3-240
blackening brass 1-502
blending overhand stroke 1-119
blocking pitch 3-54
bloom in silvering 3-286
Boehm, Joseph E. 2-257
Bonneview Observatory 1-525

Bostwick, L.G. 3-439
Bouton Observatory 1-528
Bouton, T.C.H. 1-528
Bouwers, A. 3-429
Brashear 1-70
Brashear's pitch 1-439
brasses and bronzes 2-298
Brattain, R. Robert 1-325
breakage, mirror 1-403
Broadhead, David 3-95
Brookings, Ernest 1-411
Brower, Daniel 2-379, 3-111
Brown, Earle B. 3-315, 3-346
Brown, Robert 1-528
bubble test for turned edge 1-122
bubbles, removing 1-438
bumps, raised zones, interpreting 1-234
Burch, C.R. 1-329

C

Calder, William A. 1-299
California Academy of Sciences 2-379
Callum, William 2-430
Calver 1-70, 1-476
Calver mirror focogram 1-475
Calver, 8.5-inch reflector 1-552
Canada balsam 2-93
Canfield, R.H. 3-443
Carbo grain sizes 1-421
Carborundum 3-63
Carpenter, Arthur Howe 1-220, 2-430, 3-439
Carreau, Napoleon 1-422
Case, F.A. 3-251
Cassegrainian telescopes 1-507
    collimating 2-372
    making 3-483
casting
    cope and drag 2-290
    crucibles 2-289
    furnace 2-288
    metal alloys 2-298
    molding 2-293
    molding a two-piece pattern 2-294
    molding sand 2-292
    pouring the metal 2-297
caustic test
    advantages of 1-259
    nonparaboloidal surfaces 1-280
    procedure 1-262
Cellini, Benvenuto 2-296

cellular mirror 1-413
cemented doublet design 2-127
cemented, built-up disks 1-415
cementing lenses 2-80, 3-69
center of curvature, finding 1-12
centering pitch 3-68
centering, Brashear's method 3-245
centering, lens 2-90
Chamberlain, Paul A. 3-440
chamfering disks 1-428
channeled glass tools 3-101
channeling pitch lap, carpenter's rip saw 1-124
channels in laps, purposes 1-443
channels, Laps 1-446
Chapman, C.C. 2-241, 2-300
checking progress in grinding 1-38
Chicago Amateur Astronomical Association 2-430
Chinese mirror effect 2-502
Choosing a mirror size 1-397
Christie, William 2-493
Christman, Erwin H. 2-288
chromatic aberration 2-148, 3-134
chromatic difference of magnification 3-134
Church, C.C. 3-112
circular facets 3-105
Clacey, John 2-526
Clark, Alvan 1-71
Clark, R.E. 3-27, 3-205
cleaning optics 3-339
cleaning up between abrasives grades 1-39
cleanliness, workplace 1-403
Cleveland Astronomical Society 1-56
clock stroke 1-118
cloth laps 1-439
coelostat 3-510, 3-513
coelostat telescope 3-526
cold pressing 1-461
collimation 1-53, 2-363
    Cassegrainian 2-372
coma 2-146
comatic aberration 3-151
concave tools, making 3-27
conic sections 1-12, 1-17, 1-190, 1-471
Conrady path-difference method 2-151
contamination of abrasives 1-39
continuous polishing 3-20, 3-23
contrast 3-381
convex tools, making 3-28
Cooke triplet 3-471

Cooprider, C.M. 3-195
cork insulation 2-329
correcting 1-128
    central protuberance 1-131
    color 2-63
    coma 2-59
    edge first 1-131
    HCF method 1-130
    hole 1-462
    hyperbola 1-92, 1-463
    raised or depressed ring 1-93
    spherical aberration 2-57
    turned edge 1-92
corundum or emeries 3-63
Couder, André 1-319, 2-357
Cox, H.W. and L.A. 2-479, 2-485
Crawford, J.G. 1-432
crest of the doughnut 1-197
Critchett, James C. 1-220, 1-421, 1-437, 1-448
crossed concave lens 3-214
currents, kinds of 3-365
Curtiss, Heber D. 2-334
curvature of field 3-134, 3-152
curve test during grinding 1-75
curved spider to reduce diffraction 1-465
Custer, C.P. 1-484
Cutler, Harold Nelson 3-537
cutting circular disks and holes 1-417
cutting pitch lap channels 1-8

D

Dall null test 1-315
Dall, Horace E. 1-315, 1-500, 1-507, 1-549, 1-552, 2-353, 3-353, 3-419
Danjon, M. 2-358
Darbishire, O.B. 2-263
Darling, Stephen F. 3-267
Davidson, K.F. 1-536
Davies, C.D.P. 1-467, 1-476
decementing lenses 3-72
decentering pitch lap facets 1-84
deglazing diamond tools 3-40
degreasers 3-73
Denning, W.F. 3-177
depressed zone 1-15
    appearance 1-12
    treating 1-9
DeVany test 2-460, 2-487
diagonal mirrors 2-549

grinding 1-165
    mounting 1-167
    polishing 1-166
    testing 1-166
diameter of eye lens 3-359
diamond cup wheel, principle 3-38
diamond tool diameters 3-40
diffraction 1-383
    edge 1-387
    effects 1-465
    focal 1-387
    Fraunhofer 1-386
    ring test 1-486, 2-11
Diller, C. Carvel 2-254
diopter 3-156
disk, Ramsden 3-156
distortion 3-134, 3-152
dividing
    engines 2-342
    head 2-341
dog-biscuit 1-113, 1-117, 1-244
double-star separations, list 3-384
doughnut family of curves 1-229
    mathematics 1-225
Douglass, A.E. 3-361
Draper 1-71
drives
    alarm clock 2-241
    calculating gear trains 2-247
    Dictaphone 2-242
    friction clutches 2-240
    periodic errors 2-239
    simple clutch 2-266
    synchronous motor 2-246
    universal (DC) motor 2-244
    variable ratio friction 2-251
    Victrola motors 2-242
dummy polisher 3-89
Dunlop, A.R. 3-537

E

edge diffraction 1-387
Eliason, C.W. 2-364
Ellerman, Ferdinand 3-267
ellipse 1-12, 1-81, 1-189
ellipsoid 1-12
ellipsoid testing 1-305
Ellison on refractors 2-15
Ellison's autocollimation test 1-213
Ellison, William Frederick Archdall
    1-6, 1-13, 1-20, 1-57, 1-69,
    1-94, 1-104, 1-106, 1-109,
    1-140, 2-15, 2-41
English's test for pitch hardness 2-559
English, R.E. 2-557, 2-564
Erfle, Heinrich 3-156
Euryscopic 3-156
evaporation effect 1-114
Everest, A.W. 1-63, 1-111, 1-197,
    1-223, 1-399, 1-464, 2-325
Everest, Mary A. 1-247
explosion hazard of silver nitride
    3-280, 3-282
extra-axial aberrations 1-161
    of the eye 3-159
eye
    lens diameter 3-359
    optical qualities 3-379
    relief 3-156
    resolving power 3-356
eyepiece
    cells 3-194
    constructions 3-164
    custom-built 3-163
    design procedure 3-159
    making 3-189
    prescriptions 3-169
    standard tube diameter 3-194
    testing 3-169
    war surplus 3-163
eyepoint distance (e.p.d.) 3-156
eyes, astigmatic 3-358

F

f-number, f/number, (f/No.) 3-157
facet shaving to correct a hyperbola
    1-92
fatal hyperbola 1-79
Fecker, J.W. 1-181, 1-398, 2-44, 3-15
Fernandez, José 1-106
Ferson, Fred B. 2-275, 2-287, 3-35,
    3-74, 3-77, 3-130
Field of view, apparent 3-156
figuring 1-11, 1-134, 1-462
    by machine 3-21
    Ellison's method 1-87
    small mirrors 1-134
finders 1-503
finding the center of curvature 1-12
fine grinding
    amount of pressure 1-124
    fastest method 1-123
    reducing scratches 1-124

streak of rouge test 1-431
water drop test for contact 1-431
when finished 1-77
Finn, A.N. 1-494
Flanders, Fred F. 2-335
flat, making to test refractor 2-74
flats, optical 2-523
flats, polishing 3-105
flexure 1-82, 1-398
flotation cells 2-319
focal diffraction 1-387
focal length
    back 3-156
    beware of extreme 1-99
    equivalent 3-157
focograms, making 1-484
focus of a lens 3-212
Footer, Harry 1-544
forestalling turned-down edge 1-463
Foucault knife-edge test 1-8, 1-79
    announcement 1-469
    error of observation 1-235
    unnecessarily delicate 1-70
four classes of curves 1-192
fovea centralis 3-157
Françon, Maurice 1-61
Fraunhofer diffraction phenomena 1-386
Fraunhofer type objective design 2-130
French, James Weir 1-430, 1-454, 1-458, 3-267, 3-273
Fresnel diffraction 1-386
friction effect 1-116

G

Galilean type binoculars 2-416
garden telescopes 1-22
Gardner, Irvine C. 3-247, 3-256
Garnet Fines 3-63
Garrison, Jack 3-543
Gauss condition 3-148
Gauss, Karl 3-157
Gaviola, E. 1-260
Gee, Alan E. 1-58, 2-137, 2-125, 2-254, 2-255
ghost images, elimination 2-458
glass
    for small lenses 3-240
    tables, explanation 2-71
    templates 3-45
Goddard, A.V. 2-354
graduated facets for parabolizing 1-89

Grant, William M. 3-112
Graphic method of determining,
    diaphragm stop 2-48
    focusing tube 2-48
    size of prism or diagonal 2-48
Graves, Byron L. 2-225, 2-242
Green, N.E. 2-351
Gregorian telescope 1-293
grinding 1-422
    central contract 1-429
    checking progress 1-38
    estimating fineness 1-430
    finding center curvature 1-437
    finishing with emery 1-434
    getting overall contact 1-428
    mirror handle 1-120
    nature of 1-424
    removing bubbles 1-438
    stands 1-421
    when finished? 1-434
Grubb, Howard 1-21

H

Hadley, John 1-471
Haidinger's fringes 1-374, 1-378, 2-550
Hale, George Ellery 3-513, 3-525
Halsted Observatory 2-333
Hamm, J.W. 3-37
handle, removing 1-421
Hanna, G. Dallas 2-379, 2-418, 3-111, 3-125
hardening brass 1-502
hardwood mandrels 2-303
Hargreaves, F.J. 2-351, 3-294
Hartmann's criterion 1-301
Hartness turret telescope 1-523, 2-210
Hartshorn, C.R. 3-177, 3-187
Harvard Observatory 3-361
Hastings, Charles S. 1-498, 3-133, 3-139, 3-157
Haviland, John R. 1-58, 1-493, 2-43
Hazelrigg, Hugh 1-483
honeycomb foundation (HCF) 1-32, 1-43, 1-63, 2-88, 3-32
    advantages of 1-63
    for figuring 1-448
    for zonal correction 1-66
    lap making and using 1-64, 1-446
    methods of use 1-64, 2-87
heat effects, general rules of 3-106
Hendrix, D.O. 2-493
Henry, Arthur 2-288

Herschel Condition 3-148
Herschel wedge 3-143
Herschel's mirrors 1-516
Herschel, fine grinding 1-454
Herschel, William 1-69
Herschelian telescope 1-514
Hitchcock, L.R. 3-526
Hicks, F.M. 3-7
high-power oculars 3-136
hill or a hollow, determining 2-33
hill, reducing with local polisher 1-141
Hindle test 3-503
Hindle's method of parabolizing large
    mirrors 1-195
Hindle, John H. 1-213, 1-163, 1-340,
    1-461, 1-465, 1-513, 2-319,
    3-17, 3-489
Hiner, W.B. 1-433
Hodges, Paul C. 2-507
hogging out 1-35
Hole, George 1-319
Holeman, John M. 3-162
Holoscopic 3-157
Hooker telescope 2-206
Huygenian ocular 3-135, 3-140, 3-190
Huygens' Principle 1-372
hydrant, cure for astigmatism 1-494
hyperbola 1-12, 1-189, 1-190, 1-192
hyperboloid 1-81
Hyperplane 3-157
hypocycloidal polishing 3-23

I

image motion, estimating 3-384
image spikes 1-383
Ingalls, Albert G. 1-189, 1-397, 2-353
inside-and-outside test 1-480
insufficient mixing, pitch and
    turpentine 1-124
interference filters 3-339
interference of light 1-369
interpreting shadows 1-478
Invar 1-409
inverting devices 1-418-9
Isaac Newton, erroneous conclusion
    2-55
isokumatic 3-157

J

Jacobson, Samuel 1-220
Johnson, J.W. 2-344
Johnston, H.L. 3-271

Jones, Arthur H. 2-310

K

Kalde, F.R. 2-379
Kalliscopic 3-157
Kellner, C. 3-157
kerosene in grinding 2-43
kerosene test 2-474
kerosene, use during fine grinding
    1-123
King test 1-322
King, Edward S. 2-235
King, J.H. 1-285, 1-321
Kirkham, Alan R. 1-209, 1-289, 1-293,
    1-415, 1-465, 1-513, 1-527,
    3-384
knife-edge test,
    accuracy 1-238
    allowable error 2-50
    avoiding fatigue 1-483
    description 1-10
    Hindle's method 1-215
    interpreting shadows 1-233
    learning 1-482
    measuring stick 1-242
    pattern of shadows 1-192
    pinhole separation 1-466
    precision 1-479
    residual error 1-236
    routine 1-243
    shadow appearance 1-223
    testing equipment 1-240
    use of eyepiece 2-97
    use of low-power telescope 2-97
König, Albert 3-157
Krotkov, V. 3-300

L

Lamont-Abbe 3-157
lap
    channels 1-446
    grabbing 1-459
    HCF 1-446
    in hot places 1-450
    making, dunking method 3-33
    metal 2-80
    paper polishing 1-449
    speckling 1-445
Las Estrellas Observatory 1-528
lateral chromatic aberration 3-152
lateral spherical aberration 3-151
Le Coultre, F. 1-416

lead lap finished surface 3-34
Lee, John C. 3-4, 3-275
lemon peel finish 1-154, 1-244, 1-439
    cause of 1-128
Lenart, Jr., Peter 3-35, 3-75
lens
    aberrations 2-145
    axial points 1-341
    Barlow 3-157
    Bertrand 3-157
    blocking 3-52
    centering and edging 3-67
    coating 3-337
    generating 3-35
    making 3-215
    molded blanks, defects 3-36
    production 3-35
    rough grinding after milling 3-40
Leuchinger, A.R. 1-538
Lick Observatory 3-361
Liebig, Baron Justus 3-30
limiting devices 1-419
Linde, Paul 1-141, 3-77
Littrow spectroscope 3-521
log book 1-399
long stroke 1-75
longitudinal chromatic aberration 2-146
Lord Rayleigh 1-455
low-reflection coatings 1-380
Lowell Observatory 1-531, 2-426, 3-361
Lower, Charles A. 2-251, 2-445
Lower, Harold A. 1-140, 1-485, 2-237, 2-251, 2-372, 2-443, 2-445, 2-483, 3-384
Lutz, G.H. 3-7

M

machines
    continuous polishing 3-20
    Draper 3-9
    Fecker's 3-15
    figuring 3-21
    forming the lap 3-61
    grinding and polishing 3-3
    Hindle's 3-14, 3-18
    hypocycloidal polishing 3-23
    laws of position and stroke 3-88
    Lee's 3-3
    mirrors 3-108
    polishing 3-32

    polishing and grinding strokes 3-57
    polishing mirrors 3-17
    Ritchey's 3-8
    small polishers 3-22
Mackintosh, Allan 1-401
Magnalium 1-409
magnesium fluoride 3-338
magnesium oxychloride mirror 1-416
magnifying power 1-30
Maier, Victor E. 2-266, 2-344
making
    flats 3-77
    mirrors 3-77
    pitch lap with rubber grid 1-124
    prisms 3-77
Maksutov, D.D. 3-429
Marshall, John Albert 3-282
Marshall, O.S. 3-304
Martin, L.C. 1-454
Mason, Henry H. 1-467
Mason, William A. 2-288, 2-292, 2-313
McAdam, J.V. 1-420, 1-463, 2-363
McCarthy, E.L. 1-484, 1-485
McCartney, E.B. 2-268
McDonald Observatory 2-334
McDonald telescope 2-247
McGuire, Daniel E. 1-220
McIlroy, J.L. 2-288
McLaren, D.A. 3-112
McMath Observatory 2-247, 2-250
measuring focal length using the Sun 1-51
measuring zones 1-68
Meighan, George 2-265
mercury mirror, rotating 1-412
metal
    grinding tool, size 3-51
    laps 2-80
    laps, making 3-27
    laps, using 3-27
    mirrors and flats 3-437
        Allegheny metal 3-439
        chrome steel 3-439
        stainless steel 3-440
    tools 3-49
    tools using 3-30
Michelson's equations for surface accuracy 2-51
microscope oculars 3-162
mirror breakage 1-403
mirror cell 2-226
mirror cover, making 1-103

mirror flotation cells 2-319
    3-point support 2-321
    18-point support 2-322
mirror
    astigmatism 1-494
    cellular 1-413
    copper 1-411
    glass substitutes 1-408
    grinding 1-5
    handle, description 1-120
    magnesium oxychloride 1-416
    maker's "three motions" 1-75
    materials, coefficients of expansion 1-417
    materials, relative thermal conductivities 1-417
    mirror substrates 1-405
    mounting, with sling 1-104
    polishing 1-7
    refuses to concave 1-426
    shipping 1-404
    sticks to tool 1-432
    suction 1-416
Moffitt, G.W. 2-430, 3-120
monocentric 3-158
monochromator, use of 3-593
Morse, E. H. 2-244
mount
    coudé 1-521
    double yoke 2-206, 2-224
    English (fork type) 2-203, 2-205
    equatorial 2-203, 2-207
    German 2-202, 2-204
    Gerrish polar 1-521
    Hartness turret 1-521
    importance of stability 2-215
    machining 2-230
    motor drive 2-233, 2-237
    poor man's 1-48
    Porter polar 1-522
    Porter turret 1-522
    Sheepshanks 1-521
    Springfield 2-209
    Stellafane Cassegrain 1-522
    typical Newtonian 2-200
    wooden, equatorial 2-208
mounting the mirror, Ellison's method 1-104
Muntz metal 2-298
Myers, James E. 3-34

N

Nakamura, K. 1-475, 3-14
natural corundum 3-65
natural garnet 3-65
Newton's fringes 1-374
Newton's rings 1-374
Newtonian mirror, appropriate size for first mirror 1-5
Newtonian telescope 1-4
Newtonian telescope, diffraction spike phenomenon 1-385
Nicholson, Ben L. 1-512
Nicholson, C.H. 2-429
Nobili's rings 3-289
nodal slide optical bench 1-343
nonpitch laps 1-446
null test, paraboloids 1-315

O

oblate spheroid 1-13, 1-81, 1-193
oblique astigmatism 3-134
observatory, domes 1-534
observatory, rollers 1-531
observatory, shutters 1-543
observatory, sizing 1-530
observatory, sliding roof 1-532
observatory, weatherproofing 1-531
obsidian 1-410
ocular
    Airy 1-345
    astronomical 3-139
    high-power 3-136
    Huygenian 3-140, 1-345
    Mittenzwey 1-345
    Ramsden 3-140
    solid 3-141
Odenbach, J.F. 2-346
offence against the sine condition 3-151
oil flat 2-475
one third stroke 1-5, 1-76, 1-424, 3-109
optical bench
nodal slide 1-343
    testing methods 1-349
      on the axial image 1-350
      approximate refractometry 1-363
      astigmatism 1-358
      axial chromatic aberrations 1-353
      axial critical aperture ratio or aperture tolerance 1-352

axial spherical and zonal
  aberrations 1-353
back focal length 1-351
comatic, lateral spherical or
  sinical distortion 1-358
complete telescopes 1-359
curvature of image field 1-354
distortion 1-357
extra-axial images 1-354
flange focal length 1-352
lateral chromatic aberration
  1-357
negative lenses and systems
  1-350
poorly corrected positive
  systems 1-350
radius measurement 1-365
well corrected positive systems
  1-350
working distance 1-351
optical constants, definition 2-16
optical contacting 3-92, 3-566
optical flats 2-523
  accuracy 2-553
  astronomical 2-529
  correcting 2-559
  figuring 2-540
  grinding 2-539
  one disk method 2-525
  reference flat method 2-526
  laps 2-537
  three disk method 2-523
  perforated 2-538
  polishing 2-540, 2-559
  raw materials, 2-535
    black glass 2-536
    filter glasses 2-536
    fused quartz 2-535
    grinding abrasives 2-536
    polished plate 2-535
    Pyrex 2-535
  polishing abrasives 2-536
  refiguring a perforated flat 2-542
  Ritchey's method 2-531
  suitable materials 2-557
  testing 2-544, 2-560
  turned edge prevention 2-541
optical glass 1-407
optical glass procurement 2-189
orthokumatic 3-158
orthoscopic oculars 3-152, 3-158
outstanding advantage of a reflector
  2-45

Over, Edwin 3-112
overhaul of binoculars 2-379

P

paint for inside of tube 1-502
Palomar 18-inch Schmidt 2-429
pancratic 3-158
paper polishing lap 1-449
parabola 1-12
parabolizing
  by graduating facets 1-87, 1-89
  by long stroke 1-87
  by overhang 1-18, 1-87, 1-90
  by small polisher system 1-87,
    1-90
  when finished 1-201
paraboloid 1-81
paraboloid of revolution 1-3
paraboloid, how to recognize 1-95
paraboloidal shadow, explanation of
  1-470
paraffin soaked tools 1-398
paraxial rays 2-144
parfocal 3-157, -171
Parsons, L.A. 3-112
Paul, Henry E. 2-451, 2-471, 3-547
Pease, F.G. 1-408
Peaucellier's linkage 2-83
Peck generator 3-38
Peltier, L.C. 3-413
periodic errors 2-239
Periplan 3-158
Periplanatic 3-158
Peters, C.G. 2-526, 2-528
Pettit, Edison 3-587, 3-603
Petzval surface 3-154
Phillips, T.E.R. 2-351
Pickering's seeing scale 3-371
Pickering, William H. 2-3, 2-354, 3-362
Pickering, William H. on the virtues of
  refractors 2-3
Pierce, John 1-400, 1-504, 2-233,
  2-311, 2-335, 2-529
pincushion distortion 3-152
pinhole
  cool 1-465
  determining optimum size 1-329
  illuminating 1-331
  making 1-332
  size of 1-82
  small and large usage 1-332
  tiny 1-467

　　　　too tiny 1-82
pit depth test 3-102
pitch and turpentine, insufficient
　　　　　　mixing 1-124
pitch lap
　　　　Brashear's 1-439
　　　　channels, cutting 1-8
　　　　deforming 1-131
　　　　drag during polishing 1-126
　　　　Ellison's formulae 1-83
　　　　Ellison's test 1-444
　　　　Everest's test for hardness 1-443
　　　　facet
　　　　　　size 1-126
　　　　　　thickness 1-126
　　　　　　shape of 1-85
　　　　flammability 1-446
　　　　Hindle's 1-445
　　　　making 1-6, 1-124, 1-439
　　　　modifications 1-444
　　　　properties of 1-517
　　　　Ritchey's methods 1-444
　　　　Ritchey's recipe 1-445
　　　　sinking speed 1-125
　　　　straining 1-445
　　　　substitutes 1-450
　　　　temper testing 1-111
　　　　thickness 1-84
　　　　trimming marginal facets 1-127
pits
　　　　futility of attacking with rouge
　　　　　　1-435
　　　　remedy if deep 1-109
　　　　searching for 1-435
plane-parallel glass 2-549
plano-concave lens 3-213
plano-convex lens 3-213
plastic lens cements 3-71
plate glass, how made 1-407
platyscopic 3-158
point of focus 1-3
polar alignment, photographic method
　　　　2-238
polar axis adjustment 2-366
polar heliostat 3-523
Polaroid, types 3-556
polishing 1-43, 1-127
　　　　agents 1-450
　　　　Ellison on 1-83
　　　　machine 3-88
　　　　preventing astigmatism 1-127
　　　　speed 1-127
　　　　theory 1-453

poor man's telescope 1-28
Porter turret telescope 1-524
Porter's Folly 2-220
Porter, Russell W. on mounts 2-199
Porter, Russell Williams 1-3, 1-21,
　　　　1-147, 1-290, 2-215, 2-429,
　　　　2-271, 2-275, 2-288, 2-300,
　　　　2-526, 3-189, 3-197, 3-189
positive meniscus lens 3-214
Power, Henry R. 1-421
preparing the pitch lap 1-7
Prescott, Fred L. 2-242
prescription 3-244
　　　　Achromatic 3-243
　　　　Brashear Solid Ocular 3-243
　　　　Hastings 3-242
　　　　Huygens 3-241, 3-244
　　　　Kellner 3-242
　　　　Ramsden 3-241, 3-244
　　　　Steinheil monocentric 3-242
Preston, F.W. 1-428, 1-454
Prism or Diagonal in a Newtonian
　　　　1-147
prism 3-78
　　　　commercial production 3-80
　　　　errors and tolerances 3-78
　　　　glass 1-162, 3-79
　　　　hand correcting 3-95
　　　　requisites for small-scale
　　　　　　production 3-82
　　　　testing 3-90
prolate spheroid 1-12, 1-81, 1-193
proof plates 3-45
pump diffusion 3-320
pumps, mechanical 3-318
push and pull screw cell 2-39
pyramidal error 3-114
Pyrex 1-406

Q

quantitative optical test for mirrors
　　　　1-285
quartz
　　　　cutting, grinding, and polishing
　　　　　　3-577
　　　　defects 3-554
　　　　grinding and polishing 3-560
　　　　optical 3-552
　　　　orientation 3-559
　　　　sources of supply 3-552
　　　　testing 3-554
quartz-Polaroid filter 3-548

R

Radius of Curvature, very exact 1-438
radius of equilibrium 2-324
radius template 1-5
raised zone 1-15
    appearance 1-12
Raleigh water test 2-527, 2-544
Ramsden disk 3-356
Ramsden ocular 3-135, 3-140
Ramsden, Jesse 3-158
Rank, D.H. 3-111
ray tracing
    design procedure 2-139, 2-155
    equations 2-141
    essential tools 2-140
Rayleigh limit 1-58
reciprocal of the dispersion 2-41
recognizing the paraboloidal curve 1-95
record keeping 1-473
reference sphere, in testing 1-225
reflecting autocollimator 3-111
reflecting telescope, introduction 1-3
reflector, wooden tubes 2-351
reflector-corrector 3-447
refractive index 2-40
refractor vs reflector 1-392
    accuracy in figuring 2-52
refractor
    adjustable lens cell 2-109
    advantages 2-46
    airspacing 2-132
    autocollimation testing 2-89
    calculating curves 2-16
    cell 2-108
    cemented doublet design 2-127
    cementing 2-93
    design, sign convention 2-126
    determining size of 2-102
    diaphragms (stops) 2-119
    edge thickness, maintaining 2-87
    edge tolerance 2-87
    equation for 2-127
    equation for achromatism 2-127
    equation for spherical aberration 2-127
    fine grinding 2-28
    focus control 2-112
    focusers 2-114
    Fraunhofer type objective design 2-130
    grinding the curves 2-85
    kerosene used in fine grinding 2-86
    lens cell 2-92
    lens centering 2-90
    lens polishing 2-29
    Littrow type 2-177
    measuring edge thickness 2-40
    most "foolproof" instrument 2-15
    mounting 2-120
    mounting the lens 2-36
    optical glass procurement 2-189
    precision of surface 2-49
    shaping and grinding 2-25
    stops 2-39
    supreme test 2-95
    testing and figuring 2-30
    theoretical considerations 2-44
    three disadvantages 2-46
    tripod 2-121
    tube 2-39
    tube finish 2-118
    tube materials 2-103
    under- and overcorrection colors 2-63
    use of glass with unknown constants 2-177
relative magnification 3-147
rhomboid prisms, making 3-125
richest-field telescope 3-389
right angle prism, making 1-149
Ritchey cellular mirror 1-413
Ritchey's
    sub-diameter tools 1-139
    pitch casting method 2-89
    rosin squares 2-43
Ritchey, George Willis 1-139, 1-213, 1-329, 1-413, 1-506, 2-425, 3-8, 3-272, 3-272
Ritchey-Chrétien telescope 1-505, 2-433
Rogers, H.L. 1-450, 1-463
rolled-over edge 1-45
Ronchi test
    band patterns 1-290
    band photographs 1-210, 1-213
    failure of 1-285
    gratings 1-211
    interpreting 1-213
    King variation of 1-285
    testing lenses 2-88
Roof Prism Gang 3-77
Ross, F.E. 2-425, 3-453
Ross, H.B. 1-542

rouge
    how to make 1-450
    scratches 1-460
    size 1-450
    stain 3-89
rough grinding 1-74, 1-121
    when to stop 1-76
roughing out the curve 1-74
Russell, Henry Norris 2-423
Russell, James L. 1-25, 1-56
Russell, Robert 1-310
Ryder, E.N. 2-336

S

Saeger, Jr., C.M. 3-438
saggitta
    equation for 2-127
    of a spherical mirror 1-400
    precise calculation 1-400
Scanlon, Leo J. 1-527, 2-249, 2-337
Scheffler, Lester 2-288
Schmdit camera 2-423
    f-Number, or Speed 2-452
    alignment 2-490
    angular field 2-501
    assembly 2-488
    autocollimation test 2-475
    construction 2-443, 2-455, 2-479, 2-485
    corrector plate 2-457, 2-499
        design 2-499
        figuring 2-473
            vacuum deflection formulae 2-514
        Schmidt's method of polishing 2-502
    film holder 2-464
    focal length 2-451
    fundamental types 2-496
    mirror 2-455
    mounting 2-462
    portability 2-465
    principle in optical design 2-493
    testing 2-459
    testing correcting plates 2-503
    vignetting 2-453
Schmidt, Bernard 2-426, 2-493, 2-517, 3-451
Schmidt, Bernhard, original paper on Schmidt camera. 2-517
Schroader, Irvin H. 1-251, 1-282
Schwarzschild telescope 2-433

scratches 1-42, 1-402
    cause of 1-42, 1-78
    from rouge 1-460
    prevention 1-127
secondary color 2-150
secondary spectrum 2-17
Selby, Horace H. 1-59, 1-144, 1-157, 1-305, 1-313, 1-341, 1-369, 1-450, 2-535, 2-554, 3-145
Selsi 3-158
semistroke 1-120
separating stuck disks 1-42
setting circles, graduating 1-517, 2-333
shadow bands 3-372
shadow test 1-10
    interpreting 1-478
shape of lap grooves 1-85
Sharp, Donald E. 3-273
Sheib, S.H. 1-398–9, 1-435, 2-288, 2-298, 2-335, 3-267, 3-299
shipping a mirror 1-404
Shumaker, Loren L. 1-289
sidereal day 2-233
silicon carbide 3-65
silvering the mirror 1-17, 1-101
    Brashear's process 3-262
    burnishing 3-292
    double coating 3-290
    explosive hazard 3-281
    Martin's method 3-276
    measure the thickness 3-289
    recovering silver waste 3-297
    Rochelle salts process 3-267
    silver nitrate stains, removal of 3-292
sinical error 3-151
sleeks 1-460
Slipher, V.M. 2-426
slit test 1-219
small lens work
    centering 3-232
    convex lap 3-220
    flat laps 3-226
    lap mandrel 3-219
    making a flat pitch lap 3-230
    polishers 1-90, 1-130, 3-87
    polishing 3-224
    roughing out 3-222
    spinning 3-234
    tools and curves 3-206
small lenses 3-205
Smith, Allyn G. 3-112
Smith, Nicol H. 3-298

Smith, Sinclair 2-498
solar telescope 3-513
solid ocular 3-141
sources of surface damage 3-63
Souther, B.L. 1-462, 3-267
speckling of lap 1-445
spectroscopes, uses of 3-562
speculum making, advantages of 1-72
speculum metal 1-409
    casting 3-438
spherical aberration 2-146, 3-134
spherical and zonal aberration of the
    exit pupil 3-154
spherical surfaces using milling
    machine 2-83
spherometer 2-27
    construction of 2-77
    strongly curved surfaces 3-247
spider diffraction 2-357
spit test for radius of curvature 1-112
Sprengnether, Jr., W.F. 2-312
Springfield mounting 2-211
    counterweighting 2-306
    motor drives 2-306
    casting 2-286
    pattern making 2-277
    three forms 2-272
    pier 2-306
    setting circles 2-340
stainless steel mirrors 1-409
star test 1-100, 1-486
Steavenson, W.H. 1-513, 2-330, 3-410
Steber, R.W. 1-537
steering wheel handle 1-463
Steinback, J.A. 3-112, 3-129
Stellite 1-409, 1-411
sticking mirror 1-431
stigmatic 3-158
Stokley, James 3-297
stop distance 3-154
Straight, J.W. 1-432
strain, detection 1-494
streak of rouge test 1-431
striae 1-494
    test 1-321
stroke
    blending overhang 1-132
    grinding 1-425
    length 1-34, 1-46
    long 1-13, 1-34
    one-third 1-424
    overhang 1-132
    straight 1-5

    speed 1-38
    very long 1-35
Strong, John 3-302
stuck disks 1-41, 1-460
sub-diameter tools, 1-140
    construction of 1-144
    figuring 1-140
    fine grinding 1-140
    for large mirrors 1-139
    polishing 1-140
suction mirror 1-416
Sun telescope 3-509
surface inspection 1-78, 3-66
surrounds, making for diagonal 1-164
Swenson, M.H. 3-268

T

Taylor, D. Everett 2-101, 2-288
Taylor, H. Dennis 1-458
Telaugic 3-158
telescope
    eyepieces 3-145
    housings 1-521
    magnification 3-133
    motor drive 2-233, 2-237
    mounts 2-199
    oculars 3-133
    tube 2-225
temperature during polishing 1-85
    during testing 1-99
    working 1-399
templates in lens making 3-42
    making 1-420
testing
    at focus 1-19
    convex spherical surfaces 1-321
    ellipsoid 1-305
    for complete polish 1-461
    hyperboloidal mirrors 1-310
    hyperboloids 1-310
    inside-and-outside 1-480
    mirrors 1-8
    paraboloid on near objects 1-308
    prism angles 1-150
    rigs 1-269
    star 1-486
    striae 1-321
    temperature during 1-99
    tunnel 1-468
    without masks 1-477
theory of reflecting mirror 1-4
thermal effect 1-113

Thompson, J.W. 1-419
Thomson, Elihu 1-455
three motions of grinding and
    polishing 1-35
toadstool mirror 1-45
tolerance in figuring 1-201
Tolles, R.B. 3-158
Tombaugh, Clyde 3-406
tool
    deformation 1-117
    effect 1-113
    plowing 1-118
    wooden 1-145
tools and materials, Ellison's list 1-72
transverse spherical aberration 3-151
Trott, T.B. 2-225, 2-329
tube current fix, electric fan 2-9
tube lining, sheet cork 2-225, 2-329
tubeless telescope 1-48, 1-502
turned-down edge 1-20, 1-45, 1-67,
    1-88, 1-463
    avoiding with small polishers
        1-143
    bubble test for 1-122
    correcting 1-130
    cure for 1-118
    detection 1-289
    eyepiece test for 1-88
    preventing 2-34
    probable causes 3-108
    test for slight 1-464
turned-up edge
    correcting 1-462
    cure for 1-88
    eyepiece test for 1-88
turret telescopes 1-523

U

unique shape of the parabola 1-190
using metal tools 3-30

V

Varela, F.R. 2-367
viewing interference fringes 1-148
von Arx, Wm. S. 2-242
von Liebig silvering process 3-302

W

Wade, A. 3-275
Wadsworth, F.L.O. 1-254, 1-466, 3-273
Waland, R.L. 2-473

Walkden, S.L. 3-389
walking around the barrel 1-397
Wallace, Vard B. 1-450, 2-342
Warner, G.E. 1-141, 1-335
warped mirror 1-493
washtub pedestal 1-29
Wassell 1-71
Wassell and Blacklock letters 1-517
water proofing of wooden tools 1-398
Wates, Cyril G. 2-325, 2-336, 2-366
Watrous, R.M. 1-47
Watson, A.J. 2-288
Webb, Harold B. 1-529
Webster turret telescope 1-525
Webster, J. Milo 1-523, 1-525
Weisiger, Jr., S.S. 2-337
Wells, Carl 3-112
wet, length of 1-424
why the curves develop in grinding 1-4
Williamson, Adrian 1-533
Wood, R.W. 1-412
wooden tubes 2-351
Woodside, Charles L. 2-177
Worbois, E.L. 1-31
working distance 3-158
working uphill 1-91
Wright Observatory 1-530
Wright, F.E. 1-408, 3-37
Wright, Franklin B. 1-58, 1-201, 1-206,
    1-289, 3-429, 2-433
Wyld, James H. 2-139, 2-175

Y

Yalden, J.E.G. 1-529
Young, R.K. 1-169, 3-300
Yvon, M.G. 1-260

Z

zonal testing 1-95
zones,
    accurately locating on mirror
        1-68
    bumps, raised zones, interpreting
        1-234
    measuring 1-476